Patrick Moore's Practical Astronomy Series

For other titles published in the series, go to
http://www.springer.com/series/3192

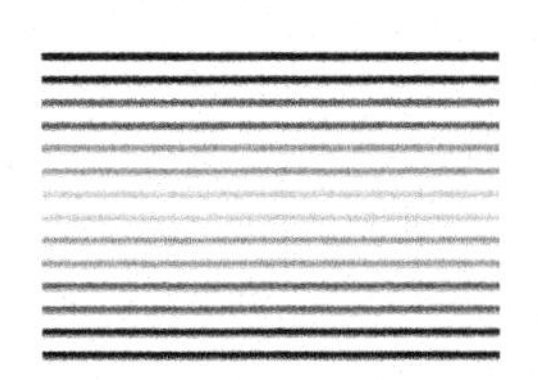

Viewing the Constellations with Binoculars

250+ Wonderful Sky Objects to See and Explore

Bojan Kambič

Bojan Kambič
Sarhova 20
SI-1000 Ljubljana
Slovenia
spikar@siol.net

ISSN 1431-9756
ISBN 978-0-387-85354-3 e-ISBN 978-0-387-85355-0
DOI 10.1007/978-0-387-85355-0
Springer New York Dordrecht Heidelberg London

Library of Congress Control Number: 2009930754

Illustrations: All pictures are reproduced by permission of their copyright owners. For further details see Picture Credits in the back matter

Printed on acid-free paper

Springer is part of Springer Science±Business Media (www.springer.com)

If Galileo had owned my binoculars,
what amazing discoveries he would have made!

All diagrams, illustrations and star charts are drawn by author, unless mentioned otherwise is written.

Whilst every care has been taken to trace copyright owners the author would like to apologise to anyone whose copyright has been unwittingly infringed.

Contents

About the Author

Bojan Kambič is from Slovenia and is well known there for his efforts in amateur astronomy. Like Sir Patrick Moore and other notable amateur astronomers, he has been honored by having an asteroid named for him (66667 Kambič) and is well known there as a popularizer of astronomy and as the author of various books, including *Zvezdni Atlas za Epoho 2000* (Star Atlas for Epoch 2000) as well as the present volume, which he translated himself with a help of Sunčan Stone. He has also written and translated over 150 articles and papers for the Slovene astronomical magazine *SPIKA* that he established in 1993.

Author's Note

This book is a guided tour through the constellations visible from mid-northern latitudes. If you want to find a certain street in an unknown city, you have to have a good map, otherwise you will get lost. This book is a kind of map of the sky! In it you will find all "important streets and houses" that you should know and all main ways and side streets that will get you to them. This book contains the particulars for more than 250 celestial objects – from double and variable stars, clusters, and nebulae all the way to the distant galaxies and quasars.

In the first part of the book, the basics of astronomical observing, choosing and using rudimentary stargazing equipment, classic astronomy, simple astrophysics, and basic cosmology are presented. The topics covered in this book include only those in astronomy that should be known to every observer before he or she begins to explore the night sky. If you do not know anything about the object that is to be observed, you will soon be bored behind the binoculars or telescope. Knowledge is the fuel for the imagination!

In the second part of the book are descriptions of all constellations visible from the mid-northern latitudes. At the beginning of each constellation there is a basic star chart and a brief description of all brightest stars. Then follow the detailed descriptions of all double and variable stars and the farthest objects, visible through the binoculars 10 × 50. Beside each description is a detailed star chart with leading stars labeled. Although it may seem that the book is written for beginners in amateur astronomy, there is a lot of interesting and useful material for experienced observers, too.

We have tried in this book not to mislead the beginner and make him or her disappointed when seeing celestial objects live for the first time. Many of the wonderful, breathtaking things we can see only when we have had many observing experiences! And each observer will react individually to what is seen. Some observers find it is wonderful to be able to see distant galaxies at all – galaxies that are millions and millions light years away – even if what they see is only a modest, featureless, few arc minutes spot of light. For other people it is exciting to separate a double star, which is seen as one with the naked eye. To yet others, it is breathtaking to observe the center of our galaxy and know what is happening there, although the clouds of gases and dust completely obscure the sight. Every observer must make those kinds of judgments for him or herself!

There's nothing really brand new in the book, nothing one can't find in other books or on the Internet. But it is hoped that what is included here is written and organized in such a way that the reader will enjoy reading and using the book!

At the end, I would like to thank Dr. Andreja Gomboc and Dr. Tomaž Zwitter from the University of Ljubljana, faculty members of mathematics and physics, for examining the manuscript and alerting me to some mistakes.

I would also like to thank the staff at Springer, who were all helpful during the preparation of the book, and especially two persons: John Watson, who was my first contact with Springer and who encouraged all along the way, and my dear editor Maury Solomon, who led me from manuscript to the finished book.

Astronomy is really a splendid science, and stargazing is a really wonderful hobby. Just think about how many miraculous things we have discovered about the universe from this little snippet of a world on the margin of a huge galaxy spinning through space! And when we turn over the leaves of this book we can see how many fascinating things there are to observe through binoculars, which cost less than dinner for two at a nice restaurant!

Ljubljana, October 2008 Bojan Kambič

PART ONE

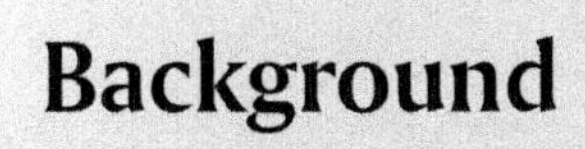

CHAPTER ONE

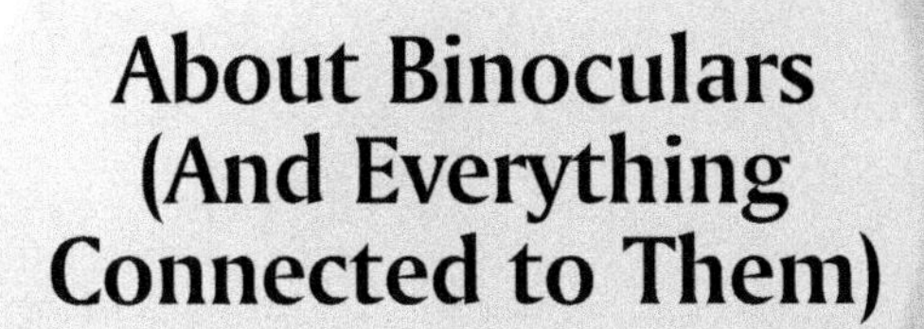

About Binoculars (And Everything Connected to Them)

The discovery of the telescope, an optical tube with lenses that magnify the observed bodies, is ascribed to the Dutch optician Hans Lippershey, who is thought to have built the first useful telescope in 1605 (Figure 1.1).

In October 1608, a book on optics was published and it described the telescope for the first time. In the same year the famous Italian physicist and astronomer Galileo Galilei made his first telescope, which had a 4× magnification (Figure 1.2). With this telescope he looked at the Moon, the Sun, the planets, and the stars and was surprised when he discovered that he could see things not visible with the naked eye.

Figure 1.1. Hans Lippershey (1570–1619)

B. Kambič, *Viewing the Constellations with Binoculars*, Patrick Moore's Practical Astronomy Series,
DOI 10.1007/978-0-387-85355-0_1, © Springer Science+Business Media, LLC 2010

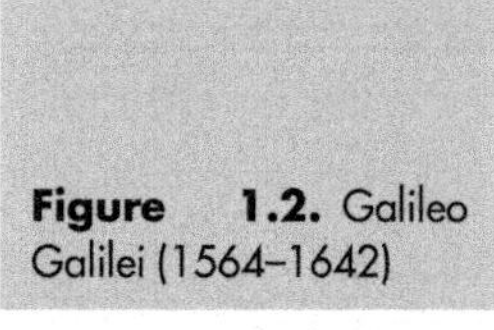

Figure 1.2. Galileo Galilei (1564–1642)

Astronomy historians are still debating whether Galileo really was the first one to think of pointing the telescope toward the night sky. He was most certainly the first to publish his astronomy findings in a book entitled *Sidereus Nuncius,* which came out in March 1610. This was also the year that most astronomy literature mentions as the decisive turning point in astronomical observation.

All important astronomers of the time, from Kepler through Newton to Huygens (to mention just a few of the most prominent names), tried to improve the telescope or construct new types, but that's another story, one that could fill a book as least as thick as this one.

As early as 1608, Dutch opticians joined two identical optical tubes and thus developed the first binoculars. They noticed that looking through both eyes is not as tiring as looking through one and is much more natural. And the binoculars could be sold for twice the price also!

The first telescopes cost a fortune and could be afforded only by the richest people. With the development of the optics industry, though, they became cheaper and more available.

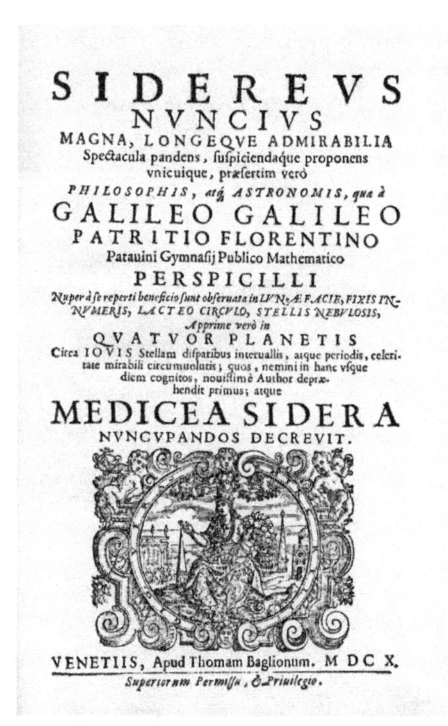

SIDEREVS
NVNCIVS
MAGNA, LONGEQVE ADMIRABILIA
Spectacula pandens, suspiciendaque proponens
vnicuique, præsertim verò
PHILOSOPHIS, atq; ASTRONOMIS, quæ à
GALILEO GALILEO
PATRITIO FLORENTINO
Patauini Gymnasij Publico Mathematico
PERSPICILLI
Nuper à se reperti beneficio sunt obseruata in LVNÆ FACIE, FIXIS INNVMERIS, LACTEO CIRCVLO, STELLIS NEBVLOSIS,
Apprime verò in
QVATVOR PLANETIS
Circa IOVIS Stellam disparibus interuallis, atque periodis, celeritate mirabili circumuolutis; quos, nemini in hanc vsque diem cognitos, nouissimè Author depræhendit primus; atque
MEDICEA SIDERA
NVNCVPANDOS DECREVIT.

VENETIIS, Apud Thomam Baglionum. M DC X.
Superiorum Permissu, & Priuilegio.

Figure 1.2A. Front page of the famous book Sidereus Nuncius

Why Binoculars?

Binoculars are refractors. Their objective consists of a gathering lens, or a number of lenses if the binoculars are of higher quality. Every pair of binoculars has a mark on the casing along the lines of 8×30, 7×40, 10×50, 15×70, etc. The first number represents the magnification and the second denotes the objective diameter in millimeters. The greater the magnification, the closer the observed bodies seem. The larger the objective, the more light gets in and the brighter the picture; in other words, with a larger objective we can see dimmer celestial bodies.

From all this we can deduce the following: the best binoculars are those that have the greatest magnification and the largest objective diameter. Although this is true, the price skyrockets along with the size of the objectives. An average pair of binoculars with an 80-mm objective diameter is

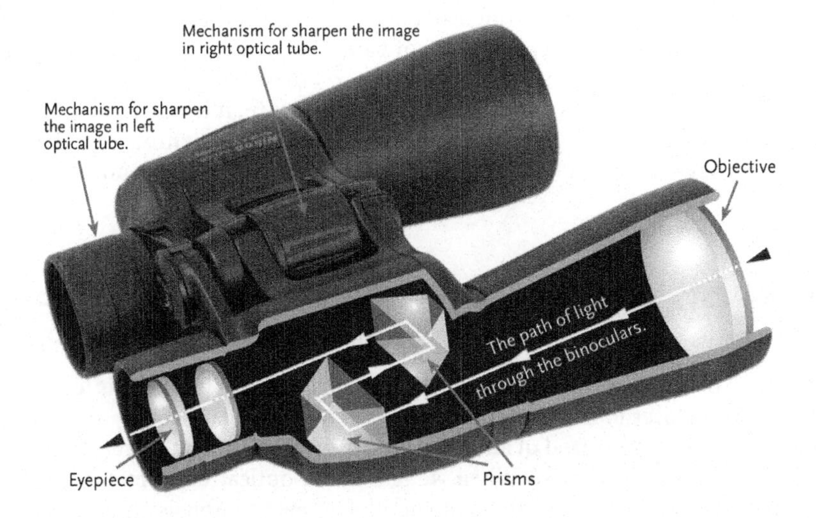

approximately 10 times more expensive than that with an average 50-mm objective (comparing the instruments of the same manufacturer, of course). The magnification is also limited by the shaking of the image when the binoculars are held by the hand. Binoculars that have more than 10× magnification have to be attached to a stable tripod.

The image that we see in the binoculars is usually right side up, as we see it with the naked eye, only much bigger. We achieve this with the two prisms that are placed between the objective and the eyepiece. The prisms shorten the tubes of the binoculars, which improves the three dimensionality of the observed image. This is a useful characteristic for land observation (which is considered the primary function of the binoculars) but unimportant for astronomy purposes.

Field of View

An important characteristic that some manufacturers print on the instrument's casing is the field of view, which is the size of the field that you can see when you look through the binoculars. This depends on the magnification and type of eyepieces. For a rough estimate, the field at 7× magnification is 7°, at 10× magnification it is 5°, and at 20× magnification it is 3°. If the binoculars are equipped with better wide-angle eyepieces, the field of view can be 25–30% bigger. It is because of the large field of view that the binoculars are an ideal instrument for beginners, and the experienced amateur astronomer also expects from it exactly this – a broad field of view that cannot be achieved with a telescope.

A large field of view enables the beginner to have a better orientation in the sky and find objects more easily, while also offering the possibility of panoramic observation of stars, star groups and clusters, and exciting journeys through the denser parts of the Milky Way – in fact, all things that cannot be viewed when looking through a telescope. Therefore, it is not at all surprising if we see a serious amateur next to his or her 20- or 30-cm telescope watching the sky with a pair of binoculars. Telescopes and binoculars often supplement each other and do not exclude each other.

Exit Pupil

An important piece of information concerning binoculars is the exit pupil. This term is not commonly known, so let's take a closer look at it here.

Every optical system is comprised of an objective, an eyepiece, and our eye(s), the eye being just as important as the first two on the list. When we talk about optical instruments' pupils we have two things in mind: the entry pupil and the exit pupil. The entry pupil is the opening in the optical instrument through which the light enters. In most binoculars this is the same as the objective diameter. The exit pupil is behind the binoculars; it is where the light exits the binoculars. We see it as a small circle of light in the eyepiece if we turn the binoculars toward a bright wall or the daytime sky and look into the eyepiece from a couple of dozen centimeters away. This small bright circle is the virtual image of the entry opening of the binoculars, and we get its size if we divide the entry pupil diameter with the magnification. The exit pupil cannot be measured with a ruler; it can only be estimated (Figure 1.3).

It is an unwritten rule that the size of the exit pupil should be the same or smaller than the size of the pupil of your eye. This should help you pick when purchasing binoculars or eyepieces. But, of course, this is not as simple in practice as it sounds.

The pupil is not always the same size. In strong light it shrinks down, and in darkness it expands as much as it can (Figure 1.4). For a while it was believed that when fully dilated a human pupil measured 7 mm. This piece of information governed all producers of optical instruments, who were convinced that the most ideal exit opening of every pair of binoculars and telescope was 7 mm. However, this is no longer considered to be absolutely true, for people are different. Some of us have owl eyes with a pupil that can be almost 9 mm in diameter. Others can achieve a maximum dilation of only 4 mm, regardless of the darkness. Generally, we can say that children and younger people have a large pupil that becomes smaller with years. However, we can find 70-year-olds who have a bigger pupil than many a teenager. So, why are we discussing this in such detail?

Figure 1.3. The exit pupil is a small disc of light that can be seen behind the eyepiece. Its size depends on the magnification and the entry pupil diameter

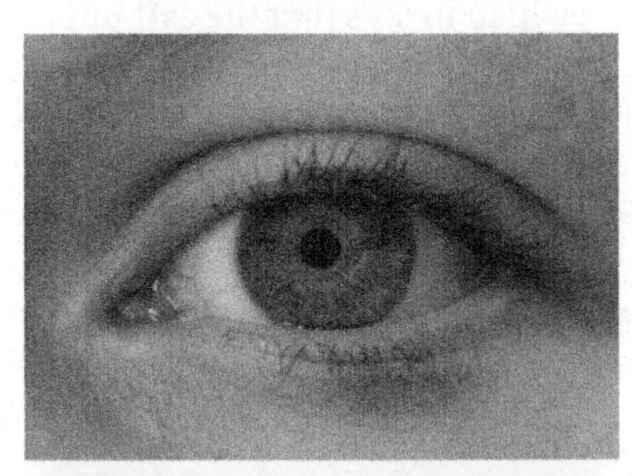

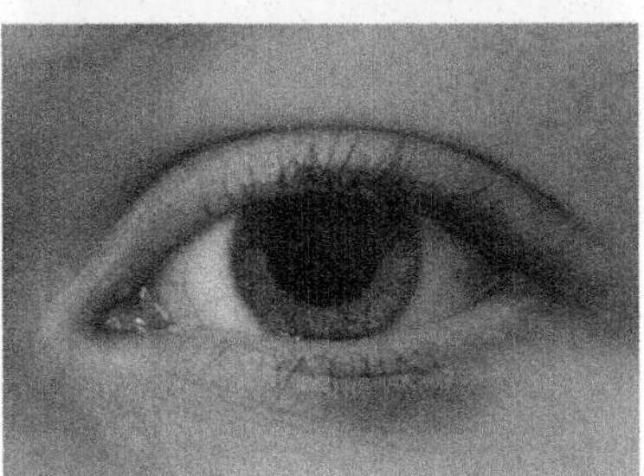

Figure 1.4. At daylight the eye's pupil is small; when dark it dilates, so that more light can fall upon the retina

In practice, it has been shown that for astronomical observations it is best not to choose the binoculars with the greatest possible pupil size. Our eye is not perfect and sees most poorly on the edges. Anyone who already has experienced watching the stars using a lowest possible telescope magnification has noticed that even if the objective and eyepiece are perfectly manufactured, the stars never appear as dots but are more or less smeared blots. This is a consequence of the imperfection of our eyes and not the eyepiece. We can demonstrate this in the following way: if we place a medium-bright star on the edge of the field of view, it will appear as a smudged dot with protrusions coming out at the sides. If we turn our head around and the protrusions spin with us, then they are a consequence of our eyes. If they turn when we spin the eyepiece, then they are a consequence of a poor eyepiece.

If we therefore wish to make the best use of the binoculars or a telescope, it is best if the exit pupil of the instrument is no bigger than between 2 and 5 mm. And it is no coincidence that these values are the same limits at which we see clearly (without optical help) in everyday life.

With the exit pupil of the binoculars we can estimate the true size of the objective. The most mistakes that occur on lenses also occur on the edges, due to the poor glass grinding. In order for these mistakes not to influence the quality of the picture the manufacturers place a diaphragm between the objective and the eyepiece, with which they limit the light that comes to the eyepiece. This, of course, means that instead of a 50-mm objective we get a 40- or even 30-mm one. Knowing only the objective diameter, therefore, does not tell us everything; we get the entire picture only in combination with the exit pupil.

Choosing Binoculars for Astronomical Observations

Because you would like to observe dim celestial bodies, the objective diameter should be as large as possible. Taking into account the different prices and availability, it is still best to select a 50-mm objective. Smaller ones are not that much cheaper, and the larger ones are much more expensive. The magnification should be 10×, which means that the exit pupil is at an optimum 5 mm and at the same time the magnification still enables us to observe the sky without a tripod. The field of view should be at least 6° (and not 5°), which means that the manufacturer used the better eyepieces, which are more appropriate for astronomical observations. Numerous experienced observers agree with this choice, and it is with such binoculars (10×50, 6°) that we have observed and described the celestial objects in this book.

Another important reason lies behind this choice of binoculars. Every good amateur telescope has a finderscope, with which we help direct the telescope in the desired direction of the sky. In better

instruments, this finderscope is an 8×50 or 10×50 auxiliary refractor. Thus, if you get used to the view of the night sky through such binoculars you will later on find it much easier to switch to observing and searching for certain celestial objects with a telescope.

In this book we will use the general term "binoculars" to mean 10×50 binoculars, which have a specific, nonvarying magnification. For binoculars of different magnifications, we will call attention to the magnification size. The word "telescope" will be used to refer to an instrument that can have different eyepieces and thus achieve various magnifications.

A Few Tips When Purchasing Binoculars

- Be sure to check while you are still in the shop whether it is possible to completely sharpen the picture in both optical tubes and whether the picture is sharp over the entire field of view.
- In order to obtain a good picture, the two optical tubes of the binoculars should be completely parallel. If the distance between the eyes has been correctly adjusted, the picture from both tubes will merge into a single round field of view. If you see two covering fields of view, there is a problem, and your observations will be disrupted to a certain degree.
- Dark objects on a light background or light objects on a dark background should not have strong rainbow edges. Rainbow edges are the sign of poor optics.
- Estimate the size of the exit pupil. If it is too small, the binoculars probably have poor objectives, and a diaphragm was used by the manufacturer to reduce the size of the entry pupil.
- A common problem with the objectives that you might miss when observing in daylight is coma. Toward the edge of the field of view the stars are increasingly distorted. Instead of bright spots we see lines or smudges. Such binoculars are fine for daylight field observations but not for astronomical observations. Pinpoint stars are the hardest test of the quality of the objective.
- With good eyepieces you can move the eye a centimeter or more from the eyepiece and not lose much of the field of view. Such eyepieces are better, and the binoculars will, in most cases, be more expensive. But if you have to lean the eye on the eyepiece in binoculars in order to see the entire field of view, you will have continual problems with lenses that will constantly smear from contact with eyelashes.
- For astronomy observations you should not use binoculars that have so-called red, green, or blue optics. Even though the picture in such binoculars is extremely sharp, it is always slightly colored (red, green, or blue), and the stars in the field of view also appeared slightly colored.
- There are also binoculars that are dedicated entirely to astronomical observations. Such instruments have zenith prisms in front of the eyepieces so you can watch the celestial objects high in the sky without breaking your neck. Of course, these are slightly more expensive.

Finally, let me answer the question that was asked above – why binoculars?

- Because you might have them already at home.
- Because they are inexpensive (at least compared with a telescope) and easy to handle.
- Because with them and this book you can immediately start your stargazing, no additional equipment required.
- Because even if at a later time you buy a telescope, you can still use the binoculars for certain kinds of stargazing or for bird-watching or even for going to the opera!

How to Measure the Diameter of the Eye Pupil at Various Light Levels
Take a pencil or other small sticks with a diameter of 7 mm. Place it vertically in front of the eye so that it touches the cheekbone and brow. Close the other eye and look toward a strong source of light. The pencil will have a dark, opaque central part and blurred edges. Now look toward the dark part of the room and wait for a few minutes. Watch the width of the dark, opaque central part. If you cannot see it any more, and at least a little bit of light is coming through the central part of the pencil, your pupil measures 7 mm.

Tripods

Is a tripod necessary or not for astronomical observations with binoculars? It's true that with the smaller magnifications you can watch with the help of your hand. But you should be aware that when doing this you will miss out on all the pleasures offered by observational astronomy.

When you are looking for a faint object and are moving with your binoculars from the bright starting star toward the object, you have to look at the star chart again and again. If you have the binoculars on a tripod, you can simply leave them pointing toward the last known star pattern, look at the chart, and continue with your search. It is not clear how you can do this without a tripod, but it is clear that an observer without a tripod will soon give up the search (and maybe even astronomy!).

Even if an object, for instance a large and bright open cluster, is easily found, you can only notice the details once you have been observing the object attentively for 5 or 10 min, sometimes even longer. There are very few people who can hold their hands still for such a long time, and the odds that you are among them are against you.

Faint objects that are on the limits of visibility with the binoculars are usually not seen without the use of a tripod. In other words, if you are not using a tripod, you will not see objects that are roughly one magnitude weaker than the limiting brightness of the binoculars. These objects are marked in this book (in the descriptions of the constellations) with a special symbol of an observer falling over. And be aware that there are a lot of them!

Stable placement of the binoculars is almost as important for astronomical observation as the optics quality. Unfortunately, a good tripod costs almost as much as a good pair of binoculars! However, there are some cheaper options. A photography camera tripod, for example, can be adapted for your purposes. If you are more skillful with your hands, you can make your own wooden or metal tripod that will enable you to observe the night sky in comfort. Some ideas can be found in the accompanying photographs (Figs. 1.5, 1.6, and 1.7).

Figure 1.5. An excellent massive, *solid*, and stable tripod, suitable for binoculars of all sizes

Figure 1.6. A special accessory allows you to place the binoculars on any photographic tripod

Figure 1.7. This tripod, which consists of water pipes, can easily be made by one person. Many different plans for constructing a tripod can be found on the Internet

A useful tripod needs to fulfill the following conditions:

- It has to be just the right weight so that it is not thrown off balance by a little breeze or does not rock backward and forward for a while every time you touch it. On the other hand, it is good if it is portable, so that you can carry it up a hill with no great difficulty to get to a better observing spot. This can be solved by making the tripod easy to disassemble into a few separate pieces.
- It must be possible to counterweight the binoculars with a weight on the other side of the axis. This is the only way to ensure that it does not tip during observation or that you do not constantly need to hold it in order to prevent it from slowly drifting downward, because if that happens, you will transfer vibrations from your hand onto the binoculars.
- The tripod or any other form of mount has to allow you to adjust the height of the binoculars, so that you can use them to observe the sky low above the horizon or high near the zenith.
- When you are watching close to the zenith you are standing under the binoculars. A good tripod is constructed in such a way that there is room for the observer underneath the binoculars. That is why the supporting sticks for binoculars are so long in the photographs.

Taking Care of the Optics

Binoculars and telescopes are optical instruments and need to be properly looked after. Even if they are small they still deserve the best of care, especially if you are going to use them at the limits of their capabilities to see faint objects or small details.

Every optical instrument gets dirty. Dirt on the lenses and mirrors scatters the light and reduces the contrast of the image. The consequence of this is that the dark sky is no longer as dark, while bright objects are not as distinctive. But taking proper care of the optical parts of your instruments does not mean that you have to clean them constantly. It is more important that you try to keep them as dust and dirt free as possible (Figure 1.8).

Figure 1.8. If you look at the lens of binoculars that you have been using for a few years under a magnifying glass, you will be shocked. However, this should not convince you to constantly clean the lenses. You will almost certainly create irreparable scratches on them that will ruin the quality of the image far more than the dust particles

Prevention

Never wipe dusty optics! Normal household dust is very abrasive because it contains miniscule stones that the wind carries around. If you wipe a dusty lens or mirror, you will rub these particles against the surface and cause the dents and microscopic scratches that you will never be able to remove. That is why it is extremely important to prevent dust from gathering in the first place.

The right tactic to use against dirt is, therefore, defensive. Whenever you are not using the binoculars, the lenses should be covered with their protective caps. If you do not have them or you have lost them, you can make new ones yourself. You can use any sort of plastic box or cap as long as it is roughly the right size. Even a plastic bag and an elastic band are better than nothing. The most important thing is that you are consistent with your covering of the lenses.

You should never touch the surface of the lenses with your fingers. The acids found on your skin can, over time, damage the optic coatings. If by accident you leave a finger mark on the optics, you should clean it immediately. We will describe the process later on.

We cannot avoid dust altogether. However, in small quantities it has a surprisingly small effect on the quality of the image. Serious research has shown that dust can cover 1/1000 of the optic surface with practically no effect on the image quality.

And here's another piece of advice. Never clean the optics just because you flashed with a battery on the objective in the middle of the night. No lens could withstand such a rigorous inspection.

Cleaning the Optics

As we said above, you should leave the dust on the optics alone for one important reason. Dirty lenses and mirrors can be cleaned anytime, but scratched optics remain scratched forever. If you do not conduct your cleaning properly, the cleaning causes big or small scratches. In some cases you may scratch the surface even when you clean properly, and scratches definitely have a worse influence on quality of image than dust. This is why you should clean the optics only very rarely (once every few years). If you make a serious effort to prevent dust from gathering on the optics, this should be sufficient. When you finally decide it is time to clean the lenses, be sure you are gentle and careful with whatever you do and be sure to follow the instructions below meticulously.

The surface that you are cleaning is not glass but an optical coating that is usually softer and thus even more vulnerable than glass. The main anti-reflex coating on the lenses is magnesium fluoride, which can be very soft if the manufacturer applied it at a low temperature. Good magnesium fluoride coatings are relatively hard. The newer and multilayered coatings are also soft, but the manufacturers toughen them. Unfortunately, you cannot know whether the coating on your optics is soft or hard.

Eyepieces and binoculars are manufactured in such a way that dust and dirt cannot penetrate inside. This means that you should never disassemble them! If you follow this rule, you will have to clean only the exterior optical surfaces.

Always remove the dust first and only then clean the surface with liquid. The traditional method for removing dust is to very gently clean the lens with a camel hair brush, which can be purchased in shops that sell photographic equipment. These brushes have very soft hair, which presses very gently on the dust particles resting on the optic surface.

The surface should be cleaned slowly, with the brush always moving in the same direction (Figure 1.9). Touch the surface with the brush and gently rotate it. You rotate the brush so that the dust particles, which will fly off, will land on the dirty and not the already cleaned surface. This also helps you to lift it away and not drag the dust particles across the lens surface. After every move you make with the brush, you should shake the dust off! It does not matter if it appears as if nothing is happening and if you feel a bit silly while performing this task. Take your time and do this job properly. When you are not using the brush it should be safely stored in a plastic bag. Before you start cleaning the lens, it is a good idea to practice using the brush on a glass surface that you have covered with a substance such as flour.

Figure 1.9. When cleaning with a brush it is very important that the dust particle stays in contact with the glass surface for the shortest time possible. Gently touch the surface with the brush and rotate it. This demands a steady hand and some practice

Even after you have cleaned the dust from the surface of the optics, some of the toughest dirt can still remain. For this, you will need to use cleaning fluid. There are various types of lens-cleaning liquids. The simplest and most effective is to use clear isopropyl alcohol or methyl alcohol (methanol). You can purchase these in drugstores or pharmacies. By no means should you use alcohol substances with any additives (for instance, to protect against misting), because they can leave stains. If you wish to dilute the cleaning liquid, use distilled water only. Stores that sell photographic equipment also usually sell liquids (ultra-clean methanol) for cleaning lenses.

You will also need sterile cotton balls or special tissues for cleaning optics. Trickle a few drops of the liquid onto the cotton ball, and with gentle circular motions spread it all across the surface. If necessary, then wipe the surface with a fresh cotton ball.

Never press against the lens; the weight of the tissue is enough. Take special care at the edge, because you do not want to get any moisture in the gap between the lens and the casing, where it could trickle between the lenses. This might cause the dirt diluted by the liquid to travel to the lower parts and leave stains there. For the same reason you should also not apply the liquid directly onto the lens but always onto a cotton ball or a tissue.

There are special tissues that do not leave threads and are dampened with methanol. These can also be bought in stores that sell photographic equipment. When you use these, you should gently pull them across the glass with a different part of the tissue only once before switching to another part. Do not press against the lens; the weight of the tissue will suffice. If you decide to clean the lens with tissues, do not use tissues designed for cleaning eyeglasses; these are usually moistened with a liquid to guard against dew (usually a silica liquid), and they will leave a coating on the lenses.

On the eyepiece you should normally clean only the lens closest to the eye, for this is the one exposed to dust, grease from the eyelashes, and finger marks. The other lenses are safely hidden deep within the eyepiece casing and should normally not be cleaned. However, if you do clean them, clean them only for dust.

When you have done everything possible to prevent dust and dirt from gathering on the lenses, it is time to forget about this problem and devote your time and efforts to observation.

The Never-Ending Battle with Dew

During observation the first objects to disappear from sight are the dim stars, followed by those that are a bit brighter. The bright stars become larger and larger, and they are surrounded by a smear of light. After a while there is only darkness in the field of view of the eyepiece. If we point at the objective with a torch, we will notice that the optics are totally misted up. Many people would put the binoculars away and go to sleep at this moment. Even though you might have been anxious to make the most of the clear night for observing or photographing, moisture can ruin all your joy. Some experienced amateurs (and even writers of telescope manuals) recommend that you dry the moist optics with a

hairdryer. This is effective, but it can also be very dangerous. Dust particles that are on the objective and are blown away by the airflow from the hairdryer can scratch the lens. A much better way to protect the lenses from moisture is to prevent them from getting moist in the first place.

What Is Dew?

When the lens (or mirror) of an optical instrument is colder than the dew point of the surrounding air, dew or white frost condenses on it. Dew forms more quickly on a dirty surface because the dust particles function as cores for condensation.

Figure 1.10. If the temperature in the room in which you keep your binoculars oscillates, the optics under the protective caps will often mist up. In order to prevent this from happening, make a few holes in the protective cap and cover it with a net plastic fabric (such as Gore-Tex™ or printing foil). Dust will not gather on the lenses, but the moisture will evaporate during the temperature oscillations

Never wipe misted lenses, because you can damage the surface and they will mist up again almost immediately anyway.

The binoculars get misted almost every time you bring them into a warm room from the outside. In order to prevent the binoculars from gathering too much dampness, you should cover the lenses with the protective caps while the binoculars are still outside. Once the binoculars have warmed up to room temperature, remove the protective caps so that the moisture, trapped underneath them, can evaporate. Only once the optics are totally dry do we cover them up again. Never let moisture remain trapped under the protective caps, because smears or something even worse can appear on the optical surfaces.

Some experts even recommend not using protective caps at all. If the temperature in the room you keep your binoculars is changing, the moisture from the air that is trapped under the protective caps condenses on the lens. You can avoid this by allowing the protective caps to "breathe" slightly, so the moisture can evaporate during the temperature changes. In order to allow for this you can cut out little holes in the protective caps and cover them with a thickly woven plastic net fabric that is permeable to air (and moisture) but does not let through dust or leave little threads all over the place. It is best to store the binoculars in an unheated dry room: in a garage, in a closed balcony, or on a terrace. Avoid damp spaces, by all means.

Binocular manufacturers put little bags with silica gel into the protective case, and these bags suck the moisture from the air. Leave the bags where they are. However, keep in mind that the silica gel capabilities of absorbing moisture are limited, so you should dry them out every now and then by putting them into the kitchen oven and heating them to 50°C.

Dew Point

Imagine a closed glass receptacle in which there is some water and upon of it mostly dry air. We know that the water will evaporate and that the air will become humid because of it. However, the air cannot absorb endless amounts of liquid. Eventually the evaporation stops. At that point we say that the air is saturated with moisture. How much water vapor can be accepted by the air before it becomes saturated with moisture depends on its temperature. The higher the temperature, the more water vapor the air can hold before it becomes saturated with moisture.

If we now cool the receptacle with saturated moist air (for instance, by placing it into the refrigerator for a while), the water vapor from the air will condense, and water drops will gather on the sides of the receptacle. This is when the moist air reaches its dew point. The dew point is the temperature at which the moisture from the air starts to fall as mist or dew.

It is also possible to measure the moisture in the air. One way is to state how many grams of water vapor are in a cubic meter of air. This quantity is called absolute humidity. Relative moisture is defined as a quotient of the absolute moisture and the saturated moisture at a specified temperature. This quantity is not expressed in units but in percentages.

Let's say that we have a glass bowl at room temperature (20°C). At first the air above the water is dry. Its absolute humidity is 0 g/m^3, while its relative humidity is 0%. We know that a cubic meter of air at 20°C can accept 18 g of water vapor before it becomes saturated with moisture. If there were 14 g of water in a cubic meter of air after 2 h of evaporation, then its relative humidity would be 78%. The water continues to evaporate. When 18 g of water evaporates into a cubic meter of air, the evaporation stops. Relative humidity reaches 100%. If we wanted the water to continue evaporating, we would need to heat the bowl. In our experiment, we cooled the bowl. Let's say that we have cooled it to 0°C. At this temperature, a cubic meter of air can accept a mere 5 g of water. That means during the cooling process 13 g of water condensed on the bowl walls.

Something similar also takes place in nature. The air always includes some water vapor. During the day, when it is warm, the water evaporates and the air becomes moist. In the evening, when the air starts to cool, it becomes saturated with moisture, and at further cooling the water starts to condense into drops of fog or dew. Of course, moisture does not fall from the sky only when the air is cooling down. It can condense from the surrounding air much before that – before the air becomes saturated with moisture – if the objects have a low enough temperature, i.e., lower than the dew point. We can check this with another experiment. When we take a cold bottle of soda from the refrigerator, it mists up quickly, if the bottle's temperature is lower than the dew point temperature. Binoculars are similar to such a bottle, and it is a rare night when the optics do not mist.

How is it possible, you might ask, that the binoculars are colder than the surrounding air? When we bring them outside from the warm room in which they kept, it slowly cools down until they reach the temperature of the surrounding air. This would be true if the heat was transferred by the process of convection only. However, because the binoculars also exchange heat with radiation, they can cool down below the temperature of the surrounding air. If you do not believe this, you should remember how the morning Sun heats us up (with radiation) when we stand in its rays, even though the air is still cold. During the day Earth and everything on it receives radiation from the Sun, while during a clear night the heat is radiated back into space.

Protective Tube

The simplest way to protect the optics from moisture is by covering the casing of the binoculars in front of the objective with a tube. The longer the tube, the better it will perform its task. Its main effect is to reduce the cooling of the optics by radiation. The tube also helps slow down the cooling of the optics due to convection and prevents disturbing light from the side to enter the binoculars. The above statement about the length of the tube should not be taken too literally.

Usually it will suffice if the protective tube is 1.5 times or at most 2 times longer than the objective diameter.

It is simple to make a protective tube at home. In a hardware or crafts store, or any place where they sell rubber and plastic products, you can usually buy an appropriate size rubber foam piece. Roll the piece into a tube and glue it together, so that it fits snugly against the binocular casing (Figures 1.11 and 1.12). Such a tube is cheap, effective, and light. You just have to make sure that whatever glue you use to attach the tube is resistant to moisture.

Figure 1.11. The protective tube for the objective of the binoculars, made from rubber foam

Eyepieces can also become misty. Even though they are heated by the warmth from your face and the cooling down time is thus prolonged, the time is shortened by the moisture in your breath and from your eyes. The most effective method for preventing moisture on the eyepiece is a rubber part that sits at the end of the lens closest to the eye, which works exactly like the tube at the end of the objective – it reduces cooling. Better binoculars have such rubber parts on the eyepieces already when you purchase them. If they do not have them, you can make them at home. All that we said about a protective tube for binoculars also holds true for telescopes, finderscopes, photographic lenses, and tracking telescopes.

Electric Warmers: Active Protection from Dew

The protective tube in front of the objective represents passive protection from dew, since it only slows down the cooling process of the binoculars. If you keep the objective constantly warm, it will mist only on rare occasions. Such warmers can easily be purchased, but it is even better if you make one yourself, for then you can adjust it to your needs and observing conditions.

The heat in a warmer comes from an electric current that runs through a resistor. We call this heat Joule's heat, after the nineteenth-century English physicist Prescott Joule. To make the warmer, you should use standard resistors, which are cheap and can be purchased at any store that sells electric parts.

Before you start with your calculations, you have to know how strong the warmer needs to be. A 3-W warmer can effectively warm a 20-cm Schmidt–Cassegrain lens, and a 1.5-W warmer is fine for binoculars, smaller telescopes, classic finderscopes, and eyepieces.

Joule's heat dissipation or the power of the warmer is

$$P = RI^2 = U^2/R = UI$$

where P is Joule's heat dissipation (its unit is watt, or W), R the resistance of the resistor (its unit is ohm, or Ω), I the current that runs through the user (its unit is ampere, or A), and U the voltage on the user (in volts, or V). The power of the warmer or Joule's heat dissipation is known (3 W or 1.5 W). The voltage depends on the source that you are going to use. **Partly because it can often be so damp outside during observations that water will drip from the binoculars, partly because you may be far away from civilization when you conduct your observations, and finally for your own personal safety, it is better not to use warmers that can be plugged into a main outlet (220 V/110 V). Instead, you should opt for a warmer that is connected to a 12 V battery.**

Now let's calculate what sort of a resistor you will need:

$$P = U^2/R \Rightarrow R = U^2/P \Rightarrow R = 122V^2/1.5\,VA = 96\,\Omega$$

So, for a 1.5-W warmer you will need a 96-Ω resistor, and for a 3-W one you will need a 48-Ω resistor.

The warmer is made up of a series of resistors that are connected together. When resistors are linked in a series, the individual resistances are added together (Figure 1.12). Because you want the warmth to be equally distributed around the optics, you will make the 96-W warmer by joining eight 12-Ω resistors in a series. If the circumference of the tube of the binocular is bigger, you will use twelve 8-Ω resistors and so on. Standard resistors may not be available for the resistance you want, so you will have to use resistors that are as close as possible to the calculated one.

Another important piece of information that you should know before you purchase resistors is the maximum power that they can withstand before burnout. If you have connected eight resistors in a series, and they jointly use 1.5 W, then one resistor uses approximately 0.2 W. If you purchase resistors that can withstand 1 W of power, the warmer will work trouble free for a long time.

You can also calculate how long you will be warming your optics with a classic 12-V car battery that has the capacity of 36 A-h. This information tells us that the battery can release electrical energy with the current of 1 A for 36 h before the battery dies.

So, what sort of current will run through our 1.5-W warmer?

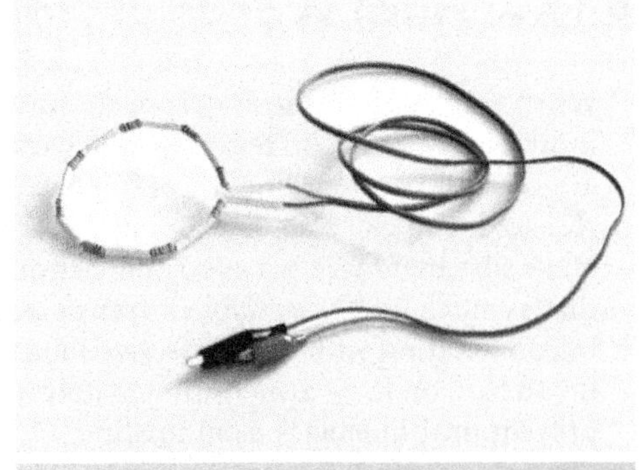

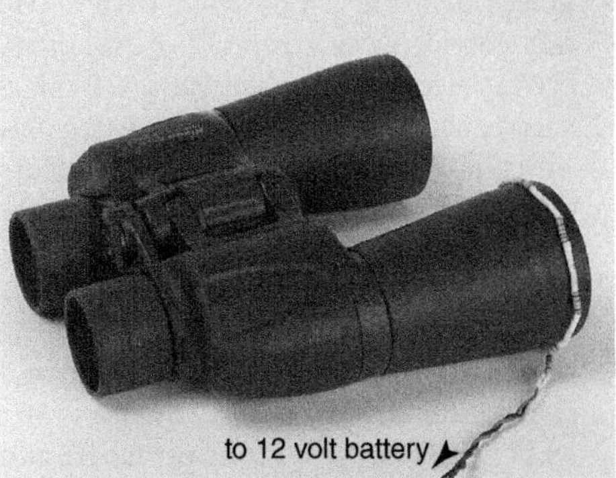

Figure 1.12. *Top*: The warmer is made from resistors that are connected in a series. All metal parts should be kept well isolated, so that you do not create a short circuit. *Middle*: The warmer is attached to the casing as close as possible to the optics. *Bottom*: The most effective is the combination of the warmer and protective tube

$$P = UI \Rightarrow I = P/U = 1.5\,W/12\,V = 0.125A$$

The car battery will allow 288 h of warmth before it will need recharging. That is more than enough, even if you have multiple warmers connected to the battery, telescope drive, and other things.

A Few Final Words of Advice

- When you solder the resistors, make sure you have carefully insulated the metal wires between the individual resistors so that they don't accidentally short circuit the warmer. The binocular casings are usually painted, but it doesn't hurt to be careful. In the store in which you purchase the resistors, you can also usually buy insulating tubes.
- Make sure that there is a good connection between the resistors and the casing of the binoculars, so that as much heat as possible is transferred to the binoculars. The warmer will function even better in combination with the protective tube.
- If you turn on the warmer immediately after you have set up the binoculars, you should be able to prevent the binoculars from misting.
- The golden rule is that with a battery you should get at least 12 h of undisturbed observation. That way you won't ever run out of electricity in the middle of the night.
- If you perform your observations in places that are extremely damp or very dry, you should try out various warmers (resistors are extremely cheap, and the warmers are simple to make) to discover which power is right for your binoculars. If you also have a voltage regulator between the battery and the warmer, you can monitor the voltage on the warmer and adjust the power of the warmer as necessary.
- Once you have mastered the technique of how to make warmers, you will find it easy and cheap to assemble them for all your needs - your binoculars, individual objectives of your photographic lenses, your eyepiece, your finderscope, and your tracking telescope or telescope objective. You can make warmers that will not be so strong as to cause air turbulence around the objective and so weak as to not prevent the optics from misting.

CHAPTER TWO

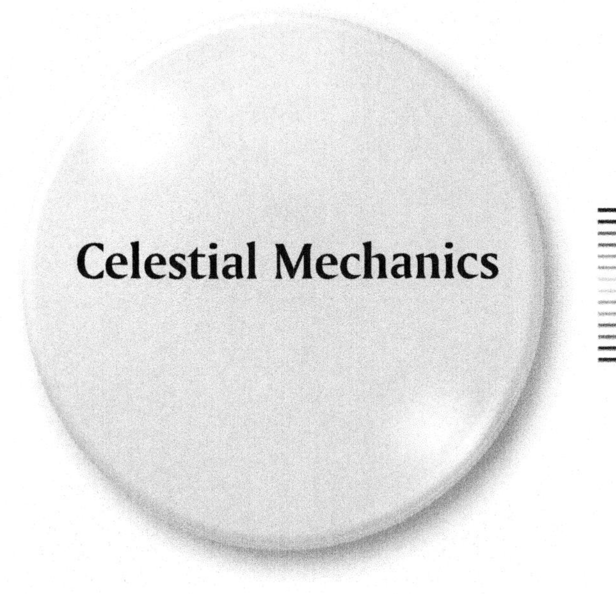

Celestial Mechanics

The Constellations

Do you remember the first night that you stood for the very first time under a clear night sky and gazed with bewilderment at its beauty? And did you think something like this as you watched: "How many stars there are! How can anyone make any sense of this mass of stars?"

An attentive observer will soon notice that individual bright stars that are rather close together in the sky seem to form simple geometric shapes – squares, rhombuses, crosses, circles, arches. Giving a name to these shapes in the sky makes them more familiar to you and easier to locate again. That is probably how the first constellations originated and obtained their names. We probably will never know who the first person was to group stars into constellations, but it must have occurred a long time ago, most likely when human beings started walking erect, looked up at the sky, and were bewildered by its beauty.

By adding to the geometric shapes the dimmer stars the constellations changed from simple figures into images of gods, heroes, animals, and everyday objects. All peoples of this world have projected their beliefs onto the sky. Modern astronomers use the constellations from the ancient Greeks, which include not only objects and animals in the sky but also ancient mythological heroes. That is why various groups of constellations tell us stories about Greek myths and legends (Figure 2.2).

By the 1930s, there was sheer chaos in the sky. Apart from the classical constellations found on sky charts, there were also all of those that had been marked in the sky throughout the long history of astronomy. Especially in seventeenth and eighteenth centuries, astronomers almost competed with each other to see who could invent more new constellations from the leftover stars (Figure 2.5).

By sailing the South Seas astronomers came to learn about stars that were not visible from Europe or North Africa, and the need to introduce new constellations for the southern celestial hemisphere

B. Kambič, *Viewing the Constellations with Binoculars*, Patrick Moore's Practical Astronomy Series, DOI 10.1007/978-0-387-85355-0_2, © Springer Science+Business Media, LLC 2010

Figure 2.1 The winter Orion is one of the richest constellations in the sky. Among its numerous bright stars are the seventh and ninth brightest stars in the sky – slivery-white Rigel and the orange Betelgeuse. Apart from them there are also some hidden treasures in the constellation that are revealed by even a small pair of binoculars – numerous nebulae and clusters – among which the biggest attraction is surely the large and bright Orion Nebula

Figure 2.2 The star chart of Orion from the wonderful Hevelli's *Uranography* that was published in 1690

appeared. Native inhabitants already had numerous suggestions for these constellations, and some of these were taken into account, while others were not.

Most of the constellations included only the brightest stars; nobody knew to which constellations the fainter stars belonged. With the huge leaps in development of modern observational astronomy at

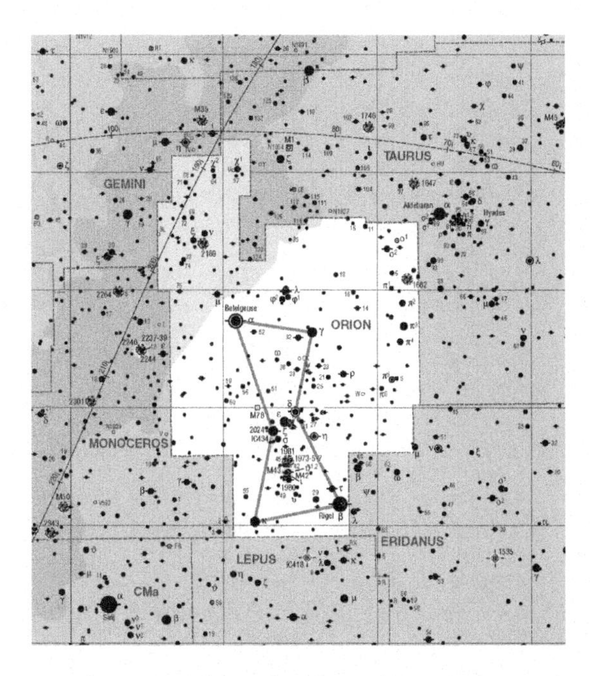

Figure 2.3 A modern star chart of Orion with the constellation borders

the beginning of the twentieth century, things began to change. In 1934, the International Astronomical Union divided the sky into 88 constellations with precisely defined borders. Today we know exactly which constellation even the faintest star belongs to. For today's astronomer, a constellation is a specific part of the sky and not merely a few bright stars that form its shape (Figure 2.3).

A constellation is thus a limited part of the sky, with stars and all nonstellar objects that usually have only one thing in common – they appear to be close together in our sky (and – as seen from Earth – seem to be roughly in the same direction). In reality some stars can be relatively close, while the others can be very far from each other and do not have any physical connection whatsoever. If we suddenly found ourselves on a different planet several hundreds light years away from the Sun, the night sky would look completely different to us.

Not all constellations are of the same size (Table 2.1), for in drawing the borders astronomers took into account historical facts. The largest constellation in the sky is Hydra, while the smallest is the famous Crux, which lies in the southern celestial hemisphere and was named in 1679.

Table 2.1 Alphabetic list of constellations

Latin name	Abbr.	Genitive	Meaning	Size [sq.°]	Page	Loc.
Andromeda	And	Andromedae	Andromeda	722	172	N
Antlia	Ant	Antliae	The Air Pump	239	182	S
Apus	Aps	Apodis	The Bird of Paradise	206	–	S
Aquarius	Aqr	Aquarii	The Water Carrier	980	183	S
Aquila	Aql	Aquilae	The Eagle	652	189	NS
Ara	Ara	Arae	The Altar	237	–	S
Aries	Ari	Arietis	The Ram	441	195	N
Auriga	Aur	Aurigae	The Charioteer	657	196	N
Boötes	Boo	Boötis	The Herdsman	906	202	N
Caelum	Cae	Caeli	The Chisel	125	206	S
Camelopardalis	Cam	Camelopardalis	The Giraffe	747	207	N
Cancer	Cnc	Cancri	The Crab	306	213	N
Canes Venatici	CVn	Canum Venaticorum	The Hunting Dogs	465	218	N
Canis Major	CMa	Canis Majoris	The Great Dog	380	228	S

Table 2.1 (continued)

Latin name	Abbr.	Genitive	Meaning	Size [sq.°]	Page	Loc.
Canis Minor	CMi	Canis Minoris	The Little Dog	183	233	N
Capricornus	Cap	Capricorni	The Sea Goat	414	234	S
Carina	Car	Carinae	The Keel	494	–	S
Cassiopeia	Cas	Cassiopeiae	Cassiopeia	598	236	N
Centaurus	Cen	Centauri	The Centaur	1060	243	S
Cepheus	Cep	Cephei	Cepheus	588	247	N
Cetus	Cet	Ceti	The Sea Monster	1231	252	NS
Chamaeleon	Cha	Chamaeleontis	The Chameleon	132	–	S
Circinus	Cir	Circini	The Compasses	93	–	S
Columba	Col	Columbae	The Dove	270	260	S
Coma Berenices	Com	Comae Berenices	Berenice's Hair	386	261	N
Corona Australis	CrA	Coronae Australis	The Southern Crown	128	268	S
Corona Borealis	CrB	Coronae Borealis	The Northern Crown	170	269	N
Corvus	Crv	Corvi	The Crow	184	271	S
Crater	Crt	Crateris	The Cup	282	271	S
Crux	Cru	Crucis	The Southern Cross	68	–	S
Cygnus	Cyg	Cygni	The Swan	804	272	N
Delphinus	Del	Delphini	The Dolphin	189	287	N
Dorado	Dor	Doradus	The Goldfish	179	–	S
Draco	Dra	Draconis	The Dragon	1083	288	N
Equuleus	Equ	Equulei	The Foal	72	287	N
Eridanus	Eri	Eridani	The River	1138	294	S
Fornax	For	Fornacis	The Furnace	398	296	S
Gemini	Gem	Geminorum	The Twins	514	297	N
Grus	Gru	Gruis	The Crane	366	395	S
Hercules	Her	Herculis	Hercules	1225	301	N
Horologium	Hor	Horologii	The Pendulum Clock	249	206	S
Hydra	Hya	Hydrae	The Water Snake	1303	307	NS
Hydrus	Hyi	Hydri	The Little Water Snake	243	–	S
Indus	Ind	Indi	The Indian	312	–	S
Lacerta	Lac	Lacertae	The Lizard	201	314	N
Leo	Leo	Leonis	The Lion	947	316	N
Leo Minor	LMi	Leonis Minoris	The Little Lion	232	325	N
Lepus	Lep	Leporis	The Hare	290	326	S
Libra	Lib	Librae	The Scales	538	330	S
Lupus	Lup	Lupi	The Wolf	334	332	S
Lynx	Lyn	Lyncis	The Lynx	545	334	N
Lyra	Lyr	Lyrae	The Lyre	286	335	N
Mensa	Men	Mensae	The Table Mountain	153	–	S
Microscopium	Mic	Microscopii	The Microscope	210	342	S
Monoceros	Mon	Monocerotis	The Unicorn	482	343	NS
Musca	Mus	Muscae	The Fly	138	–	S
Norma	Nor	Normae	The Set Square	165	332	S
Octans	Oct	Octantis	The Octant	291	–	S
Ophiuchus	Oph	Ophiuchi	The Serpent Holder	948	352	NS
Orion	Ori	Orionis	The Hunter	594	364	NS
Pavo	Pav	Pavonis	The Peacock	378	–	S
Pegasus	Peg	Pegasi	The Winged Horse	1121	376	N
Perseus	Per	Persei	The Victorious Hero	615	382	N
Phoenix	Phe	Phoenicis	The Phoenix	469	438	S
Pictor	Pic	Pictoris	The Painter's Easel	247	–	S
Pisces	Psc	Piscium	The Fishes	889	391	NS
Piscis Austrinus	PsA	Piscis Austrini	The Southern Fish	245	395	S
Puppis	Pup	Puppis	The Stern	673	397	S

Table 2.1 (continued)

Latin name	Abbr.	Genitive	Meaning	Size [sq.°]	Page	Loc.
Pyxis	Pyx	Pyxidis	The Compass	221	182	S
Reticulum	Ret	Reticuli	The Net	114	–	S
Sagitta	Sge	Sagittae	The Arrow	80	408	N
Sagittarius	Sgr	Sagittarii	The Archer	867	411	S
Scorpius	Sco	Scorpii	The Scorpion	497	428	S
Sculptor	Scl	Sculptoris	The Sculptor	475	438	S
Scutum	Sct	Scuti	The Shield	109	441	S
Serpens	Ser	Serpentis	The Serpent	636	444	NS
Serpens Caput	SCa	Serpentis Caput	The Serpent's Head	428	444	NS
Serpens Cauda	SCd	Serpentis Cauda	The Serpent's Tail	208	447	NS
Sextans	Sex	Sextantis	The Sextant	314	453	NS
Taurus	Tau	Tauri	The Bull	797	455	N
Telescopium	Tel	Telescopii	The Telescope	252	–	S
Triangulum	Tri	Trianguli	The Triangle	132	468	N
Triangulum Australe	TrA	Trianguli Australis	The Southern Triangle	110	–	S
Tucana	Tuc	Tucanae	The Toucan	295	–	S
Ursa Major	UMa	Ursae Majoris	The Great Bear	1280	473	N
Ursa Minor	UMi	Ursae Minoris	The Little Bear	256	482	N
Vela	Vel	Velorum	The Sails	500	485	S
Virgo	Vir	Virginis	The Virgin	1294	487	NS
Volans	Vol	Volantis	The Flying Fish	141	–	S
Vulpecula	Vul	Vulpeculae	The Fox	268	495	N

abbr. Genitive of the constellation's name is used with names of the stars, double stars and variable stars: Alpha Orionis, Delta Scuti ... Many times it is abbreviated: instead of Alpha Orionis we write Alpha Ori, instead of Delta Scuti we write Delta Sct ...

size Size of the constellation in square degrees.

page Page in this book, where the constellation starts.

location If the constellation lies in the north celestial hemisphere (above the celestial equator) it is labeled as N; if it lies on south celestial hemisphere, it is labeled as S. If the constellation lies along celestial equator, so that one part is on north celestial hemisphere and the other is on south celestial hemisphere, it is labeled as NS.

Maybe some of you might be wondering why we can use the constellations from the ancient Greeks, who lived roughly 2,500 years ago. Haven't things changed since then? Today we know that everything in the universe is moving. Our Solar System and all other stellar systems travel around the center of our Galaxy, and every star also travels in its own direction at its own specific speed. However, great distances divide us from the stars. It is true that a star may whizz across the universe with the fantastic speed of 50 km/s, but if we observe this star, for instance, from 100 light years away (approximately 1 million billions kilometers), we will be able to recognize its movement only with the most precise instruments. Centuries will pass before we will notice with our own eyes that a star has changed its position in the sky. And for this example we have chosen a rather fast star that is relatively close to us.

Asterisms

The ancient astronomers often divided the constellations into smaller parts that were, in turn, given their own names. We call such a part of a constellation that has its own name but does not have the importance of a constellation an *asterism*. The best known asterisms that many wrongly consider to be constellations are the Big and Little Dipper. In reality, the Big Dipper is a part of the constellation Ursa Major, while the Little Dipper is part of the constellation Ursa Minor (Figure 2.4). Another well-known asterism is the Hunter's Belt, which is formed by three rather close and bright stars in the constellation of Orion.

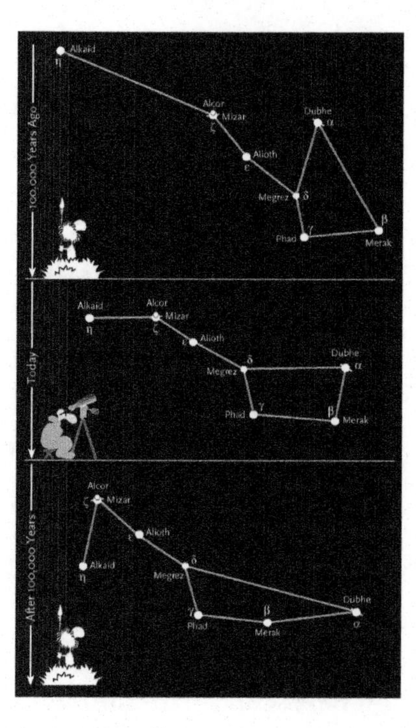

Figure 2.4 The shape of the Big Dipper 100,000 years ago, today, and in another 100,000 years. The appearance of constellations changes through time; however, millennia must pass for us to be able to notice these changes with the naked eye

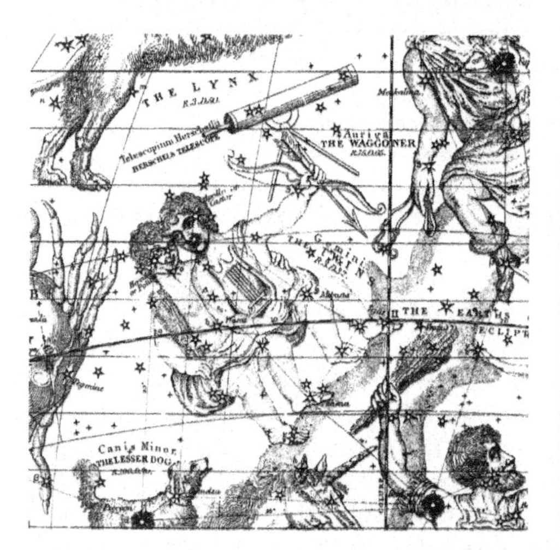

Figure 2.5 A part of a star chart from 1835 (author Elijah H. Burritt), in which we can see the classic constellations as well as Herchel's telescope. It was created from the dimmer stars between Gemini, Lynx, and Auriga by the Austrian astronomer Maximilian Hell at the end of the eighteenth century. Today the constellation no longer exists

The Celestial Sphere

When we look into a clear night sky it seems shaped like a ball, with Earth at its center. At any given moment in time we can see half of the sphere from our observing point. The second half lies below the horizon. The stars all appear equally distant from us and seem to be "pinned up" onto the inside of this celestial sphere. However, this sphere does not remain motionless. If we memorize the position of a certain bright star in relation to a nearby house or tree and then look at the same star from the same point an hour or two later, we will notice that it has moved to the west. The sky rotates from the east to the west (Figure 2.6). Of course, the rotation is not real; it just appears to be happening, for in reality Earth is rotating around its axis and we are rotating with it.

Figure 2.6 Here's how you can tell that the sky is really rotating. Place a photographic camera on a stable tripod and point it toward the sky. If you expose the image for a long period of time, say, for a couple of hours, arcs of the stars will be shown in the image, in this case the northern part of the sky. The star in the center is Polaris

Every sphere that rotates around its axis has two points that do not change their position. These are the poles, in our case represented by the north and south celestial poles. Because the rotation of the sky is a consequence of Earth rotating around its axis, the celestial poles are exactly above Earth's geographic poles – the north celestial pole is directly above Earth's North Pole and the south celestial pole directly above the Earth's South Pole.

In the same way as Earth the celestial sphere has its celestial equator, which lies directly above Earth's equator, or, to put it another way, exactly in the middle between the two poles. The celestial equator is a great circle that divides the celestial sphere into two equal hemispheres: the Northern and Southern Hemisphere.

Wherever we stand in Earth's Northern Hemisphere to observe the sky, the north celestial pole and celestial equator are always above the horizon, and the south celestial pole is always below the horizon. Wherever we stand in Earth's Southern Hemisphere to watch the sky, the north celestial pole is always below the horizon, and the south celestial pole and celestial equator are always above the horizon. If our observing point is exactly on the equator, the northern and southern celestial poles are on the (mathematical) horizon, and the celestial equator is in the zenith (Figure 2.7).

Celestial Coordinates

Soon after people started observing and describing stars in the celestial sphere, the need for a coordinate system arose. The system that was adopted made it simple to describe the position of the star (or any other celestial body) using only two coordinates. This was possible because in general stars do not change much in position relative to each other over time.

The celestial coordinate system is similar to the system we use for places on Earth. Geographic latitude is represented in the sky by declination, while geographic longitude is represented by right ascension. The choice of starting point for declination seemed to be obvious. It is measured in degrees from the celestial equator to the north (+) or south (–) celestial pole. The declination of the stars on the celestial equator is 0°, on the north celestial pole +90°, and on the south celestial pole –90°. For

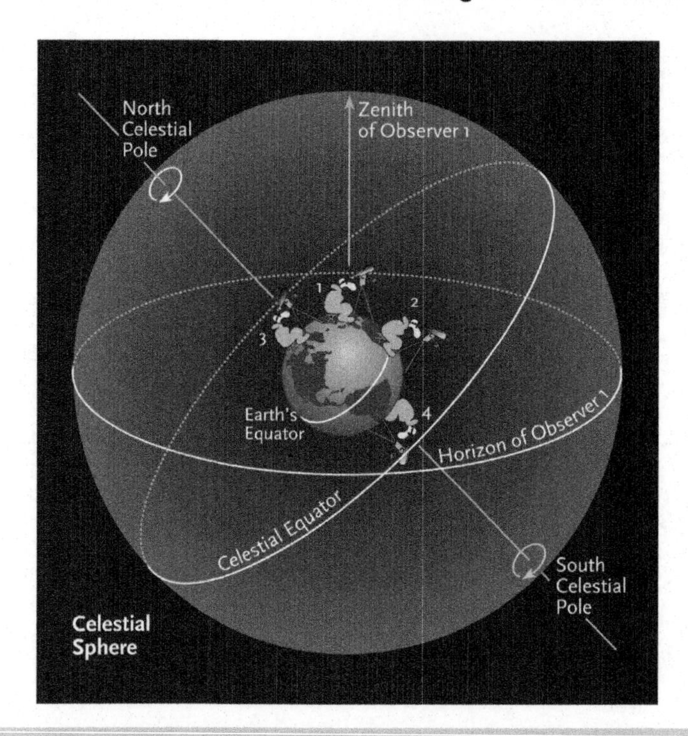

Figure 2.7 For all observers on Earth the sky appears to be a sphere with Earth (the observer) at the center. At any given moment of time we can see one half of the sphere (if there are no high hills there); the other half lies below the horizon. In the illustration we can see only the horizon for observer 1, that is, the observer watching from mid-northern geographic latitudes. With a bit of imagination we can guess how observers on other parts of Earth can see the sky. Observer 3 is sitting at the North Pole. In his zenith is the north celestial pole, and along his (mathematical) horizon runs the celestial equator. Observer 4 is sitting at the South Pole. In his zenith lies the south celestial pole, and along his (mathematical) horizon also runs the celestial equator. Observer 2 is sitting on the equator. In his zenith is the celestial equator, and on the horizon on opposite sides lie the north and south celestial poles

more precise measurements of declination the degree is divided into 60 arcmin ($'$), and an arc minute is divided into 60 arcsec ($''$) (Figure 2.8).

The selection of the coordinate starting point for the right ascension is – the same as on Earth – a matter of choice. In the same way as all geographic meridians are equal among themselves (they are all great circles on the sphere), so are the celestial meridians. Due to historic reasons we have agreed on Earth that we will start counting geographic longitude from the meridian that runs through the observatory in Greenwich, London. The coordinate starting point of the system on Earth is therefore at the point at which the Greenwich meridian crosses the equator. At that point the geographic latitude and longitude are both 0°. In the same way, astronomers needed to select a point on the celestial equator that would represent the starting point of the celestial coordinate system. They agreed that this would be the point where the ecliptic crosses the celestial equator and in which the Sun is at the spring equinox. This point is called the vernal point, or point γ (gamma), and it lies in the constellation of Pisces.[1]

Right ascension is measured in hours, from 0 h to 24 h. Its starting point is the vernal point, and it grows toward the east. The advantage of such a division (hours instead of degrees) lies in the fact that for every

[1] The ecliptic is the apparent annual path of the Sun across the celestial sphere. In one year the Sun crosses the ecliptic only once. The ecliptic is not parallel to the celestial equator but makes an angle of $23^\circ 26'$. The intersections are two points. We have already mentioned the vernal point, and the second point is called the autumnal point, which is where the sun is in the autumn equinox.

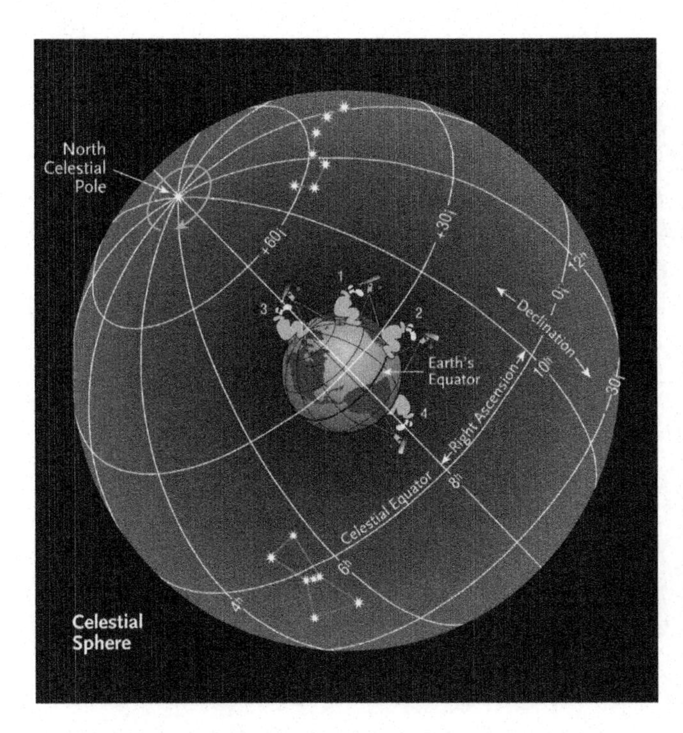

Figure 2.8 The sky coordinate system can best be understood if you visualize the coordinate system of latitudes and meridians on Earth and project it onto the sky. Once again four observers are shown on Earth. The first (1), who is observing the sky from mid-northern latitudes, can see the complicated net of the coordinate system in the sky, depending of the geographical direction of the sky he is looking toward (see Figures on pages 32 and 33). The remaining three have better luck. Their sky is extremely simple. Detailed drawings of what they see are shown on figure 2.10

hour the sky rotates for approximately one right ascension hour. That is why it is simple to ascertain from star charts when a certain constellation or star will be above the horizon. If, for instance, Orion is at its highest above the horizon and we are interested when (in how many hours) we will be able to see Cancer, we look at the central right ascension of the Orion (5.5 h) and the central right ascension of Cancer (8.5 h). From this we can immediately see that Cancer will be at its highest in approximately 3 h. Looked at it this way, the sky becomes a giant 24-h clock. But beware! On the equator the arc of 1 h measures 15°, and the closer we get to the poles, the coordinate system lines converge and the arcs, which represent 1 h, are shorter and shorter (see illustration above).

A right ascension hour is divided into 60 minutes (min) and a minute is divided into 60 seconds (s). The way you write minutes and seconds at the declination and right ascension is different, for they are different. If 1 h of right ascension on the celestial equator equals an arc of 15°, 1 min equals the arc of 15′, and 1 s the arc of 15″.

Because the celestial sphere rotates, the coordinate system rotates with it. The position of the celestial body on the celestial sphere can therefore always be given with two coordinates. To put it another way, if we know the two coordinates, we can always find the celestial body on the celestial sphere.

Rotating Sky

At the beginning, we mentioned that the night sky rotates. In fact, it only appears to be rotating, for in reality Earth is rotating around its axis. Because we know that Earth takes 24 h to make one complete rotation, we would expect that the same star that is rising on the horizon at this moment will appear on the horizon again in 24 h. However, observations show differently.

If we choose a bright star and note down the time this star sets behind the neighbor's roof (or some other clearly distinguishable and nonmoving object) we will discover that the next day (if viewed from the same point), the star will set behind the roof 4 min earlier. This phenomenon is even more

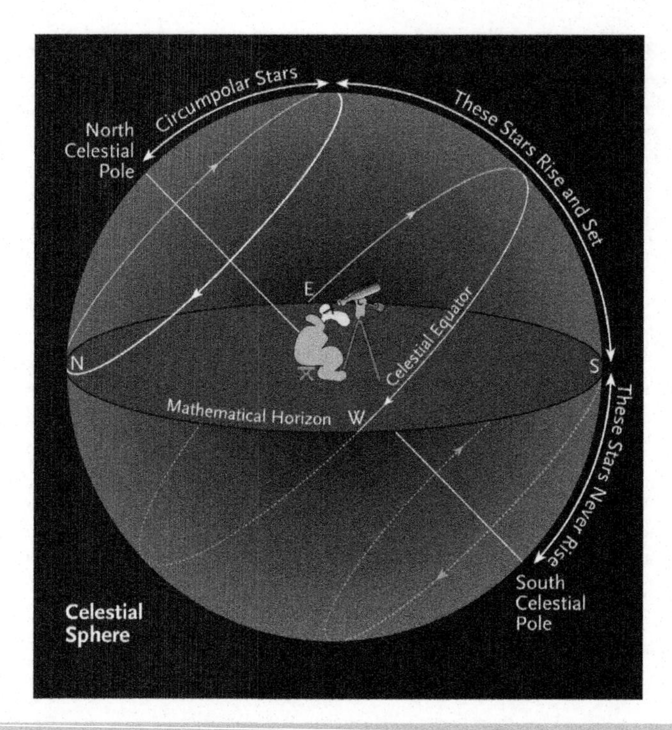

Figure 2.9 Which stars will be circumpolar, which will rise and set, and which will never appear above horizon depends on the geographic latitude of the observing point. If we move toward the equator, the north celestial pole moves toward the horizon, and the area of the sky around the pole, where the stars never rise and set, becomes smaller. If we move towards the north geographic pole, the north celestial pole moves toward the zenith, and the area of sky around the pole, where the stars never rise and set, increases. What the sky looks like at its extreme points (exactly on the pole and on the equator) is shown in the illustrations on the next page

noticeable if there is a longer period between the two observations. In other words, a star that is setting behind the neighbor's roof at the beginning of January at 10 p.m. will set behind the same roof at the beginning of February at 8 p.m. and at the beginning of March as early as 6 p.m. Over the year the constellations slowly drift toward the west, and on the east new ones appear. These changes in the sky are a consequence of Earth revolving around the Sun. That is why in the spring there are different constellations in the sky than in summer, autumn, or winter.

Imagine a celestial sphere with Polaris on it (or the position of the north celestial pole). We know that the sky seems to rotate around Polaris. The stars that are close to the pole travel in smaller circles, and those that are further travel in larger circles. Because Polaris lies at 45° above the horizon (when viewed from mid-northern latitudes), the stars that are close to it are visible throughout the night and throughout the year, for they rotate around the pole and never fall below the horizon. We call these *circumpolar stars.* Stars that are far from the North Pole rise and set during their daily and annual movements across the sky.

Of course, there are also stars that are close to the south celestial pole (from mid-northern latitudes it is 45° below the horizon). These never come above our horizon during their daily and annual movements (and can thus never be seen from mid-northern latitudes). Whether a star is considered circumpolar or that it rises and sets or is never above the horizon depends on the geographical latitude of the observing point and the declination of the star. On the north geographical pole, where we have Polaris in the zenith and the celestial equator on the horizon, all stars above the celestial equator (with a positive declination) are circumpolar stars, while those below the horizon (with a negative declination) can never be seen. On the south geographic pole the situation is exactly the opposite.

If our observing point is on Earth's equator, we have the celestial equator in the zenith, while the poles lie on the opposing sides of the horizon. In this case none of the stars is circumpolar, and throughout the year we can see all stars (and constellations) that can be seen from Earth. (This is why large observatories are built as close as possible to the equator.) If our observing point is somewhere in between, certain stars always appear above the horizon, others rise and set, while some other can never be seen. Stars that rise and set at a certain geographical latitude φ have the declination δ within the following limits:

$$(90^\circ - |\varphi|) < \delta < (90^\circ - |\varphi|)$$

Constantly above (below) the horizon is the star on a positive (negative) geographical latitude φ, if its declination δ fulfils the condition:

constantly above on $\varphi > 0$
constantly below on $\varphi < 0$ $\delta > (90^\circ - |\varphi|)$

Constantly below (above) the horizon is the star on the positive (negative) geographical latitude φ, if its declination δ fulfils the condition:

constantly below on $\varphi > 0$
constantly above on $\varphi < 0$ $\delta < (|\varphi| - 90^\circ)$

For all of the above-stated conditions we have to take the absolute value for the negative (south) geographical latitude.

On mid-northern latitudes, the circumpolar stars (constantly above the horizon) are around the north celestial pole to the declination +45°. These stars can be seen on any clear night throughout the year. Constantly below the horizon (can never be seen) are stars with a declination below –45°. Stars with a declination of between +45° and –45° rise and set and can be seen only at certain periods of the year (Figure 2.10).

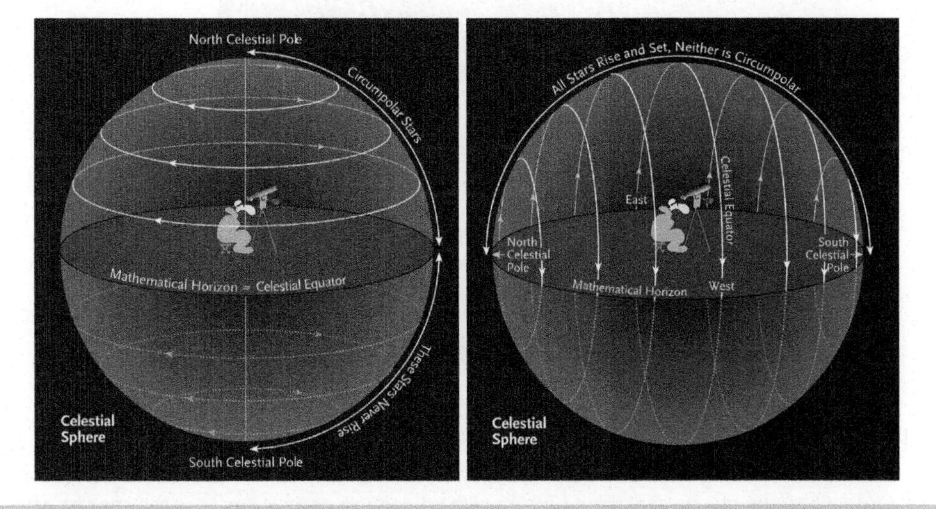

Figure 2.10 The appearance of the sky for the observer on the north geographic pole (illustration left) and on the equator. For an observer sitting on the north geographic pole, the stars rotate around the vertical axis and therefore never set during the night or during the year. However, this observer can see only stars in the northern celestial hemisphere (those with a positive declination), and he can never see those in the southern hemisphere. Sitting on Earth's South Pole, the observer can see the movement of stars just like the observer on the North Pole, except that he has the south celestial pole in his zenith and can see all the stars in the southern celestial hemisphere (with a negative declination). However, he can't see the ones in the northern celestial hemisphere. The observer on the equator is in the best position. During the night and throughout the year he can see all of the stars that can be seen from Earth

The Directions of the Sky and Using Star Charts

Where is north, south, east, and west in the spherical sky? It is defined by the celestial coordinate system. North is always in the direction of Polaris, regardless of which direction the binoculars are pointing and how they are turned. Let's look at the example.

All charts (in Part II of this book) are oriented so that north is up, east is on the left, south is down, and west is on the right. In our search, for example, of the globular cluster M 13 in Hercules we can use the chart on page 304, and if the constellation at the moment is above the eastern horizon, we have to know the orientation of the coordinate system in this part of the sky. If the sought-after cluster is on the chart above the Zeta, you will search for it in the sky in vain in that place. You have to turn the chart approximately 45° counterclockwise, as seen in the figure below. The cluster is thus left of Zeta. What about the other parts of the sky?

The coordinate sky net in various directions in places with median geographic latitude 45° has been shown on schematic figures. When we are turned toward the south (see Figure 2 on next page), you do not need to turn the charts at all. North truly is upward, south is downward, east is on the left, and west is on your right. For the other parts of the sky, the position of the coordinate system is more complex, as seen in Figures 1, 3, and 4.

Figure 1 – View East

If we observe the stars above the eastern horizon, we have to rotate the chart for about 45° counterclockwise. If we are observing stars above the western horizon, we have to rotate the charts for about 45° clockwise. Only when the charts are turned in the right direction can we say that the sought-after object is, for instance, above the selected star or below it or left or right of it. This is why we try to avoid such descriptions, and instead describe the position of the celestial bodies as east of..., northwest of..., etc.

If we are observing the stars in a northerly direction, it depends on the position of the observed body in which direction and to how many degrees we have to rotate the chart so it will show what we can see in the sky. Experienced observers, who are well acquainted with the constellations, look at the position and orientation of a constellation in that part of the sky they are observing and then rotate the chart so that the orientation in the sky matches the chart. You do always need to know where the north celestial pole or Polaris is, for this tells us which way is north, and then the other

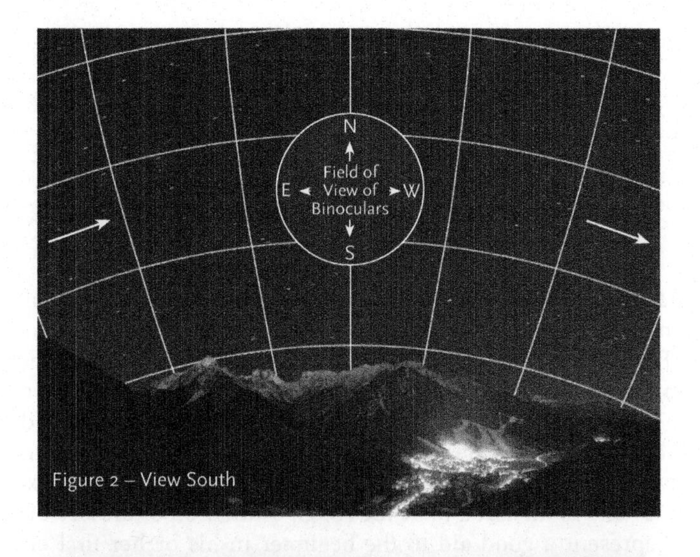

Figure 2 – View South

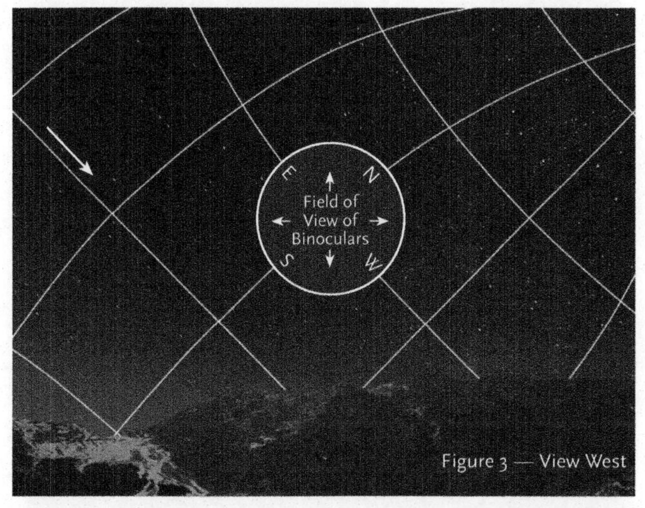

Figure 3 — View West

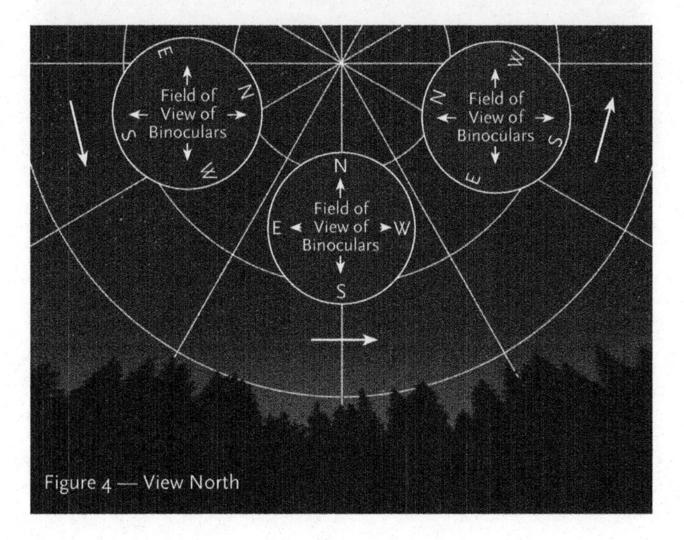

Figure 4 — View North

directions in the sky are easy to figure out. Let us here also point out that this seemingly complicated orientation of the field of view becomes easier the more experienced in observation you become. An experienced observer always knows how the field of view in the binoculars is oriented, without needing to think about it.

Seasonal Charts

In the seasonal charts that follow numbered 1 (January) to 12 (December), we can see how the constellations in the mid-northern latitudes change during the year. Every chart covers a piece of the sky along the meridian[2] from the southeast to the southwest (90°) and from the northern to the southern horizon. The charts depict the sky as it appears around midnight of the 15th of each month. The times that they can be seen hold true for the local time zone. Various conditions (daylight saving time, etc.) are not taken into account. This means that we have to add an hour to the times next to the dates in which daylight saving time applies. The circle with the stretched out hand next to each chart represents approximately 25° and is intended to be the rough estimation of the size of the individual constellation. The charts represent a good aid to the beginner in his or her first encounters with constellations. More on this in the First Steps section later.

[2] A meridian is an imaginary great circle on the celestial sphere. It passes through the northern point on the horizon, through the celestial pole, up to the zenith, through the southern point on the horizon, and through the nadir, and is perpendicular to the local horizon.

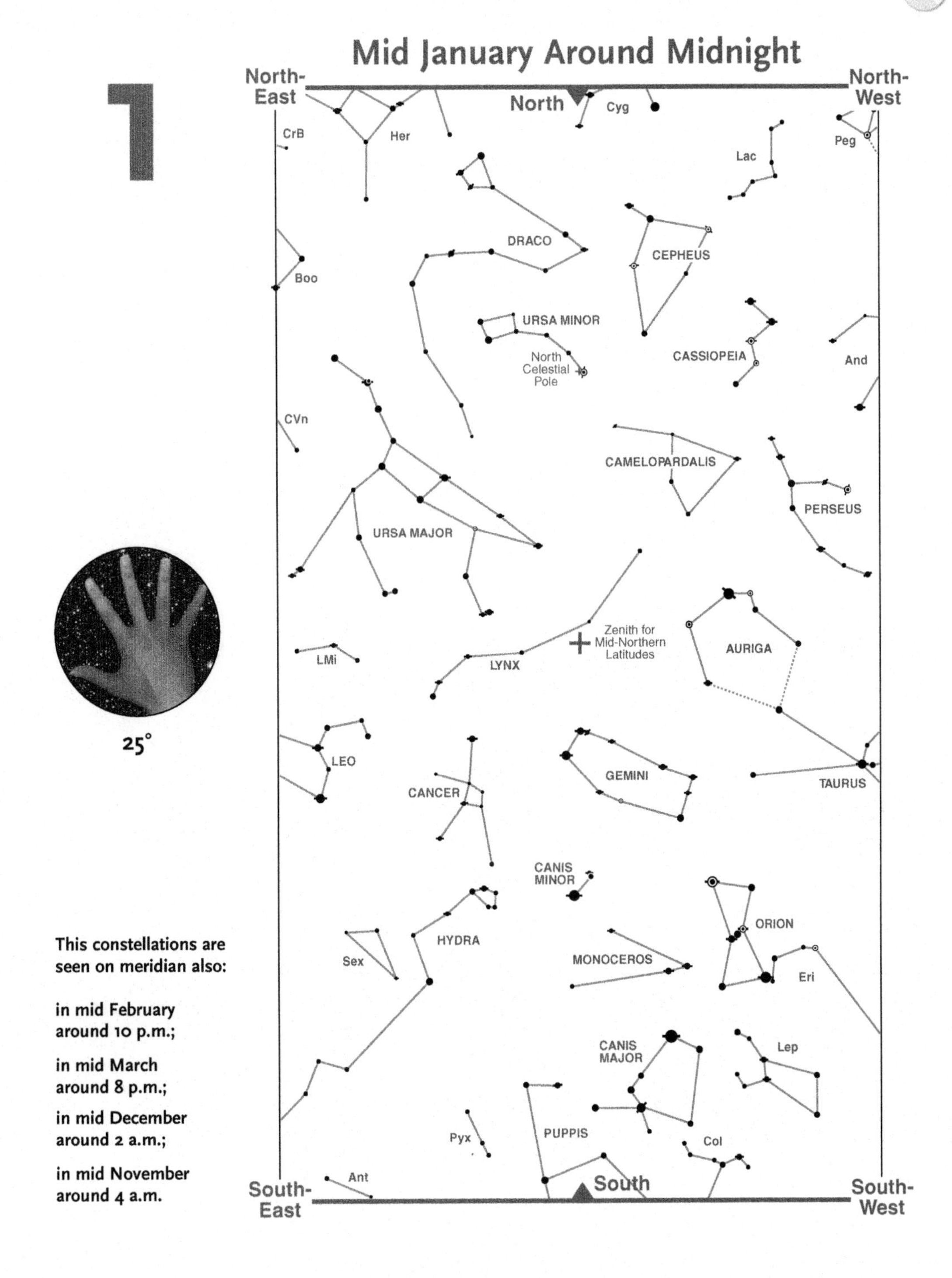

This constellations are seen on meridian also:

in mid February around 10 p.m.;

in mid March around 8 p.m.;

in mid December around 2 a.m.;

in mid November around 4 a.m.

Mid February Around Midnight

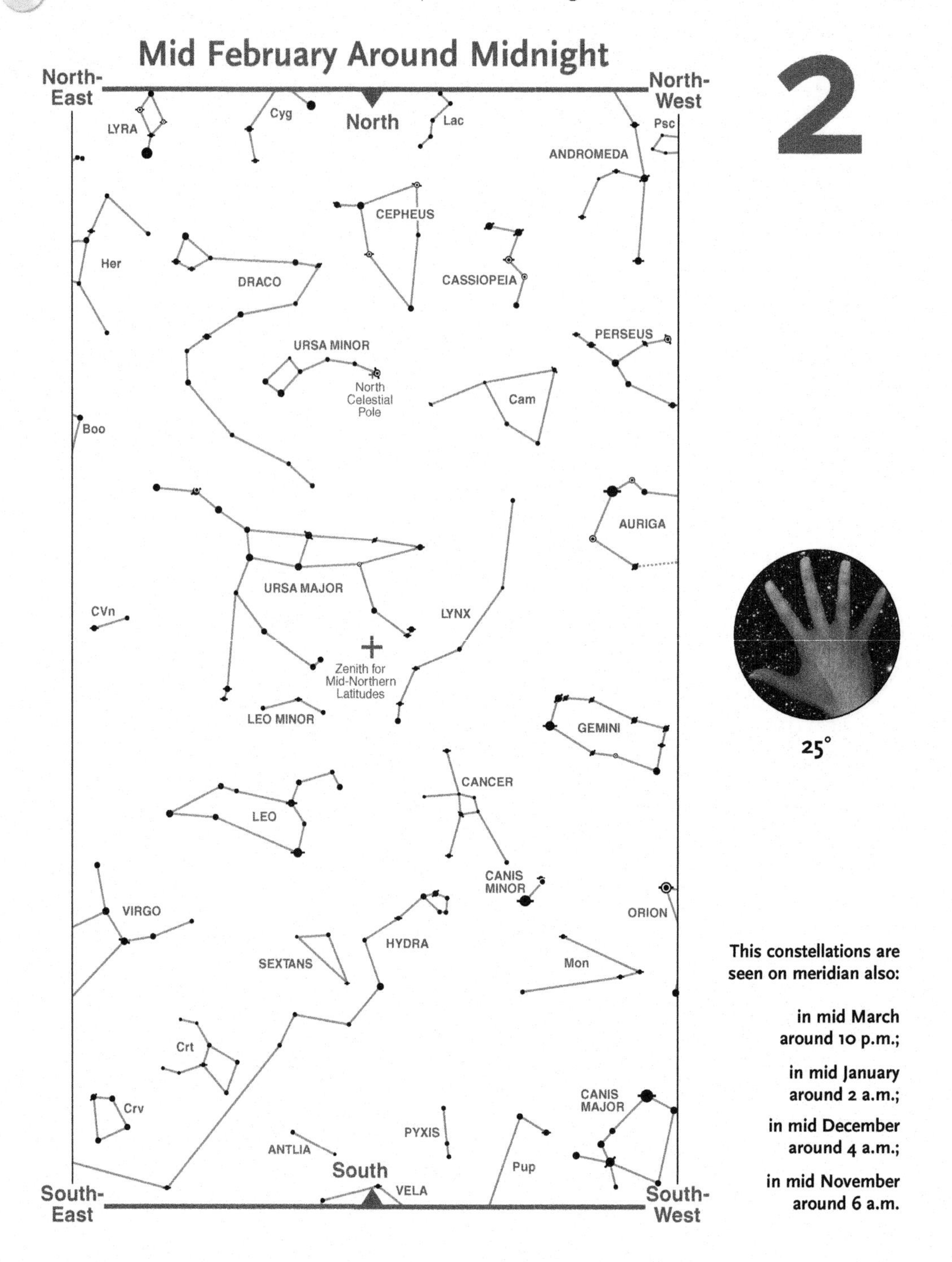

This constellations are seen on meridian also:

in mid March around 10 p.m.;

in mid January around 2 a.m.;

in mid December around 4 a.m.;

in mid November around 6 a.m.

3

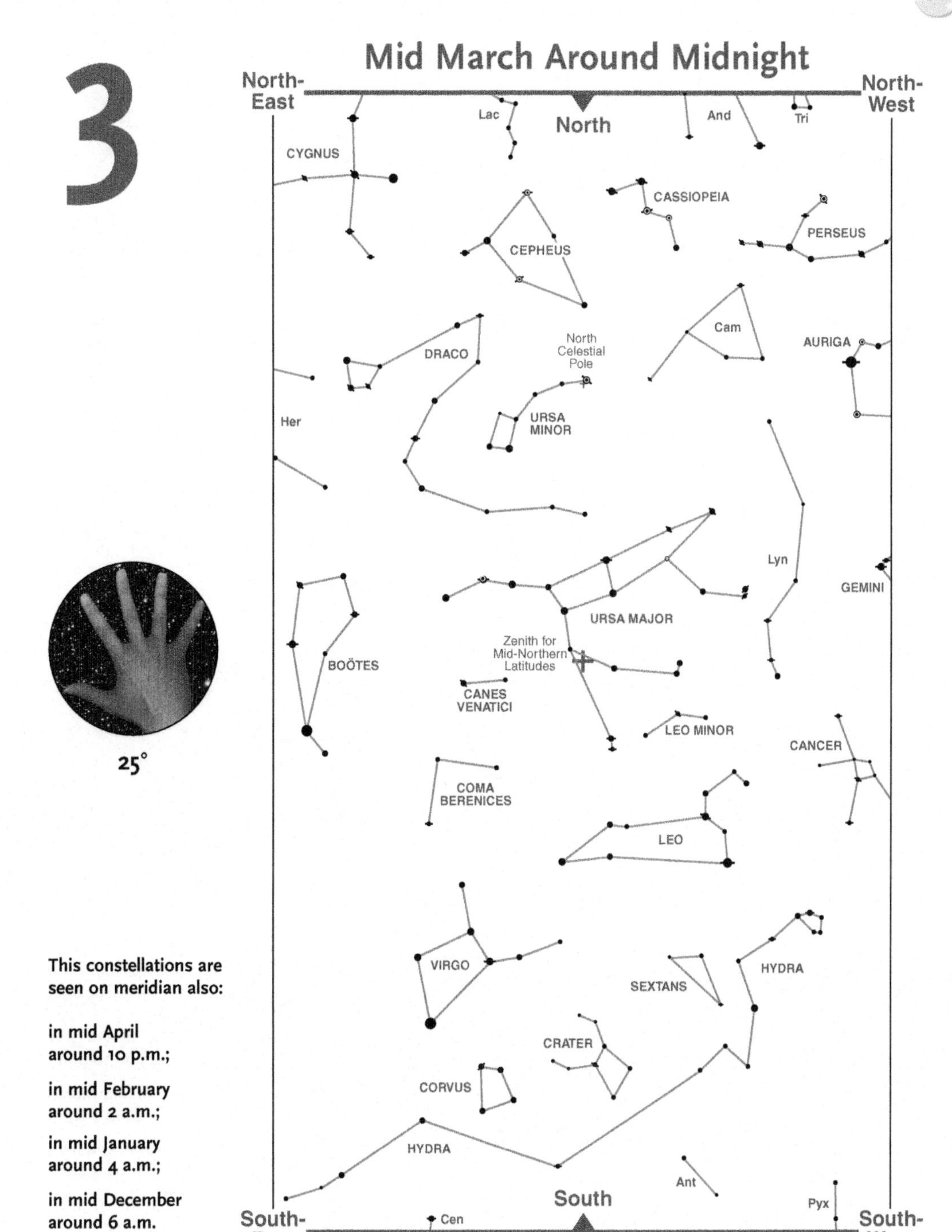

This constellations are seen on meridian also:

in mid April around 10 p.m.;

in mid February around 2 a.m.;

in mid January around 4 a.m.;

in mid December around 6 a.m.

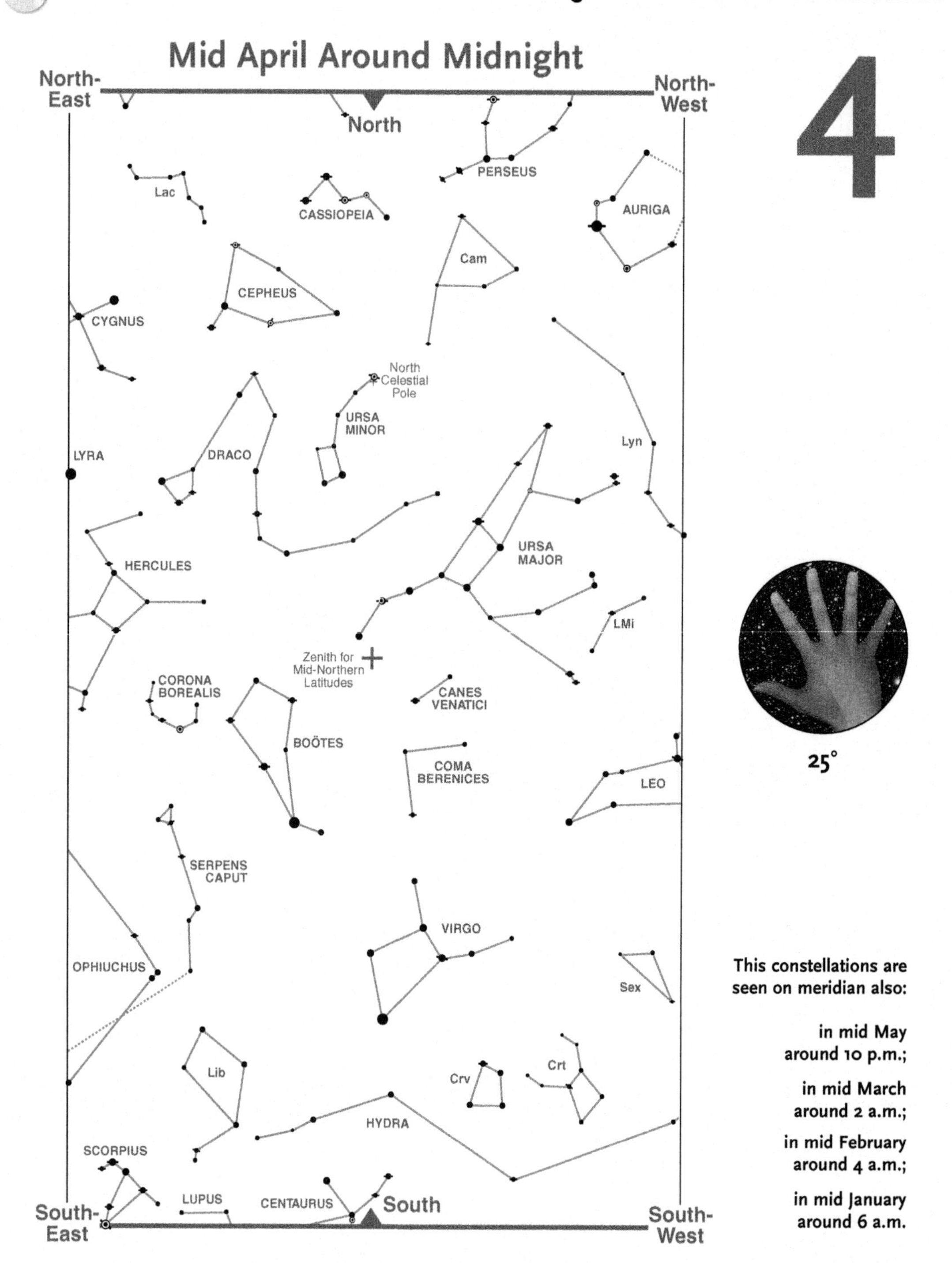

This constellations are seen on meridian also:

in mid May around 10 p.m.;

in mid March around 2 a.m.;

in mid February around 4 a.m.;

in mid January around 6 a.m.

5

Mid May Around Midnight

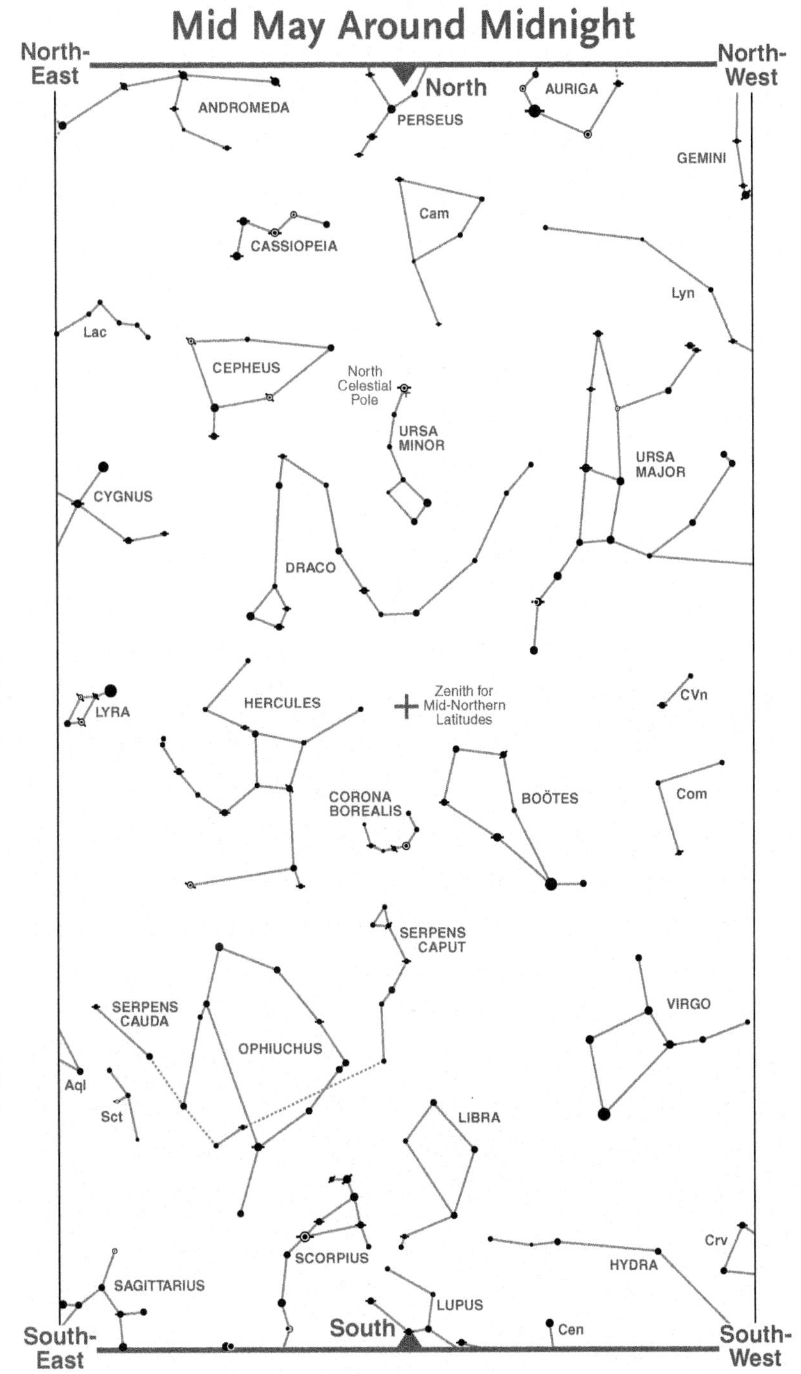

25°

This constellations are seen on meridian also:

in mid June around 10 p.m.;

in mid April around 2 a.m.;

in mid March around 4 a.m.;

in mid February around 6 a.m.

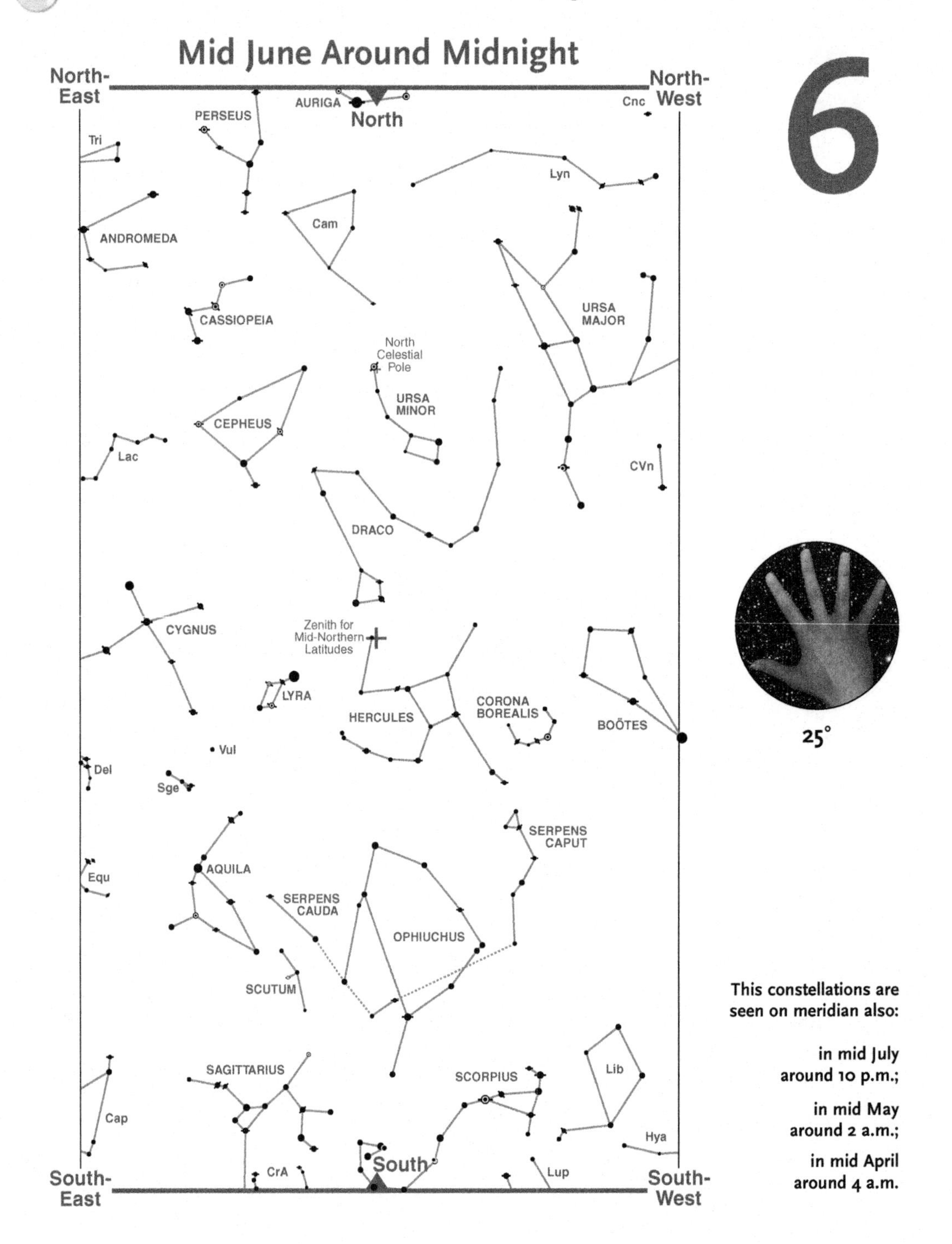

This constellations are seen on meridian also:

in mid July around 10 p.m.;

in mid May around 2 a.m.;

in mid April around 4 a.m.

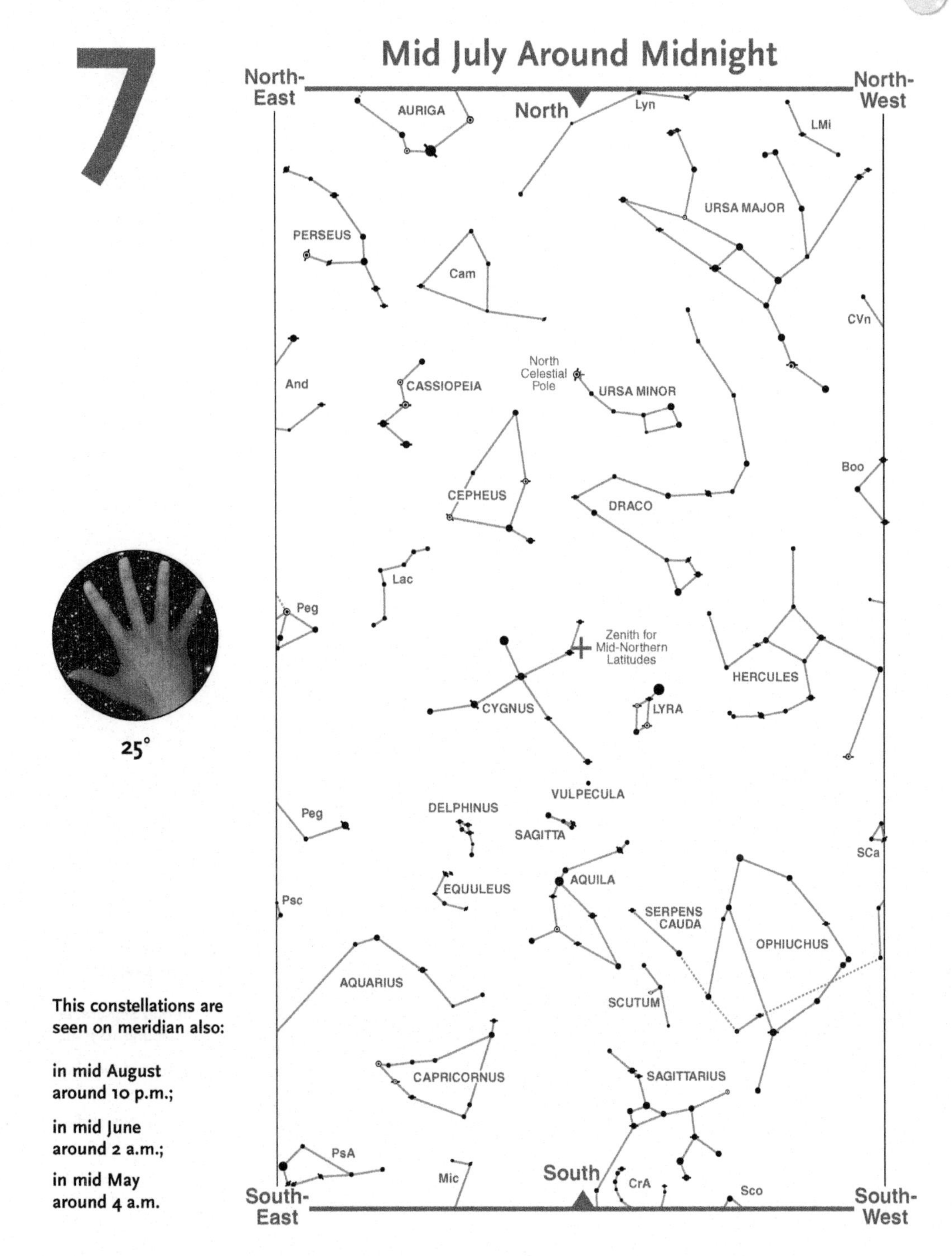

This constellations are seen on meridian also:

in mid August around 10 p.m.;

in mid June around 2 a.m.;

in mid May around 4 a.m.

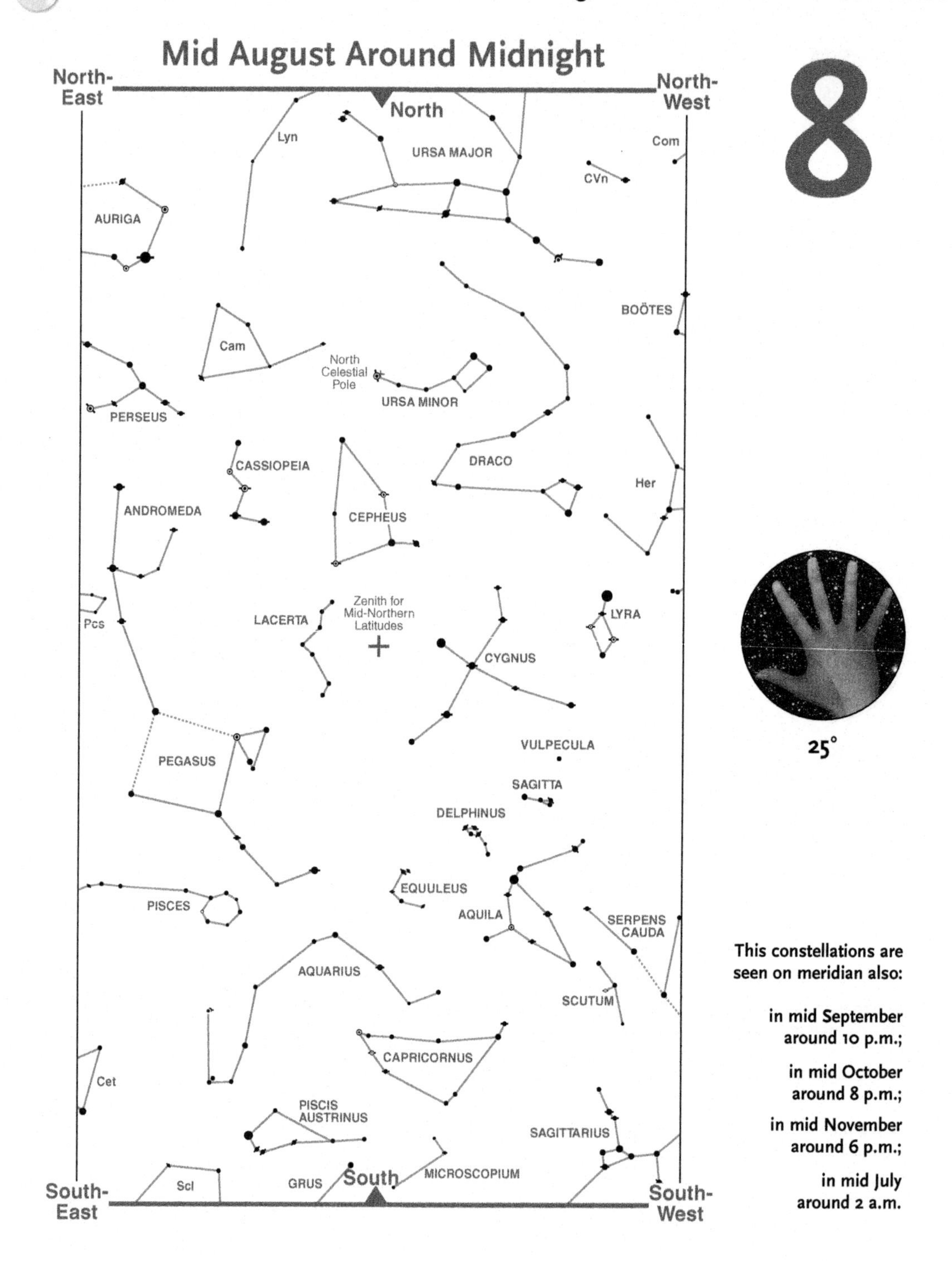

This constellations are seen on meridian also:

in mid September around 10 p.m.;

in mid October around 8 p.m.;

in mid November around 6 p.m.;

in mid July around 2 a.m.

9

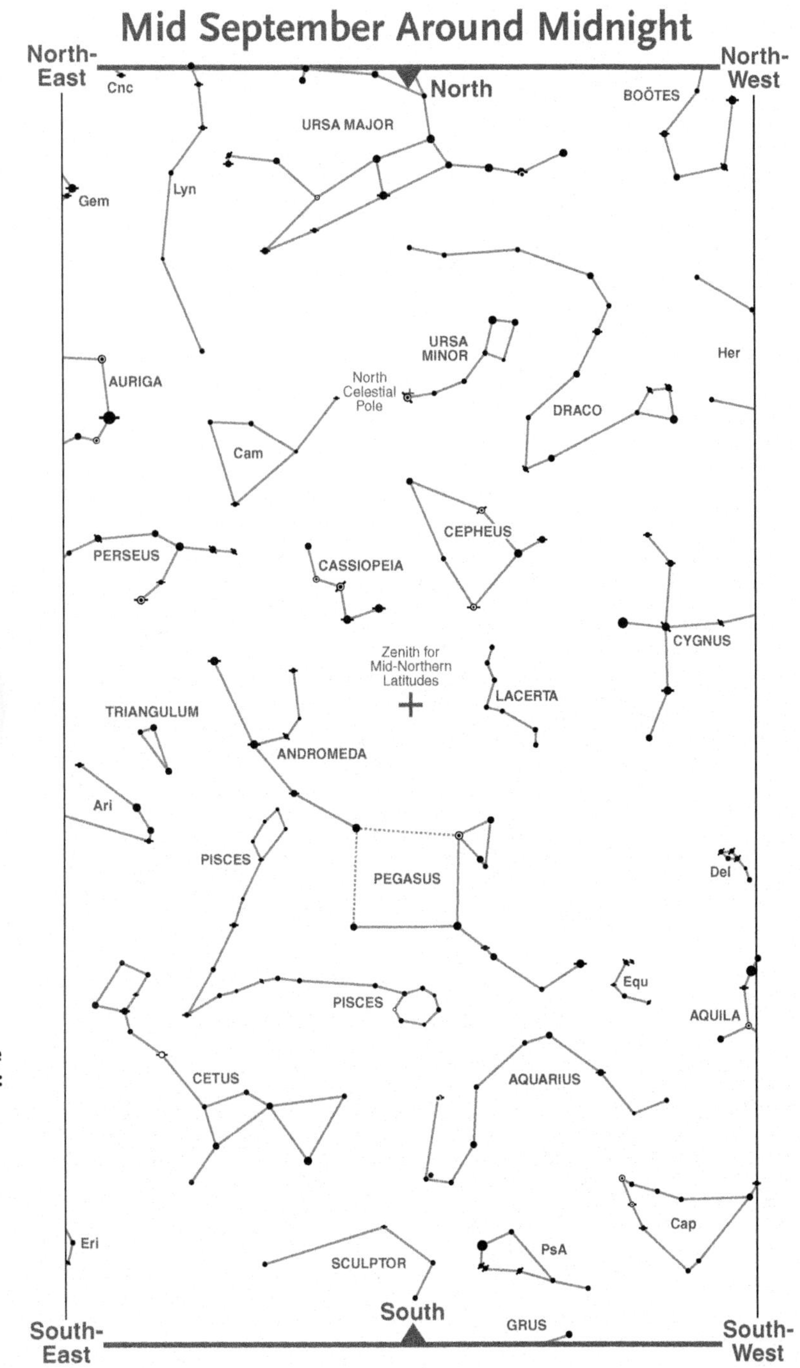

25°

This constellations are seen on meridian also:

in mid October around 10 p.m.;

in mid November around 8 p.m.;

in mid December around 6 p.m.;

in mid August around 2 a.m.

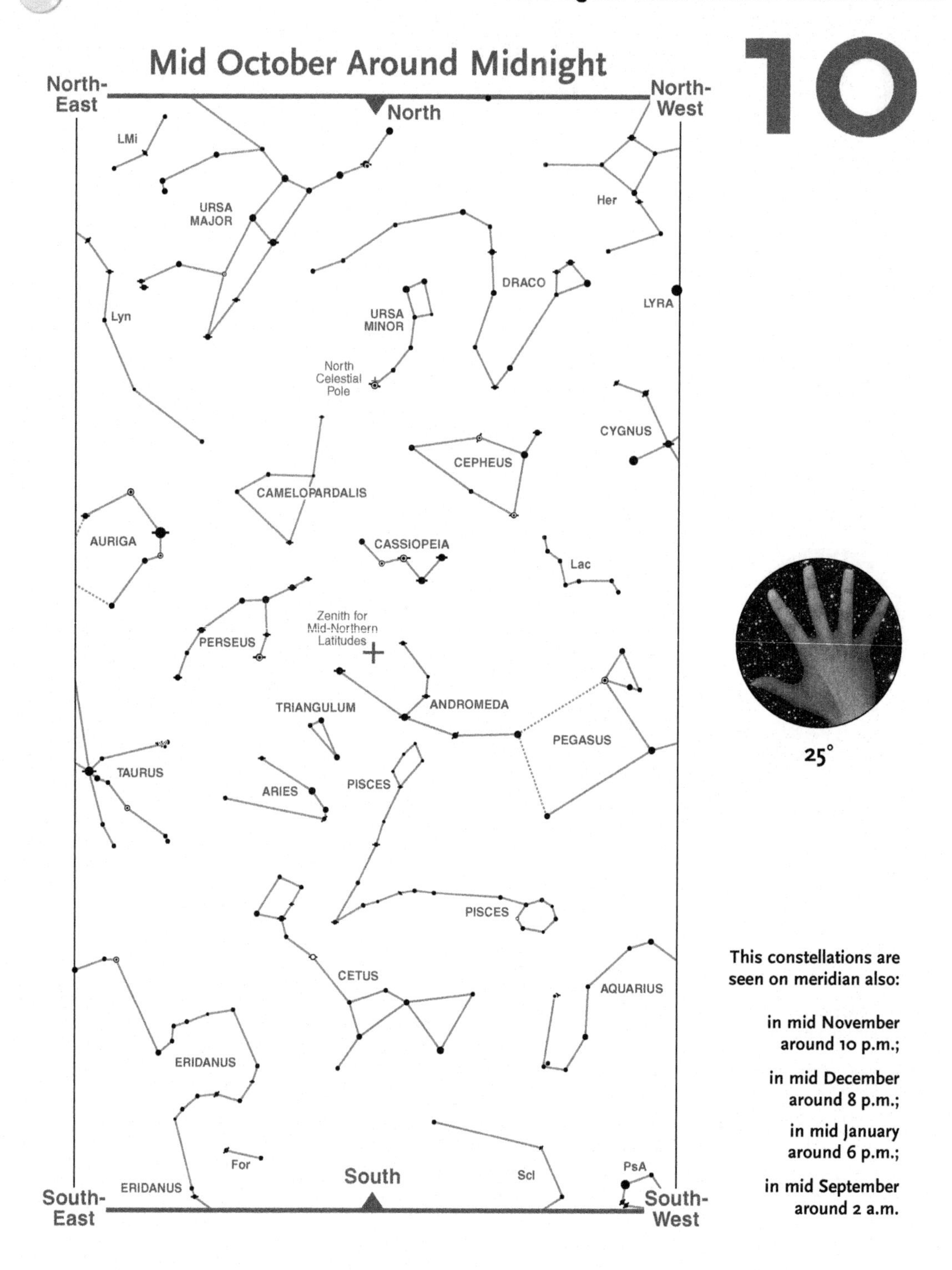

This constellations are seen on meridian also:

in mid November around 10 p.m.;

in mid December around 8 p.m.;

in mid January around 6 p.m.;

in mid September around 2 a.m.

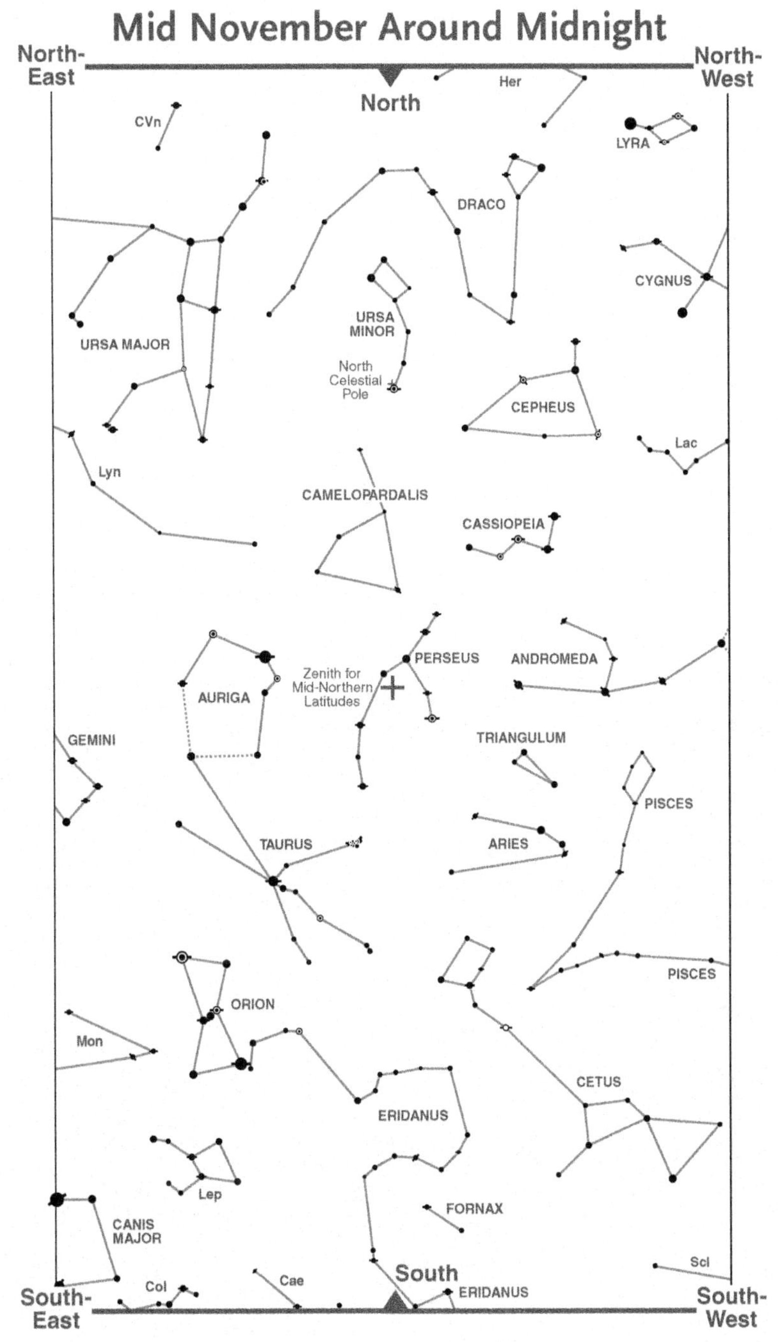

25°

This constellations are seen on meridian also:

in mid December around 10 p.m.;

in mid January around 8 p.m.;

in mid October around 2 a.m.;

in mid September around 4 a.m.

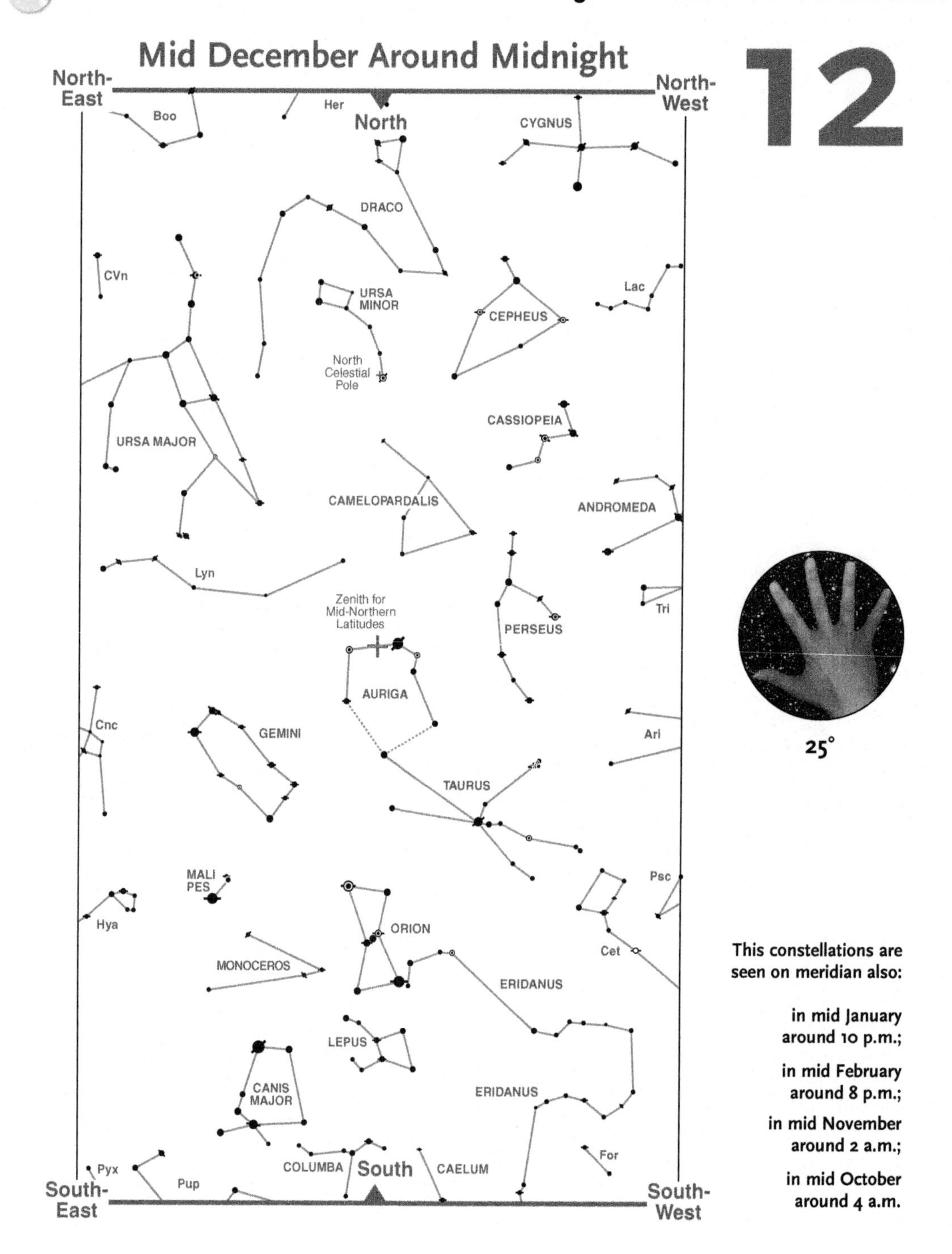

This constellations are seen on meridian also:

in mid January around 10 p.m.;

in mid February around 8 p.m.;

in mid November around 2 a.m.;

in mid October around 4 a.m.

Measuring Angles in the Sky
Astronomers measure the apparent distances between the stars in the sky with angles. (We say "apparent" because these are not the true distances between the stars.) Betelgeuse and Rigel in the constellation of Orion are 18.5° apart. From Betelgeuse to Gemini it is 33°. The apparent diameter of the Moon and Sun is approximately 0.5°. The comet tail is 90° long. The star is 15° above the horizon.

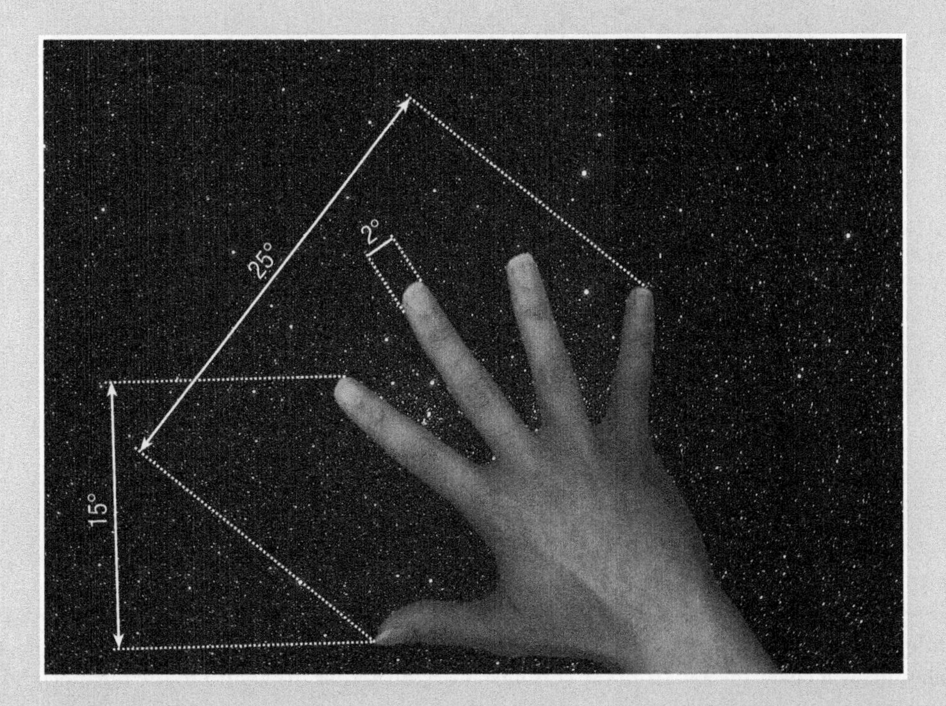

For a rough orientation - especially as we take our first steps across the sky - we can utilize something we always have on us, our hand. If we stretch it out and spread the fingers apart, we have a protractor; with this we can estimate the angle distances in the sky.

First Steps

If you have a friend or an acquaintance who is already familiar with the constellations, it is best if he or she helps you take your first steps across the sky and points out a few of the brightest stars and the constellations associated with them. Once you are familiar with a few constellations, you can use the seasonal charts (found in this book), a planisphere, a star atlas, or some other aids to find and recognize the remaining constellations.

A lot of people are familiar with the asterism called the Big Dipper. If you are among them, you can use the seasonal charts and first locate the neighboring constellations of the Great Bear and then their neighboring constellations and so on across the sky. But if you do not know a single star and a single constellation and want to learn how to recognize them by yourself using this book, then read the following paragraph very carefully.

In general, recognizing what is in the sky is pretty simple. What you need to know to begin is the rough direction of north–south from your observing point. You can define this direction with a

compass or by looking where the Sun is at midday – roughly south. At a certain date, say in mid-April at midnight, when you are standing under a clear night sky and turn toward the south, the spring constellations will cover the sky from the southern horizon across the zenith to the northern horizon. The sky is dominated by three bright stars (see chart P1 on page 51): Arcturus in Boötes, Spica in Virgo, and Regulus in Leo. It is enough to recognize one constellation, the one with the greatest number of bright stars. From this starting point you can then simply find the neighboring constellations, then their neighboring constellations, and so on across the celestial sphere. But beware! When you think you have found, for instance, Regulus and Leo, have a look at the chart that depicts stars up to magnitude 5 (we will speak about stellar brightness and its unit magnitude in the next chapter), which is found in the description of this constellation in the second part of this book (Figure 2.11A). In addition to the brightest stars that make up the shape of constellation Leo, you also should try to recognize all of the fainter stars. Only then can you be sure that you are truly looking at Leo. (It happens all too often that beginning amateur astronomers search for too small patterns of stars and are satisfied with the first grouping that is roughly similar to the one they are looking for.)

When you have established the location of Leo, you should look at seasonal chart 3 (a cutout from this chart can be found on Figure 2.11B), which includes the constellation Leo, and notice that the following constellations surround it: to the west is Cancer, north is Leo Minor, northeast is Coma Berenices, southeast is Virgo, south is the Sextans, and southwest is the head of the Hydra. Once you recognize these constellations, with the help of the descriptions that are included in the second part of this book, you will already know seven constellations. And then you can travel ahead across the celestial sphere.

Charts P1–P4 depict the spring, summer, autumn, and winter skies in the Northern Hemisphere, with only the brightest stars and thus only the most visible constellations or asterisms. Alongside the charts are the dates and hours of visibility. Novices in sky gazing should first – depending on the season and time – find one of these constellations. This should be the starting point. For easier

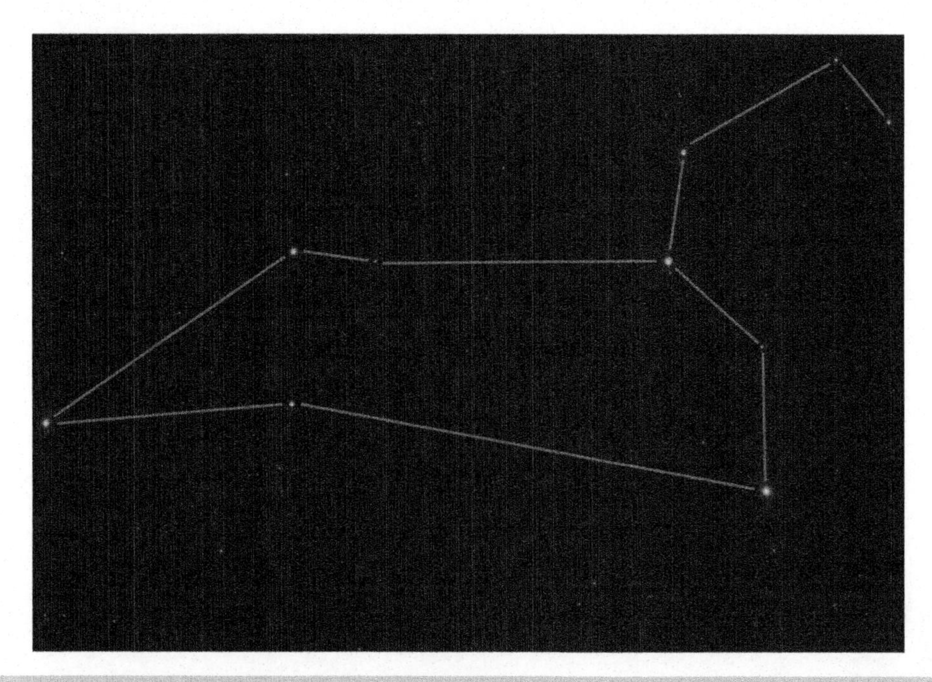

Figure 2.11 Photography of Leo with stars up to magnitude 8

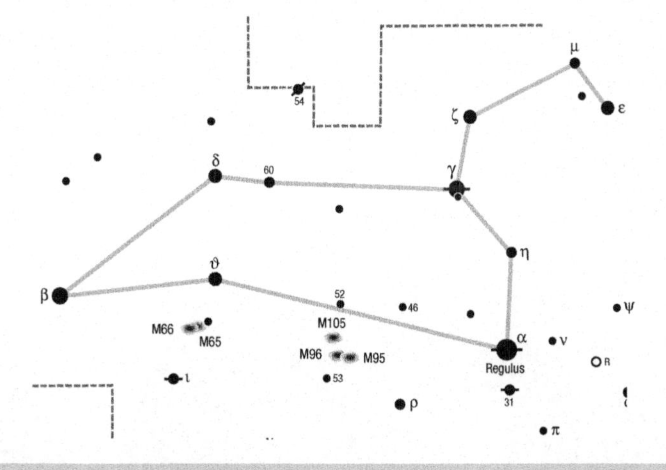

Figure 2.11A Map of Leo with stars up to magnitude 5

orientation we have also drawn in some of the more important angle distances and an open, stretched out hand, which represents approximately 25°. The zenith is also a useful point for orientation.

Places with no light pollution or smog are the most suitable for observing the night sky through binoculars. But on a clear, moonless night there are so many stars in the sky that sometimes even the more experienced observers can be mislead, let alone beginners. Thus, while you are still inside in a bright room, take a good look at the brightest stars and the angles between them on the chart. Only then should you step out under the night sky, turn toward the south, and for the first few minutes

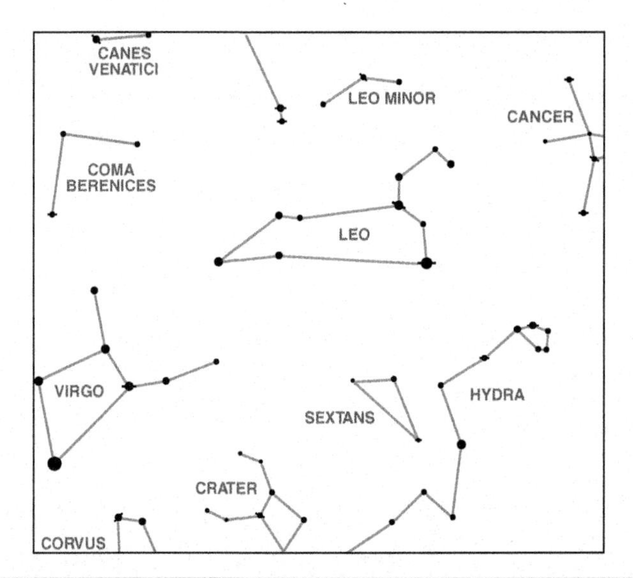

Figure 2.11B Leo and surrounding constellations

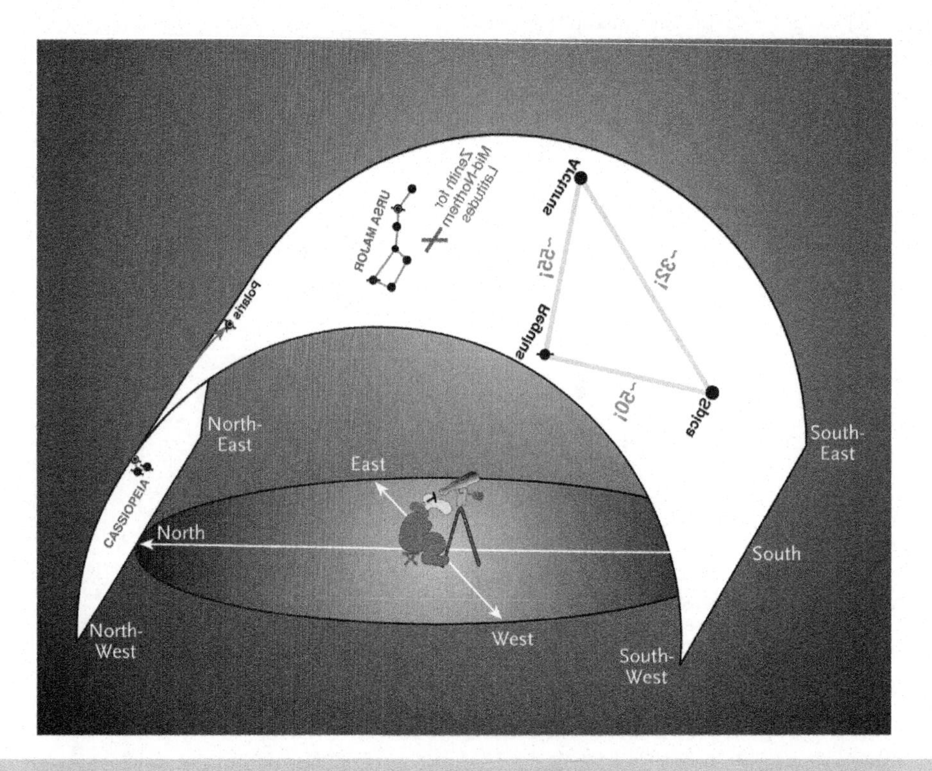

Figure 2.12 It helps to picture the charts for learning about constellations (P1 to P4, as well as the seasonal charts) folded over the meridian of the observing point from the south across the zenith to the north. If you have a problem with this, you should photocopy the chart from this book and then bend it while using it during observation

P1

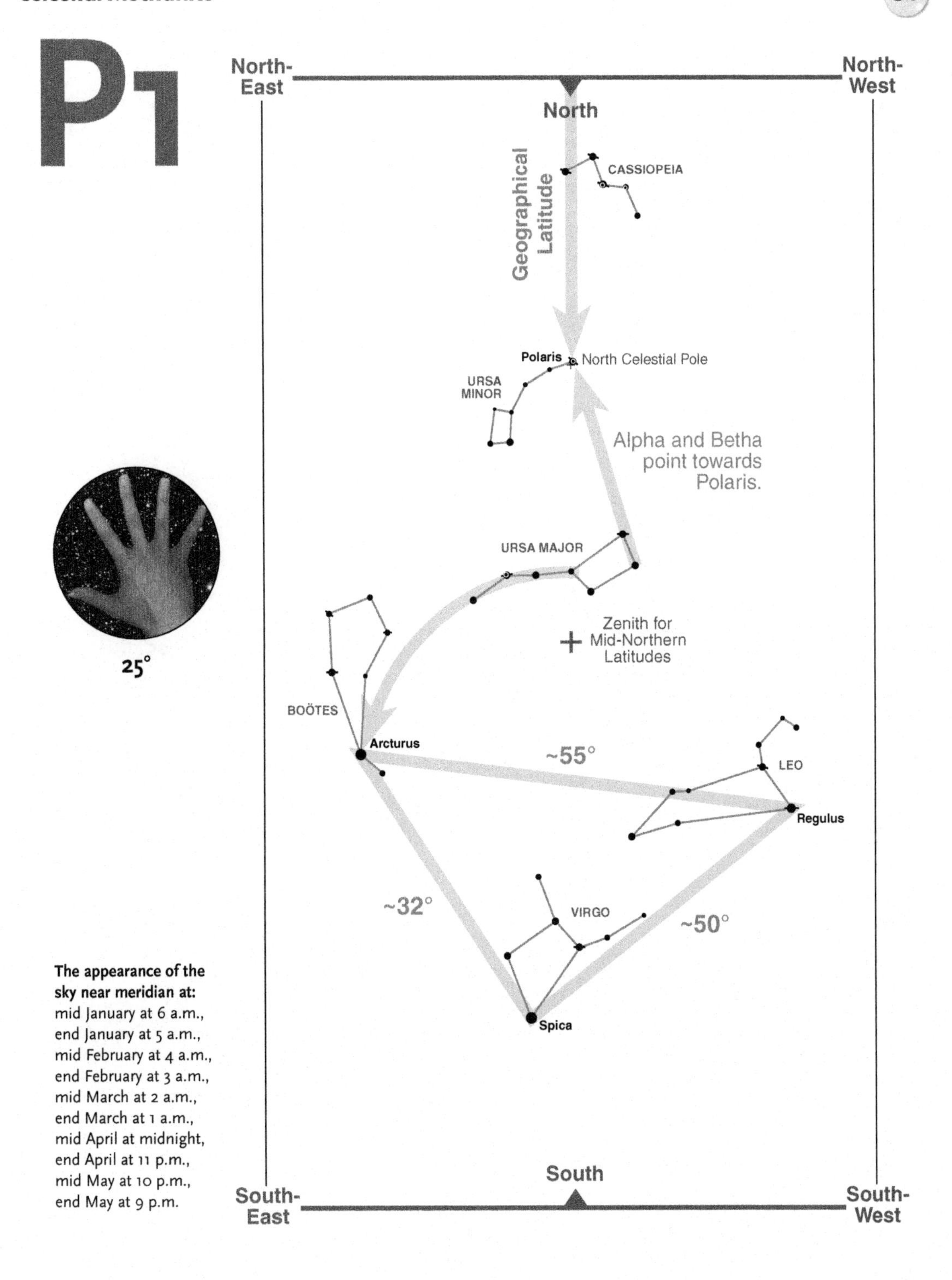

The appearance of the sky near meridian at:
mid January at 6 a.m.,
end January at 5 a.m.,
mid February at 4 a.m.,
end February at 3 a.m.,
mid March at 2 a.m.,
end March at 1 a.m.,
mid April at midnight,
end April at 11 p.m.,
mid May at 10 p.m.,
end May at 9 p.m.

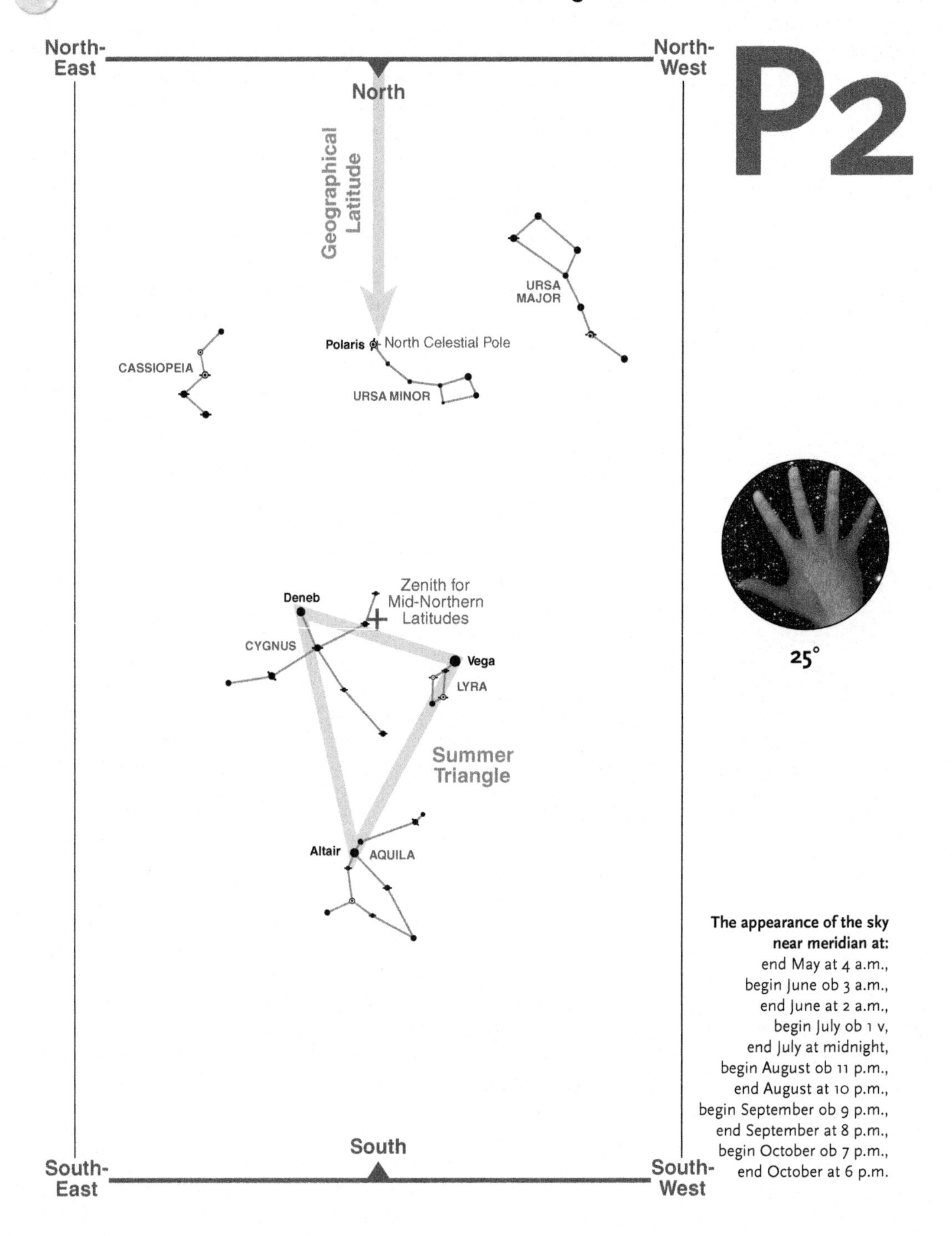

The appearance of the sky near meridian at:
end May at 4 a.m.,
begin June ob 3 a.m.,
end June at 2 a.m.,
begin July ob 1 v,
end July at midnight,
begin August ob 11 p.m.,
end August at 10 p.m.,
begin September ob 9 p.m.,
end September at 8 p.m.,
begin October ob 7 p.m.,
end October at 6 p.m.

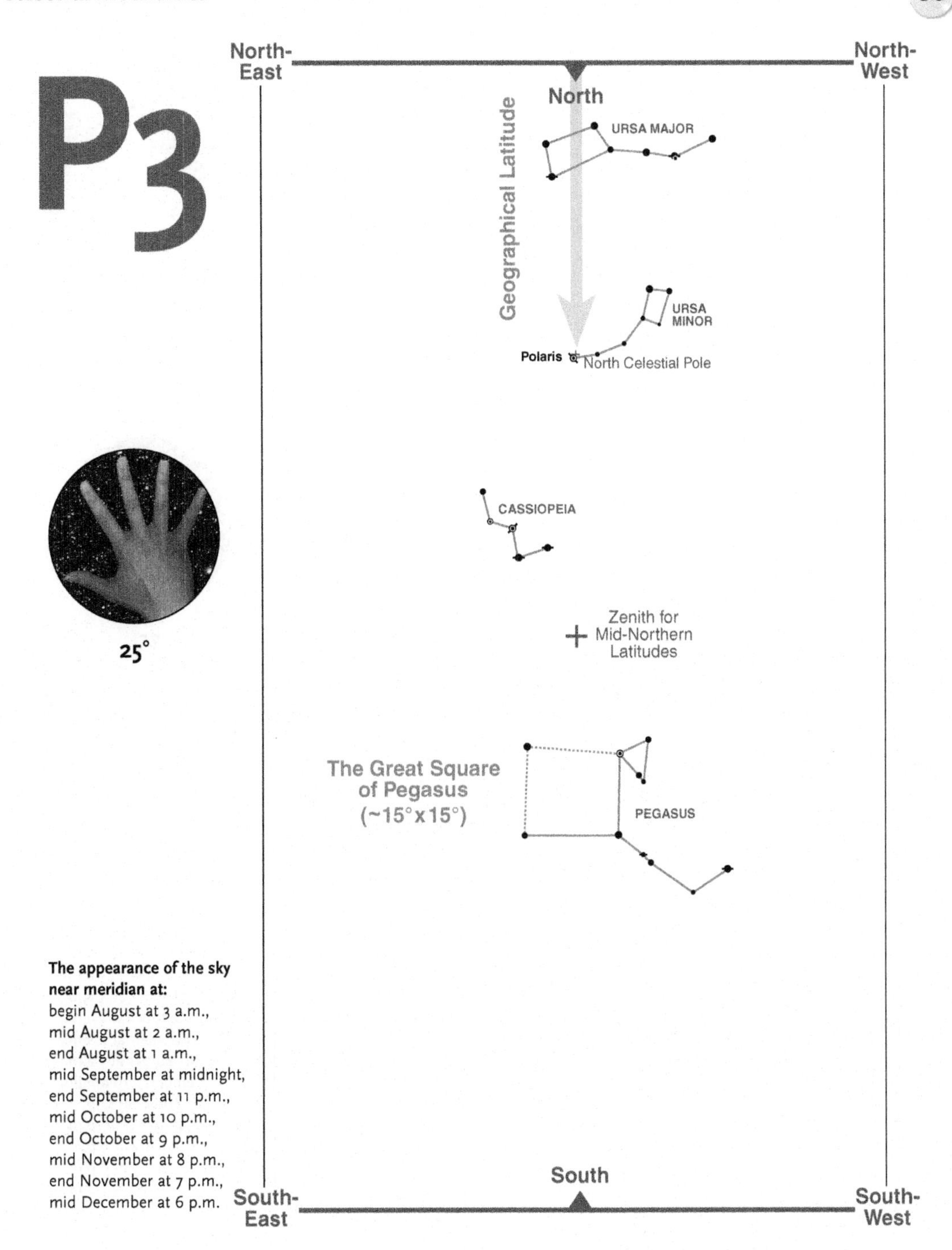

The appearance of the sky near meridian at:
begin August at 3 a.m.,
mid August at 2 a.m.,
end August at 1 a.m.,
mid September at midnight,
end September at 11 p.m.,
mid October at 10 p.m.,
end October at 9 p.m.,
mid November at 8 p.m.,
end November at 7 p.m.,
mid December at 6 p.m.

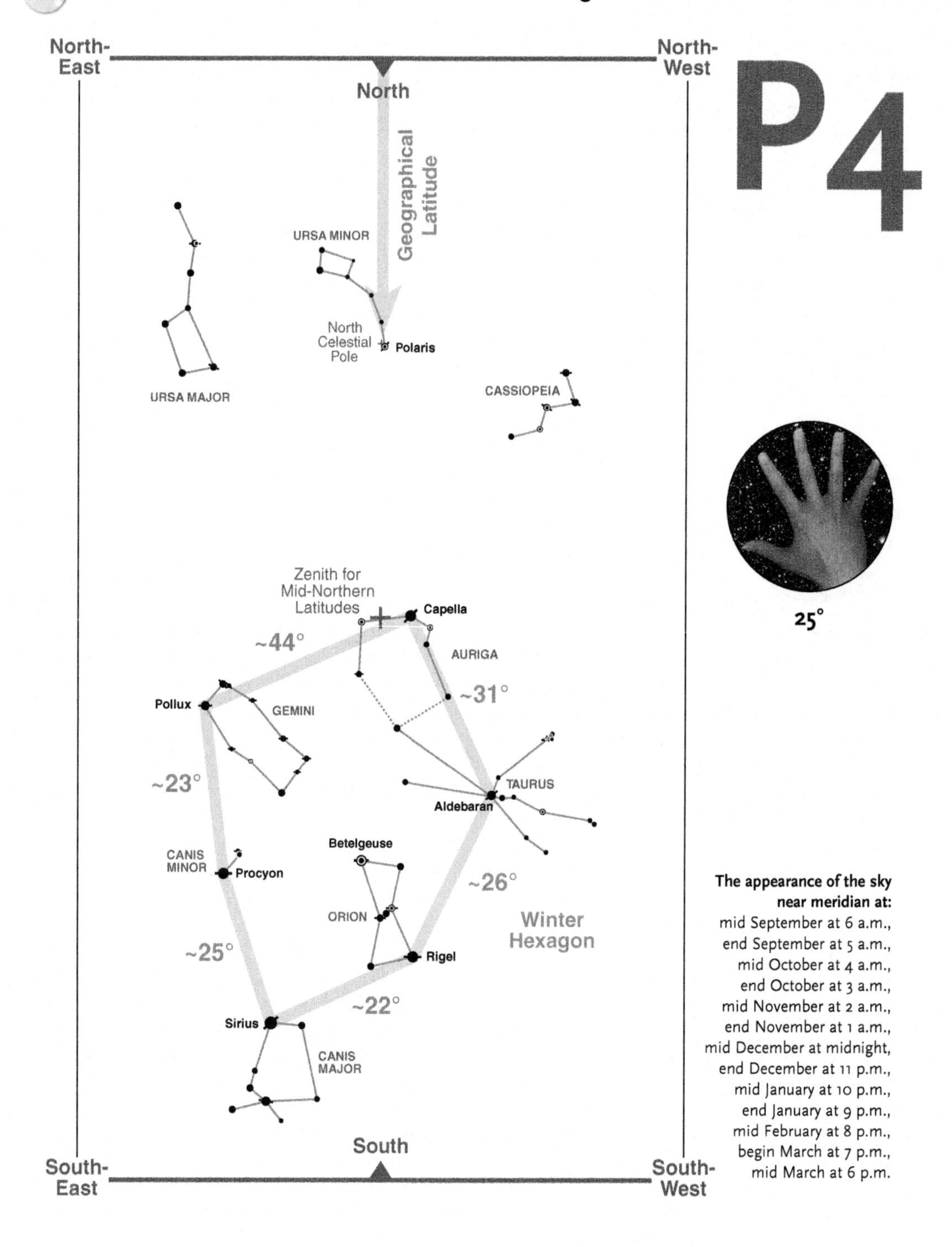

The appearance of the sky near meridian at:
mid September at 6 a.m.,
end September at 5 a.m.,
mid October at 4 a.m.,
end October at 3 a.m.,
mid November at 2 a.m.,
end November at 1 a.m.,
mid December at midnight,
end December at 11 p.m.,
mid January at 10 p.m.,
end January at 9 p.m.,
mid February at 8 p.m.,
begin March at 7 p.m.,
mid March at 6 p.m.

while your eyes are still getting adjusted to the dark, you will see only the brightest stars and will certainly recognize them.

You can also bring a flashlight outside with you. When you turn it off, your eyes will not have had time to adjust to the dark, and you will only see the brightest stars in the sky. You can also try and learn about the constellations by observing them from light polluted places, where even under the best conditions you will only see stars up to magnitude 3, which, for the novice, is almost ideal. However, after you are able to recognize the brightest stars and want to learn about the entire constellation, you should find an observing point with dark, clear skies.

A final possibility is to start learning about the constellations at twilight, when the Sun has already set but it is not yet night, and only the brightest stars are visible in the sky. Such conditions exist every day for approximately half an hour.

CHAPTER THREE

Those Wonderful Distant Suns

The Brightness of Stars

Some stars shine brighter than others. We say that they are of different magnitudes, or levels of brightness. Hipparchus, who lived around 150 B.C., classified all stars visible to the naked eye into six classes according to their brightness. The writings in which this ancient Greek astronomer described the magnitude system, almost certainly for the very first time, are long gone.

Claudius Ptolemy, who lived three centuries later, knew Hipparchus's work and probably used it as the basis for his two stellar catalogs, which were published in Book VII and Book VIII of the famous *Almagest.* In the catalogs, Ptolemy classified 1,022 of mostly Northern Hemisphere stars in six groups according to brightness, which he named *magnitudes* (in Latin, *magnitudo* means "greatness").

The scale used in *Almagest* was a kind of tool for identification in which stars were classified into various classes according to the visual impression they made on the observer. The brightest stars made up the first class, and he named these stars of the first magnitude (1 m). Stars that were barely visible to the naked eye could be found in the sixth class. He named these stars of the sixth magnitude (6 m). This classification system remained unchanged until the invention of telescopes in the early 1600s. With telescopes we are able to see stars that are otherwise too faint to be seen by the naked eye. Therefore, the scale for brightness had to be expanded. Celestial objects that are weaker than magnitude 6 are of magnitude 7 (7 m), 8 (8 m), 9 (9 m), etc. Objects that are brighter than magnitude 1 are of magnitude 0 (0 m), -1 (-1 m), -2 (-2 m), etc.

With the advance of photometry and instruments for measuring brightness, by the end of nineteenth century astronomers were capable of measuring the brightness of objects in the sky very accurately. The basic magnitude classes remained, but they also gained decimal places. Here we are

B. Kambič, *Viewing the Constellations with Binoculars*, Patrick Moore's Practical Astronomy Series,
DOI 10.1007/978-0-387-85355-0_3, © Springer Science+Business Media, LLC 2010

talking about visual brightness (m_v) in visible yellow-green light with a wavelength around 550 nm, where the human eye is the most sensitive (see text in the box on the next page).

Nowadays, astronomers view the magnitude scale as one of the anachronisms of modern science. At the same time the moment for a successful reformation of the scale has long since passed. The magnitude scale has several peculiarities. First, it is inverse, which means that the higher the number, the fainter a star. And second, it is logarithmic, with the base of the logarithm not 10 or *e* but 2.5, a number without any special significance. After the battle it is easy to be a general, but still, it was a great mistake for astronomers to have allowed this clumsy system to take root! For contemporary observers who encounter magnitudes for the first time and are interested only in how bright a star is, the scale is – to put it mildly – awkward. However, astronomers are used to it and so do not strive to change it.

In this book, the charts for individual constellations include bright stars up to magnitude 5. Magnitude 6 stars are shown only when they are an important part of the constellation's figure or if they form a close naked eye pair with a brighter star. Different stellar magnitudes are indicated by the different sizes of the black dot. In size they line up as follows:

- ● Stars of magnitude —1 (brightness from magnitude —1.6 to —0.5),
- ● Stars of magnitude 0 (brightness from magnitude —0.6 to 0.5)
- ● Stars of magnitude 1 (brightness from magnitude 0.6 to 1.5),
- ● Stars of magnitude 2 (brightness from magnitude 1.6 to 2.5),
- ● Stars of magnitude 3 (brightness from magnitude 2.6 to 3.5),
- • Stars of magnitude 4 (brightness from magnitude 3.6 to 4.5),
- • Stars of magnitude 5 (brightness from magnitude 4.6 to 5.5) and
- · Stars of magnitude 6 (brightness from magnitude 5.6 to 6.5).

Such a classification is necessary in order for the charts to be clear and understandable; however, it has its drawbacks. When viewed in the sky, stars of, for example, magnitudes 3.6 and 3.5 appear to be of practically equal brightness, since the difference in their magnitude (one tenth) can only be detected by an experienced observer. However, in the charts they are shown with symbols of two different sizes, because the first stars belong in the class of magnitude 4 stars, and the second one in the class of magnitude 3 stars. On the other hand, in the case of two stars with, for example, 2.6 and 3.5 magnitudes (differing therefore by 0.9 magnitude), we can easily see the difference in the brightness with the naked eye, but they are depicted by the same symbol in the chart, since they both fall in the class of magnitude 3 stars.

In the more detailed charts in this book, which are given to help you find fainter objects, stars up to magnitude 9.5 are shown, i.e., all stars that are visible with binoculars in good observing conditions. Only in open clusters, which are among the nonstellar objects most suitable for observing in binoculars are stars up to magnitude 11 shown.

Limiting Magnitude

As mentioned earlier, on a clear, moonless night we can see stars with the naked eye as faint as magnitude 6.5. With binoculars we can see even fainter stars, since its lenses are much larger than the lenses in our eyes or, more accurately, have a much larger diameter than the pupil of our eye, and therefore collect more light. However, the binoculars also have a limiting magnitude that depends on the size of the objective: the greater the diameter, the fainter the objects we can observe. The approximate value of the limiting magnitude can be estimated with the simple equation:

$$M_{\text{limiting}} = 6.5 + 5\log D$$

where D is the objective's diameter expressed in centimeters. Using binoculars with a diameter of 5 cm, we can see stars as faint as magnitude 10. If we have an objective with a 10 cm diameter, we can see stars up to magnitude 11.5. With the currently most popular amateur telescope (which has an objective's diameter of 20 cm), the limiting magnitude of stars is 13.

The value for the limiting magnitude of stars that can be seen with the naked eye (6.5 magnitude) depends on numerous factors. Probably you have already noticed that in cities we can usually see only stars to magnitude 3 or possibly 4. Because of light pollution, which has also made its way into smaller towns, villages, and even the countryside, the limiting magnitude is most often only about 5! Only on the clearest and driest nights, usually in winter, the limiting magnitude drops to magnitude 6.5 or 7. However, there are places on Earth where the sky is still really dark. On the best nights, when there is no Moon and air humidity is low (in deserts), the limiting magnitude can drop to magnitude 8! Can you imagine how the night sky looks like from such a place? The sky is jam packed with stars, and one can see the zodiacal light, *gegenschein* (or counter shine) and the zodiacal belt; these are phenomena that most of us only know from books. An experiment with an artificial star in a laboratory environment, where the disturbances caused by humidity and atmospheric blur were removed, also showed that the limiting magnitude of the human eye is around 8!

In practice the limiting magnitude also depends on the altitude of the observed object and on the observing conditions, such as the transparency and tranquility of the atmosphere. It might happen that we get excellent observing conditions when the limiting magnitude of the binoculars in the vicinity of the zenith drops as low as 11, but mainly the conditions are such that we can only get more or less close to magnitude 10.

When we talk about limiting magnitude, we cannot ignore the problem of light pollution. In fact, conditions are getting worse each year, and in some places they are already catastrophic. A while ago, you could read in books that all you have to do is drive to a nearby hill to observe the night sky, but nowadays this is no longer enough. The combination of light pollution and atmospheric haziness, which is a consequence of air pollution, affects practically all densely inhabited parts of the world. There are only a few places left that have a truly dark night sky and not an orange horizon plus a glow at night that makes you wonder whether the Sun has truly set!

More on Magnitudes

When we are talking about brightness or magnitudes of the celestial bodies in books like this, we are talking about visual magnitudes, i.e., magnitudes that we see with our eyes, normalized to the value they would have in the absence of the atmosphere. But celestial bodies radiate in all wavelengths of electromagnetic spectrum. We know that hot stars radiate mostly ultraviolet light and cool stars mostly infrared light. That's why astronomers use a few different types of magnitudes. Besides visual magnitude (m_v) the most important is bolometric magnitude (m_{bol}), which measures the total energy output of the celestial body. In older references you can find photographic magnitudes (m_P), which depend on sensitivity of photographic emulsions on films or plates. With the development of modern CCD detectors in the last few decades photographic magnitudes are no longer in use.

Extinction

Before light from a star reaches the eye, it has to travel through the atmosphere. This causes extinction – weakening of the light due to the absorption and dispersion of it in the atmosphere. Whenever we talk about limiting magnitude or use formulae to calculate it, we have to be aware that the result is valid only for the zenith, i.e., the direction directly above our observing site. In this case we are looking at the stars through the thinnest possible atmospheric layer. As we move further away from the zenith, the layer of air gets thicker, and this increases the absorption and dispersion of the light, which causes the limiting magnitude to rise. The atmospheric extinction becomes especially important when we are observing celestial bodies less than 20 degrees above the horizon (see table below).

Angle from zenith	Reduction in magnitude	Angle from zenith	Reduction in magnitiude
0°	0.00	60°	0.23
10°	0.00	70°	0.45
20°	0.01	75°	0.65
30°	0.03	80°	0.99
40°	0.06	85°	1.77
50°	0.12	87°	2.61

Let us digress for a moment to draw attention to something we usually do not pay much attention to. The field of view of the binoculars is 6 degrees. This means that between the stars at the top and the bottom of our field of view there is up to 6 degrees of height difference; that is what causes the atmospheric extinction to vary. Two stars that are otherwise equally bright will appear variously bright at different heights low above the horizon: both will appear dimmer than they truly are, but the lower-placed one will be even dimmer than the higher-placed one. Extinction also depends on the altitude above sea level, humidity, and level of aerosols in the air. After a cold weather front has passed, when the sky is crystal clear (especially if we are in the mountains), the values are much smaller when compared to the ones in the table, while on humid summer nights they can be even bigger. In star atlases (and in the charts in this book) extinction is, of course, not taken into account.

Integral Brightness and Surface Brightness

When talking about brightness we should also explain two terms that are more related to other celestial objects than they are to the stars.

Stars, whose light travels to us from enormous distances, are seen by us as tiny points of lights (see the box on p. 62). With vast celestial entities such as diffused nebulae, clusters, and galaxies the situation is different. The light comes from various parts of its spread out surface. The brightness given for these objects tells us in reality how bright the object would be if all of its light was gathered into one point, but it is not! That is why a star of magnitude 10 is relatively easy to see, but it is much harder to notice a galaxy or globular cluster of magnitude 10. This is also the reason that with galaxies we usually see only the central core and not also the surrounding (dimmer) parts, for instance, the spiral arms in galaxies of magnitude 8, even though our binoculars can detect objects up to magnitude 10! This inconsistency is even more obvious with nebulae that are apparently much larger than galaxies. A good example is the North America Nebula in Cygnus, which is approximately magnitude 4. Based on this number we should be able to find it easily with the naked eye. However, the size of the nebula is 120 times 100 arcmin, which is 2° by 1.7° (the apparent size of the Moon is 0.5°). When we compare the data for the brightness of the nebula and its size, we can see that there is relatively limited brightness per unit of surface; we thus say that this is a nebula with a small surface brightness. In reality, this nebula can only be noticed with the naked eye in perfect observing conditions. When considering whether we will be able to see a vast celestial body with our binoculars, we always have to take into account two pieces of information: its brightness and the apparent size. The greater the brightness and the smaller the apparent size, the more likely it is that we will see the body. That is why the book also carries the information (next to every nonstellar celestial body's name or designation) on the integral brightness and apparent size. Of course the best help to us is, as always, observing experience.

Figure 3.1. The Andromeda Galaxy (M 31) and the globular cluster Omega Centauri are approximately of equal brightness. The first shines with an integral magnitude of 3.5, and the second with an integral magnitude of 3.7. But because the apparent size of M 31 (180′×60′) is much bigger than the apparent diameter of the cluster (36′), we can see the galaxy only on clear nights and in moonless skies as a gentle cloud of light, while the central part of the cluster is so bright that it was mistaken for a star of magnitude 4, which is also the reason for its name, Omega Centauri

Luminosity and Absolute Brightness

We have already mentioned that a star's magnitude (its symbol is m) tells us how bright a star in the sky is. Magnitudes measure the density of the light flow that reaches us from the star but tells us nothing about the stars themselves. Are they distant giant stars, nearby dwarfs, or maybe just plain average stars? To emphasize this fact we often talk about apparent magnitude or apparent brightness. If we lived in a different part of the galaxy, the brightness of these same stars would be totally different, because their distances would be totally different.

Luminosity tells us what the total power or how much energy the star emits per time unit. The luminosity is therefore linked directly to the star – its mass, core, surface temperature, actual size, etc. It is expressed in watts (W). The luminosity of the Sun is 3.9×10^{26} W. The largest and hottest stars surpass the Sun in their luminosity by several hundred thousand times! The luminosity can be calculated from the brightness, if we know the distance to the star.

There is another term, related to the brightness of stars, that is called *absolute brightness* or *absolute magnitude* (its symbol is M), and this tells us how bright the star would be in the sky if it were at a distance of 10 parsec (32.6 light years). So absolute magnitude is a certain measure of luminosity.

All three terms are not related just to the stars but to all kinds of celestial bodies. Let's see some examples. The brightest star in the sky, Sirius, has magnitude −1.44. Its luminosity is 10^{28} W and its absolute magnitude is 1.4. Its celestial neighbor Rigel in Orion has a magnitude of 0.1, a luminosity of 2.6×10^{31} W (2,600 times bigger than Sirius), and an absolute magnitude of −6.7. When we compare all these data, we see that although Sirius in the sky looks much brighter than Rigel, in fact this is only because Sirius is much nearer to us. If we put both stars to 10 parsec distance, Rigel would be much, much brighter than Sirius, because Rigel's luminosity is much larger. And, just to compare, our Sun's magnitude is −26.8, while its absolute magnitude is only 4.8. If we put it at 10 pc distant, it would be only a 5th magnitude star.

Star Names

Only the brightest stars have proper names, and in most cases they obtained them in ancient times. It would be impossible to memorize the names of all of the 9,100 stars that are visible with the naked eye, not to mention the names of billions of stars that are visible through telescopes or on long-exposure images.

In 1603, the German astronomer Johann Bayer (1572–1625) came up with a much more practical system of delineating stars. He named the stars in an individual constellation with the letter of the Greek alphabet so that the brightest was given the name Alpha, the second brightest Beta, the third Gamma, and so on. Only rarely did he use a different criterion from star brightness. An example of this exception is the constellation Ursa Major, in which the stars in the asterism known as the Big Dipper were named from Alpha to Eta, and the remaining stars in the constellation were named according to their waning brightness. To differentiate the stars from different constellations we have to follow the Greek letter with the name of the constellation (in genitive) the star belongs to. A few examples are Alpha Lyrae, Alpha Ursae Majoris, Alpha Orionis or Zeta Crucis, Zeta Centauri, Zeta Hydrae, etc.

All constellations have many more stars than there are letters in the Greek alphabet. Once we reach Omega, we start denoting stars with numbers. This system was introduced to astronomy by John Flamsteed (1646–1719), which is why we usually call it the Flamsteed system. Thus, when you look at the charts you will see, for example, 29 Orionis (on the actual charts we have left out the name of the constellation for practical reasons, so you will see only the number 29 alongside the star), 61 Cygni, 89 Virginis, 52 Leonis, 139 Tauri, etc.

Diffraction Pattern

There is an interesting fact related to star's images seen in the telescopes, which many amateur astronomers are not aware of. The image of the star in the telescope is not the image of its surface or disk. Stars are too far to be seen as disks, no matter how big the objective is or how high the magnification is. In fact, the image we see in the eyepiece is a diffraction pattern of the star's light, produced in the optical instrument we observe with, and it occurs because of the wavelike nature of light. The diffraction pattern is composed of a bright circular region in the center, surrounded by concentric dark and light circles. We can see this pattern in the eyepiece if we slightly blur the image of a star. The central peak of the pattern of perfect optics contains 84% of the total light. If the quality of optics is poor, the first diffraction ring is so strong and broad that it mingles with central peak.

The diffraction pattern is named the "Airy disc" after George Biddell Airy (1801–1892), the English Astronomer Royal who first described the phenomena.

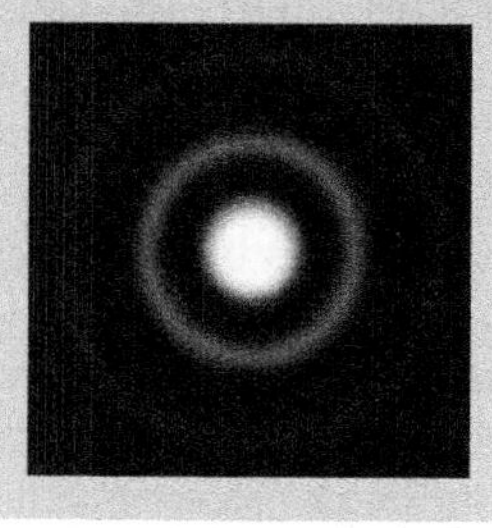

Double Stars

Every now and then an attentive observer can see two stars that seem extremely close in the sky. Such a couple is called a double star, and if there are three or more stars we call them triple or multiple stars. When looking through binoculars, the number of visible double stars increases significantly.

The astronomers who observed double stars through the first telescopes were certain that these only appeared to be together. After long and careful observations William Herschel (1738–1822) published in 1793 that there are actual double stars among them. The stars in a true double star (also called a binary star) are gravitationally bound, and they rotate around a common center of mass. From Earth we see that the distance between the stars periodically changes. The changes are usually slow, and only astronomers with professional equipment can measure them. This was a truly great, we could say epochal, discovery, for it proved that Newton's gravitational law was universal and that it holds true also in other parts of the universe. Using Kepler's law we can, with a binary star, define the total mass of the system by measuring the time it needs to orbit and the orbit radius, and in some cases we can also determine the masses of the individual stars (Figure 3.2).

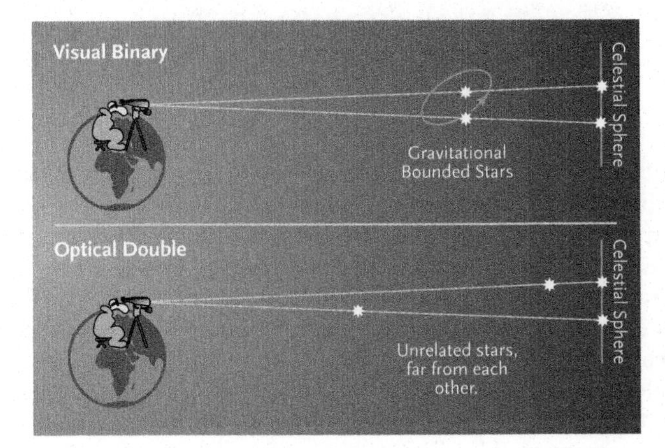

Figure 3.2. A true and an apparent double star. In a true double star the two stars are close together, are gravitationally bound, and orbit around a common center of gravity. With a virtual double star the stars only appear to be together, which means that when viewed from Earth they seem to be in almost the same direction

With apparent double stars the two components are not gravitationally linked and are usually very far apart. They only appear to be close when viewed from Earth, for they are in the same general direction. Examples of such stars are Mizar and Alkor (80 Ursa Major) that are 11.8 arcmin apart and can be easily split even with the naked eye. In apparent double stars, the distance between the stars can also change, but this is only due to the stars' proper motions.

In this book, we will often mention astrometric and spectroscopic binaries. An *astrometric binary* is a star pair in which we cannot see the dimmer member of the pair, but we can conclude it is there due to the periodic declinations in the proper motion of the brighter companion. We cannot see the accompanying star because it is too dim or too close to the brighter star. However, from the measurements we can still estimate the orbital period and thus the total mass of the binary system. In the past, when professional astronomers had poorer equipment than most amateur astronomers have today, there were numerous astrometric binaries. Later on, with larger and higher-quality telescopes, they could separate and see many of the accompanying stars. The best known example is probably the brightest star in the sky, Sirius, with its accompanying star, a white dwarf. An astrometrical binary star is best understood by looking at the illustrations on Figure 3.3.

With a *spectroscopic binary* the stars are so close together that we cannot separate them even with the largest telescopes. Astronomers can see only that there are two stars and not just one when they study the spectral lines of elements such as hydrogen and helium in the incoming light from the star.

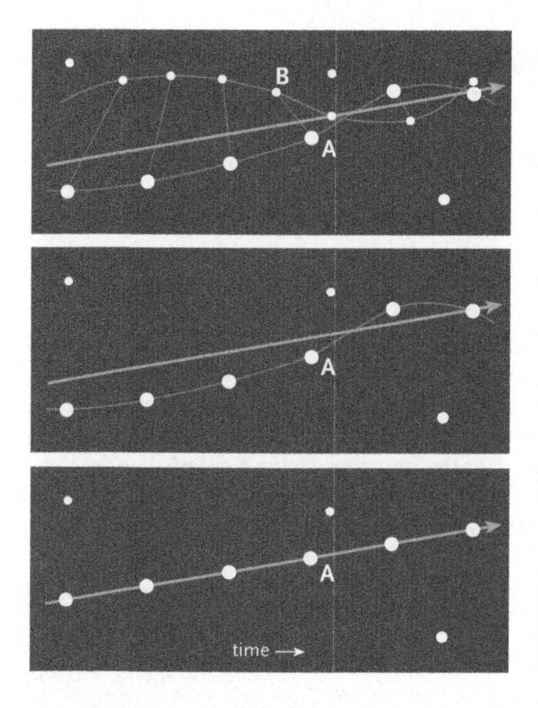

Figure 3.3. *Top:* Stars A and B revolve one around the other, while at the same time having a common proper motion through space that we can recognize when compared to other stars. From Earth we can therefore see them travel in the way depicted in the top illustration. *Center:* Even if we cannot see star B, we can infer its presence from the periodic movements of star A. *Bottom:* If this star had no accompanying star, it would move along a straight line

Due to the Doppler effect, the lines periodically shift from their basic position, because the two stars are periodically approaching and receding from us while they circle around each other.

The orbital periods of spectroscopic binary stars are extremely short; we measure them in days. If the stars when seen from Earth circle each other in such a way that they never cover one another, the brightness is constant. In rare examples the orbit – as seen from Earth – is such that there are regularly occurring partial or total eclipses. The brightness of such a binary star varies for the observer on Earth, and we call the phenomenon an *eclipsing binary star* or an *eclipsing variable star.* Such binary stars can be found in tables of variable stars. Astronomers in the past noticed the changes in the brightness of the star, but they did not know the reason behind them.

From the 1990s onwards, astronomers have been using a new observation technique called optical interferometry. They observe a binary star with two or even more telescopes that are far apart and combine the light from all the telescopes into an interferometric image. This system has a resolution equivalent to a mirror as large as the greatest distance between the telescopes, which could be 100 m or more. In this way they have already managed to separate some spectroscopic binary stars such as the famous Capella in the constellation Auriga.

With many double stars both components are the same color, while some others are famous due to the color contrast of the two stars. One of the most beautiful binaries is the Albireo (Beta Cygni), in which the brighter star is a nice golden-yellow color, while the dimmer one is blue. For more on colors of the stars see later in this chapter.

Limiting Resolution

Whether you are going to be able to separate a certain double star with your telescope (and see two stars) depends on the diameter of the objective (D) and the wavelength (l) of the observed light (star color). The larger the diameter, the greater is the resolution. Limiting resolution tells us the minimum separation between two stars that can still be seen as split in specific telescope:

$$\text{Resolution (in radians)} = 1.22\,\lambda/D$$

Amateur astronomers started using an even simpler formula, which gives us the limiting resolution in arc seconds and in which we assume that we are observing at a wavelength of 550 nm, the range in which the eye is the most sensitive:

$$\text{Resolution (in arc seconds)} = 12/D$$

where D is the diameter of the objective in centimeters.

In both examples, the result is a theoretical limiting resolution in which we assume that we have ideal optics (objective and eyepiece) and excellent observing conditions that enable unlimited magnifications. In practice we can only get more or less close to the theoretical limiting resolution.

The theoretical limiting resolution of a 5-cm objective is 2.4 arcsec, a 10-cm objective 1.2 arcsec, a 20-cm objective 0.6 arcsec, etc. For comparison we can say that the limiting resolution of the eye is approximately 1 arcmin (60 arcsec) (Figure 3.4).

With binoculars, where we have a defined and unchanging magnification given in advance (for instance, 7 times at 7×50 or 10 times at 10×50), assuming that we cannot change the eyepieces, we

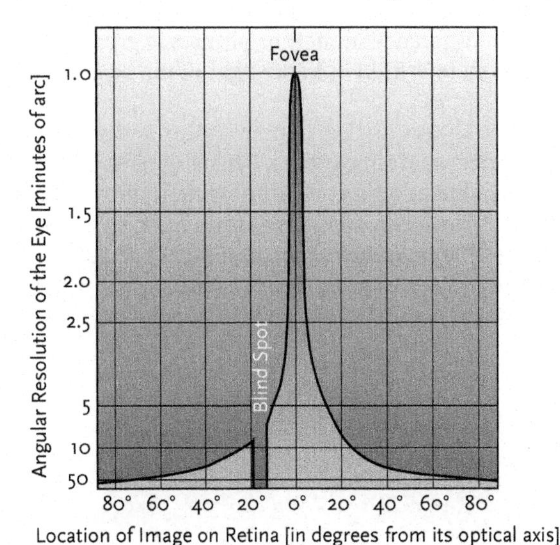

Figure 3.4. Angular resolution of the human eye is approximately 1 arcmin only very near the fovea or near the axis of the eye's lens

cannot even come close to the limiting resolution of the objective. In the best possible observing conditions we can, with 10×50 binoculars, come closer to a resolution of 20 arcsec, even though the theoretical limiting resolution of the objective is 2.4 arcsec. If we wish to make the best use of the objective to its extreme limits, we would need at least 150 times magnification.

In this book, we will therefore focus only on the crème de la crème of the double stars. Special mention was given only to those in an individual constellation that are renowned due to their strong color contrast, a few of the "ordinary" ones that are especially useful for testing the quality of optics of the binoculars, and those that played an important role in the history of astronomy.

Double Star Names

If we can resolve the double stars with the naked eye then the stars were usually ordered according to their brightness as all other stars have been, except that numbers 1 and 2 were added to them, for instance, the double star Alpha-1 and Alpha-2 of the Capricornus (also written as Alpha^1 and Alpha^2), Mi-1 and Mi-1 of the Boötes, etc. With triple (3) or multiple stars (n), the numbers go to 3 or n.

If a pair can be separated merely by an optical aid, it is most common that the stars received the additional letter A, B, C, etc, according to their decreasing brightness. One of the most beautiful multiple stars in the sky (which is unfortunately impossible to separate with binoculars) is the Iota Cassiopeiae, which is composed of Iota A, Iota B, and Iota C.

Observing Double Stars

As we have already mentioned, amateur astronomers are interested mainly in those double stars that can be admired for their interesting or beautiful color contrast and those that are useful for testing the quality of the optics of their instruments. We will encounter simple pairs in the descriptions of double stars that appear in later sections of this book, including pairs that can be easily resolved in any binoculars and some difficult pairs. There are two types of the latter. In the first case they are close double stars, with the difference being that the two stars are on the limit of the resolution of the binoculars. Another difficult pair is one in which the stars are rather close together (though we can still separate them easily), but the difference in their brightness is greater than 2 magnitudes. In such examples, the dimmer of the two stars is hidden in the light of the brighter one. In both cases we will get better results if we use better optics.

There is something that almost always disturbs observations and makes them harder to conduct, regardless of the quality of the optics – the atmosphere. Because our atmosphere is restless, the stars in the field of view are more or less blurred and constantly seem to move about here and there, which makes it difficult to observe them. However, sometimes for a fraction of a second the picture freezes, and at that moment we might be able to see two stars that before appeared as one. You have to wait patiently for those rare moments when observing tight double stars or when testing the optics of the binoculars (Figure 3.5).

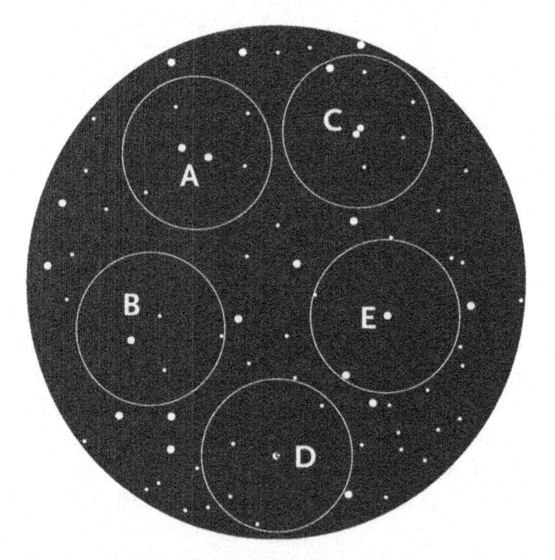

Figure 3.5. The first thing that we need to know before we start observing a double star system is the size of the field of view in comparison to the separation between the stars and the position angle. In the binoculars that we are looking through, do the stars appear far apart, close together, or can we not resolve them at all and see only a single star? There are a few examples of these situations in the illustration, but we will learn mostly through observations. *A* is a double star with similar-magnitude stars that are far apart; *B* is a double star with different-magnitude stars that are far apart; *C* is a close double star with similar-magnitude stars; *D* is a close double star with different-magnitude stars (we might not be able to see the fainter one or we will see it merely when there are excellent observing conditions); and *E* is a double star that cannot be split

Position Angle

Sometimes the dimmer star in the pair is very close to the brighter one and becomes lost in the strong shine of its neighbor. Or, it might be so faint that we can hardly see it. This is why we need to know where we should be looking for the fainter star. Included in the general information about a double star, such as the magnitude of both stars and their separation, is always their position angle (p.a.). This piece of information tells us how many degrees – measured from the north toward the east – the dimmer star is offset from the direction northward. The position angle is thus between 0° and 360°. In this book, after the position angle the year of the last measurement is given. This data refer to measurements of separations as well as the position angles (Figure 3.6).

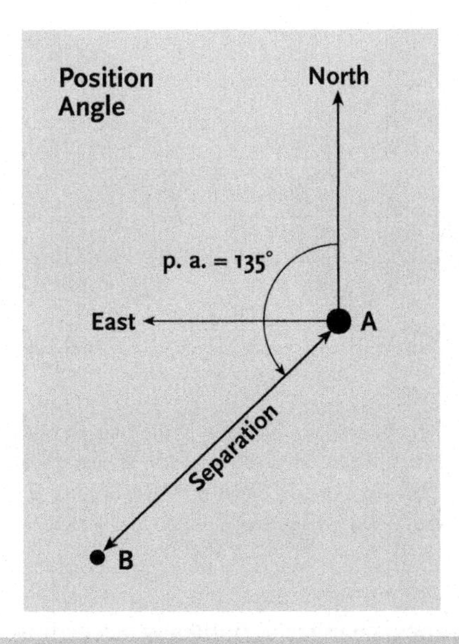

Figure 3.6. The position angle tells us the position of the fainter star in relation to the brighter in degrees. The angle is measured from the north toward the east, from 0° to 360°

Variable Stars

The magnitude of some stars is not constant, it changes over time. We call these stars *variable stars* or *variables.* Variable stars can be separated into two large classes: apparently variable stars and true or physically variable stars.

The apparently variable stars are those in which we notice a difference in the magnitude even though it never really changes. How is this possible? Let's imagine a close binary star, in which the stars are so close together we cannot resolve them in any telescope. If the plane of orbit of this double star is in line of sight as seen from Earth, then one star periodically covers the other during their orbiting. We could say that we see eclipses. Such double stars are called *eclipsing binaries.* The brightness changes are due to perspective effect. The eclipsing variable is the brightest when we can see both stars, and it reaches its minimum when one star covers the other. The best known variable of this kind is Algol (Beta Persei), which is reduced in magnitude by 1.3 for a few hours every three days (Figure 3.7).

In the second class of variables are stars in which the brightness really does change due to variations of their luminosities. We further divide these into regular and irregular variables.

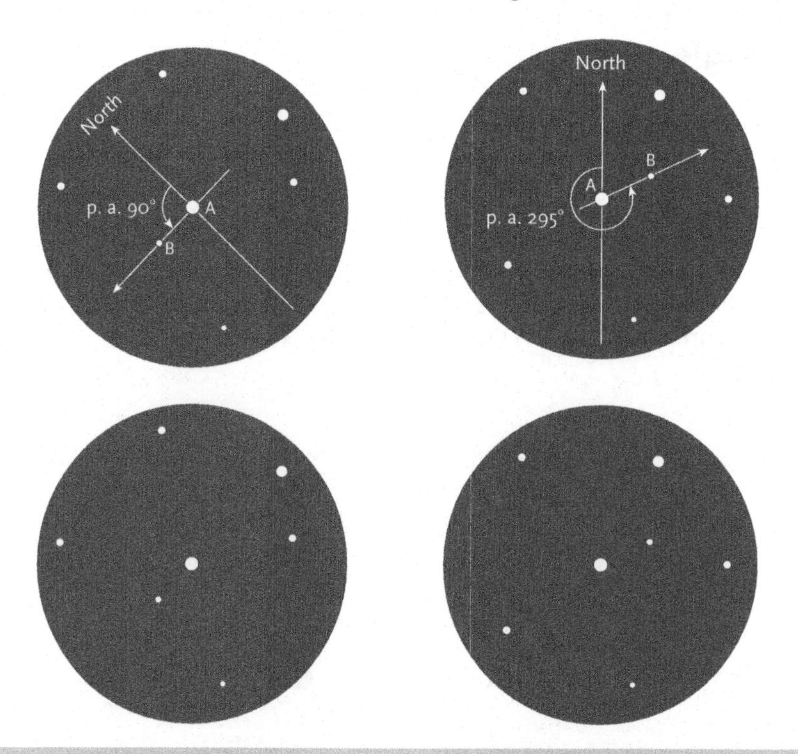

Figure 3.6A. Let's take a closer look at a two examples. If the position angle of the pair is, for instance, 90°, the fainter star is east of the brighter (see illustration upper left). Second illustration depicts position angle 135°. The upper illustrations are a schematic depiction, while the bottom ones are the views in the binoculars. The most important thing is (as we have already mentioned) that we know how the field of view is oriented, i.e., that we know where north, south, east, and west are

For regular variables the brightness changes periodically, according to some physical law. For these stars we can predict in advance their changing magnitudes. The periods of the regular variables are different – from an hour all the way to a year. The reason for the changes in brightness lies in the periodic expansion and contraction of the star, which causes the surface temperature and therefore the luminosity to rise and fall. The representatives of this class are Cepheides, which got their name from the first variable star of this kind discovered – Delta Cephei.

The periods of Cepheides are in usually between one day and one week long. Variable stars of the RR Lyrae type are very similar to Cepheides, only for them the period of oscillating is much shorter, usually shorter than one day. Cepheides and variable stars type RR Lyrae are also called short-period variable stars.

In the sky, we can see stars in which the brightness periodically changes over a longer period of time. Their cycles can last from a year up to several years. These are the long-period variable stars of the Mira type or Mira variables, which got their name from the variable star Mira in Cetus (Omicron Ceti), which is typical of this type. All stars of this kind are red giants with various dimensions and luminosities. Their luminosity changes because they have used all of their reserves of nuclear fuel (hydrogen) and have become unstable. Their changes are not as regular as with Cepheides and can vary from cycle to cycle. That is why some of them are placed in the class of semi-regular variable stars. An example of such a star is Betelgeuse in Orion.

For irregular variable stars – as the name tells us – neither the period nor the maximum and minimum brightness from cycle to cycle is the same. For this type of variable star, there are also a number of subclasses. We will describe only a few here.

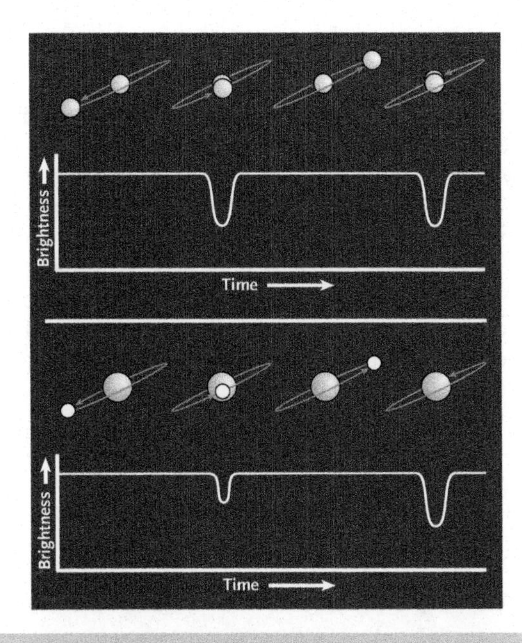

Figure 3.7. Eclipsing variables are, in fact, binary stars in which the orbital plane of the stars roughly or exactly coincides with the direction of our viewing sight. In the illustration, two typical examples are depicted. *Top:* A binary star composed of similar stars (for instance, Beta Lyrae); *bottom*: binary star composed of an orange giant and a white hot star (as in Algol)

The variable stars of the R Coronae Borealis type are usually bright, but on occasion suddenly and without warning their luminosity fades dramatically. The stars of the U Geminorum type, that are also called dwarf novae, behave in a completely different way. They are usually very faint, and every now and then they suddenly become extremely bright. For the stars of the RV Tauri type the luminosity at maximum and minimum oscillate greatly, but the period is unpredictable.

Novae

Novae are a special type of variable star. These are stars that, with a sudden, momentary outburst become extremely bright and quickly reach their maximum. Then their brightness fades until they reach their original level of brightness after a certain period. A nova is thus not a new star that is born the moment it increases its luminosity, as we might deduce from the name.

In reality, novae are binaries in which the stars are close together and one of them has already finished its life and ended it as a white dwarf. This star, with its strong gravitation, pulls on the hydrogen-rich outer layers of its neighbor. The material gathers on the surface of the white dwarf and compresses, due to the strong gravity. When there is enough material, a thermonuclear explosion occurs at the surface, which we see as a rapid increase of brightness, or the nova.

All novae erupt many times; however, the period between two outbursts is different from star to star. Those novae where we have already detected two or more eruptions are called *recurrent novae.* A typical representative of this type is T Coronae Borealis, which usually shines with magnitude 9 but has outburst as a nova at least twice – in 1866, when its brightness level rose to magnitude 2 and in 1946, when it reached magnitude 3.

Graph of Brightness

Astronomers measure the brightness of numerous variables and follow their cycles. In this endeavor they are helped to a great extent by amateur astronomers, for there are a lot of variable stars and the measurements are relatively simple. Thus it is considered a waste of precious observing time to use the large telescopes for such observations. This is one of the rare fields of astronomy where amateurs can truly work side by side with professionals.

The measurement results are displayed in a curve that shows how the brightness changes during the time observed (Figure 3.8). From the curve we can decipher the period, note maximum and minimum brightness, and see how the curve changes (or remains the same) from one cycle to another – and with this we check whether something unusual is happening with the star, perhaps something that will need more serious observation.

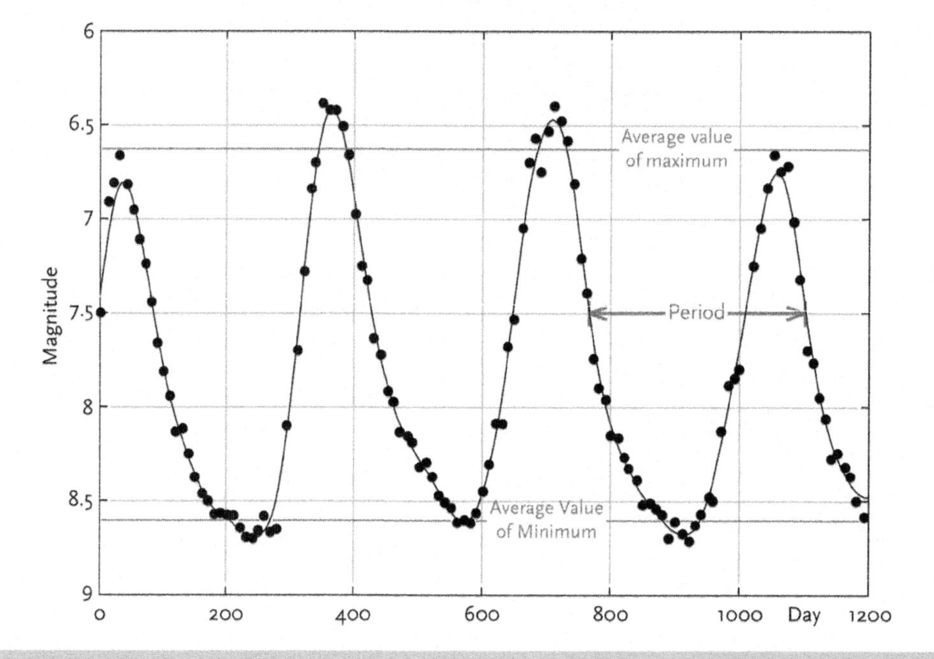

Figure 3.8. Here is an example of a curve of the brightness of an irregular long-period variable. From it we can see the average value of brightness that the star achieves at maximum and minimum and the period. In this book, all light curves are provided by the American Association of Variable Star Observers, with visual observations taken from the AAVSO international database. The data span the 1,200 days prior to June 30, 2008. The data points are 10-day averages of visual magnitude estimates. The solid lines are spline fits through the data (not mean curves)

From this data, astronomers try to create models of variable stars and ascertain the reason behind the changes in luminosity. That is how, for instance, they discovered for Cepheides the connection between its period and luminosity and used what they learned to measure distances to these stars and even to the neighboring galaxies in which the Cepheides are seen.

Variable Star Names

If the variable star is bright enough to have a name from the Greek alphabet, then it has no additional name. Thus, we have variable stars such as Alpha Herculis, Delta Capricorni, Omega Virginis, and so on.

As mentioned earlier, the fainter stars in a constellation have their own numerical designations (Flamsteed designations), while the variable stars – so that we can differentiate them immediately – are given different designations. Usually this is by a capital or double capital of the alphabet: S Virginis, RR Lyrae, etc. This system allows 334 different designations. If there are yet more variables in a constellation, we mark them as V335, V336, etc.

Every constellation contains a lot of variable stars. In this book, we have described just a few of the brightest, the most typical, or those that have played an important role in the history of astronomy.

Star Colors

The science that studies the nature of color and the way people perceive it is called colorimetry. Its findings are vital for the manufacturing of photographic films and light detectors, television and computer screens, color printing, etc. In this book we will focus only on how we perceive star colors.

With the naked eye we can see approximately 9,100 stars in the sky, out of which only about 150 are not usually white. Among the brightest stars the most common colors are yellow-orange, yellow, and light blue. Only rarely are stars truly red, and even these are seen only with the help of optics. Among the white stars we can differentiate between three nuances: "cold white," which have sort of a blue tint, the "normal white," which have no color tint, and "warm white," which leans toward yellow. The variations of colors and tints are reduced when we look at fainter stars. It is very difficult to recognize the blue tints at stars with a magnitude 2. At magnitude 3 the orange ones lose their color, become less noticeable, and are usually seen more as brownish-yellow. From magnitude 4 onward, almost all stars are more or less white. If you are observing with binoculars, the perception of colors extends to roughly magnitude 7, and from there onward most of them are white again. In order for us to understand this, it is necessary to understand how people perceive colors.

In normal daylight we can see the world in a multitude of colors. Three characteristics are important for every color: hue, saturation, and brightness. Hue is the term used for the dominant wavelength within the spectrum. Saturation tells us how clear the hue is; the fewer other colors mixed in, the clearer and more saturated a hue is. Brightness is linked to the intensity of light that falls on the retina in our eye. There are two types of cells (photoreceptors) that respond to light in our retina. Cones are for daytime vision, rods are for nighttime vision.

We have three types of cones, or color receptors, each of which is most sensitive to either red, green, or blue light. The cones with the highest sensitivity at 450 nm give the impression of blue; cones with highest sensitivity at 550 nm give the impression of green; and those with maximum sensitivity at 600 nm give the impression of red (Figure 3.9).

The colors with all their nuances emerge as a combination of various excited cones, with our brain also playing an active role. The proportion of signals from the three types of receptors determines the thousands of color impressions that we can differentiate between at normal levels of lightness. On color prints or TV screens, we can also create any millions of colors and their tints merely by mixing the three basic colors. The basic colors can be well differentiated, pure, spectral colors, but they are usually selected so they match the red, green, and blue peak sensitivity of the receptors in our eyes.

Color-blind rods for night vision come in only one type; thus we can only see black and white with them. It takes about half an hour to achieve a complete adaptation for night vision (Figure 3.10).

Physicists and astronomers look upon light in a more physical way. They describe it using a spectrum, which is a graph of light intensity in relation to wavelength. Today, astronomers know how to routinely and precisely measure the spectra of stars and other celestial objects.

Over 100 years ago, astronomers ascertained that the star colors are extremely similar to the colors of black bodies at various color temperatures. (A black body is a "perfect emitter," which means that the light intensity that it emits at every wavelength depends only on its temperature.) The spectrum of the star is thus compared to the spectrum of a black body at a given temperature. Such a graph represents the color as we would see it if we had thousands of different types of cones – thousands of basic colors, one for each wavelength. Our red, green, and blue cones are sensitive each in their own way to a broad span of wavelengths; roughly each one is responsible for half of the visible spectrum (see Figure 3.9).

People with the most common form of color blindness have only two types of cones (as a number of animals have). They see fewer colors because they see only a combination of two basic colors. If we had four types of cones, we could see additional colors, and we could differentiate between those that we now consider to be the same.

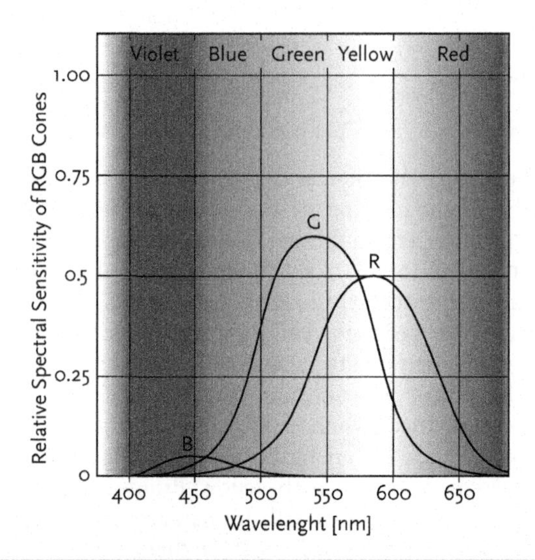

Figure 3.9. The responses to light of the three types of cones. The spectral areas overlap each other. The red cones also respond to green and even blue light, though they are less sensitive in these parts. The brain interprets any combination of responses as a different color. From the graph we can see that the eye is sensitive in different ways to various colors. This means that the yellow light that falls onto the eye causes a stronger reaction than red or blue light. When we compare different colors the eye deceives us, because it does not have the same sensitivity to all colors. Therefore, astronomers use instruments that do not have these faults to define star color (and with it also surface temperature of stars, as we will see later)

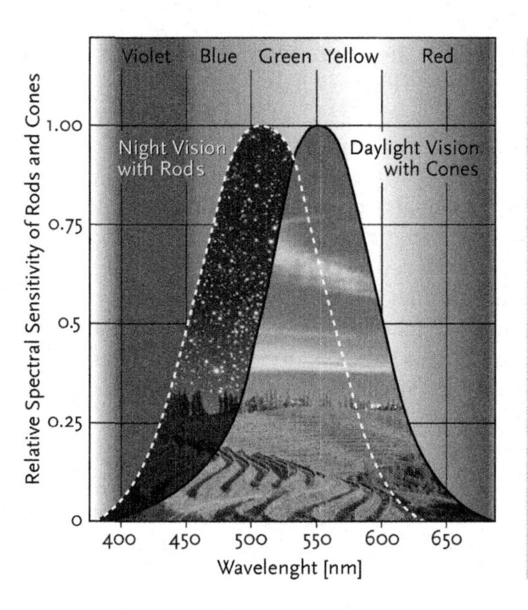

Figure 3.10. The spectral sensitivity of the cones (daytime vision) has its peak in the yellowish-green color with a wavelength of 550 nm, which is also the peak of the Sun's spectrum, while the spectral sensitivity of the rods (nighttime vision) has its peak in the green-blue light with a wavelength of 505 nm, which is the peak of clear night sky radiation

Influence of Brightness on Colors

If it is brighter than usual, a color hue seems brighter and less saturated. At the most extreme level of brightness we see everything in white. When normal brightness increases the yellows are the first to change into white and the red and violet the last. The reason for this lies in the overburdened receptors, which are operating at more than 100% efficiency. This is a state that our brains understand as white color. If the brightness is lower than normal, the white becomes grayish, while the brighter colors appear to be purer, and the normally pure colors become darker. Yellow and orange tones become brownish.

Illusions

The word *illusion* might seem a bit odd here, but as we will see later, it is appropriate. The most common illusions occur when the stars are too faint for normal color vision. At that point the color-blind rods that were until then hidden in the retina rise from it and contribute their share in the creation of the image. But let's take one step at a time.

Figure 3.11. Which observer in this picture is larger? Our brains play strange tricks on us during astronomical observations, too

Stars of magnitude 0 and 1 are still just about bright enough for normal recognition of colors. Thus, we can see them in their natural colors. The first changes appear in stars that shine with a magnitude of between 2 and 3. At magnitude 2, a white star appears to be slightly greenish, while at magnitude 3–3.5, it is of a gentle blue-greenish hue.

Between magnitude 3 and 4 the color-blind rods play an even greater part in the creation of the image. Stars that are in reality blue appear to be colorless or at most somewhere between brown and yellow.

The dimmest stars that we can see with the naked eye do not show any colors, because most of the image is created by the rods that respond best to the green, blue, and violet wavelengths. However, our cones still respond to the light, and they contribute to the creation of the image but not enough to give the impression of color. The consequence of this is that our limiting visible magnitude differs for blue, orange, and yellow stars. If the lowest magnitude that we can still see is 6.5 for yellow, we can still see blue stars at magnitude 6.7, while orange stars can be seen at a mere 6.1 magnitude.

For celestial objects that are too bright for normal color vision, the changes are simpler to describe. As we have already mentioned, all colors move toward white. The spectroscope would show Sirius with magnitude –1.44 as blue, even though we see it as totally white. Venus is in reality as yellow as the Sun, for it only reflects light, but due to its great magnitude (approximately –4) we see it as completely white.

Our brains also play tricks on us when we are trying to differentiate between the colors of close double stars. When colors from one part of the spectrum are missing, our brain shifts the remaining colors into a different part of the spectrum. If the brighter component in the double star is extremely yellow or orange, its otherwise white neighbor can obtain a hint of green, blue, or even violet. The color contrast will be most noticeable when the image of the stars is not completely in focus, and we have to watch at such magnification that the stars are as close as possible in the eyepiece. Later in this book you will find quite a number of such double stars.

You can also see how the brain tricks you by adjusting colors in everyday life. If you are driving in a car at night along a road that is lit by bright orange streetlamps, the otherwise yellowish lights of the cars driving in the other direction appear to be bluish in the prevailing orange light. Check it out next time!

You are also experiencing an optical illusion at this very moment if you are reading this book under a reading lamp. Doesn't the paper look white? But the light emitted by the light bulb is light orange, which you can verify by taking a photograph of the room with film that is intended for taking photographs outdoors. Our brains interpret any dominating color as white.

Observing Star Colors

There are plenty of clear nights when sky conditions are not entirely suitable for serious observing. During such nights you can practice observing star colors. In the descriptions of the constellations later in this book practically all stars that are not usually white are mentioned. Check if you can see them in the described colors and tints. It is very interesting to compare similarly colored stars. Unfortunately, there are limits to how much you can do this, because the chosen stars need to be above the horizon at the same time, and it is also best if they are at roughly similar heights; otherwise, the atmosphere will influence their color. When you become a master at observing bright stars, you should try observing fainter ones. Choose stars that are similar to the brighter ones. Do the fainter stars appear to be browner in color, even though the spectrum does not show this?

Following are a few pieces of advice that should come in handy when observing star colors:

- Eye cones are most densely scattered on the fovea, a roughly 2-degree wide area of the retina on the optic axis of the eye lens. For good color perception you should look directly at a star and not past it.
- With eyes that are totally adapted to night vision, we don't see colors as well as when they are nonadapted. Relatively mild light pollution therefore helps us slightly by preventing the rods from completely adjusting to the dark and thus seeing only gray tones.
- Observe stars when they are at least 30 degrees above the horizon, if possible even higher. The thick layer of the atmosphere shifts their color hues toward the red part of the spectrum, which becomes most obvious when the Sun is setting. Even in the zenith, the stars are slightly less blue than we would see them if they were viewed above the atmosphere. Moist air on a humid night causes redder colors than dry air does.
- To observe true star colors you must have a neutral chromatic adaptation. This means that prior to observing you should not expose your eyes to a strong single color light. Our perception of colors depends on what sorts of combinations our brain accepts as "white." You should avoid reading using artificial light directly prior to your observation, because this tricks the brains to take the

yellowish orange (the color of the glowing thread in the light bulb) as white. Yellow and yellowish-orange stars will thus lose a lot of their hue, while the blue will be more emphasized.

- For checking star charts while outside you should never use red light, even though this is recommended by numerous writers. Orange and reddish stars will thus lose a lot of their hue, while the blue will be emphasized. It is wiser to use a normal but dim white light.
- People with normal sight (or properly adjusted sight) see stars under a view angle of 1.5 arcmin. Such a small source lights only a few cones on our fovea – maybe only twice as many as are needed for any color perception at all. Thus, it can actually help to be short-sighted when identifying star color. Check if the colors of the very bright stars become more obvious when you take your glasses off. If you are observing with binoculars, you can achieve the same effect if you slightly blur the image.

More on Star Colors

Science fiction artists are very keen on drawing foreign worlds with a blue, yellow, or red sun, or even all three at once! Such works certainly inspire our imagination and take us in our minds to alien planets that we will most probably never be able to visit. But as we are discussing star colors we must ask ourselves: Are those blue, orange, and red stars in the skies of hypothetical planets realistic? Would we see them like this if we stood on one of the planets that orbits around them? The answer is NO!!!

All stars are suns, even the coldest red giants, such as Betelgeuse. Their levels of surface brightness are so high that if we saw them up close, they would overburden our cones to the point at which they would appear white. If you doubt this, you should realize that the temperature of the glowing thread in an incandescent light bulb is approximately 2,500°C, which is similar to the surface temperature of the colder red giants. And in the same way as the thread in the light bulb is blindingly white, the surface of a red giant would also appear as white.

The actual orange-reddish color of Betelgeuse's light would become slightly more apparent if we looked at the lighted landscape of its hypothetical planet. But even here the difference would be hardly noticeable to our eyes. The color difference between the sunlight on Earth and on Betelgeuse's planet would be roughly the same as the difference between outdoor (daytime) and interior (artificial) light. The planet would most certainly not be bathed in a ruby red color.

The Lives of Stars

Observing stellar colors is a pleasant bonus in amateur observations. Professional astronomers use light received from the stars to obtain a lot of useful information besides color.

Everything that we know about the stars and other objects in universe we have learned from the electromagnetic radiation they emit. Only the Sun and some close-by supernovae emit enough neutrinos to be picked up by modern detectors. You can see the electromagnetic spectrum by passing light through a prism or a diffraction grating. Nowadays astronomers use spectrographs mounted on large telescopes to routinely record and study the spectra of celestial objects. In fact, they do not build large telescopes to directly observe the objects but to obtain their spectra!

Color and Temperature

Let's start by a quick recapitulation of basic physics.

Heated objects emit electromagnetic radiation, including visible light. A piece of iron, for example, at room temperature emits only infrared light. Heated to 600°C, it glows red. As we increase the temperature, the color changes to yellow and then white. Physicists invented the idea of a perfect black body, which absorbs all incoming light and re-radiates light, depending only on its temperature.

In the black body's spectrum there are all of the single color components of visible light, from violet, with a shorter wavelength, to red, with a longer one (Figure 3.12). The spectrum is uninterrupted; colors change gradually and seamlessly into the neighboring color. From the spectrum, we can deduce surface temperature of the body. The most powerful emissions of the black body occur at certain wavelengths, which is in accordance to Wien's law, inversely proportional to the temperature. If we looked at a glowing piece of iron with a spectroscope, we would see a rainbow spectrum; it would depend on the temperature which color would glow the brightest (Figure 3.13).

Figure 3.12. With a hint of imagination we can spot all the colors of the rainbow in this continuous spectrum: from red on the right, through yellow, green, and blue, to violet on the left

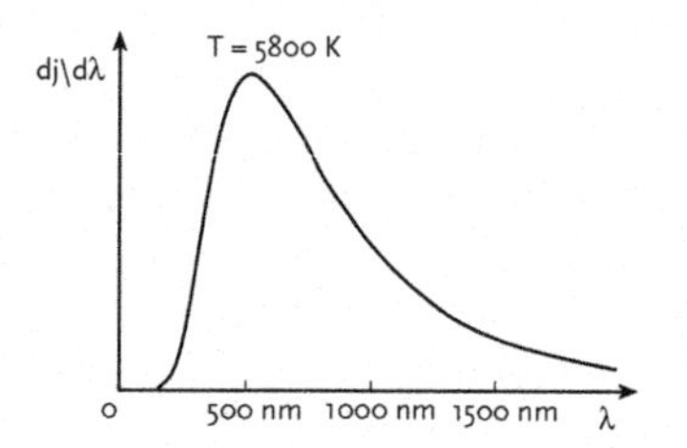

Figure 3.13. Physicists like to talk about the spectral radiance as a function of wavelength

However, atoms in rarefied hot gas do not emit light in a continuous spectrum. As you might remember from high school physics, electron states in atoms have a precisely determined energy. During a transition of an electron from one state (or level) to another with a lower energy level, a photon of light is emitted, which carries away the energy, equal to the difference in energy between the two electron states. The electron state with the lowest energy is known as the ground state, the state with the next higher level of energy as the first excited state, followed by the second excited state. The light emitted by the atom during such a transition shows up in the spectrum as a narrow bright line at a very precise wavelength. If we look at a specific gas with a spectroscope, we may see several spectral lines that are characteristic for that type of radiating gas. This is known as the discrete or emission spectrum (Figure 3.14).

If, on the other hand, light with a continuous spectrum travels through gas, the atoms of gas will absorb light of particular wavelengths. These are the wavelengths that would otherwise appear in the emission spectrum of this gas if it were hot.

As we saw above, a photon is emitted when an atom goes from an excited state to a state with lower energy, such as the ground state. Absorption is the inverse phenomenon. A photon is absorbed, and the atom goes from the ground state or an excited state to an excited state with higher energy. Atoms that absorb certain wavelengths of light with a continuous spectrum emit the same light – with a major difference being that they emit it isotropically in all directions. The observer will not notice this light coming in his or her direction. Narrow dark lines will therefore appear in the continuous spectrum. This is an *absorption spectrum.* Atoms of a given chemical element have a characteristic group of spectral lines, from which we can recognize the element. We meet all kinds of spectra in astronomy (Figures 3.15 and 3.16).

dj/dλ

λ

Figure 3.14. Emission spectrum, emitted by a rarefied hot gas (upper panel) and a graph showing energy density distribution over different wavelengths

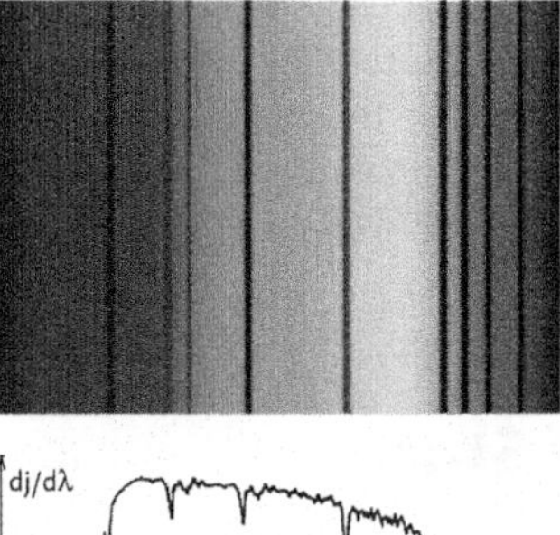

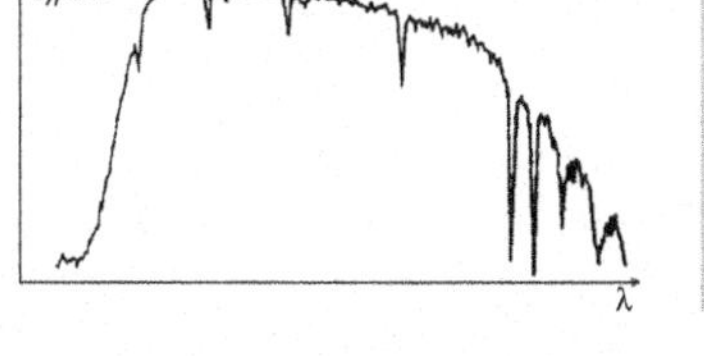

Figure 3.15. Absorption spectrum... and spectral radiance as function of wavelength

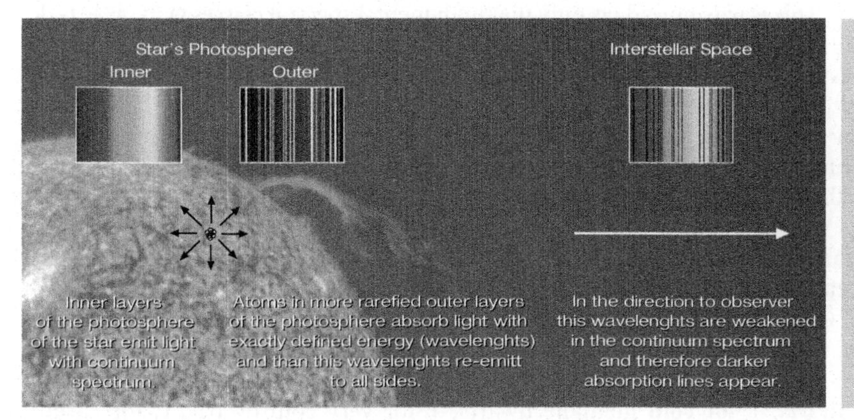

Figure 3.16. Absorption lines in the stellar spectrum appear at exactly the same wavelengths, at which the hot gas in the star's photosphere would produce emission lines

Stars are hot gaseous spheres that have emitting atoms. We can see the light emitted from the surface. As we have already mentioned, a star's spectrum is very similar to the spectrum of a black body. This means that using Wien's law, we can determine the stellar surface (effective) temperature. The Sun emits most strongly at a wavelength of 500 nm. Its surface temperature is 5,800 K. Antares emits the most light at the wavelength of 830 nm and therefore has a lower effective temperature, about 3,000 K.

A star surface is not like the surface of a solid planet. Stars are gaseous, so their surface is composed of a layer of gas several hundreds kilometers thick and becomes thinner with increasing height. It is from the surface that most of the light is emitted and eventually reaches us. Inner layers of the surface emit a continuous spectrum. Atoms in more rarefied outer layers absorb certain wavelengths from the continuous spectrum. Dark absorption lines appear in the stellar spectrum. From these, astronomers determine the chemical composition and even abundances of certain chemical elements. And this is by far not all!

We can learn a lot about emitting gas also from the shape of the spectral line. If it is narrow then the gas is rare and has a moderately low temperature. If the line becomes broader at high temperatures, it means the atoms are moving faster, toward and away from the observer, and because of the Doppler effect the observed wavelength is shorter and longer. The line also broadens during the rotation of the star around its own axis, due to the collisions in the dense gas between atoms and to a strong magnetic field, if there is one. The width and shape of the spectral lines can therefore be used to deduce the density of gas in a stellar atmosphere or in interstellar nebulae; it can also be used to determine the rotation velocity of a star and strength of the magnetic fields. Regular monitoring of spectral lines and their eventual periodic oscillation due to the motion of the star toward and away from us can reveal spectroscopic binaries as well as more massive cold companions – brown dwarfs and planets.

Classification of Stars

Toward the end of nineteenth century, astronomers at Harvard Observatory began to systematically take photographs of the stellar spectra. Taking into account detailed spectral characteristics, Edward C. Pickering (1846–1919) classified stars into classes, which he labeled alphabetically from A to Q. The basis for his classification system lay in the absorption lines of hydrogen and helium, with the addition of some other elements, for instance, iron in the case of colder stars. His coworker Annie Cannon, who classified almost a quarter of a million stars by their spectra, established that certain classes were unnecessary. This is why nowadays only the fundamental spectral classes O, B, A, F, G, K, and M remain. Spectral class O corresponds to the highest surface temperatures of stars, while M corresponds to the lowest. We can easily remember the classes with a help of an old mnemonic, "Oh, Be a Fine Girl, Kiss Me."

With the advance of spectroscopy, which soon became one of most important branches of astronomical observation, it was discovered that stars within the same spectral class may differ. So each main class was further divided into subclasses, designated by Arabic numbers, for example, G2 for our Sun.

For an even more detailed definition of the stellar spectrum, prefixes and endings were added, and we write these with a small or capital letter of the alphabet. Here's an illustration: normal dwarf stars obtained the prefix "d," while white dwarfs obtained "D."

For the purposes of designating special groups of stars, a new and expanded classification system was introduced. Fundamental classes were joined by class W for Wolf–Rayet-type stars, L for red and brown dwarfs, T for cold methane dwarfs, Y for very cold dwarfs, C for carbon stars (the older classes R and N were joined to form this class), P for planetary nebulae, Q for novae, and so on.

In addition to this, it was soon discovered that the stars within the same spectral class can substantially differ in their luminosities. Because of this, new luminosity classes were introduced, and these are designated by a Roman numeral: Ia are super giants with intense luminosities, Ib are super giants with less intense luminosities (compared to class Ia, of course), II are bright giants, III are giants, IV are subgiants, V are stars on the main sequence of the H-R diagram (which we discuss in the next section), and VI are subdwarfs.

The complete spectroscopic notation for our Sun is, therefore, G2V. It tells us that according to the surface temperature, the Sun belongs to the spectral class G2 and that it is a star on the main sequence of the H-R diagram. The complete denotation of Deneb is A2Ia. This tells us that it is a white star of the spectral class A2 and a super giant with intense luminosity.

The Hertzsprung–Russell Diagram

In the beginning of the twentieth century, astronomers took a big step forward in understanding the evolution, or life, of stars. This was due to the introduction of a diagram that has the spectral class or surface temperature of the stars on the horizontal axis, and luminosity or ratio between the stars' and Sun's luminosity on the vertical axis. What we are talking about here is the famous Hertzsprung–Russell diagram or, in short, the H-R diagram (Figure 3.17).

Around the turn of the century, the Danish astronomer Ejnar Hertzsprung (1873–1967) was studying stellar spectra. He discovered that some orange stars, for example Betelgeuse and Antares, have narrower absorption lines, while certain other orange stars within the same spectral class have wider lines. He concluded that this was because the former have intense luminosities, while the latter have less intense luminosities. Stars with a similar surface temperature but different luminosities should differ only in their sizes. The former are therefore giants, while the latter are dwarfs. The common belief until then was that stars within the same spectral class, i.e., stars with approximately the same surface temperature, also share the other parameters. Hertzsprung started investigating open clusters, in which all members are equally distant from us, and so we only need to determine their magnitudes, which is also a measure of their luminosity. He entered the stars into a diagram, with spectral class on its horizontal axis and magnitude or luminosity on its vertical axis.

Around this same period, the American astronomer Henry Norris Russell (1877–1957) was studying the origin and evolution of stars. In his research work, he used similar diagrams. He also discovered in 1909 that orange stars can be very big or small. After several years of work and collaboration with Hertzsprung he established that the best way to capture characteristics of stars is with diagrams like these. In 1914, he published a diagram in *Nature* magazine, a diagram that is now called the Hertzsprung–Russell diagram.

Astronomers were initially amazed that stars are not more or less evenly distributed over the diagram, but that about 9/10ths of them lie on a diagonal line (today called the main sequence), which stretches from the upper left part of the diagram toward the lower right part. In the upper left corner, there are stars with intense luminosities and high surface temperature. As we go to the right along the main sequence toward the lower surface temperatures, luminosities decrease. Below the main sequence are white dwarfs, stars with weaker luminosities and high surface temperatures; and above it are red giants, that have intense luminosities in spite of their low surface temperatures.

The Hertzsprung–Russell diagram is also a useful tool for estimating the distance of stars. From the shape of the spectral lines we can figure out if a star is on the main sequence, a giant, or a dwarf. Then we can infer its luminosity from the temperature, and from its apparent magnitude we can calculate the distance. However, the most important role played by the H-R diagram was in studying the life of stars.

The Birth of Stars

In the beginning of the twentieth century, when astrophysics was still in its early stages, there was very little known about what goes on inside stars. From 1928 onward, scientist began to have an

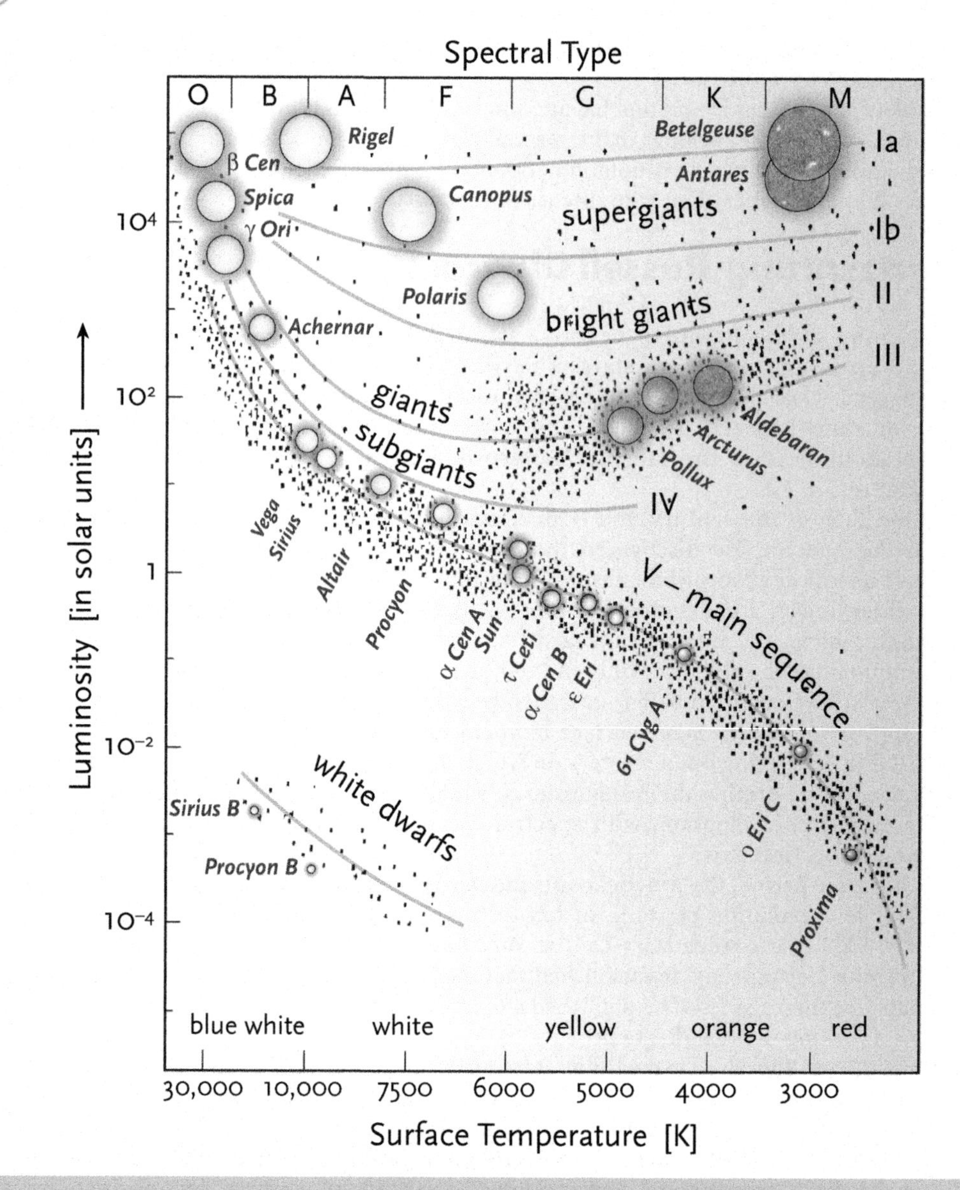

Figure 3.17. The Hertzsprung–Russell diagram here shows the spectral and luminosity classes of some of the better known stars

inkling that stars produce energy during the nuclear process of fusing light nuclei. Before that, it was thought that stars contract continuously and use their own gravitational potential energy to compensate for their radiation losses.

In the 1930s, it was discovered that stars are mainly composed of hydrogen. In the beginning of the 1940s, the new findings from nuclear physics began infiltrating the study of astrophysics. In 1938, Hans Bethe and Carl-Friedrich von Weizsäcker realized that stars acquire energy by merging (fusion) their hydrogen nuclei. In the 1950s, the first stellar models appeared. Initially these were calculated by hand and later by computers. Nowadays, stellar models have become highly sophisticated. They tell us what is happening in the center of a star, how energy is transported to the surface, how the temperature and pressure change with depth, and how all this affects the surface temperature and luminosity, which are quantities we can measure. Before long, the Hertzsprung–Russell diagram began to play an important role in the development of the science of astrophysics, since it was ideal for testing results.

Stars are born in cold interstellar clouds of gas and dust. The typical diameter of such a cloud is somewhere between 30 and 60 light years, and it contains a mass of 10,000–100,000 Suns. Although these interstellar nebulae appear completely static and unchanging, when we observe them through a telescope, we know that they are in fact dynamical systems in which changes – in an astronomical sense – occur fairly quickly. The dynamic is influenced by two forces. Gravity is an attractive force and tries to gather gases and dust together. During the contraction the temperature of the cloud rises. The result is a repulsive force (pressure), which tries to blow the cloud apart. If such a cloud was isolated from the other stars in the galaxy, nothing exciting would ever happen in it, because of the equilibrium between gravity and pressure.

Different external influences, such as an explosion of a nearby supernova, cause disturbances similar to sound waves. Matter becomes denser in some parts of the cloud and rarer in others. If the density of a cloud gets higher in an area, the gravitational attraction increases in that area. Only when gravitation prevails over the repulsive force will the cloud start to locally contract and condense. The random denser clumps become star embryos, while the cloud develops into an open cluster (Figures 3.18, 3.19 and 3.20).

When the first clumps of matter appear in the cloud, matter around them sinks and starts sticking together faster and faster. In the center of each clump a massive core starts to form, the density of which constantly increases and therefore also increases the attractive force. The clumps in turn become increasingly denser and hotter. When the core density reaches 10^{-8} kg/m^3, matter becomes opaque for light.

A further increase in density causes even faster heating of the core, since the cloud loses less energy through radiation. A dense and hot central core is produced, which is surrounded by a thinner and colder envelope in which matter is still falling toward the center. Such a cloud is called a protostar. Its mass is typically around 10^{30} kg, it has a diameter of about 10^{10} km, and its surface temperature is roughly 170 K. The diameter of a protostar is comparable to the size of the entire Solar System.

The protostar radiates in infrared light. It compensates for the loss of energy with continuous contraction. Due to the released gravitational energy it heats up, the central pressure increases, and the star radiates more and more. It takes a few tens of millions of years before a protostar with a surface temperature of about 1,500 K is created from the first embryo, emitting most of its light in the infrared part of the spectrum. Over the next 20 million years, with continued contraction the protostar will heat up to the point at which nuclear reactions in its core will start and it will enter the main sequence of the H-R diagram.

With the increase of the surface temperature, the power of the stellar wind[1] also increases. Wind rarefies the surrounding nebula and blows it away, so eventually the matter falling on the protostar runs out, and its growth stops. From the matter that remains in the close vicinity of the star, an accretion disc is formed (if it is not a binary star), and from this planets may begin to coalesce.

How much the interior of the protostar heats up depends on its mass. The higher the mass, the higher the central temperature and pressure. The temperature needed for hydrogen to burn is roughly 10 million K. Protostars with smaller than a 0.08 mass of the Sun never reach the temperature needed for maintaining nuclear reactions. These objects are called brown dwarfs.

---

[1] A stellar wind is a stream of charged particles that are ejected from the upper atmosphere of a star. It consists mostly of a high-speed protons and electrons that have enough energy to escape the star's gravity.

Figure 3.18. The most famous and certainly the most beautiful stellar nursery in our galaxy is the big Orion Nebula

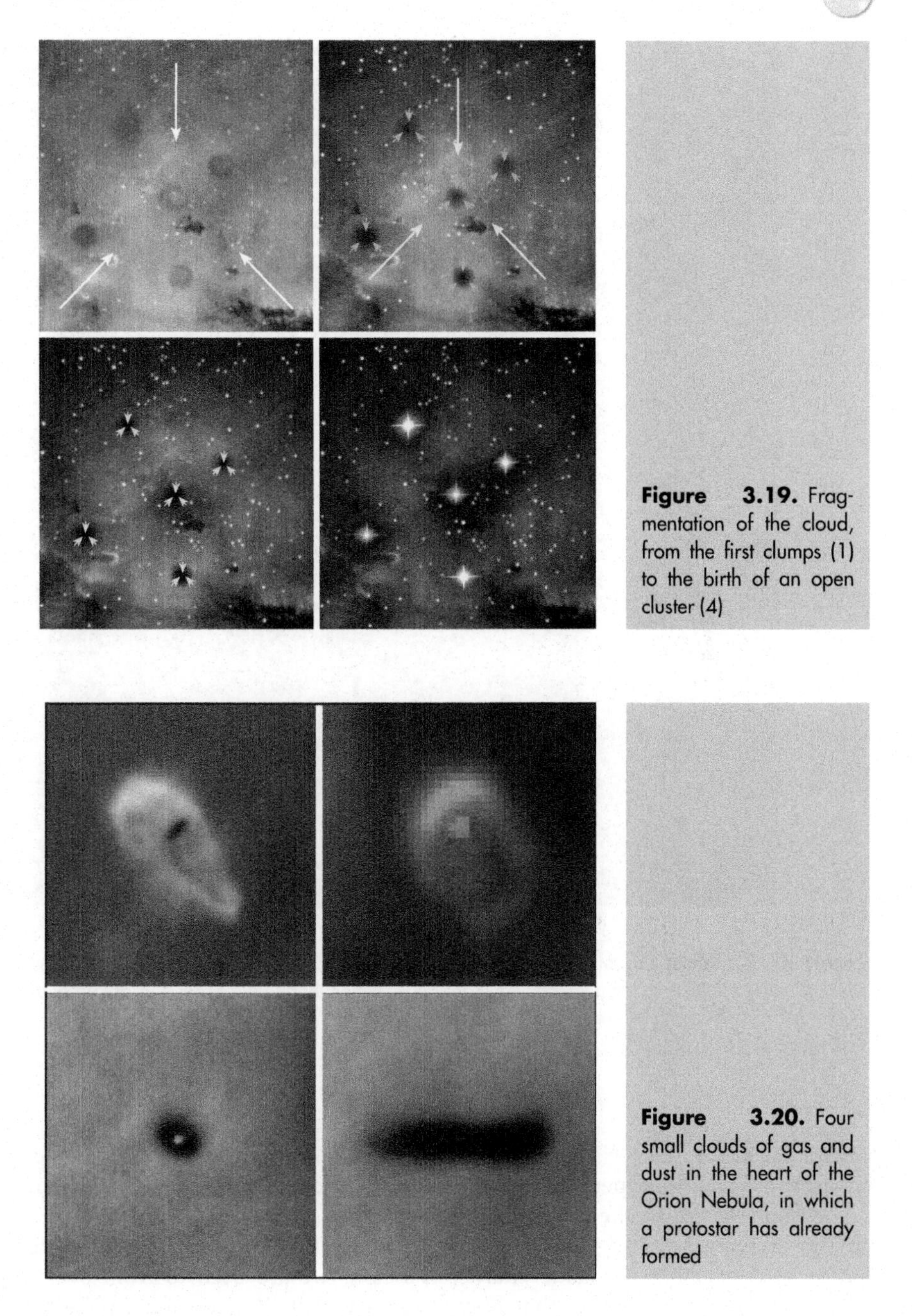

Figure 3.19. Fragmentation of the cloud, from the first clumps (1) to the birth of an open cluster (4)

Figure 3.20. Four small clouds of gas and dust in the heart of the Orion Nebula, in which a protostar has already formed

Life on the Main Sequence

When nuclear reactions take place in the core of the star, contraction stops, since equilibrium is established; the pressure of the gas in the core equals the gravitational attraction of the external layers. This is the moment when we can say that a star is born (Figure 3.21). If we were to put that star on an H-R diagram, it would be located on the main sequence. This is the beginning of the long and peaceful period in which the star will likely live up to 90% of its life. Where on the main sequence it will appear,

Figure 3.21. The star is born! In the heart of the famous nebula M 16 in the constellation Serpens dense clouds of gas and dust called the Pillars of Creation are beautifully visible. In their interior, young, hot stars are born. But only when the biggest and most massive among them clean the surroundings with their fierce stellar winds can their light break through the nebula and reach us

which nuclear reactions will take place in its core, how luminous it will be, its life span, and the way it will grow old and die depends mainly on its mass and its chemical composition when it is born.

The temperature in the core of the Sun reaches 15 million K, the pressure is 3×10^{14} Pascals, and the density is 150 kg/dm^3 (150 times the density of water). Under these conditions, the matter no longer has the same characteristics as we are used to from everyday life. Atoms are totally ionized, which means that they have lost all their electrons. In the cores of the stars, we can therefore find a hot and dense brew of atomic nuclei, and among them electrons roam freely. Such a gas of charged particles is called *plasma.* Due to the high temperature, the nuclei move very fast and can collide regardless of the repulsive force present as a result of the corresponding positive electric charges.

At lower temperatures, hydrogen atoms merge through the so-called p-p chain (see Figure 3.22) which is divided into branches. This chain gets its name from protons (p), which are the main players. At the lowest temperature reactions of the p-p I branch are dominant. At somewhat higher temperatures reactions of the p-p II branch are dominant, and at even slightly higher temperatures reactions of the p-p III branch are dominant. After one cycle of p-p chain reactions has taken place, we get a

Nuclear reactions of p-p chain	
Nuclear reaction	Released energy [MeV]
p-p I branch	
$p + p \rightarrow d + e^+ + \nu$	+1.442
$d + p \rightarrow {}^3He + \gamma$	+5.493
${}^3He + {}^3He \rightarrow {}^4He + 2p$	+12.859
p-p II branch	
${}^3He + {}^4He \rightarrow {}^7Be + \gamma$	+1.586
${}^7Be + e^- \rightarrow {}^7Li + \nu$	+0.861
${}^7Li + p \rightarrow {}^4He + {}^4He$	+17.347
p-p III branch	
${}^3He + {}^4He \rightarrow {}^7Be + \gamma$	+1.586
${}^7Be + p \rightarrow {}^8B + \gamma$	+0.135
${}^8B \rightarrow {}^8Be + e^+ + \nu + \gamma$	+18.074
${}^8Be \rightarrow 2{}^4He$	

Nuclear reactions of CNO cycle	
Nuclear reaction	Released energy [MeV]
CN cycle	
${}^{12}C + p \rightarrow {}^{13}N + \gamma$	+1,944
${}^{13}N \rightarrow {}^{13}C + e^+ + \nu$	+2,221
${}^{13}C + p \rightarrow {}^{14}N + \gamma$	+7,550
${}^{14}N + p \rightarrow {}^{15}O + \gamma$	+7,293
${}^{15}O \rightarrow {}^{15}N + e^+ + \nu$	+2,761
${}^{15}N + p \rightarrow {}^{12}C + {}^4He$	+4,956
NO cycle	
${}^{15}N + p \rightarrow {}^{16}O + \gamma$	+12,126
${}^{16}O + p \rightarrow {}^{17}F + \gamma$	+0,602
${}^{17}F \rightarrow {}^{17}O + e^+ + \nu$	+2,762
${}^{17}O + p \rightarrow {}^{14}N + {}^4He$	+1,193
OF cycle	
${}^{17}O + p \rightarrow {}^{18}F + \gamma$	+5,609
${}^{18}F + e^- \rightarrow {}^{18}O + \nu$	+1,655
${}^{18}O + p \rightarrow {}^{19}F + \gamma$	+7,993
${}^{19}F + p \rightarrow {}^{16}O + {}^4He$	+8,115

Figure 3.22. Two sets of fusion reactions, by which stars on the main sequence convert hydrogen to helium and produce energy

helium nucleus, two positrons (e^+) and two neutrinos (ν) from four protons. The energy that is released is divided among the X-ray photons (γ), particles, and neutrinos. The energy of photons and particles is carried through collisions to the surrounding matter and thus heats it and preserves the high temperature in the core of star. The neutrino can escape undisturbed, and its energy for the star is lost (Figure 3.23).

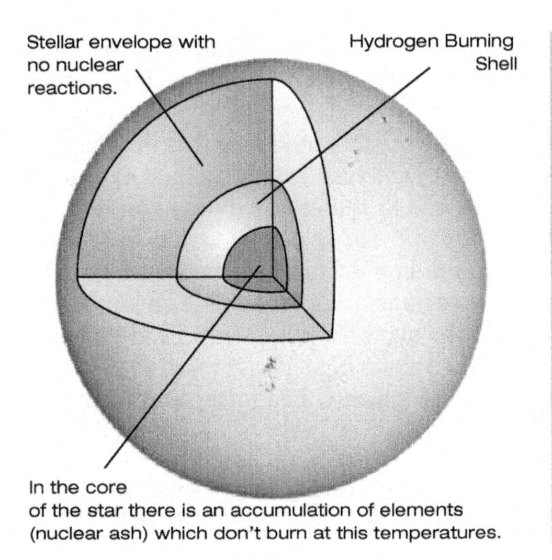

Figure 3.23. As the temperatures rise, nuclear reactions become even faster. In stars this happens in the center, which is why the first element to be used up is hydrogen. Elements that do not burn at these temperatures start collecting, forming the material called nuclear ash

At even higher temperatures in the cores of massive stars, reactions of the CNO cycle take place during which even heavier nuclei are created. These reactions run much faster than the previously mentioned ones.

Over 600 million tons of hydrogen nuclei in the core of our Sun merge into helium nuclei every second. Regardless of this terrifying number we can, with a rough calculation, estimate that the Sun (1.99×10^{30} kg) will have enough mass to survive for a 100 billion years in peace. Of course, this estimate is at least 10 times too high, for the Sun is only 75% hydrogen, and nuclear reactions do not take place across the entire star.

Neutrinos are the only particles that can escape out of the core of the Sun undisturbed, and they tell us much about what is happening there, such as what sort of nuclear reactions are taking place and how efficient is the hydrogen fusion process. This is why astrophysicists are trying their utmost to catch them and count them with neutrino detectors.[2] At this moment we can only say that the detectors register far fewer neutrinos than are predicted in the theoretical models of the Sun. Only time will show whether the detectors are not sensitive enough or something happens to the neutrinos on the way from the Sun to us (nowadays this seems to be the most likely solution to the neutrino problem). Another alternative, of course, is that we simply do not know enough about conditions in the core of our Sun.

As long as a star has enough hydrogen in its core it remains stable. It spends most of its life in such a state. It is interesting that stars with a smaller mass live longer than the ones with greater mass, even though we would expect it to be the other way around. In the stars with greater mass the core pressure is higher, so the temperature is higher and thus the nuclear reactions run faster, which is why the luminosity is more intense and nuclear fuel reserves are consumed sooner. A rough estimate of the connection between the life span of a star on the main sequence and its mass is

$$T \propto 1/M^{2.5}$$

A star with a mass 10 times the mass of our Sun shines approximately 3,000 times brighter but will leave the main sequence approximately 300 times sooner. A star with one-tenth of the Sun's mass shines with a brightness of 3,000 times less than that of the Sun but will live 300 times longer!

Aging

In the previous section on life on the main sequence, we said that the star is stable during this period. However, we should not take this statement too literally. Throughout its life while the star is gathering nuclear ashes, it is constantly shrinking and increasing its core temperature and density as well as luminosity. However, in comparison to other periods in the lifetime of a star, these changes are relatively slow.

For the Sun, we have very precise models and know that the changes take place over a long period, over 11 billion years. When the Sun emerged from the cloud of gases and dust and nuclear reactions started in its core, the temperature in its center was some 12 million K, the density 80 kg/dm^3, and the luminosity about 70% of today's value. Today the Sun is 4.55 billion years old and is approximately in the middle of the period it will spend on the main sequence. One can find today's data in the previous chapter. Right now the Sun has consumed almost half of the hydrogen stock in its core. At the age of 7.7 billion years, our Sun will reach the highest surface temperature it will have during its life on the main sequence (6,000 K) and will have 1.3 times higher luminosity than it has today. At the age of 10.9 billion years it will have 2.2 times higher luminosity than today, and hydrogen will run out in its core. It will

[2] Neutrinos are particles that can easily escape the dense and hot core of the Sun. About 16,000 billions neutrinos from the Sun pass through this book every second. However, it would take a million years for one to become trapped and start a reaction.

leave the main sequence, and a period of aging will start. For stars with a higher mass, the described changes will occur faster, and for ones with a lower mass, they will occur slower. How a star gets old, what happens to it during the aging process, and how will it end its life, depends mainly on its mass.

Stars with Masses Far Greater than Our Sun

The results from the models show that when the hydrogen fusion in the core stops, a massive star has approximately the following composition: most of it is still hydrogen (60.2% by mass); this is followed by helium (35.4%), and all other elements represent 4.4%. In this phase, the core of the star is mainly composed of helium, but it is surrounded by hydrogen-rich outer layers.

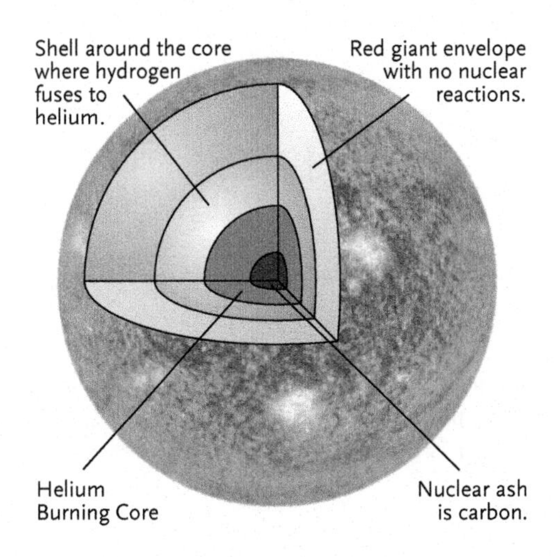

Although in the core energy is no longer being produced, the star still radiates and starts cooling. The pressure in the core starts falling, and the star begins to collapse, condensing and warming up under the gravity of its own outer layers. But even before the temperature in the core reaches the value at which helium starts burning, nuclear fusion of hydrogen in the inner layers of the envelope starts. When this happens, the radiation of the released photons press down on, condense, and warm the core even more, while the outer layers of the stars are blown out. The core becomes smaller and hotter, but the size of the star and its luminosity increase! The surface temperature falls. The star turns into a red giant.

When the temperature in the core reaches 100 million K, fusion of helium starts. The shrinking of the core stops, and the star reaches equilibrium once again. This whole phase lasts from a few hundred thousand to a few million years.

During the thermonuclear burning of helium in larger quantities carbon (carbon-12) and oxygen (oxygen-16) are produced. The first originate from three helium nuclei, which collide at the same time. When there are enough carbon nuclei, the next important reaction can take place, from which oxygen is produced from carbon and helium. But even helium eventually runs out, and the star loses its internal source of energy. It starts to shrink and heat up, until in the core carbon (at a temperature of 500 million K) and oxygen (at a temperature of a billion K) start to burn, while at the same time helium continues to burn in the surrounding inner layers. The core gets smaller and hotter, but the outer layers expand outward, and the star becomes a red super giant. In this phase the energy is produced in many layers; hydrogen continues to burn in the outermost layers of the core, where it is rich, hot, and dense enough to do so. In the center of the core, heavier elements are being converted into even heavier elements, as far as the temperature permits. When all the fuel there is burned to ashes, the star shrinks and heats up, and the burning moves to the surrounding layer, which is still rich with fuel. But in the now denser and hotter core the nuclear ash starts to burn. In this phase, the red super giant resembles an onion.

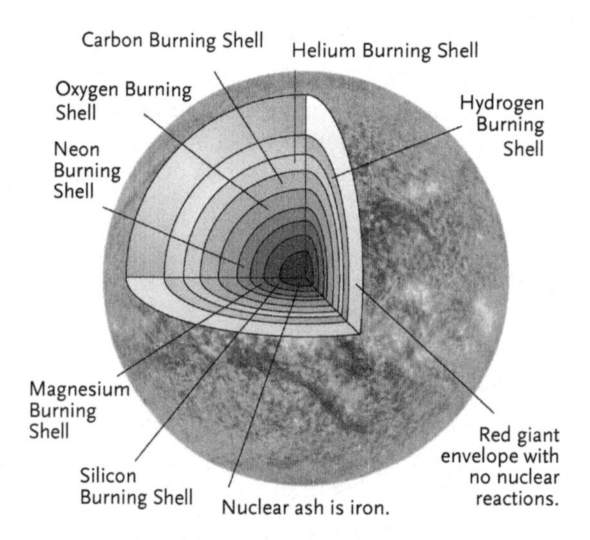

In the table below, we can see the duration of an individual phase and the average temperature and density at which these phases take place. The data are for a star 25 times more massive than our Sun. The life span of such a massive star is only 8 million years!

Phase	Temp. [K]	Density [kg/dm^3]	Duration
Hydrogen burning	4×10^7	5	7 million years
Helium burning	2×10^8	700	700,000 years
Carbon burning	6×10^8	2×10^5	600 years
Neon burning	1.2×10^9	4×10^6	1 year
Oxygen burning	1.5×10^9	10^7	6 months
Silicon burning	2.7×10^9	3×10^7	1 day
Collapse of the core	5.4×10^9	3×10^9	0.35 seconds
Explosion	app.10^9	–	10 seconds

How big are the red super giants? The biggest we know are VY Canis Majoris (1,950 times bigger than the Sun), VV Cephei (1,750 times bigger than the Sun), and V354 Cephei (1,520 times bigger than the Sun). All three are so huge that it is almost impossible to imagine! If we put them into the center of our Solar System, the stars would extend far beyond the orbit of Jupiter, almost to Saturn! Among the brighter stars, which are visible with the naked eye, Mu Cephei (1,420 times bigger than the Sun) is currently sixth largest, while Antares (700 times bigger than the Sun) and Betelgeuse (650 times bigger than Sun) are tenth and eleventh (data are from March 2008). The surface temperatures of the red super giants are around 3,500 K, so all of them have a distinct orange color. (There is more on the average density of these stars in the section on Mira later in this book.)

Most of the red giants and super giants are pulsating long-period variable stars. These stars do not have a constant period nor a maximum nor minimum brightness. The brightest and most interesting representatives are presented in the latter part of this book, in the descriptions of the individual constellations. The changes in the brightness of these stars appear to be due to the expanding and shrinking of their very rarefied outer layers. This takes place because of the eternal fight between the gravitation, which wants to shrink the star, and the pressure in the center, which wants to blow matter outward. In fact, all stars pulsate continuously; but in the red giants, the end effect is much greater than in normal stars that are on the main sequence of the H-R diagram. Astrophysicists say that if we had precise enough instruments, we could see all stars as pulsating variables.

Nuclear reactions in stars do not continue indefinitely. In order for the core, composed of carbon and oxygen, to start burning, it has to be at least as massive as 1.4 Suns in order to reach the proper temperature. Of course, this means that the initial mass of the star has to be much greater. Today we do not know exactly how much stellar matter is lost during a star's evolution, but astrophysicists estimate that a star should probably have an initial mass of at least 11 Suns.

Why does a star lose its mass in the first place? When the transformation into a red giant begins and the star blows out, gravitation on its surface weakens, to the point where the outer layers are no longer strongly bounded to the core. Any outburst on the photosphere or other disturbance and even tidal forces from eventual planets cause a continuous flow of mass into interstellar space.

Even if the mass of a star is very great, the thermonuclear reactions will stop at a certain point. The heaviest element that can be produced in stellar cores is iron (Fe). The next reaction, in which cobalt is produced from iron, is endothermic. This means that instead of producing energy, the star is losing it. The death of the red giant is approaching!

Stars with Masses Lower than 2.25 Suns

Now we will explore what kind of destiny awaits stars with less mass after a rather peaceful life on the main sequence. The beginning of the aging process is similar to that of stars with greater mass, but the ones with less mass never reach the temperatures at which carbon and oxygen start to burn in the core. As an example of a star with lesser mass we will take our Sun, for that is the star we are certainly most concerned about for the future.

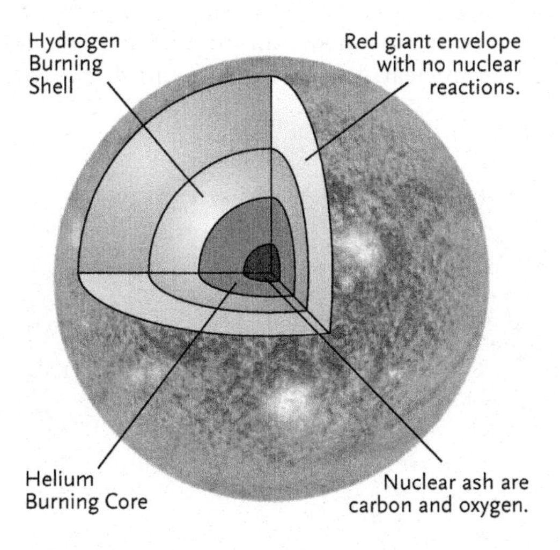

As we have already mentioned, at the age of 10.9 billion years the Sun will leave the main sequence of the H-R diagram. By that time the hydrogen in its core will run out. During the cooling process the gravity of the outer layers will start to shrink the Sun, and because of this the internal temperature will increase and the burning of hydrogen in the layer close to the core will become more efficacious. This will cause additional shrinking of the core and at the same time cause the outer layers to blow outward and cool down. The surface temperature will fall to 4,500 K, but due to the enlargement of the diameter of the Sun the luminosity will increase. The Sun will become a red giant, with a diameter of 9.5 times bigger than today and a luminosity that surpasses today's value by 34 times.

Because of the expansion gravity on the surface will decrease, and the Sun will not be able to retain most of its outer layers. Gradually it will thus lose a considerable part of its matter, which will float off into interstellar space. Its mass will decrease to 0.72 of today's value. When the temperature in its core comes close to 100 million K, the helium nuclei will start to burn in thermonuclear reactions. The product of this will be carbon nuclei, and in the later phase oxygen will be produced from the carbon. The helium burning will take place at higher temperatures and pressures, which is why the luminosity of the Sun will be constantly increasing. At the end of the red giant phase, at the age of 12.2 billion years, it will be 2,300 times that of today's value. Its diameter will increase to 2 astronomical units, and it will therefore be 200 times bigger than today's Sun!

During this entire period the Sun will constantly lose its outer layers, and its mass will decrease to half of today's value. The amount of helium will decrease in the core, while the amount of the carbon and oxygen nuclei will increase. Gradually the helium in the core will run out, the Sun will shrink and warm up, and helium burning will start in the neighboring outer layer. In the next outer layer, hydrogen will still be burning. But in stars with a smaller mass, the burning of helium in the layer close to the core is neither long lasting nor stable, for these stars never reach high enough temperatures. In the Sun's core, the nuclear ash will be composed of carbon and oxygen. Through the course of time, the nuclear reactions will cease completely and the Sun (or better say its remnant) will begin the last phase of its journey.

Before we examine the last phases in the lives of stars, we have to say a bit more about red giants. We have divided aging stars into two groups by mass and described their evolution. This division is rather artificial. The universe is much more variegated! Among the red giants and super red giants at the end of their aging we can find stars in all intermediate phases, as we have already mentioned. The stars with the lowest mass might never even start burning helium in the core, so they end their life when they run short of hydrogen. Stars that are slightly more massive than our Sun can start burning oxygen in the core (beside helium and carbon), and the star ends its life with nuclear ash composed of helium, carbon, oxygen, and neon. Even more massive stars can burn neon, and the nuclear ash in their core is magnesium. Only the most massive can use nuclear fuel to the end and finish their lives by creating silicon or even iron in their core.

At the end of a red giant phase the diameter of the Sun will be some 200 times bigger; its outer layers will reach the orbit of Earth.

Star Death

The initial mass determines the life of stars from birth to the end of the aging process, but for how they end their lives we must divide them into three groups: stars with an initial mass up to 11 solar masses, stars with initial mass between 11 and 50 solar masses, and those with initial mass greater than 50 solar masses. These three groups can be differentiated by what remains after their death.

Stars up to 11 Solar Masses

Let's start with the smaller stars and continue the story of our Sun, which we started to weave in the previous section.

When the thermonuclear reactions in the core of the Sun run out and the internal sources of energy dry up, our star will continue to radiate into space and lose its energy. This means that it will start cooling down. The pressure in its center will not be able to resist the gravity of the outer layers, so the Sun will start to shrink, heat up, and become increasingly dense. But it will not reach the temperature needed to start a new cycle of thermonuclear reactions because its mass is too small. Such a star, which radiates only because of its internal store of heat, is called a white dwarf.

What is the state of the matter inside a white dwarf? During the life of the star its core constantly shrinks and eventually becomes so dense and hot that the atomic nuclei (mainly carbon and oxygen) are almost joined together without electrons, which are freely moving around. Such a state is called degenerate gas, and we find it only at very high densities. We cannot find it on Earth, and it has also not been created in laboratories yet. The pressure of electron-degenerate gas depends not on temperature but merely on density. Thus, the electrons are responsible for stopping the further contraction of the star. Calculations show that the star shrinks and becomes increasingly dense before the electrons start interfering. The typical diameter of a white dwarf is only some 12,000 km. These stars are therefore approximately as big as our planet Earth! Their central density is from 2×10^5 to 2×10^9 kg/dm^3. A sugar cube made of matter from an average white dwarf would weight 40 tons on Earth!

During the contraction the star heats up. When the white dwarf phase begins, its surface temperature is 100,000 K and more! Such a hot star radiates a lot of ultraviolet light and X-rays. These highly energetic photons excite atoms and molecules in the outer layers of the star, which were previously blown away. The gas starts to emit light. A planetary nebula appears, and it stays visible for approximately 10,000 years, until the gases, which are still expanding, become so rarefied that they dissipate into interstellar space (Figure 3.25).

Because the white dwarf still radiates, it cools down slowly. The process lasts for several hundred million years. At the end, a dark and cold ball of nuclear ash remains (Figure 3.24).

Although the white dwarf is a very hot star at the beginning, it is extremely small and so has low luminosity. A typical value is some 0.0025 luminosity of our Sun.

The nearest white dwarfs circle Sirius (in constellation Canis Major) and Procyon (Canis Minor). They are not visible through average amateur instruments. For amateur observers the most interesting is the white dwarf that is a member of the triple star Omicron–2 Eridani. This is the only white dwarf seen through smaller telescopes.

Earth's Fate?

What is the destiny of our Sun, and how is it linked to the destiny of Earth? Will the expanding Sun burn up Earth or will something else happen?

Models show that Earth will lose all liquid water on continents when the Sun's luminosity reaches 1.1 of today's value. The Sun will reach this luminosity at the age of 5.7 billion years. The oceans will evaporate at the luminosity 1.4 of today's value, at the age of 9.1 billion years. Models have been made for a cloudless Earth. But because Earth has clouds, this will happen a little bit later. This means that life on Earth will end even before the Sun leaves the main sequence.

The Sun will most probably not capture and burn up the inner planets. At the end of the red giant phase it will indeed expand as far as today's Earth orbit, but due to its loss of mass and hugely greater diameter its gravity will be much weaker, so the inner planets will continue to move around it but in greatly enlarged orbits.

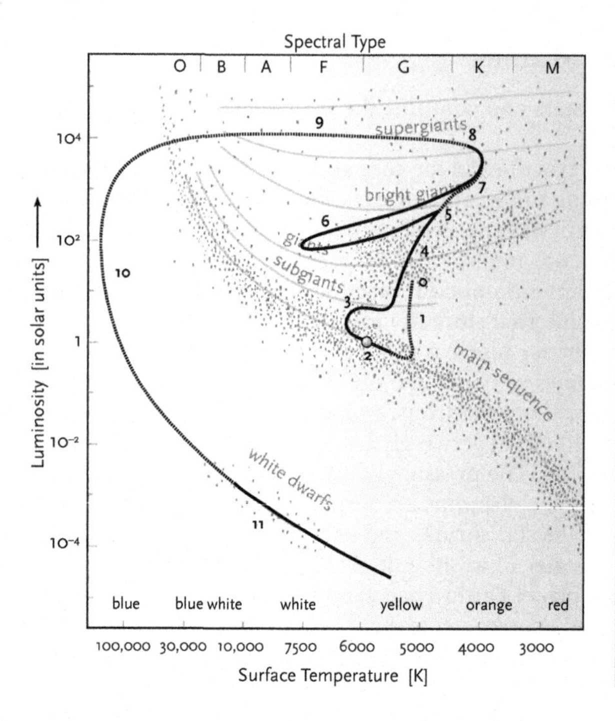

Figure 3.24. The life cycle of a star with one solar mass is shown in the H-R diagram. Only the most important turning points in its life are marked: 0–1, evolution from protostar to star; 2, period on the main sequence; 3, period when hydrogen is running short in the core; 4, red giant phase (hydrogen is burning in the shell around the core); 5–7, helium burning phase; 8, red super giant phase; 9, planetary nebula appears; 10, the core is cooling down and shrinking; and 11, white dwarf

Stars with Masses Between 11 and 50 Solar Masses

When of the last nuclear ash starts accumulating in the center of these stars (during their aging process), the mass of the inactive core increases, and the gravity of the outer layers squeezes it more and more. The core becomes so dense that only the electron-degenerate gas can resist the gravity (as we mentioned in the description of the white dwarfs). But massive stars started their lives with greater mass, so now there is enough matter for the inactive core to reach the Chandrasekhar limit of 1.4 solar masses. When this happens, gravity prevails over the electron-degenerate gas pressure, and the core collapses into itself. The velocity of the collapse is amazing – 70,000 km/s, or 0.23 the speed of light! The temperature and density increases in an instant so that the atomic nuclei decay into protons and neutrons, while the protons and electrons melt into neutrons.

This collapse lasts only a few tenths of a second, and in this really short time an immense gravitational energy of 10^{46} Joules is released. Neutrinos that carry this energy away leave the core within 10 s of the collapse. Most of the neutrinos escape into space undisturbed; a small number of them are absorbed by the outer layers of the star. But even this small amount of energy is enough to blow up the outer layers of the star in a tremendous explosion, which is called a supernova explosion. After the explosion, the remains are so hot that they radiate as brilliantly as 100 billion Suns for a few weeks, comparable sometimes to the luminosity of the entire galaxy!

Only during a supernova is there enough energy to allow for elements heavier than iron to be produced in larger quantities in nuclear reactions. The explosion spreads them all across interstellar space, where

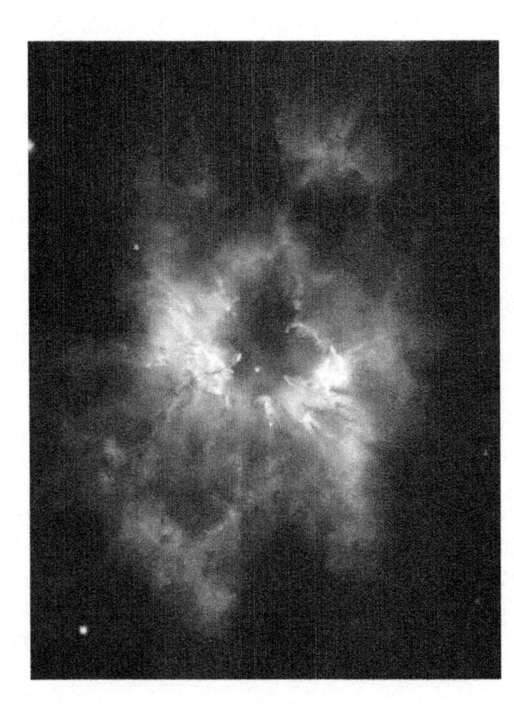

Figure 3.25. During the aging process the star loses its outer layers, which are blown into interstellar space. When a hot white dwarf originates from a stellar core, it radiates most of its light in the ultraviolet and X-ray part of the spectrum. The highly energetic photons excite atoms in the gases that were previously blown away, and now they emit light. We see the planetary nebula. The image shows planetary nebula NGC 2440 in constellation Puppis. Its central star is one of the hottest white dwarfs known. Its surface temperature is as high as 200,000 K

they enrich gases that the star has blown away during its life. The shockwave of the explosion mixes and whirls all the matter surrounding the star and creates ideal conditions for the formation of new stars. So the next generation is born from an enriched mixture of gases and dust. Our Solar System originated from the remains of a supernova explosion. The initial velocity and power of the shockwave from a supernova is so great that it spreads across interstellar space for up to 200 years after the explosion. Gradually it slows down, cools, and fuses with interstellar matter. This process can take up to 10,000 years.

The explosion of the supernova completely destroys the entire star. The outer layers are blown away into interstellar space, and the core collapses into a small ball - even smaller and denser than a white dwarf - called a neutron star or a pulsar. The ball is composed mainly from neutrons. The equilibrium in a neutron star is maintained by the neutron-degenerate gas pressure. Neutrons are so compressed that there are 10^{44} of them in a cubic meter. Their density is comparable to the density of an atomic

nucleus! The typical mass of a neutron star is between 1.35 and 2.1 solar masses, while its diameter is only between 20 and 10 km! The central density may reach an unbelievable value of 10^{15} kg/dm^3.

We will say more on the unusual properties of pulsars in the latter part of this book, when we reach the discussion of the Crab Nebula in the constellation Taurus.

Chandrasekhar Limit

Subrahmanyan Chandrasekhar (1910–1995) was born in Lahore, India. At the age of 19, he graduated from a university in Madras and in that same year he boarded a ship to England to attend postgraduate studies at Cambridge University. During the long voyage, the young man passed his time by researching the theory of white dwarfs, proposed by Ralph H. Fowler in 1926. By the time the ship docked in England, he had almost completely formulated his theory, which brought him a Nobel Prize for physics 53 years later.

Chandrasekhar discovered that the greatest mass the white dwarf could have and not collapse into a much denser neutron star was approximately 1.4 solar masses. If its mass is lower than this so-called Chandrasekhar limit, it is stable and therefore represents the final stage in the evolution of stars with a lesser mass. But if its mass is greater, the collapsing continues until a neutron star or even a black hole is created.

Chandrasekhar has written many excellent books on hydrodynamics, relativity, and energy transportation with radiation. He embarked on a career at the University of Chicago, where he stayed for over 50 years. His important contributions to twentieth-century physics were honored with the Nobel Prize in 1983, which he shared with William A. Fowler from Caltech.

Stars with Masses Greater than 50 Solar Masses

The luminosity of these really massive main sequence stars surpasses our Sun by 100,000 times or even 1,000,000 times. They are all in the luminosity class Ia. In order to maintain such immense luminosity (and therefore loss of energy), they have to have a very powerful internal source, which is why they must spend

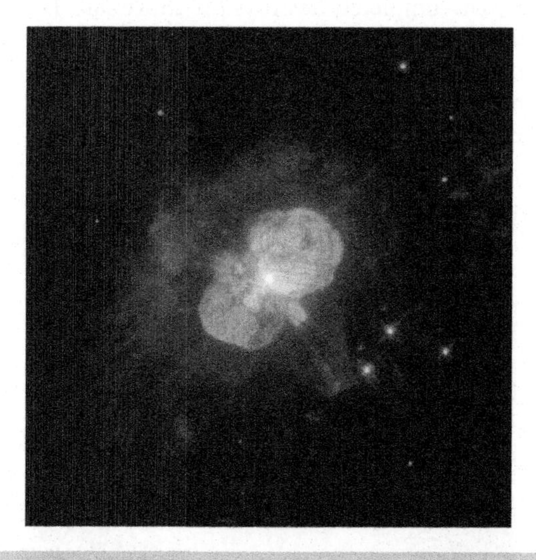

Figure 3.26. One of the most massive stars in our galaxy is Eta Carinae. Its mass is estimated to be more than 100 solar masses! Around the star we can see its blown-out outer layers (the last major outburst occurred 150 years ago), which have spread into interstellar space. The lifetime of such a massive star is only around a million years, and its end is known – a tremendous supernova explosion and the origin of a black hole. Eta Carinae is some 10,000 light years from us

their stores of nuclear fuel so quickly. Their life is exceptionally short; instead of billions it is measured in millions of years. Due to the stormy processes and strong stellar wind they lose their outer layers very quickly and sometimes even in the phase of hydrogen burning. When hydrogen runs short and the star begins to leave the main sequence of the H-R diagram, almost all outer layers are stripped from it, and only a huge helium core remains. This group of rather rare celestial objects are called Wolf–Rayet stars. Their life is not very different from the previously described massive stars, but it occurs much faster.

However, there is another important difference between them. The iron core, which develops during the last phase of the red super giant stage within the star's center, is much more massive. When the nuclear reactions run out, and the star starts to shrink, its collapse cannot be stopped by the pressure of the electron-degenerate gas, so the core collapses into itself. A tremendous supernova explosion occurs. The outer layers are blown away into interstellar space, and the core collapses into a neutron star.

But now nothing can resist gravity, and now is the time to show who is the chief of the universe. This amazing force, which pervades the entire universe, holds us gently chained to Earth so that we do not feel it at all, drives planets around their suns and suns around the centers of their galaxies; this incredible force that forms clusters and super clusters of galaxies has to fight the entire life of the star with its matter. When gravity squeezes, the pressure of matter increases and reestablishes the equilibrium. More and more nuclear reactions occur, and they run at increasingly higher temperatures and densities. When gravity squeezes even more, the degenerated electrons and neutrons appear and resist it. Only now, with these gigantic masses, gravity is the victor!

The core, which was as massive as a few Suns, was already squeezed to a ball with a diameter of 10 km. The degenerated neutrons try to stop the collapse. But the gravitational vice does not stop; it gets even stronger, and finally even the neutron star fractures under its own gravity and implodes. This is the show stopper! The star, just a few seconds before so shiny and boastful, shrinks into a point, into nothing, into something that mathematicians call a singularity, into something that we cannot even describe with today's physical laws! And when the dense clouds of the supernova explosion blow away and we look at where the star shone not so long ago, we do not see anything! Darkness! But this darkness has such a powerful gravitational field around itself that even light cannot escape! From the star originates a black hole, which is definitely the most exotic object astronomers know of.

The black hole does not radiate anything; it just swallows everything that comes its way. This is why we do not see it! If we are lucky and it is close to a star, we can see how matter from the star spirals toward the black hole and then disappears.

It is difficult to even imagine the density of the matter inside a white dwarf. But if we envision a black hole being created from our Sun, we would have to squeeze it down to a diameter of 6 km, and the Earth to 1.6 cm! Try to imagine this!

The Lives of Stars in Binary Systems

So far we have described the life cycles of single stars or of those that live in widely separated binary systems. The evolution of stars in close pairs is a bit different. Let's briefly describe the two most interesting examples.

If a binary system is composed of stars that have individual masses greater than 11 solar masses, the one with the greater mass soon uses up its hydrogen stores and starts its transformation into a red giant. The gravity on its surface weakens, and the hydrogen-rich outer layers start to flow toward the neighboring star. If there is plenty of mass, a common envelope is created around the two stars. Because both stars still circle each other, the envelope obstructs them and absorbs their rotational energy. This additional energy causes the envelope to be blown away from the system. So now the binary star is composed of an aging stripped helium core, which is extremely hot, and a normal star on the main sequence of the H-R diagram. If the mass of the helium core is big enough, it will explode as a supernova in the course of time. Its remains will collapse into a neutron star or a black hole, depending on the mass (Figure 3.27).

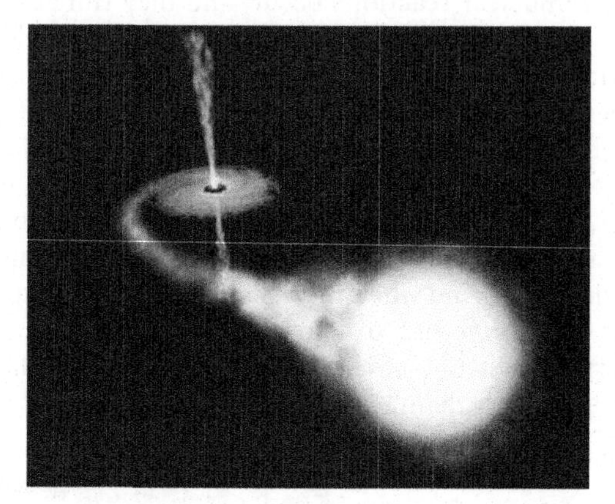

Figure 3.27. This binary system is composed of a super giant and a black hole. The latter pulls matter from the star, which spirals into it. During this fall, the matter heats up so much that it radiates outward in X-ray wavelengths. An example of such a binary star is Cygnus X–1

If the binary system survives the first explosion of the supernova, it will continue to evolve. When the second star leaves the main sequence and starts to transform into a red giant, the strong gravitation of its neighbor pulls on its outer layers. Around the neutron star or black hole an accretion disk of matter forms, and the falling matter gets so hot that it radiates in the X-ray part of the electromagnetic spectrum. Such objects are known as X-ray binaries. At the end of its life the second star also explodes as a supernova, and what remains is a gravitationally bound system of two pulsars, a pulsar and a black hole, or two black holes, depending on the stars' initial masses. Computer models of such binary stars show that the objects could also collide and merge over the course of time, which could explain one type of extremely strong gamma-ray bursts that are seen in the depths of the universe.

If stars in the binary system have individual masses lower than 11 solar masses, the one with the greater mass will also end its life sooner and transform into a white dwarf. Its close neighbor will be at that time in the red giant phase, so its surface gravity will weaken. The matter from its outer layers will accumulate on the white dwarf (Figure 3.28). If the growing mass of the white dwarf reaches the Chandrasekhar limit, it will explode as a supernova type Ia (Roman numeral 1 and letter a).

This special kind of supernova is of great importance to astronomers. The explosion always occurs in the same manner (when the mass of a white dwarf reaches 1.4 solar masses), so the power of the explosions is pretty much the same. That is why this kind of supernova is often used as a "standard candle" for measuring great distances in the universe.

Figure 3.28. A binary star composed of a normal, Sun-like star in a red giant phase, and a white dwarf (as RS Ophiuchi). The strong gravity of the white dwarf pulls the matter from its neighbor. When the mass of a white dwarf reaches the Chandrasekhar limit, a supernova Ia explosion occurs

Why Is a Black Hole Black?

If you want to shoot a rocket into space, you have to give it sufficient initial velocity to stay in space. Otherwise it might take off but fall back to Earth after a while. This is called *escape velocity,* and for Earth it is 11.2 km/s.

The escape velocity from a celestial body depends on its mass and size. The greater the mass and the smaller its diameter, the higher the escape velocity. Take our Sun as an example. The escape velocity from its surface is 618 km/s. If our Sun were to become a white dwarf, it would be the size of Earth, and the escape velocity would be 6,700 km/s. If the Sun were to become a neutron star (with a diameter of 24 km), the escape velocity would reach as much as 150,000 km/s, which is half of the speed of light! But when we squeeze it even more to become a black hole (with a diameter of 6 km), the escape velocity would be equal to the speed of light! Even light would not have enough velocity to escape the black hole's gravity!

Using Newton's mechanics Pierre-Simon de Laplace and John Michell (at the end of the eighteenth century) considered about bodies that have such strong gravity that even light cannot escape from them. But it was at the beginning of the twentieth century that black holes got a solid theoretical basis with Einstein's General Theory of Relativity. At the beginning, it seemed that these exotic objects were only theoretical models, and that they did not exist in real space. Astronomers found the first observational evidence of actual black holes in the 1970s. Today we know that black holes are not rare at all and that they settle in the centers of most galaxies. We will discuss this in greater detail in the chapter on the Milky Way Galaxy.

The Nearest Stars

In November 2006, an international group of astronomers under the leadership of Todd J. Henry announced that they had discovered 20 new stars close to the Sun. The measurement of the parallaxes revealed that these include the 23rd and 24th closest stars.

How is it possible that we know so little about our vicinity that even today we are still discovering stars that are so close? And this is at a time when powerful and sophisticated space telescopes are helping us to observe the edge of the universe, billions of light years away?

Anyone who has aimlessly looked through a telescope at the sky and observed the thousands and thousands of stars gliding across the field of view should not be surprised by the question posed

Figure 3.29. Which of the countless stars is the closest?

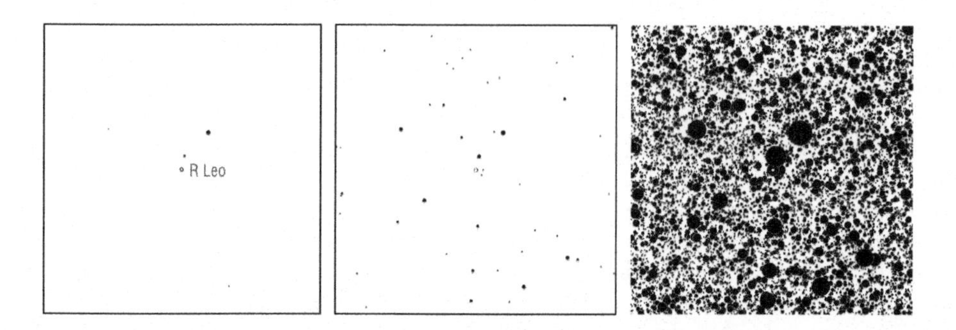

Figure 3.30. As we look for ever fainter stars, their numbers increase drastically and with this also the number of possible candidates for nearby stars. The field in charts measures only 4×4°, which is an insignificant part of the entire sky. On the first chart there are stars up to magnitude 6, on the second up to magnitude 10, and on the third, which is almost covered with dots, up to 18. The charts depict the vicinity of variable star R Leonis

above. If we wanted to know all the stars in our vicinity, we would have to carefully examine and measure all stars that we can see with the largest telescopes. Their number is truly astronomical. If we attempted do this in a systematic way, we would not be done for ages (Figures 3.29 and 3.30)!

Of course, not all discoveries of close stars are coincidental. Astronomers have their own sieve, with which they can roughly separate the closer from the more distant stars, and they check and measure only those that pass through the sieve.

Parallaxes

When Earth travels around the Sun, the close stars change their position in the sky in relation to those further away that lie in the same part of the sky. This phenomenon is called the parallax. The parallax principle was already known by the ancient Greeks, but only the development of modern astronomical instruments allows us to actually measure the parallax of the stars.

If a star is 3.26 light years from Earth (1 parsec), then it moves, on average, around its position for 1 arcsec over a period of one year. This angle is called the parallax. It is a very small angle. If we place a quarter 4.5 km away, it is seen under the angle of 1 arcsec. However, we cannot see the coin with the naked eye. We need a telescope in order to see that.

Measuring parallaxes is not simple. We have to follow a star night after night and precisely measure its position in relation to a few stars in its vicinity. It is impossible to measure the parallax for all of the stars. This is why astronomers started off by measuring the parallax of the brightest stars. They thought that the brightest stars were the closest to us. This would hold true if the stars were all at least roughly the same; however, they are not. They quickly discovered that most of the bright stars are so far away that their parallax cannot be measured at all. But the list of the closest stars began emerging: 61 Cygni, Sirius, Procyon, Alpha Centauri, and so on.

The bright stars were followed by less bright ones. With careful long-term measurements, we have so far managed to measure the parallax, and with this the distance to, a few thousand stars. This took much hard work and many sleepless nights for astronomers dedicated to this work! Large observatories have been systematically dealing with measuring parallaxes for over a century.

After all of the bright and slightly less bright stars had been measured, astronomers started looking at some of the faintest ones. There are many more of those, and astronomers could not just start at one end and finish at the other. They became picky. Which of the millions of stars would they dedicate their valuable observation time to? It could easily happen that an eager beginner would waste his entire professional career measuring weak stars, which would, after years of measurements, show that they are so far from us that their parallaxes could not be measured (Figure 3.31).

If only there was a way we could sieve this mass of stars in such a way that only the most promising candidates would remain...

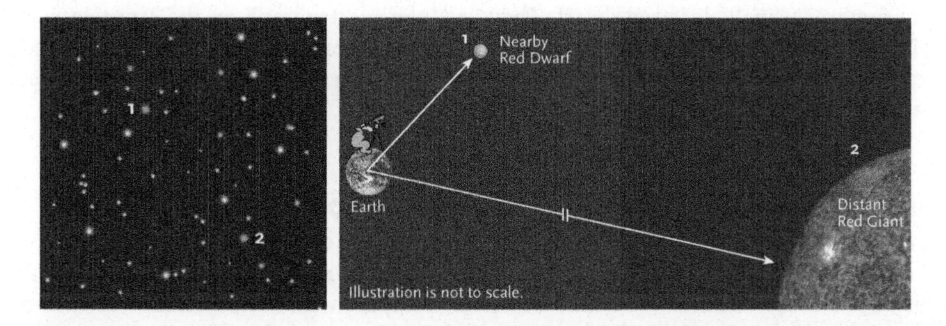

Figure 3.31. Two stars that appear to be exactly the same in a telescope or on images can in reality be totally different. The first is a red dwarf, a small star with low luminosity that is close to us. The other is a red giant, much further away. Its dimensions and luminosity greatly surpass the first. In order to establish the difference between them, we need to record the spectra of both stars

The Sieve

Stars from the main sequence of the Hertzsprung–Russell diagram, which live out their lives quietly, have a more or less well-defined correlation between the color and luminosity. The greater the luminosity, the bluer the star; the lower the luminosity, the more reddish it shines. If we compare the brightness of the star with its color or luminosity, this can help us decide which star we should select for the measurement or which star could possibly be in our vicinity. Let's take a look at an example.

Let's assume that a small reddish magnitude 12 star was our close neighbor. We can also assume that this is a star from the main sequence of the H-R diagram; thus we can roughly estimate its luminosity based on its color. From the estimation of the luminosity and the measured brightness (the density of the light flow that reaches Earth), we can get the estimate for the distance of the star:

$$j = L/4\pi R^2 \rightarrow R = (L/4\pi j)^{1/2}$$

where j is the density of the light flow that reaches Earth, L the luminosity of the star, and R its distance from us.

Our line of thought is reasonable; however, nature is never that simple. In the H-R diagram we can find stars that have the same surface temperature and thus the same color as our selected star, but a thousand times higher luminosity. These are red giants. And if such a red giant is much further from us than a small neighboring red star from the main sequence, the two stars will appear absolutely the same in the sky. We thus also need a measure with which we can separate the stars on the main sequence from the giants and super giants. This method is spectroscopy. The spectrum of the star reveals which one is from the main sequence and which one belongs to the class of giants. Before we start measuring the parallax, we therefore have to record the spectrum of the selected star.

The stars in our galaxy are never still; they are in constant motion. Some travel through the Milky Way in groups; others prefer to wander about alone. We call this their proper motion. The greater the proper motion of a star, the greater the possibility that it is close to us. The conclusions are not exact, but they are fairly reasonable for a rough selection of candidate stars. This is the same sort of conclusion as when we are looking out of the window of a speeding train. We assume that the things that travel through our field of view faster are closer to us than objects that move more slowly; of course, we could be wrong in assuming this. Still, in most cases, we are right.

Let's take a look how Todd Henry and his coworkers found the previously mentioned close stars with the help of sieves. They were looking for stars that showed an appropriate relationship between their brightness, color, luminosity, and motion. They found a number of perspective candidates. First of all, they had to ascertain if there were any faraway red giants among those that would appear to be exactly the same as a small close red dwarf. They used the spectrograph on the 4-m telescope at the Cerro Tololo observatory in Chile to check the spectrums of these stars and eliminate the red giants. Only after they had performed this task could they start measuring the parallaxes. The final result was the identification of 20 new star systems in our vicinity.

A Closer Look at Our Nearest Neighbors

Table 3.1 in this chapter listing the 25 closest star systems to us reveals not only the name of the star and its distance but a whole array of data that we can compare with our Sun and with each other.

As we have already described, the absolute magnitude is the measurement for the star's luminosity. It tells us how bright the star would be if we observed it from a standard distance of 10 parsecs (32.6 light years). The absolute magnitude of our Sun is 4.8. In the table the absolute champion is Sirius, which has 25 times greater luminosity than the Sun, while the faintest star is the newly discovered one in position 24, with an absolute magnitude of 17.4, the luminosity of which is merely 1/700,000 of the Sun's (Figure 3.32).

Table 3.1. The table shows 25 of the stars (or star systems) closest to the Sun. Stars that can be seen with the naked eye are emphasized and those that are visible in binoculars are printed in italics. Data is from January 1, 2008. (Source: Research Consortium on Nearby Stars, or RECONS: www.chara.gsu.edu/RECONS/index.html)

Star	Constellation	R.A. h m s	Dec. ° ′ ″	Proper motion ″/year	Dist. l. y.	Spectral type	Bright. Mag.	Absol. Bright. Mag.	Mass
Sun	–	–	–	–	–	G2 V	–26.72	4.85	1.00
1. Proxima		14.29 43	–62 40 46	3.85	4.24	M5.5 V	11.09	15.53	0.11
Alpha Centauri A	Centaurus	14 39 36	–60 50 02	3.71	4.36	G2V	0.01	4.38	1.14
Alpha Centauri B		14 39 35	–60 50 14	3.72	4.36	Ko V	1.34	5.71	0.92
2. *Barnard's star*	Ophiuchus	17 57 48	+04 41 36	10.4	5.96	M4.0 V	9.53	13.22	0.17
3. Wolf 359	Leo	10 56 29	+07 00 53	4.70	7.78	M6.0 V	13.44	16.55	0.09
4. *Lalande 21185*	Ursa Major	11 03 20	+35 58 12	4.80	8.29	M2.0 V	7.47	10.44	0.46
5. **Sirius A**	Canis Major	06 45 09	–16 42 58	1.34	8.58	A1 V	–1.43	1.47	1.99
Sirius B		06 45 09	–16 42 58	1.34	8.58	DA2 N	8.44	11.34	0.5
6. BL Ceti A	Cetus	01 39 01	–17 57 01	0.337	8.73	M5.5 V	12.54	15.40	0.11
BL Ceti B		01 39 01	–17 57 01	0.337	8.73	M6.0 V	12.99	15.85	0.10
7. Ross 154	Sagittarius	18 49 49	–23 50 10	0.666	9.68	M3.5 V	10.43	13.07	0.17
8. Ross 248	Andromeda	23 41 55	+44 10 30	1.62	10.32	M5.5 V	12.29	14.79	0.12
9. **Epsilon Eri**	Eridanus	03 32 56	–09 27 30	0.977	10.52	K2 V	3.73	6.19	0.85
Epsilon Eri P1		03 32 56	–09 27 30	0.977	10.52	planet	–	–	–
10. *Lacaille 9352*	Piscis Austrinus	23 05 52	–35 51 11	6.90	10.74	M1.5 V	7.34	9.75	0.53
11. Ross 128	Virgo	11 47 44	+00 48 16	1.36	10.92	M4.0 V	11.13	13.51	0.16
12. EZ Aquarii A		22 38 33	–15 18 07	3.25	11.27	M5.0 V	13.33	15.64	0.11
EZ Aquarii B	Aquarius	22 38 33	–15 18 07	3.25	11.27	–	13.27	15.58	0.11
EZ Aquarii C		22 38 33	–15 18 07	3.25	11.27	–	14.30	16.34	0.10
13. **Procyon A**		07 39 18	+05 13 30	1.26	11.40	F5 IV–V	0.38	2.66	1.57
Procyon B	Canis Major	07 39 18	+05 13 30	1.26	11.40	DA N	10.70	12.98	0.5
14. **61 Cygni A**	Cygnus	21 06 54	+38 44 58	5.28	11.40	K5.0 V	5.21	7.49	0.70
61 Cygni B		21 06 55	+38 44 31	5.17	11.40	K7.0 V	6.03	8.31	0.63
15. *GJ 725 A*	Draco	18 42 47	+59 3749	2.24	11.52	M3.0 V	8.90	11.16	0.35
GJ 725 B		18 42 47	+59 37 37	2.31	11.52	M3.5 V	9.69	11.95	0.26
16. *GX And A*	Andromeda	00 18 23	+ 44 01 23	2.92	11.62	M1.5 V	8.08	10.32	0.49
GX And B		00 18 23	+44 01 23	2.92	11.62	M3.5 V	11.06	13.30	0.16

Table 3.1. (continued)

Star	Constellation	R.A. h m s	Dec. ° ′ ″	Proper motion ″/year	Dist. l. y.	Spectral type	Bright. Mag.	Absol. Bright. Mag.	Mass
17. **Epsilon Ind A**		22 03 22	−56 47 10	4.70	11.82	K5 Ve	4.69	6.89	0.77
Epsilon Ind B	Indus	22 04 10	−56 46 58	4.82	11.82	T1.0 M	–	–	0.04
Epsilon Ind C		22 04 10	−56 46 58	4.82	11.82	T6.oM	–	–	0.03
18. DX Cancri	Cancer	08 29 49	+26 46 37	1.29	11.83	M.6.5 V	14.78	16.98	0.09
19. **Tau ceti**	Cetus	01 44 04	−15 56 15	1.92	11.89	G8 Vp	3.49	5.68	0.92
20. GJ 1061	Horologium	03 36 00	−44 30 45	0.826	11.99	M5. 5 V	13.09	15.26	0.11
21. YZ Ceti	Cetus	01 12 31	−16 59 56	1.37	12.13	M4. 5 V	12.02	14.17	0.14
22. *Luyten's star*	Canis Minor	07 27 24	+05 13 33	3.74	12.37	M3. 5 V	9.86	11.97	0.26
23. SO 0253+1652	Aries	02 53 01	+16 52 53	5.11	12.51	M7. 0 V	15.14	17.22	0.08
24. SCR 1845–6357 A	Pavo	18 45 05	−63 57 48	2.66	12.57	M8. 5 V	17.40 J	19.42	0.07
SCR 1845–6357 B		18 45 03	−63 57 52	2.66	12.57	T	–	–	–
25. *Kapteyn's star*	Pictor	05 11 41	−45 01 06	8.67	12.78	M1. 5 V	8.84	10.87	0.39

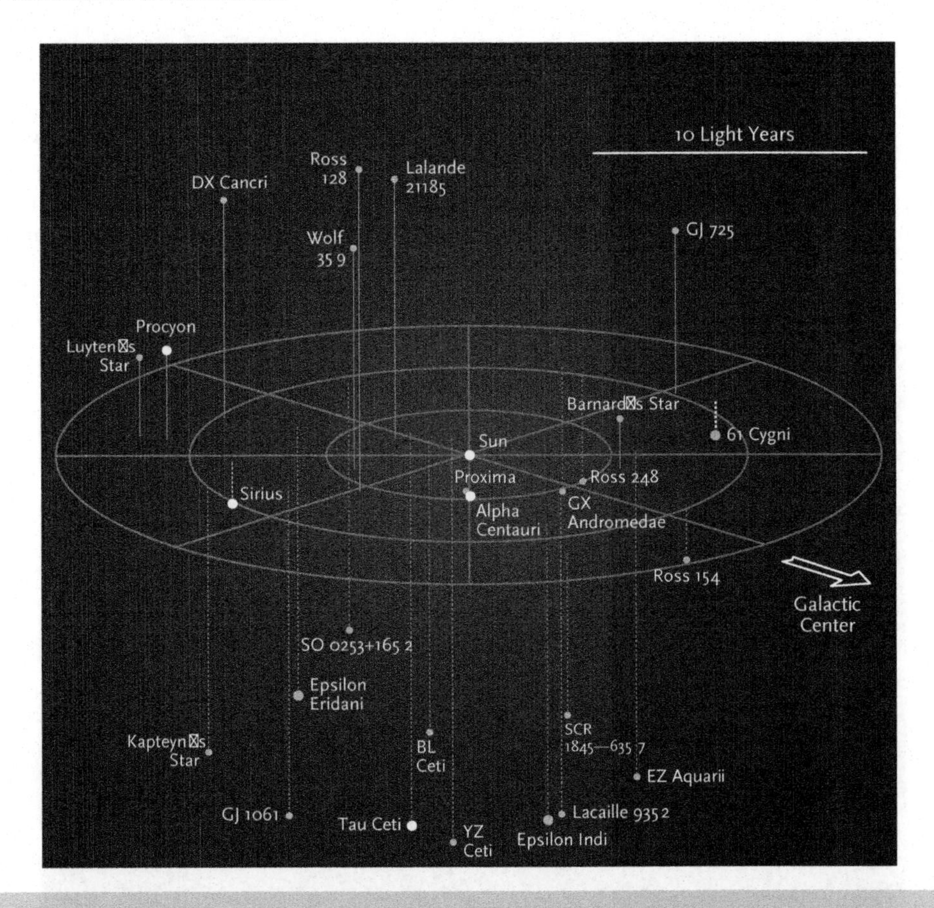

Figure 3.32. The 25 closest star systems. To make the illustration clearer, the double and multiple stars are shown with a single dot

The spectral types of the nearer stars can be roughly divided into a few groups: hot white stars of the spectral class A, such as, for instance, Sirius; yellowish stars similar to our Sun of the spectral classes F and G; orange-red stars of class K; red dwarfs of class M; and brown dwarfs of class T. In addition, we also have two white dwarfs in the vicinity, both of which belong to the spectral type DA and accompany Sirius and Procyon.

The distance tells us how much time it takes for the star's light to reach us. When we compare it with the distance between Neptune and the Sun, which is 4 light hours, we get a rough idea of how much empty space divides us from even the nearest stars.

The Search Is Not over Yet

We are well aware of the brighter stars in our vicinity. Most of the ones that are still hidden are red dwarfs, which are usually only 10–50% of the Sun's diameter and mass, and less than 1% of its luminosity. The last piece of information tells us that they are so faint in the sky that we do not notice them even through binoculars, let alone with the naked eye – even if they are as close to us as, for instance, Sirius. And red dwarfs are very common in the universe. If we look at the within 30 light years of the Sun, an area that includes approximately 300 known stars, we can see that over 70% of them are red dwarfs. Astronomers are convinced that there are at least 100 still to be discovered within this same area.

Parallax measurements taken by Earth-based observatories continue even though we now have space-based telescopes. During its successful mission, the European astrometric satellite *Hipparcos* measured 120,000 stars up to magnitude 12.5. It could measure the parallax up to one-thousandth of an arc second, and yet the faint red dwarfs remained hidden. Until ESA's next space astrometric mission, named *GAIA* (on illustration below), which will start at the end of 2011, we will continue to search for our neighbors from Earth and update the star charts of our vicinity in space, charts that we only started drawing 170 years ago.

How Far Are the Stars?

The ancient Greeks and even the Babylonians knew how to find the answer to this, but they did not have precise-enough measuring instruments. The principle of measuring the distance of celestial bodies was clear. It is called parallax.

Stretch out your arm and raise your thumb. Close your left eye and look toward the wall on which there is, say, a light switch. Remember the position of the thumb in relation to the switch. Now close your right eye and look at the thumb with the left eye. You can see that the position of your thumb moves in relation to the light switch. This happens because the observation point has changed (from the right eye to the left). This is the parallax principle. The distance between the left and right eye is called the base. If you know what the base is and measure the angle by which the thumb has moved in relation to the switch between the two views (left and right eye), it is a simple matter to calculate how far the thumb is. The movement of the thumb in relation to the switch is greater if the thumb is closer to the eyes and if the distance between one and the other observation point is greater.

In astronomy, calculating the distance of the thumb is the same as calculating the distance of the observed celestial body, say, moon, planet, or nearby star. Instead of the wall with a light switch, we have the background of very distant stars, which serve as motionless points, from which we can measure the movement of the selected nearby body.

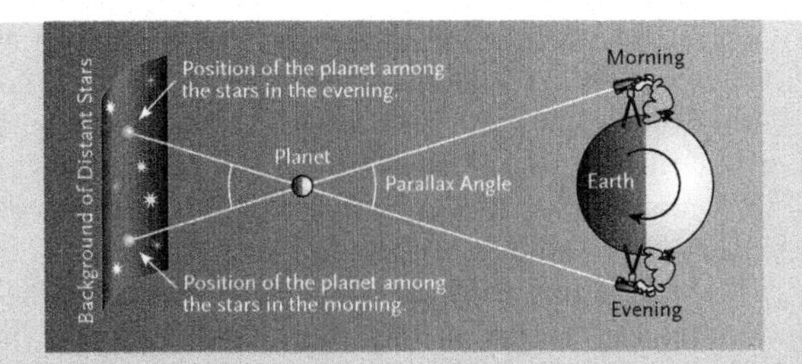

How Far Are the Moon and Planets?

The closest target to us is the Moon. Astronomers ascertained very early on that the distance between the left and right eye is not big enough to note the Moon's parallax. The left and right eyes were thus substituted for by two observers, who were a few hundred kilometers apart and were watching the Moon at the same time. By measuring the parallax of the Moon and comets Tycho Brahe (1546–1601) ascertained that comets are faraway celestial bodies and not meteorological phenomena in the upper layers of our atmosphere, as Aristotle had thought.

Kepler's laws on the movements of the planets (the first two introduced in 1609, while the third was formulated in 1619) represent the basis of all Solar System studies. In the seventeenth century, astronomers already knew quite precisely how long it took for each planet to orbit the Sun. In order to calculate all of the distances in the Solar System, you also need to know the distance between the Sun and at least one of the planets.

It was soon realized that Europe is too small for measuring the parallax even of the closest planets. So instead of sending astronomers with precious instruments to the other side of the world, Giovanni Domenico Cassini (1625–1712) had an ingenious idea. He waited for 12 hours for Earth to rotate around.

When the measurements are performed on the equator, the distance between the observation points is the same as the diameter of Earth. Cassini, who was measuring the parallax of Mars from a Parisian observatory, had a base of roughly 8,400 km. He measured the position of the planet in relation to the stars in the background in the evening and then again in the morning. The measurements showed a movement of a mere 30 arcsec, and a distance between Mars and Earth in opposition of roughly 60 million km. From these measurements and the aid of Kepler's laws, the astronomers could calculate the distance between Earth and the Sun and the distances between all the other planets and the Sun. This was the first time the vastness of our Solar System was truly recognized.

Measuring Distance to the Stars

Regardless of how exact the astronomers were at measuring the positions of the stars in the evening and morning, they could not calculate their parallax. The distance between the "left and right eye" had to be increased from the diameter of Earth (12,700 km) to the diameter of Earth's orbit around the Sun (300 million km). Because Earth revolves around the Sun, the observer on Earth changes his or her position in space. If we measure the positions of the selected stars, for instance, in March and then for the same stars in six months time, we would be able to detect the apparent shifts of the closest stars in comparison to those that are much further away from us. The first person who managed to measure the parallax of any star was Friedrich Wilhelm Bessel. In 1838, after an in-depth analysis of his measurements, he published the fact that the star 61 Cygni in one year circumscribes in the sky an ellipse with a radius of 0.3 arcsec. In this way, astronomers could tell for the first time the distance of any star. The result shocked them – the star 61 Cygni is 11.1 light years away from Earth, which is approximately 700,000 times further from Earth than the Sun! The year 1838 represented a great turning point in our knowledge of the universe, and Bessel's success was the peak of the 300-year endeavor to measure the distance to the stars.

The parallax of the stars is the pillar stone for defining the distances in the universe. However, we can only reach a few thousand light years distant with it, which is negligible in comparison to the size of the universe. So how do we measure greater distances?

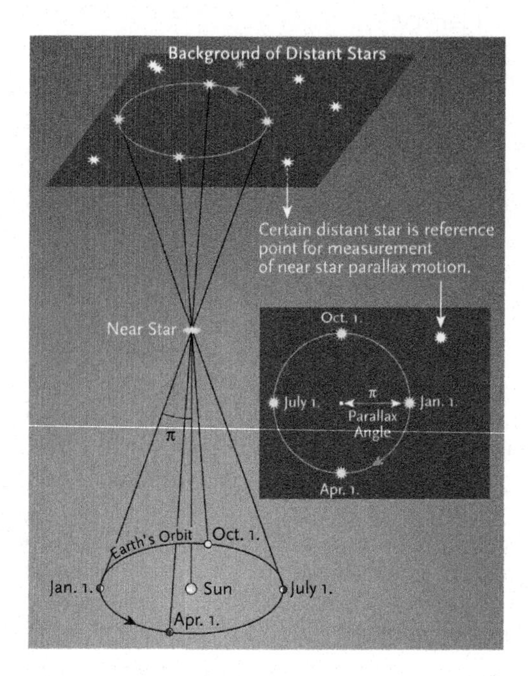

Standard Candles

Once more there is a simple principle we can utilize! The distance of a body can be calculated from its brightness – if we know its luminosity. Thus, if we find typical bodies with defined luminosities, we can use these as standard candles. To measure vast distances today, we use a class of variable stars called the Cepheides and the special type of supernova explosion called the supernova Ia.

At the beginning of the twentieth century, astronomer Henrietta Leavitt recognized that there was a direct correlation between the period of changing brightness and the luminosity of the Cepheid variable stars. Harlow Shapley was the first to realize the importance of this discovery. The Cepheids revealed the distance to the star groups in which those stars were members. Today, we define distances across the entire galaxy using the Cepheids, which are bright enough to see them in the neighboring galaxies all the way to a distance of several hundred million light years.

In very remote galaxies, we cannot differentiate between individual stars; not even the bright Cepheids are seen. That is why we have to find a brighter standard candle. These are the explosions of supernovae Ia that we discussed earlier, involving double stars consisting of a white dwarf and a red giant. The white dwarf with its strong gravitation pulls matter off its neighbor and explodes when its mass surpasses the point of stability (Chandrasekhar limit), at approximately 1.4 of the Sun's mass. All supernovae Ia explosions have therefore similar luminosity, so by observing them it is possible to ascertain distances up to one billion light years.

Hubble's Law and the Expansion of the Universe

In the 1920s, Edwin Hubble studied the spectrums of galaxies. He noticed that the spectral lines in the light coming from the galaxies, except for the few closest ones, were without exception shifted toward the red part of the spectrum. In other words, all galaxies are moving away from us, or to put it differently, the universe is expanding. When he compared the redshifts of those galaxies, for which the astronomers have measured the distance in one or more of the previously described ways, he concluded that the connection between the redshift and the distance is linear. The redshift, or the speed of receding, is greater the further away the galaxies are from us:

$$v = H_0 \times d$$

This is the famous so-called Hubble's law that Hubble published in 1929. H_0 is Hubble's constant and is 71 km/s/megaparsec. The law is merely a linear approximation that holds true as long as the speed of receding is low compared to the speed of light.

From Hubble's constant and the measured speed of a receding galaxy we can determine the galaxy's distance. The law is constantly checked using those galaxies for which we have managed to measure the distance with Cepheids and supernovae Ia. So far we have managed to verify the law up to the distance of approximately one billion light years, but most probably it also holds true for galaxies and quasars further away.

Defining the distance of celestial objects is a long and tough job. The exactness of the parallax measurements, which is the first step on the scale of distances in the universe, depends predominantly on technological development. The exactness of specifying greater distances with Cepheids depends on our knowledge of interstellar space between us and the chosen Cepheid. The interstellar dust and gas can weaken the brightness of the Cepheid, which, in turn, appears fainter in our sky, which again means that it will appear to us as further than it is in reality. We encounter the same problems also with supernovae Ia.

To know exact distances in the universe is important. Believe it or not, a more precise measured parallax of a single Cepheid could in the end influence many cosmological theories as well as modify our current estimates of the size of our entire universe.

The Battle for Parallax

As we mentioned earlier the first person to measure the parallax of any star was Friedrich Wilhelm Bessel. Bessel was an excellent mathematician and astronomer who, between 1821 and 1833, precisely measured the positions of 75,000 stars! He was particularly interested in star 61 Cygni because of its large proper motion in the sky. Bessel anticipated that the star must be quite close to us.

The battle for the supremacy at measuring parallax was very fierce; astronomers were well aware that the discovery could bring them eternal glory. Only one person could win, but the competition has given us some of the most important astronomical discoveries to date.

While he was trying to measure parallax, Edmund Halley (1656–1742) discovered the proper motion of stars at the beginning of the eighteenth century. Nineteenth-century astronomers rightfully predicted that the large proper motion of the individual stars in the sky must, to a certain extent, also be a consequence of their closeness. That is why they chose the candidates for measuring the parallax of those stars for which they had recorded large proper motions. Bessel stumbled across the star 61

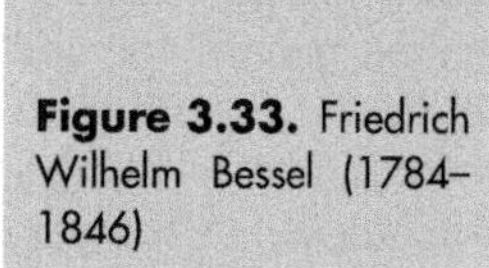

Figure 3.33. Friedrich Wilhelm Bessel (1784–1846)

Cygni, which changes its position by 5.2 arcsec per annum, when he was preparing a new star catalog. Halley thus helped Bessel to win.

During his search for a star parallax in 1728, James Bradley (1693–1762) discovered an aberration. This is a phenomenon in which the observer, who is moving, does not see the celestial body in the same direction as he or she would standing still. Earth revolves around the Sun, and so do we. Due to the aberration the virtual position of the star that lies in the pole of the ecliptics in one year draws a circle with a radius of 20.5 arcsec around its true position. All other stars in the sky describe more or less flattened ellipses. For stars that lie in the plane of the ecliptic, the ellipse becomes an arch with a length of 41 arcsec. The aberration is proof that Earth revolves around the Sun.

When William Herschel at the beginning of the nineteenth century decided that he would join the parallax hunters, he focused especially on close double stars and precisely measured the changes in their positions. He did not discover parallax. However, he did notice that some of the double stars changed their position in relation to each other as if they were circling one around the other under the influence of gravitation. This was an epic discovery in itself. He found true double stars and showed that the laws of gravity are universal and hold true throughout the universe.

CHAPTER FOUR

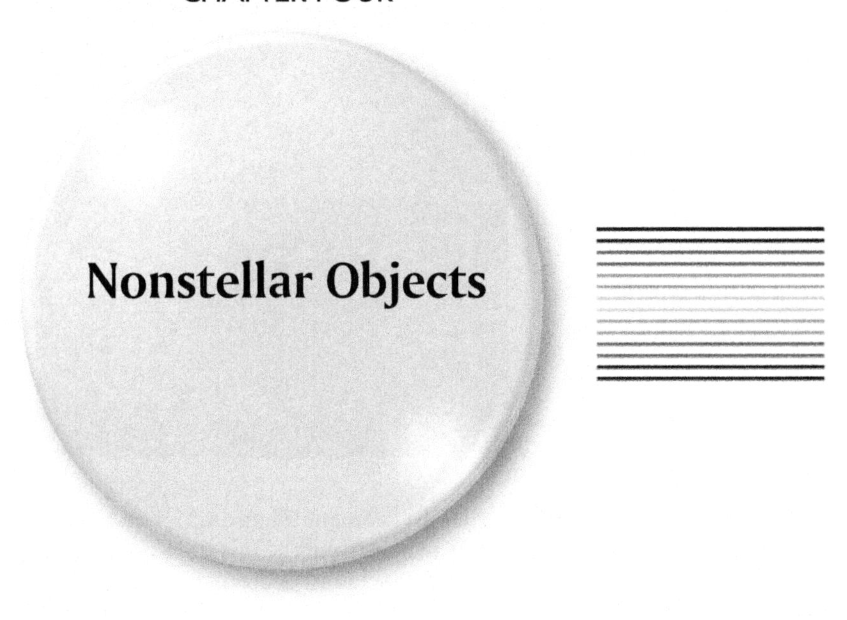

Nonstellar Objects

Star Clusters

Not all parts of our home galaxy are densely populated with stars. We are 4 light years away from the closest star, Proxima Centauri, and in a sphere measuring 20 light years in diameter we would encounter a mere dozen or so stars. Using binoculars and, in some cases, the naked eye we can see in places here and there a more dense group of stars in the sky. We call these star clusters and divide them into two general groups: open and globular.

Open Clusters

Open clusters are groups of between around 10 and a few hundred stars that appear in the sky close together and are also truly close together. Some of the clusters are visible with the naked eye, for instance, the Pleiades and Hyades in the constellation of Taurus and Praesepe in Cancer. Open clusters have irregular shapes, and the stars are gravitationally bound to each other. This tells us that they were born from the same cloud of gas and dust. For astronomers who study the evolution of stars, open clusters are especially important because the original chemical composition is the same for all of them, and they are all approximately the same age and distance from us. Studying them can show how the characteristics of the star and its evolution are influenced by its mass (Figure 4.1).

Open clusters are young groups of stars that usually include hot blue stars at the beginning of their life journey. Yellow and red giants are rare, while red and yellow super giants have not been seen at all so far. In numerous clusters, the remains of the embryo cloud is still clearly visible as a soft nebulosity that wraps itself around the stars. Some clusters are buried in still denser nebulosity, where new stars

B. Kambič, *Viewing the Constellations with Binoculars*, Patrick Moore's Practical Astronomy Series,
DOI 10.1007/978-0-387-85355-0_4,

Figure 4.1. The large and wonderful open cluster M 41 in the constellation Canis Major. As a gentle speck of light it can be seen with the naked eye, while the binoculars separate the hazy speck into numerous faint stars. The field on the image is approximately 1×1° in size

are in the process of forming at this very moment (Figure 4.2). You can read more about these clusters in the descriptions of open clusters in a later part of this book.

Because most of the open clusters can be found in the disk of the galaxy they are sometimes referred to as galactic clusters. Today we know of approximately 300 open clusters that can be seen from Earth.

Open clusters are not stable star groups. Interactive gravitational forces and the forces affecting other nearby stars constantly disturb them. That is why the stars in an average cluster in most cases become dispersed (lose their gravitational bond) in less than a billion years. Our Sun also used to be a member of an open cluster that has already been scattered throughout the galaxy.

Figure 4.2. Open cluster NGC 2244 in Monoceros is still wrapped in a gentle nebula of gas and dust from which its stars were born. Although the cluster of stars is clearly visible in binoculars, the nebula can only be seen on long-exposure images. For a comparison of the apparent sizes, the globular cluster M 13 from page 112 is placed into the image in scale proportion

Some star groups are already so dispersed and so weakly linked gravitationally that we no longer call them clusters but associations. Usually these include from 10 to 100 stars that are scattered across an area measuring a few hundred light years across. If they are blue and white giants within the spectral classes O and B, they are called OB-associations. Apart from OB-associations we also know T-associations that are made up of very young variable stars of the type T Tauri. Because these stars have a lower luminosity, we know only some of the groups closest to us. The stars that form associations are still receding from each other, and the ties break off completely over time. In numerous cases they are still surrounded by the cloud of gases and dust from which they emerged.

Open clusters are a favorite target of amateur astronomers, mainly because we can separate the stars with small telescopes. It is different with the nebulosities that surround the stars. Except for rare exceptions we can observe these only with larger telescopes, and even then only with the aid of special filters for observing nebulae. But they show their true beauty on long-exposure images. Due to their large apparent size they are also suitable for amateur astrophotography.

Open clusters are ideal for observing with binoculars. Most can be separated into individual stars, but even those that remain merely spots of light are usually large and bright, thus clearly visible. Due to the large field of view of the binoculars you will have a beautiful panoramic view of these star clusters, a view that can usually not be seen even through better telescopes.

In the latter part of this book, you can find alongside the name or designation of the open cluster information on the integral brightness and apparent size (in brackets). Brightness is expressed in magnitudes (m), and the apparent size in arc minutes (′) or arc seconds (″).

Globular Clusters

Globular clusters, as the name tells us, are symmetrical and usually have the form of a globe or sphere. They include hundreds of thousands of stars, which is much more than an average open cluster, and the stars are also much closer together. The stars bunch up around the center, where they can be found so close together that they are hard (if not impossible) to resolve even with the largest telescopes. The stars in the globular

Figure 4.3. Globular cluster M 13 in Hercules. A true comparison between the apparent size of clusters in images of this and the previous page can be made only once we are aware that the field in the image of M 13 is approximately $0.5 \times 0.3°$ big, while in the image of M 41 it is $1 \times 1°$, and in NGC 2244 it is approximately $2.5 \times 3°$

clusters are also gravitationally bound. Among them we find mainly older stars. Globular clusters contain the oldest objects in the universe and are believed to have emerged together with the galaxies.

In binoculars and smaller amateur telescopes, globular clusters are clearly visible as soft, misty spots of light in which the most that we can see is the slightly brighter core or a slight inconsistency in the form. The larger amateur telescopes can already resolve individual stars at the edge of a cluster. The best known and among amateur astronomers of northern latitudes favorite globular cluster is most certainly M 13 in Hercules (Figure 4.3).

So far we have discovered about 160 globular clusters; there are perhaps only 20 more to be discovered. They are not evenly spread across the sky. Because they inhabit the galactic halo in most cases we find them outside the equatorial plane of the galaxy.

The globular clusters discussed in this book always include the name, integral brightness, and apparent diameter. However, this is the size that we can achieve visually only by using the largest telescopes or on images. Using binoculars we can see only the brightest, central part of the cluster. In most cases the speck of light measures merely a few arc minutes. This holds true for all globular clusters described in this book. In binoculars they appear as relatively bright and clearly visible specks of light that can easily be differentiated from the stars. But this is also all that can be seen with a 5-cm objective and a 10× magnification.

Planetary Nebulae

Planetary nebulae have nothing in common with planets; they obtained their name based on their appearance in early telescopes. In the sky, they look like small, faint discuses or rings of light. In reality they are vast shells or rings of thin gas that was blown away into space by a star in the center of the nebula. The emergence of planetary nebulae is an unavoidable consequence in the evolution of "normal" stars, the mass of which can reach up to 11 times the mass of the Sun (Figure 4.4).

Figure 4.4. Planetary nebula M 97 in Ursa Major is too faint to be seen through binoculars

Planetary nebulae are – except in few rare cases – extremely small in our sky. That is why they are seen in binoculars as objects similar to stars, i.e., points of light and not even very bright points of light. In order for us to be sure that we are looking at a planetary nebula and not a star, we need to have a good star chart. In this book, we presented only the brightest examples of planetary nebulae. We can only attempt watching such objects once we have some observing experience (Figure 4.5). As already stated, the planetary nebulae are more appropriate for those observers who wish to see as many objects as possible, regardless of their appearance. However, knowing the positions of the brightest planetary nebulae will come in handy to all observers, especially those who will continue their observations with a larger telescope.

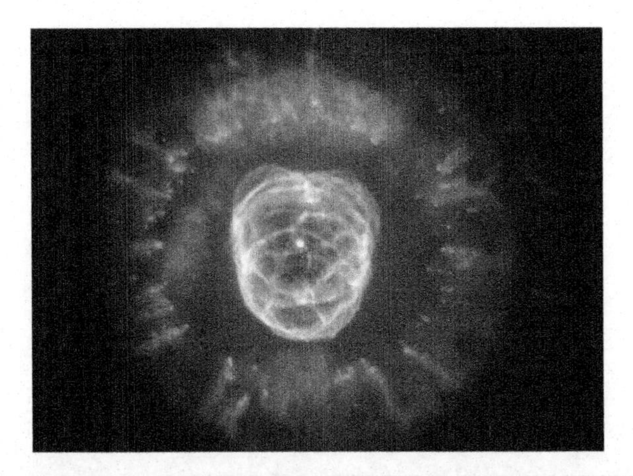

Figure 4.5. Planetary nebulae, photographed through large telescopes and using long exposure times, are very diverse. In the expanding gas, one can easily see the influences of their interstellar surroundings and the consequences of the strong stellar wind from the central star. The image shows planetary nebula NGC 2392, called the Eskimo Nebula, in Gemini. The star in the center was once similar to our Sun; at the end of its life (approximately 10,000 years ago) its outer layers were blown away and a wonderful planetary nebula appeared

In the latter part of this book you can find alongside the name or designation of the planetary nebula information on the integral brightness and apparent size (in brackets).

Nebulae

We can see some nebulae in the sky with just the naked eye, but we can see many more with telescopes or on long-exposure images. Nebulae are vast areas of gas and dust. We differentiate them into reflective, emission, and dark nebulae.

Reflective nebulae are those that do not emit their own light but reflect the light from the stars in the vicinity. An example is the Pelican Nebula in Cygus. We can see it because the nearby stars shed light on it (Figure 4.6).

Emission nebulae emit their own light (Figure 4.7A). The nearby hot stars with their strong ultraviolet light ionize the gas atoms in the nebula, which causes them to emit photons of specific wavelengths. Examples of famous nebulae are the North America and the Rosette.

The remnants of supernovae form a special type of emission nebulae; in these the nebulae shine due to fast-moving electrons that excite the gas. The best known nebula of this type is the Crab Nebula (M 1) in Taurus, which can easily be seen in amateur telescopes; it is at border visibility in binoculars. Of course, some nebulae shine in combined light; they reflect the light from the nearby stars as well as emit their own light. Such is Orion Nebula (M 42).

If there are thick layers of dust in the nebula that do not let light come through and there are no bright stars in the vicinity, the nebula is dark. Such a nebula prevents us from seeing into the depths of the universe behind it; that is why we see it as a dark patch in the starry sky or against the background of a brighter nebula. In the southern celestial hemisphere, an especially dark nebula that lies within the Milky Way has the name of Coalsack Nebula. The best known dark nebulae seen from mid-northern latitudes are the famous Horsehead Nebula in Orion (Figure 4.8) and the Pipe Nebula in Ophiuchus.

It probably does not need to be stressed that we can observe nebulae only in a very dark sky, on a moonless night far from distracting artificial lights. To get a better view of nebulae through binoculars one has to use special filters during observation, since these increase the contrast of the image in the

Figure 4.6. The unbelievably turbulent gas and dust clouds of the reflective nebula Pelican in Cygnus can be seen only on long-exposure images

Figure 4.7. NGC 6888, known as the Crescent Nebula, is an emission nebula in Cygnus

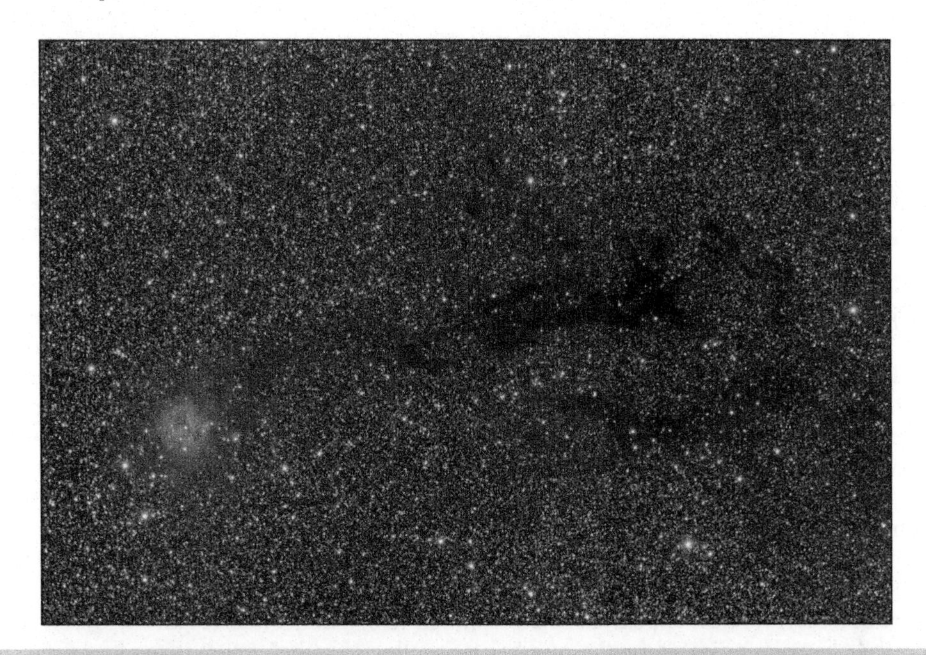

Figure 4.7A. Fantastic combination of reflection nebulae Sh2-125, known as the Cocoon Nebula, which is involved with open cluster IC 5146, and a dark nebulosity designated as B168, which spreads west of it

Figure 4.8. The famous dark Horsehead Nebula in Orion is only visible on long-exposure images. It measures only 6.5 × 8 arcmin in size

eyepiece. Filters are adjusted to the type of nebulae – either reflective or emission. You can buy special filters for some of the brightest and best known nebulae that are specially adjusted to the light emitted by that specific nebula.

Galaxy Types

If we could observe the universe from a great distance we could see that the space is filled with countless galaxies. (The word *galaxy* is of Greek origin – *galactos* – and means "milk.") Galaxies are giant independent star systems that include billions of stars. Today we believe that in the whole universe there are over 100 billion galaxies. We can find them all the way to the edge of the visible universe. Galaxies are not at all randomly scattered around the universe, however; they are merged in larger and smaller groups, called *clusters*. One of the biggest, closest clusters is the one between the constellations of Leo, Coma Berenices, and Virgo, which includes thousands of galaxies and is called the Virgo Cluster (Figure 4.9). Our galaxy also belongs to a smaller cluster that we call the Local Group, which includes approximately 30 members.

According to their appearance, galaxies are divided in the Hubble classification scheme into four large groups: elliptical, lenticular, spiral, and irregular. Elliptical (E) galaxies are typical for their misty, more or less featureless appearance that is similar to bright nebulae (Figure 4.10). The extent of their elliptical shape is indicated by the number that follows the letter E: 0 is totally round, while 7 depicts a very flattened shape.

Lenticular galaxies (S0 – the letter S and the numeral 0) are an intermediate form between elliptical and spiral galaxies. They look like elliptical galaxies, but in the center is a slightly denser core (Figure 4.11).

A spiral galaxy (S) is typically recognizable by its spiral arms, which more or less tightly wrap themselves around the central bulge. Those with tightly wrapped arms are marked with Sa, the ones with slightly looser arms Sb, even looser Sc, and so on. The last in line is the mark Sm, which designates a practically completely diffused spiral galaxy, in which the spiral arms are already so loose that we can no longer separate them from the rest of the galaxy (for example, the two Magellanic Clouds).

Galaxies that do not belong in any of the three groups according to their appearance are called irregular and are marked by the letter I.

Figure 4.9. The closest big cluster of galaxies in Virgo is approximately 60 million light years from us. It includes over 2,000 members. This image shows the bright elliptical galaxies M 86 and M 84 that are on the border of visibility using binoculars. The field in the image is approximately 1 square degree in size

Figure 4.10. Various types of elliptical galaxies: E0–1 (M 87), E3 (M 86), and E6 (M 110)

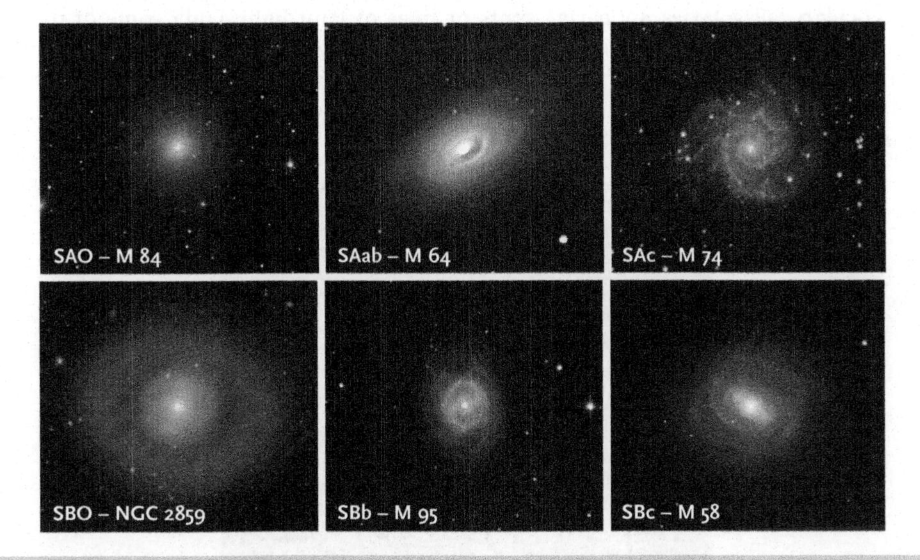

Figure 4.11. Lenticular, spiral, and irregular galaxies can be found in two forms: as normal galaxies and as barred galaxies. The image depicts various examples of lenticular and spiral galaxies

Spiral, lenticular, and irregular galaxies can also appear in another form: as barred galaxies. The bar is comprised of stars and usually runs across the center of the galaxy. If a spiral galaxy is nonbarred, we denote it as SA; if it is barred, we denote it as SB. Nonbarred lenticular galaxies are therefore indicated with an SA0, and those barred by an SB0. If we catch a hint of a bar in an irregular galaxy, we mark it with IB if not it is IA.

Looking at a few examples should help the seemingly complicated classification system become much clearer. If a barred spiral galaxy has tightly wrapped arms, we call it an SBa, and if its arms are wider open, it is marked SBc. If it is a nonbarred galaxy, we call it an SAa and SAc. The two Magellanic Clouds are spiral galaxies with practically completely opened arms and a hint of a bar, which is why they are marked SBm. If a lenticular galaxy without a bar appears to be closer in form to an elliptical galaxy, it is marked $SA0^-$, and if it is more reminiscent of a spiral galaxy, it is marked $SA0^+$. If the same galaxy had a bar, it would be marked $SB0^-$ or $SB0^+$, depending on its appearance.

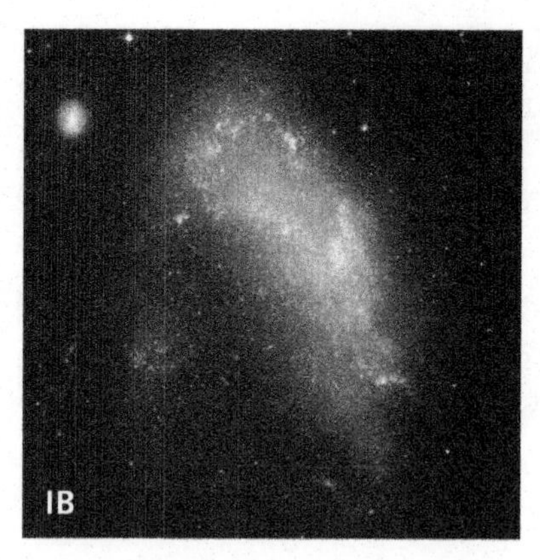

For observers with binoculars and small telescopes, the previous discussion is merely for edification. In binoculars, all galaxies appear as spots or lines of faint light. Only some of the brightest examples can hint toward an irregularity in the otherwise equally bright (or better to say faint!) spot of light seen only under perfect observing conditions. Unfortunately, when using a 5-cm objective almost all galaxies are too far away for us to see anything more than the galaxy's core, which is usually the brightest part (Figure 4.12).

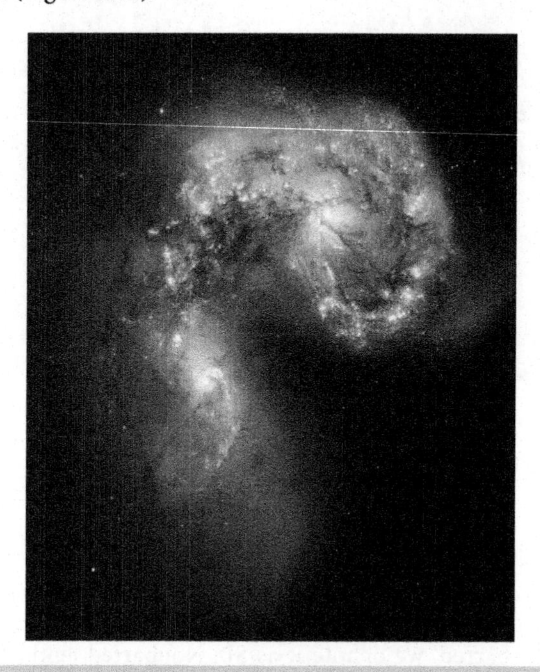

Figure 4.12. In galaxy groups, there are often close encounters or even collisions between individual members. Such colossal events usually result in the fusion of the two galaxies and cause a disturbance in the surrounding interstellar gases and dust. The consequence of this is the vigorous birth of stars. On the image we can see the collision of galaxies NGC 4038 and NGC 4039 (in the constellation Corvus)

The Milky Way Galaxy

In the summer and winter, far away from the light-polluted skies of the cities (and lately even towns) on a moonless night we can notice a veil-like strip of faint light, an irregular shape, that spreads across the sky from one side of the horizon to the other. This is the Milky Way. The main building blocks of

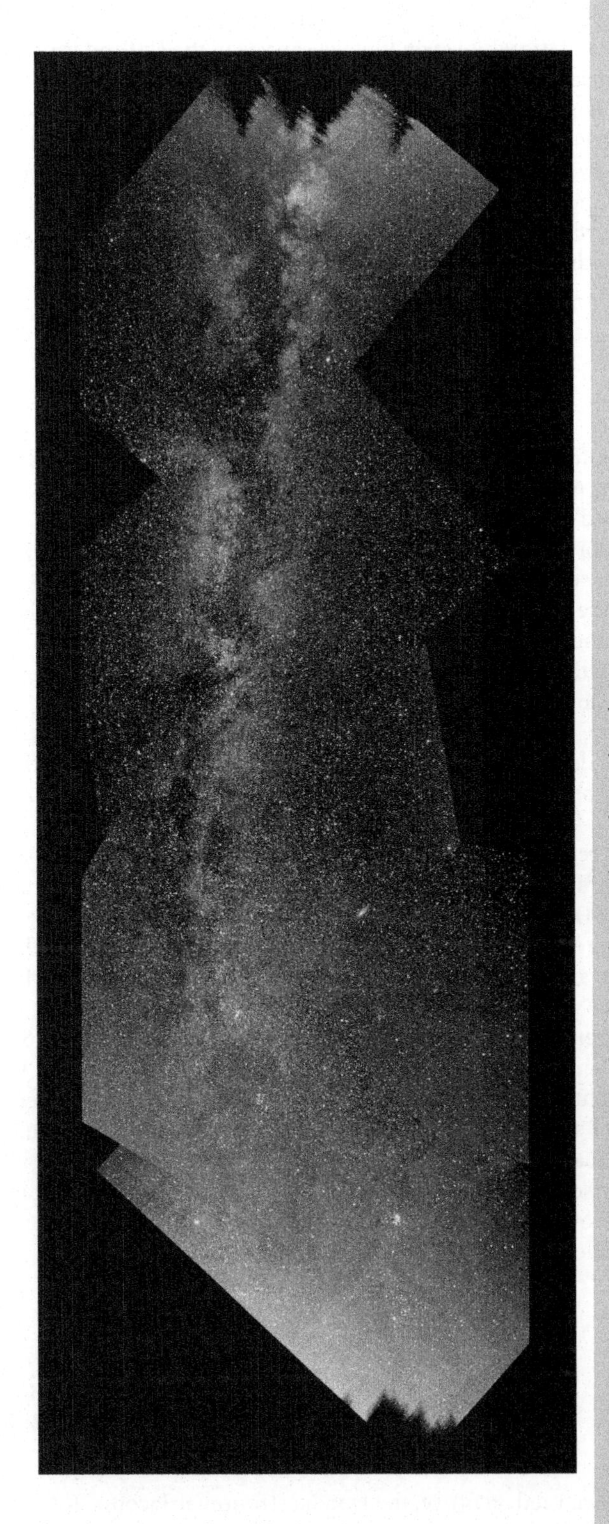

Figure 4.13. A mosaic of the summer Milky Way from the southern (right) to the northern horizon

the universe are the galaxies – giant star cities, populated by billions of stars. All stars that we can see with the naked eye and those that we can see through telescopes are members of our own galaxy. The Milky Way is also the name of the white strip that we can see in the sky and is our view (the view from Earth) on our galaxy (Figure 4.13).

Young people who do not know what a real night looks like, due to light pollution, cannot even imagine what a clear starry night sky looks like. Maybe you can come closest to this image in the middle of a desert or in the winter high up in the mountains during a temperature inversion, when the valleys (and the light pollution) are hidden under a thick layer of fog. Only there can we get a hint of what we have lost with the widespread placement of public lighting. Under such conditions, the Milky Way can be seen so clearly that it is no wonder that all peoples of the world have their legend about it. But we will leave the myths and legends to other writers.

A Brief History

Until the discovery of the telescope, astronomers could only roughly describe the Milky Way and more or less guess about its nature. Claudius Ptolemy (second century) from Alexandria, the last great astronomer of antiquity and the author of the famous *Almagest,* described it as "a strip, white as milk, of an irregular shape – in some places broader, in some narrower. In some parts it branches out and if we look at it closer we can see that it changes in its brightness as well as color."

Galileo Galilei gave us the first hint regarding what this veil-like strip could be. When he directed his first telescope toward the Milky Way, the gentle nebula split into numerous faint stars of various luminosity and colors, exactly as we can see elsewhere in the sky with the naked eye. Galileo was most probably the first person to have a hunch that the universe was much bigger and richer than was previously thought.

Numerous astronomers dedicated themselves to the study of the galaxy. Their hypotheses were, of course, a reflection of the observation tools and techniques available at that time. We will mention only a few here.

The first assumption as regards the galaxy being a great family of stars went back to the eighteenth century, when the British astronomer Thomas Wright (1711–1786) and the German philosopher Immanuel Kant (1724–1804) came to the conclusion that the Milky Way is a great flattened group of stars. Their thoughts can be roughly summed up as follows: because the stars in the sky are condensed in a narrow strip, there is a greater chance that the galaxy has the form of a flattened disk than a globular shape.

At the end of the eighteenth century, the German-born British astronomer Sir William Herschel (1738–1822), through systematic checking of the sky and counting the stars in the field of view of a telescope, came to the conclusion that the galaxy was sort of an ellipsoid with a denser central area and a somewhat denser ring of stars around it, while the Sun was supposedly in the central part in the vicinity of the galactic equator. At the time astronomers still did not know how to measure the distance of the stars, so Herschel could not even roughly estimate the size of the galaxy. His estimate as regards the form was also not correct; however, this still remains the first time someone had given any real thought to it from a scientific viewpoint (Figure 4.14).

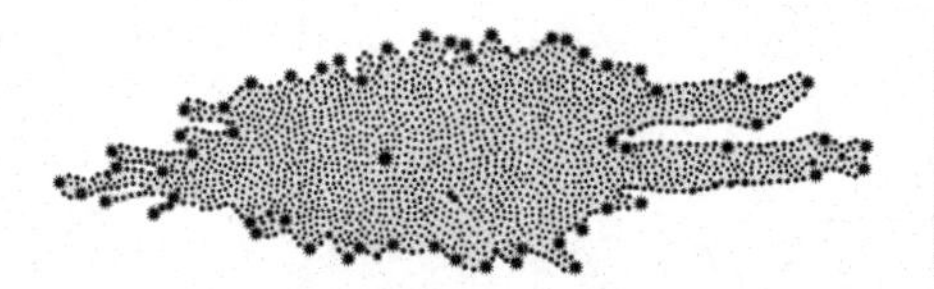

Figure 4.14. Herschel's model of the galaxy

At the beginning of the twentieth century, the Dutch astronomer Jacobus Kapteyn (1851–1922) and the American astronomer Harlow Shapley (1885–1972) had at their disposal large modern telescopes for observing the sky. They were among the first to try to establish the size and shape of the galaxy through measurements. Kapteyn concluded that it had the shape of a disk measuring 50,000 light years across, and the Sun was close to its center (Figure 4.15).

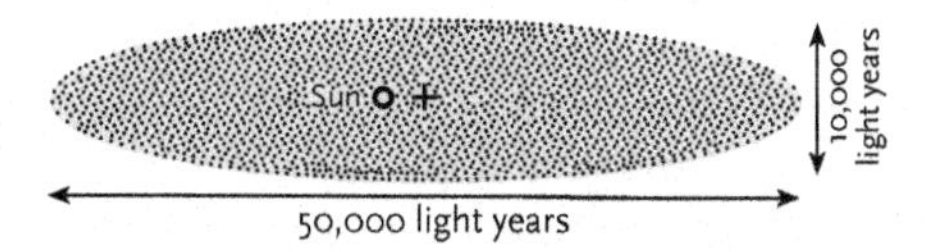

Figure 4.15. Kapteyn's estimation of the shape and size of the galaxy

Shapley, on the other hand, was convinced that the galaxy incorporated a much bigger group of stars. He estimated that it had a diameter of 320,000 light years and that the Sun was not in its center but two-thirds of the way toward the edge of the disk. Shapley reached his conclusions by measuring the positions and distances of globular clusters, which are so bright that they can also be seen from great distances. In the individual cluster, he measured the luminosity of the variables type RR Lyrae which – similar to the Cepheids – can prove useful for establishing distances. In this way, he got the distances of clusters and with that their true position in the sky (Figure 4.16).

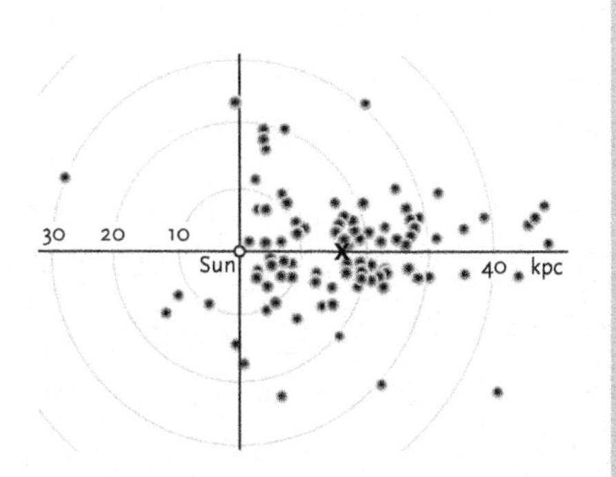

Figure 4.16. Shapley's diagram of the pattern of globular clusters in the galaxy; X denotes its center. A kiloparsec (kpc) is 3,260 light years

His reasoning was as follows: globular clusters are not equally spread across the sky. There are many more of them in the direction of the constellations of Sagittarius and Ophiuchus. If the Sun was in the center of the galaxy, globular clusters would be more equally dispersed across the sky. (At the time astronomers logically assumed that the clusters were more or less equally spread across space and that there was a sort of center of gravity for all clusters somewhere in the center of the galaxy.) From this, Shapley concluded that the Sun is not in the center of the galaxy.

The models of these two astronomers were completely different. Kapteyn's estimation of the size of the galaxy was too low, for he did not include star extinction (the weakening of the light from the stars due to interstellar dust) in his calculations. Shapley rightfully concluded the position of the Sun in the galaxy but greatly overestimated its size. He also did not know the true values of the absorption effects of the interstellar dust.

The 1920s brought with it a great new discovery. Edwin Hubble (1889–1953) ascertained that spiral nebulae such as Andromeda (M 31) are so far away from us that they are independent space bodies, most probably similar to our galaxy (more on this in the section on the Andromeda constellation). This was a huge step in the right direction. Thousands of galaxies in the sky became models with which astronomers could compare their findings as regards our own galaxy.

In the 1940s, Walter Baade (1983–1960) worked at the Mount Wilson observatory, which at the time housed the largest telescope in the world, its mirror measuring 2.5 m. Baade photographed the nearby galaxies and studied their stars. He concluded that the galactic disks are mainly populated by blue stars, which he called Population I stars. In the central parts of the galaxies, he mainly found red and orange stars, which he called Population II stars. Population I stars are relatively young (from one million to a few billion years old). Their chemical composition is similar to that of the Sun: hydrogen and helium dominate, with about 2–3% of heavier elements. Population II stars are older (typically around 10 billion years) and were supposedly created at the same time as the galaxy. This population also includes the stars in globular clusters.

Galaxies emerged soon after the Big Bang from the large spherical gas clouds (the proto-galaxies) that started condensing. The first stars (Population II) consisted of mostly hydrogen (75% of the mass) and helium (25% of the mass). The most massive Population II stars quickly ended their life cycle and after 10 million years or so exploded as supernovae. Elements heavier than helium are created mainly during nuclear fusion reactions in the cores of the massive stars and during explosions of supernovae, which have enriched interstellar space with heavier elements. Younger stars, such as the Sun, were thus born from hydrogen and helium with a mixture of heavier elements. The presence of two so different star populations in the same galaxy is therefore a logical consequence of the galaxy's evolution.

By studying other galaxies, Baade thus ascertained that their spiral arms are populated mainly with Population I stars, the young, hot stars of the spectral types O and B. He logically concluded that this must also be the case with our own galaxy. Due to interstellar dust, optical observations are limited to the parts of the galaxy in our vicinity (roughly up to 10,000 light years away). By measuring the positions of the blue giants in our galaxy, optical astronomers partially recognized its spiral structure, which was the first true evidence that we live in a spiral galaxy (Figure 4.17).

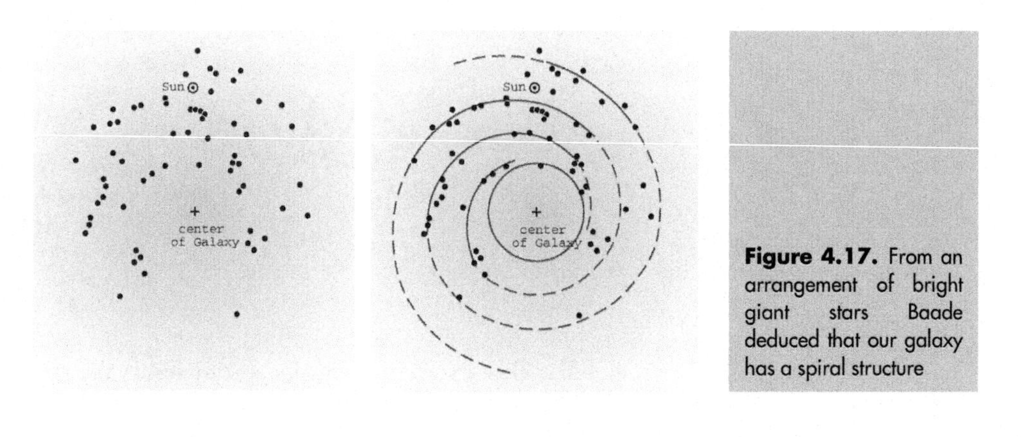

Figure 4.17. From an arrangement of bright giant stars Baade deduced that our galaxy has a spiral structure

Interstellar Space

The space between the stars is not empty. It is filled with interstellar matter that is comprised of gas and specks of dust. On average, a cubic centimeter of interstellar space consists of only a few atoms. For comparison, a cubic centimeter of air, which we breathe in, has 10^{19} molecules, and a cubic centimeter of an ultra high vacuum, which we can create in the best laboratories on Earth, contains a few million molecules. Compared to conditions on Earth, interstellar space is practically an ideal vacuum.

Interstellar gas can radiate in various parts of electromagnetic spectrum, depending on its temperature. At the top end of the temperature scale and at low density it radiates in the X-ray wavelengths. If it is cold, it emits infrared light and radio waves. If the atoms in it are in an excited state due to the ultraviolet light being emitted from a hot star nearby, we can see it as radiating emission nebulae or planetary nebulae. As a point of interest you should note that the density of gas in the densest nebulae that shows up in our images is as low as approximately 100 atoms/cm^3!

There is even less interstellar dust than there is gas, roughly about 1% of the total mass of interstellar matter. Dust particles were flung into space by stars that in the later phases of their lives simply blew away the outer layers of their atmosphere into their surroundings. When old massive

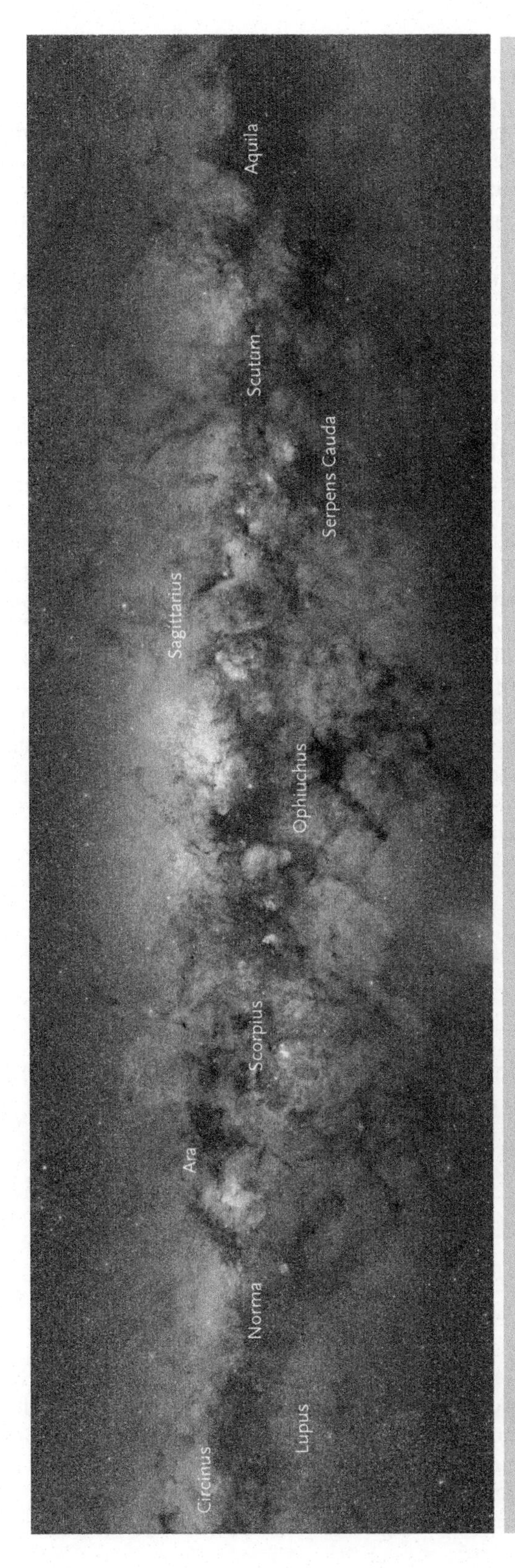

Figure 4.18. The part of our galaxy with the densest gas and dust clouds

stars explode as supernovae, atoms of oxygen, carbon, iron, etc., are scattered into interstellar space. Silicon and iron form very small crystals and oxygen, carbon, and nitrogen atoms stick to them.

Dust particles that pervade space are very small. Their diameters are somewhere between 1 μm and 1 nm (roughly the size of a molecule), and they are very far apart when compared to conditions on Earth. The average distance between two dust particles in the equatorial plane of the galaxy, where the clouds are the densest, is about 100 m. Thus, the average density of dust is even lower than the average density of interstellar gas. Regardless of this, the dust still manages to greatly disperse, absorb, and polarize the light coming from the stars, which causes great difficulties with our observations. When we in visible light look along the plane of our galaxy, we cannot see much further than a few thousand light years away. Even though the dust particles are extremely rare, there are quite a lot of them on a path that covers a couple of thousand light years. It is only when we watch above or below the plane of the disk that we can see the celestial bodies further away

The most common interstellar matter is hydrogen. Its average density in the galactic plane is around 2 or 3 atoms/cm^3. The unevenly distributed gas in some places creates parts where the density can be as much as 10 times higher than average. This is where hydrogen molecules (H_2) can emerge, as well as molecular clouds, which are comprised of more complex molecules. As we get further away from the galactic plane, the average density of the interstellar gas falls drastically.

Hydrogen atoms can be found in a neutral (unexcited) state, an excited state, or ionized. In order for a hydrogen atom to reach an excited state, it has to absorb a photon that has high energy. This is why there are very few atoms of this type in interstellar space. In the vicinity of hot, young, bright stars of the spectral type O and B, which strongly radiate in the ultraviolet part of the spectrum, the atoms are ionized all the way to the distance of a few tens of parsecs. By this criterion, we can divide the entire galactic vastness into areas in which hydrogen is not ionized – we call these areas H I areas (H and Roman numeral I) and areas of ionized hydrogen, which we call H II areas. The key step in revealing the true structure of the galaxy was made by astronomers after they discovered the fact that neutral hydrogen radiates radio waves with the wavelength of 21.1 cm. And because clouds of hydrogen are dispersed all across the disk, especially in the spiral arms, it was the radio astronomers who "photographed" the entire galaxy (Figure 4.19).

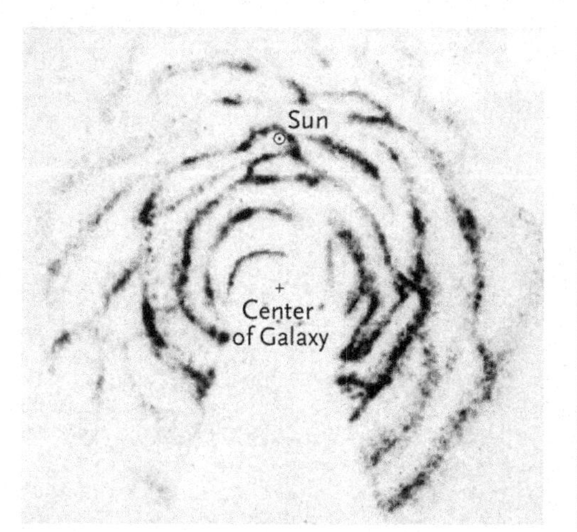

Figure 4.19. On an early radio wave chart of our galaxy that showed the pattern of neutral hydrogen, it was proven that we live in a spiral galaxy

In 1956, astronomers erected a 25-m radio telescope in Dwingeloo in the Netherlands. For a while this was the world's largest radio telescope. Its receivers were tuned so that it could watch the 21-cm radio waves that are transmitted by neutral hydrogen atoms. In this way it could study the structure of our galaxy. Optical astronomers had determined that it was likely that the Milky Way had a spiral structure. The charts in the radio wavelengths, such as Figure 4.19, showed without a doubt, that as much as 80% of our galaxy consists of spiral arms.

Seeing Without Light

The development of science and technology in the second part of the twentieth century brought with it telescopes that gather light from wavelengths other than visual ones, detectors that detect photons with high energies, and satellites with which we can observe from above Earth's impenetrable atmosphere. Thus, today we have charts of the galaxy in all parts of the electromagnetic spectrum. No two such charts are completely the same, and every one of them tells us something about the celestial bodies and phenomena totally different from the bright stars and the ionized gas that we can see in the visual light.

When we move from optic to shorter wavelengths, we first come across ultraviolet light, which is ideal for researching astronomical bodies at temperatures of around 100,000 K; in this light the hottest stars are seen. Thus we see the bright, young, massive stars and the hot white dwarfs that emerged after the death of less massive stars, such as our Sun. But also the cold stars, which do not emit ultraviolet light themselves, could be observed because of the radiation emitted from their chromospheres and coronae, the same as our Sun.

Some common elements, such as carbon and oxygen, have the strongest spectral lines in these wavelengths; thus the view of the Milky Way in ultraviolet light reveals more about the frequency of these elements in stars and gas clouds than we can learn merely from the data found in the visible light. Interstellar gases and dust absorb ultraviolet light, which is why our view in this part of the electromagnetic spectrum is somewhat restricted.

X-rays, which have even shorter wavelengths and are emitted by even hotter gases and stars, can also be used to learn about the galaxy. The brightest sources of the X-ray sky are double stars in which the star from the main sequence star or the red giant transfers its matter to a neutron star or a black hole that was formed from a previous massive star. The brightest clouds of gas in X-ray wavelengths are not nebulae of ionized hydrogen (H II), but the remains of supernova explosions that have erupted matter so violently that during the collisions with interstellar gases temperatures increased to a million degrees or more. Lately, as the sensitivity and angular resolution of detectors in satellites has drastically improved, astronomers have begun locating even fainter X-ray sources, and on these images some normal stars have appeared, mainly due to their coronary radiation.

The photons that have the greatest energy of them all are called gamma photons. With satellites, which have taken the detectors above the atmosphere, astronomers have charted the sky in the gamma spectrum and discovered that the galaxy is a strong diffusing transmitter of this radiation. The main sources are cosmic particles – protons and other matter accelerated with magnetic fields to a speed close to the speed of light. When the cosmic particles hit the interstellar gas, gamma radiation is emitted. Gamma radiation can thus tell us two important things: where the interstellar gas is the thickest and where there is the highest count of interstellar particles – typically in spiral arms.

If we move from the visible wavelengths in the other direction, we step into the kingdom of infrared radiation, which is emitted by bodies with temperatures of between 20 and 4,000 K. The sources of near-infrared radiation (shorter wavelengths, higher temperatures) are mainly stars that are cooler than the Sun. These are older red giants that are already on their way to exploding as supernovae or to finishing their lives as white dwarfs. But also the youngest stars, which have not yet stabilized on the main sequence of the H-R diagram, emit this light.

Interstellar gas is totally transparent for infrared light, too. So we can see the entire galaxy in infrared, including the events that have taken place in its very center.

In the far-infrared light, we can see interstellar matter, heated to between 10 and 100 K. This is mainly dust that was heated by a star's light. The hottest, brightest stars are the most efficient heaters, and thus the far-infrared radiation is an excellent indicator of the current level of star births.

Radio waves reveal interstellar gas and dust. Radiation at certain radio wavelengths reveals cold molecular gas. Other radio wavelengths reveal neutral hydrogen. This is the most common hydrogen phase and has already been charted for the entire galaxy. As with the infrared and X-ray wavelengths, radio wave observations also allow us to see the entire galaxy.

THE ELECTROMAGNETIC SPECTRUM

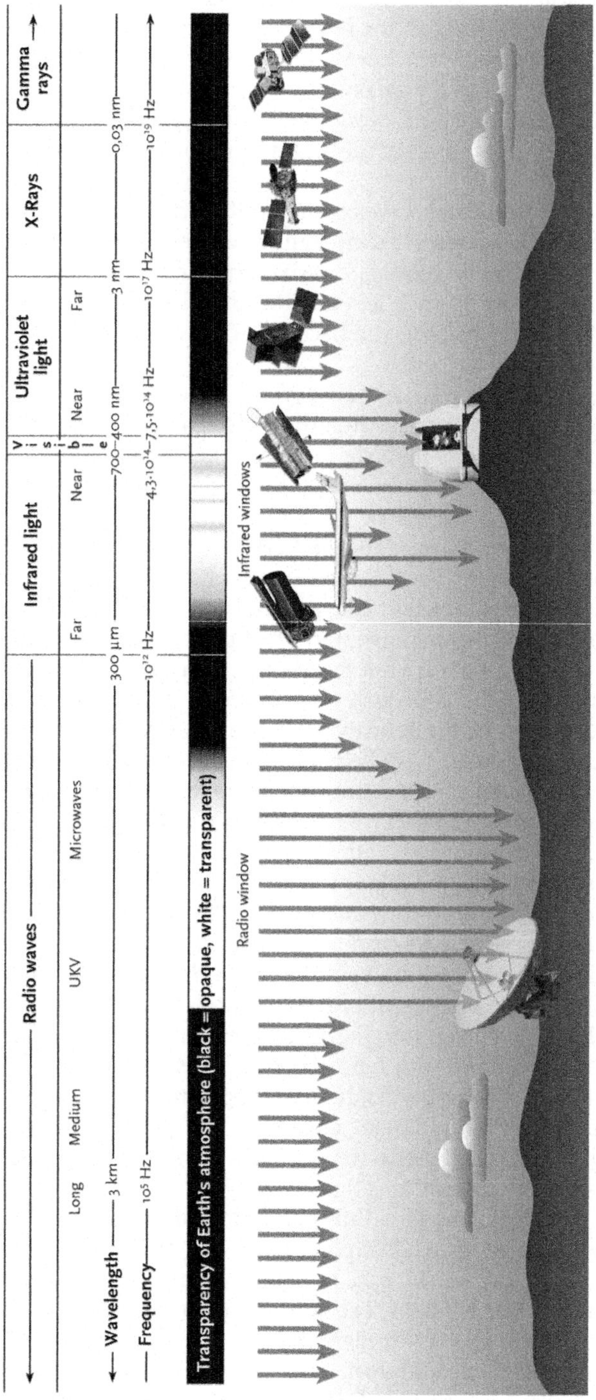

Radio waves
Infrared light
Visible
Ultraviolet light
X-Rays
Gamma rays

Long
Medium
UKV
Microwaves
Far
Near
Near
Far

Wavelength — 3 km — 300 µm — 700–400 nm — 3 nm — 0,03 nm

Frequency — 10^5 Hz — 10^{12} Hz — $4{,}3\cdot10^{14}$–$7{,}5\cdot10^{14}$ Hz — 10^{17} Hz — 10^{19} Hz

Transparency of Earth's atmosphere (black = opaque, white = transparent)

Radio window

Infrared windows

Thus, today we have a relatively good idea as to what our galaxy looks like, its shape and size, and where the most notable objects lie.

The Structure of Our Galaxy

The Milky Way is a spiral galaxy. It has the shape of a strongly flattened disk with a diameter of roughly 130,000 light years and a central bulge with a diameter of 15,000 light years. The estimated mass is somewhere between 750 and 1,000 billion solar masses, and it is estimated to have between 200 and 400 billion stars. If we compare it to other spiral galaxies, it is one of the bigger and more massive representatives of its kind.

The newest observations with the NASA's infrared Spitzer Space Telescope confirmed that through the central bulge lies a bar of stars and that our galaxy therefore belongs among the barred spiral galaxies.

The galaxy can be disassembled into three components (the figure below depicts a side view): galactic disk, central bulge, and galactic halo.

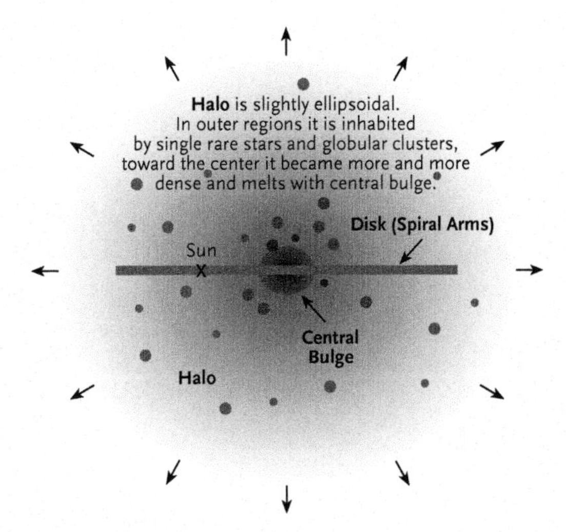

Disk

The disk is comprised of the two main spiral arms, few smaller ones, and a few spurs. The Sun, which is approximately 26,000 light years from the center of the Galaxy and 20 light years above its equatorial plane, is close to the inner edge of the smaller spur, called the Orion or Local arm, which is approximately 15,000 light years long. Our neighbors are the inner part of the Scutum-Centaurus arm and the outer part of the Perseus arm. They are approximately 6,500 light years from us, the first closer to the center of the galaxy and the latter on the outer side. These main arms are approximately 80,000 light years long. The furthest part of the Scutum-Centaurus arm was the harder to distinguish, for it lies on the other side of the galaxy, and observation of it was disturbed by the radio waves from the center (Figure 4.20).

The spiral arms start at the edge of the central bulge and wrap around it. In comparison to the diameter, the thickness of the disk is much, much smaller. The disk is thickest at the bulge, where it is estimated to be a few thousand light years across. In the place where the Solar System is located, it is only a few hundred light years thick.

Most stars in the disk belong to the Population I class. These are mainly young stars that were born from the mixture of gas and dust, enriched with heavier elements. There is a lot of interstellar matter – gas clouds and dust – which represents as much as 15% of its mass. Many of these clouds can be seen

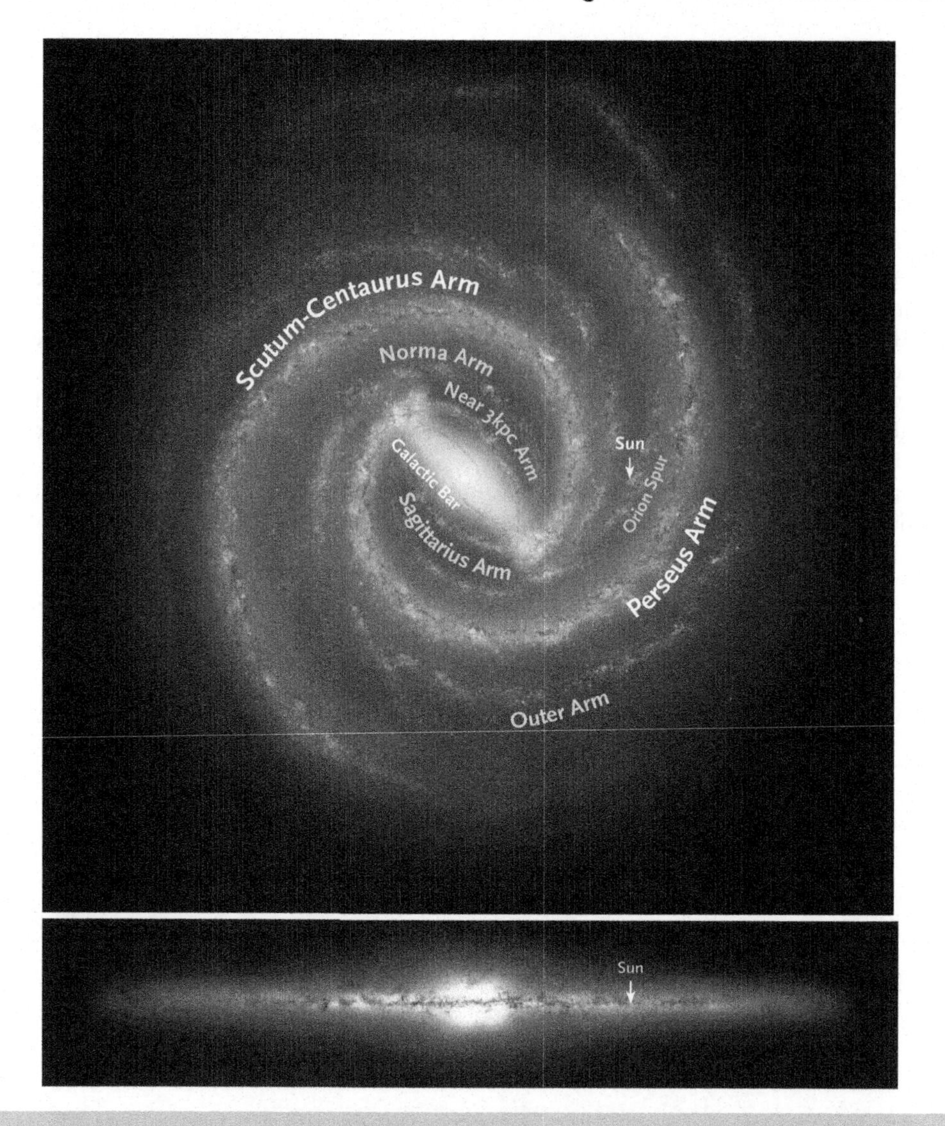

Figure 4.20. This illustration of our galaxy is based on today's knowledge and the newest observations. The top image could be seen by an observer in a galaxy far away in the direction of Leo or Boötes, while the bottom image could be seen by someone watching from between the Cyngus and Scutum. The Milky Way Galaxy is depicted as a barred spiral galaxy. Today, it is marked as SAB(rs)bc II. The SAB means that the galaxy does not have a well developed and visible bar. The (rs) means that there is a ring of gas and dust around the center. The bc tells us that the spiral arms are not tightly wrapped around the center. This classification is, of course, not final. Astronomers predict that new evidence will show that the bar is well developed and thus easily visible, and the classification will move from SAB to SB

in visual light as emission or reflective nebulae or as dark clouds that restrict the light from the stars behind. All open clusters also lie in the disk.

It is very hard to define the outer borders of the galaxy. Toward the edge of the disk the concentration of stars slowly drops (similar to the density of Earth's atmosphere with altitude); thus the galaxy does not have a clearly defined outer border. Most stars in the disk lie within a diameter of 100,000 light years, while individual ones and interstellar gas can be traced as far as 65,000 light years from the center.

The flatness of the disk is proof that the galaxy rotates around its axis. Stars, gas, and dust clouds move around the center along almost circular orbits, similar to the planets around the Sun; the further they are from the center, the slower their motion. Calculating the speed of the Sun is a very complex task, and it is difficult to be precise about it. Astronomers have estimated that the Sun and the stars close to us move at approximately 220 km/s, which means that the disk, in the position where we are, makes one rotation in 240 million years; this is what we call a galactic year. In order to illustrate this unimaginably long period, let us note that since its birth the galaxy has made only some 50 rotations. From the time of the extinction of the dinosaurs, roughly 60 million years ago to today, only a quarter of a galactic year has passed.

The Galactic Halo

The diameter of the halo is over 300,000 light years; however, over 90% of its population is gathered within a globe that measures 200,000 light years across. These are mostly old Population II stars and globular clusters; there is no gas or dust. The stars with larger masses exploded a long time ago as supernovae; thus the older, orange-reddish cold stars prevail. In globular clusters, which measure between 50 and 150 light years across, there are a few hundred thousand stars apiece; in some even a million can be found. These stars are also very old – as old as the galaxy.

Due to the reciprocal gravitational disturbances among the stars, globular clusters over time lose stars from the edge, which move into the surrounding space. Astronomers are of the opinion that a large portion of the stars in the halo migrated there from the globular clusters. Individual globulars can be found as far as 200,000 light years from the center of the galaxy, so we can also say that the halo does not have a clearly defined border. Apart from their age, the stars in the halo are also similar in their chemical composition, in which helium and hydrogen represent 99.9% of the matter and heavier elements a mere 0.1%.

As in the disk the stars and globular clusters in the halo circle the center of the galaxy; however, their orbits are not as regular. Numerous stars move in a retrograde motion (in the opposite direction to most); the orbits are large and irregular, with a large inclination toward the galactic plane; and many stars move in a chaotic manner. Astronomers think that in this case gravitational disturbances during close encounters with neighboring galaxies of the Local Group have pushed the stars into their orbits or they have sailed into our halo from galaxies that were swallowed up by ours.

The Central Bulge

At its edge the halo contains few stars, but toward the center it becomes more densely populated, and it eventually ends as it merges into the central bulge. The diameter of the central bulge measures approximately 15,000 light years. Here the density of stars is at its greatest in the entire galaxy. The stars are similar to those from the halo – old, orange-reddish, with a slightly higher percentage of heavier elements. The difference is that here there is also a lot of interstellar gas and dust and a lot of stars that are at the end of their lives – white dwarfs, neutron stars, and black holes. As recently as a decade ago, we imagined the central bulge as some sort of a massive globular cluster in which the stars were getting denser, the closer to the center we got. Observations in the last few years and the measurements of the positions of over 30 million stars in the central bulge increasingly point toward the fact that the stars are somewhat more condensed in the bar, which measures approximately 27,000 light years in length and is 7,000 light years wide. The bar is surrounded by a ring of gases in which most of the galactic molecular hydrogen is gathered. In the ring, we can witness the highest rate of star births from anywhere in the entire galaxy. Astronomers say that this would be the most visible part of our galaxy, if we could watch it from, say, the Andromeda Galaxy (Figure 4.21). The birth of new stars in the halo and the galactic bulge ended ages ago, except for in this ring and in the gas clouds surrounding the core itself.

Figure 4.21. The central bulge with the bar and the gas ring in which new stars are being born. This is how we would see the central part of our galaxy from the direction of the Andromeda Galaxy

Center of the Galaxy

At the very center is the galactic core. Observations with radio interferometers make it clear that the core is extremely small; in diameter it measures a mere 13 astronomical units. The core, though, is a strong source of infrared radiation, and in the surrounding space violent and extremely powerful energy processes take place.

To our great disappointment, thick clouds of gas and dust obstruct our view in visual light reducing the light by a trillion times. We have learned almost everything we know about the galactic core from astronomers who observe the universe in infrared light. But radio waves, X-rays, and gamma radiation also emerge out of this area.

In the 1930s, Karl G. Jansky (1905–1950) discovered that in the constellation of Sagittarius, exactly where Shapley predicted that the center of the galaxy would be located, lies a strong source of radio waves. In the beginning of the 1950s, astronomers used radio telescopes to discover (step by step) that the center has a very complex structure. Apart from the clouds of ionized gases they also recognized a small but strong source of radio waves from the area named Sagittarius A (Sgr A). This source is not a normal cloud of ionized gases, since a part of its radiation is of nonthermal origin. The radiation is emitted by electrons with great energy that move in a spiral fashion in a strong magnetic field. It has only been in the last few decades that research in all fields of electromagnetic radiation has revealed how truly complex the center of the galaxy is (Figures 4.22 and 4.23).

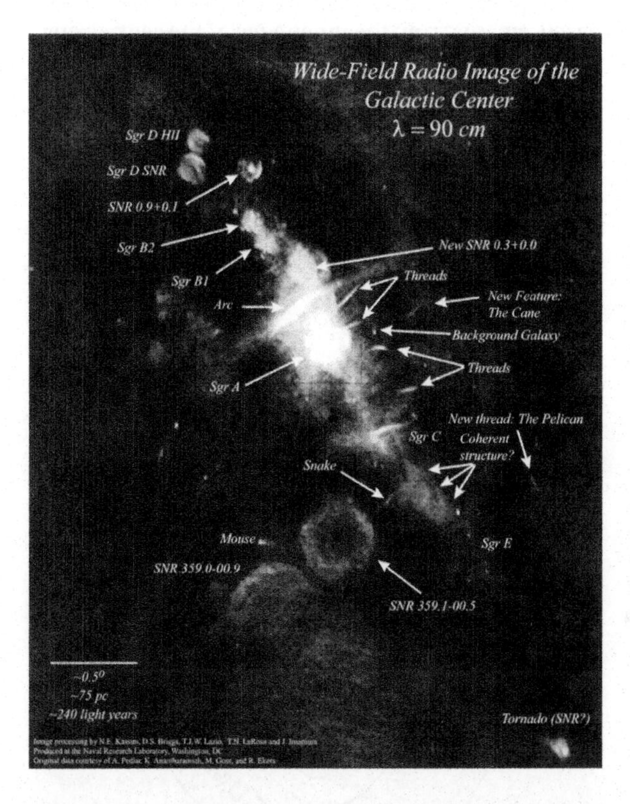

Figure 4.22. This radio image of the center of the galaxy covers an area 1,500 times 1,700 light years in size. We can see the entire diversity of the central area: numerous gas filaments and bubbles, the remains of supernovae with neutron stars or black holes in their centers, and even active areas in which stars are born. In the very center of the picture lies Sagittarius A, which is the strongest radio source on the image

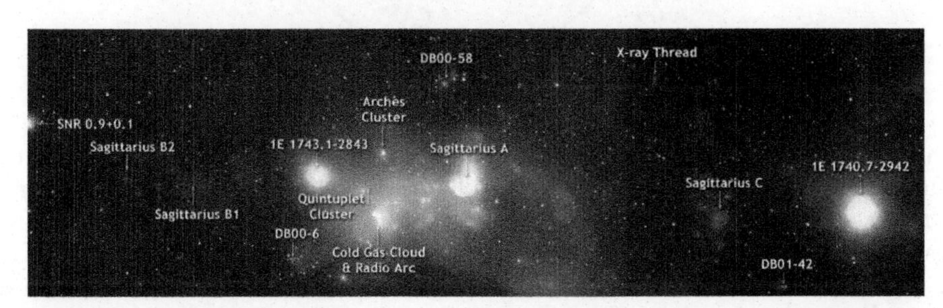

Figure 4.23. An X-ray image of the center covers an area 900 times 230 light years in size. On it are hundreds of stars, white dwarfs, neutron stars, and black holes, all buried in a cloud of gas with temperatures in the millions of degrees. Due to the frequent explosions of the supernovae in this area, the gas is enriched with heavy elements that will slowly be spread around the entire galaxy. In this way, the active galactic center influences the development of the entire galaxy. For comparing this image with the radio wavelength image (above) use Sagittarius A (Sgr A) and Sagittarius B2 (Sgr B2) as guiding points

As we have already mentioned the stars in the central bulge get denser and denser, the closer to the center we get. In 1982, astronomers discovered that at a distance of 15 light years from the core lies a ring of warm (a few hundred K) molecular hydrogen, which is mixed with dense areas of ionized hydrogen (H II areas). Called the Galaxy Central Ring (see illustration below), this structure includes gases and dust that have the mass of some 10,000 Suns, and their luminosity is 20 billion times greater than the luminosity of our Sun! It is interesting to note that the ring has a sharp inner edge, 5 light years from the center. The clouds of gases in the central ring are in a strong turbulent motion and move faster the closer they are to the center. The average speeds are somewhere around 100 km/s. Within this area there are no areas of denser gases and dust, so astronomers have named it the Central Cave. There are only numerous scattered stars, from which we can only see the infrared light of the brightest giants. Spinning gas jets emerge from the inner edge of the central ring, moving at speeds of up to 1,000 km/s. The gases spiral toward the center, where they are pulled by strong gravity. Here they heat up considerably, with an additional helping of energy from the mighty stellar winds of the surrounding giants.

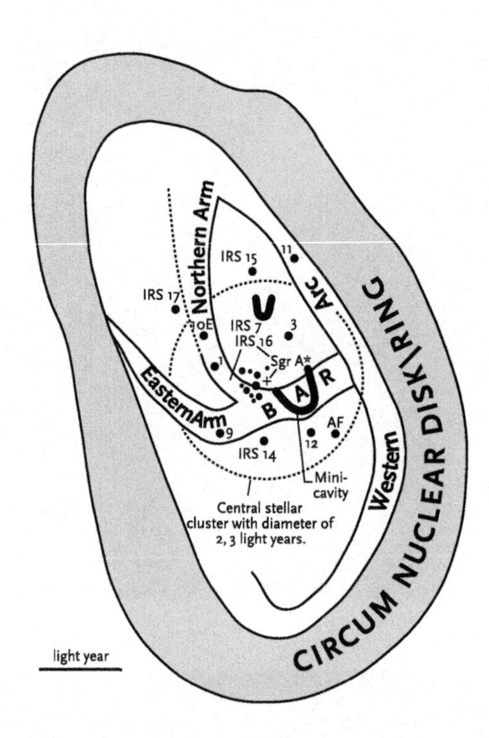

All of this speaks to the very dynamic events in the center of the galaxy. But the greatest mysteries can be found in the heart itself, in an area less than 1 light year across. Only recently have astronomers with infrared interferometers managed to break through to this area and can now view individual stars in the central cluster. At the moment they can see only the giants, for only their light can penetrate through the thick layers of gases and reach us. The stars, moving at speeds of up to 1,000 km/s and even showing signs of acceleration, are an excellent indicator of the shape and size of the gravitational potential in the center. Astronomers have noticed that in the very center lies a small yet extremely powerful source of infrared and radio emissions, known as Sagittarius A* (asterisk denotes that it is a part of the larger source Sgr A, which was mentioned earlier). This extremely unusual object covers a very small area, one that measures a mere 45 astronomical units (roughly the size of Uranus's orbit around the Sun), while its mass has been estimated to be approximately 3.7 million solar masses! Out of all objects known to astronomy only black holes have such a great mass gathered into such a small area. Yes, there is no doubt. In the center of our galaxy lies a black hole that gobbles up everything that crosses its path. All electromagnetic radiation that reaches us from there comes from the spinning of the accretion disk, a spiral vortex that emerges when a black hole pulls toward it the gases from the central ring and the surrounding stars (Figure 4.24).

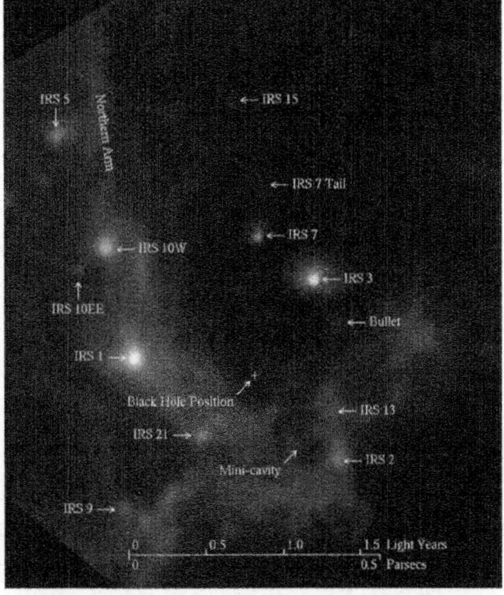

Figure 4.24. The central part of the galaxy shown in infrared light; the image was taken with the 10-m telescope called the Keck II. It is orientated the same as the illustration on the previous page

However, this is not all! Observations with the X-ray satellite Chandra revealed that, in the vicinity of the galactic black hole there are a myriad of double X-ray stars in which one of the two stars is a neutron star or a black hole. Astrophysicists have estimated that in the direct vicinity of the galactic black hole there are at least another 20,000 stellar black holes!

Unfortunately, the resolution of today's interferometers is not great enough for us to see directly the vicinity of the black hole or within the black hole itself. When advances in technology allow interferometric observation in the submillimeter wavelength spectrum, we will be able to see the event horizon of our galactic black hole. And this will be an event that should not be missed (Figures 4.25 and 4.26).

Figure 4.25. An area the size of a few light years in the central cluster of stars surrounding the center of the galaxy was photographed in infrared light at the European Southern Observatory (ESO) with the 8.2-m telescope. The marking in the center indicates a black hole

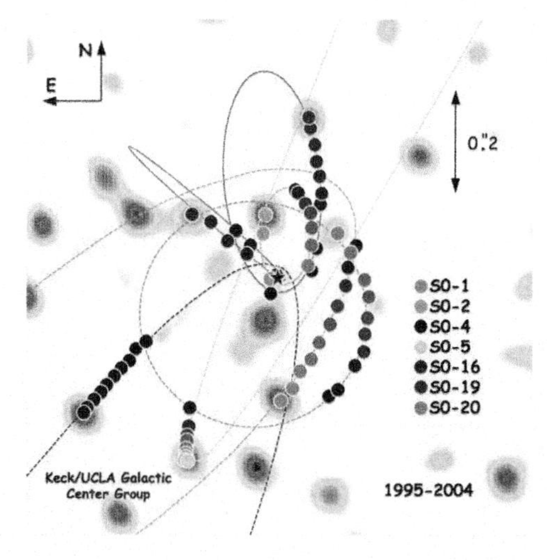

Figure 4.26. The black hole in the center of our galaxy is not a fairy tale or a product of the rich imagination of astronomers. Our knowledge of its presence is based on the observations of movements of the stars in the cluster within the central 1.0 ×1.0 arcsec of our galaxy. These orbits provide the best evidence yet for a black hole made up of an astounding 3.7 million solar masses

Dark Matter in Halo

If in the center of our galaxy unusual events happen that we find hard to understand but can still be explained with astrophysical theories, the halo holds even more interesting and to date impossible to understand surprises. Something is there that we cannot see, something that we cannot even theoretically explain. But we know it is there and that, with its gravitational pull, it influences the evolution of the galaxy and in fact the entire universe. This is the still mysterious dark matter (Figure 4.27).

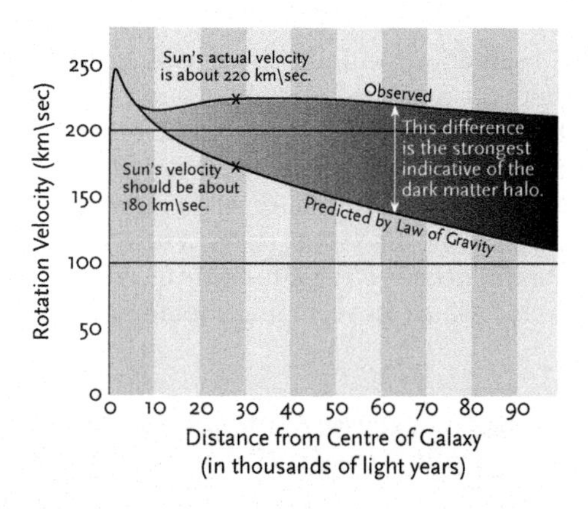

Figure 4.27. This diagram shows measured orbiting speeds and their expected values. Astronomers blame the discrepancy on mysterious dark matter

In the 1930s, the Swiss astronomer Fritz Zwicky and the Dutchman Jan Oort suspected that the universe includes much more mass that it appears to hold at first glance. Zwicky studied the large cluster in the constellation Coma Berenices and discovered something unusual: the speeds of galaxies within the cluster are too great, and there is not enough visible matter for gravity to hold together the cluster members that should, taking all physical laws into account, have dispersed ages ago.

Oort compared the number of stars in the vicinity of the Sun and their orbiting patterns around the center of the galaxy and reached a conclusion similar to that of Zwicky. Our galaxy also hides much more than meets the eye. Only in the last decades has it really become clear how great the divide truly is between what we can see and what we cannot see. We know that this something exists, because it exerts a gravitational force on normal matter. Astronomers call this invisible substance *dark matter.*

Let's take a quick look at how Oort reached his conclusions. He measured the mass of the galaxy at various distances from the center and determined that it is extremely large and getting even bigger when we measure it further away from the galactic disk. How do we even measure the mass of an object such as our galaxy? We analyze the speed of galactic objects (stars, nebulae, and interstellar gas clouds) in relation to the center of the galaxy. We should remember Kepler's laws. The faster a star circles at a certain distance from the center, the greater the mass within the orbit. The same law is used for determining the mass of double stars or the central star around which planets orbit. The speed of the planets depends on their distance from the star and its mass. When we measure the Sun's speed around the galactic center and determine the mass within our orbit, we get 100 billion solar masses, which does not agree with the estimation of the mass of the visual stars and gas clouds. But even more unusual things happen further away from the Sun.

We would expect that stars move slower and slower the further away they are, as predicted by Kepler's law, since most of the galaxy mass should be gathered in the central bulge. However, we notice that their speed remains the same with distance. This shows that a large part of the galactic mass is not gathered in the vicinity of the center. If we continue with our measurements even further from the center and measure, for instance, the movements of the globular clusters in the halo or the interaction of the entire galaxy with the Andromeda Galaxy and other galaxies, we get the mass of over a thousand billion Suns. The ratio between the mass and luminosity is thus between 20 and 30 to 1 and cannot be ascribed to the stars that we cannot see – not even the countless small ones.

The discrepancy between the mass that is measured and the mass that we can see (that radiates) is so great that it can mean only one thing: in the halo and among the globular clusters there have to be between all the visible stars great quantities of some hidden and invisible matter that does not radiate in any part of the electromagnetic spectrum that we can measure. This is most certainly not normal star material, nor is it gas, because that could be detected in the ultraviolet light or in radio waves.

The dark matter also represents most of the mass in other galaxies, galaxy clusters, and the universe in general. And what is this dark matter? Its true nature is still unknown, but there are plenty of theories as to what it is composed of. Some think that it is comprised of brown dwarfs, unsuccessful stars with masses not big enough to sustain nuclear reactions in their centers. Others are of the opinion that these are white dwarfs that have gone extinct ages ago or that they are massive black holes. The boldest theorists speak about an unknown exotic matter that we don't know anything about.

Can we say anything with certainty regarding the beginning, evolution, and composition of the galaxy if we do not know its most massive component? We have to be aware that we are not looking merely for a few small stars that have hidden themselves from our telescopes but something that most of our universe is made of.

The Origin and Evolution of Our Galaxy

When we start researching events as complex as the origin of the galaxies, we soon find ourselves on slippery ground. There are just too many questions that we have to answer with, "We cannot say with certainty." Galactic astronomy does not have a long tradition; we only first learned about galaxies 80 years ago, and the first serious research started only half a century ago. In the last decade, new observational data have been accumulating with such speed that theoreticians can hardly keep up with it. The script of the origin and evolution of the galaxies is far from completed.

Did the Milky Way Galaxy emerge from a single gas cloud that was shrinking and merging into stars or by the merge of a number of smaller galaxies? Up until 2006, astronomers were opting for the first choice. Then astronomers at the European Southern Observatory conducted an extensive study of the stars in the disk and the central bulge and realized that the stars in each area differ greatly one from another. The disk and the central parts were developed separately, and the stars did not mix. Before they thought that the stars were born in the disk and then slowly moved to the core; today, it is the general opinion that all the stars in the center were created at the same time and from the same matter. If the galaxy was formed through the merging of a number of smaller galaxies, the center would also include stars from disks. On the other hand, our galaxy is at this time adding to it at least three satellite dwarf galaxies, as you will see in the next chapter. In the future they will merge completely with our galaxy. Why couldn't this have happened before?

Because we do not even know what dark matter is made from, we cannot state with certainty that we know how it influenced the origin and evolution of the galaxy. And dark matter is its predominant component.

Is there a black hole in the center of the galaxy? Today we can answer this question affirmatively, but only a few years ago astronomers would only shrug their shoulders. But we still do not know for certain when and how was it created and how its presence has influenced the evolution of the galaxy.

For a long time astrophysicists were baffled as to why the stars in our galaxy were from only two populations and not more. And why do the young stars have a similar composition to the much older stars, even though the first were created from gases and dust that were enriched numerous times and should therefore include much more heavier elements than their predecessors. This question has been, to a certain extent, answered in the last 10 years. In the galaxy halo there is still a lot of hydrogen that can only be spotted if it is gathered into a cloud. And such an enormous cloud, which is made up of a few million solar masses, was discovered in the mid-1960s, and in 1999 its distance and position were measured. It was shown that the cloud above the galaxy, between 10,000 and 40,000 light years away, is coming closer at a speed of 160 km/s. Such clouds supposedly ensure a fresh influx of hydrogen that rains down on the galaxy and brings new matter without adding any heavy elements; thus the concentration of heavier elements does not change significantly.

We can talk about the age of the galaxy with greater certainty. It can be estimated on the basis of the age of its stars. The Milky Way was born less than a billion years after the Big Bang. The first stars definitely emerged from clouds of hydrogen and helium. The more massive ones evolved quickly and in the end exploded as supernovae. During their life and especially at the end of it, they scattered a significant part of their mass into their surroundings in the form of planetary nebulae and novae and supernova remnants. This is the matter from which new stars were born, enriched by heavier elements.

Today we can find all types of stars: giants and dwarfs, hot and cold, young and old, stable and variable. The smallest have a mass that equals a mere 0.08 of the Sun's mass, while the largest can be more than 100 times more massive than the Sun. However, there are not many of the latter ones left, for these have a very short lifespan. Most stars are faint, cold, orange-reddish stars from the main sequence of the H-R diagram. A typical star in the Milky Way is thus very similar to our Sun: small, faint, and not very hot.

As you might have gathered by now, mass is the most important factor in the evolution of stars and galaxies. If, for instance, our galaxy consisted merely of stars that had approximately 10 solar masses, it would be extremely bright, but it would shine for only 10 million years, and then the stars would explode one after the other as supernovae. After a few cycles of such galactic recycling, a dark galaxy would remain, made from only stubs – neutron stars and black holes. If it was comprised of only small stars, the size of, say, 0.1 solar mass, it would be extremely faint, but it would shine for at least a few hundred billion years. How long our galaxy will still shine depends on how many new stars are being born in it. Today we estimate that approximately 3–5 stars are born every year. Such a number of newly born stars appears to be small at first glance, but it is enough to keep the galaxy shining with the same brightness for billions and billions of years to come.

A part of a star's mass stays forever trapped in the star remains, whether it is a white dwarf, a neutron star, or a black hole, and will never be used again to create new stars. This means that there is less and less gas available and at some point – perhaps in 20 or 30 billion years – there will be no hydrogen left and stars will no longer be born. The brightness of the galaxy will start slowly but surely to dim. In the far distant future, in many billions of years, only dead stars and long-lived red dwarfs will be found in the galaxy. But in the end even these stars will switch off their nuclear reactors, and the galaxy will turn into a dark disk of extinguished stars.

This would be the certain scenario if our galaxy were the only one in the universe. But because it is a member of the Local Group and is gravitationally linked to other galaxies, unpredictable things can happen, such as a collision with the large and massive Andromeda Galaxy. We will discuss this in greater detail in the next chapter.

Satellite Galaxies and the Local Group

So far we have discovered approximately 20 satellite galaxies that are gravitationally linked to our galaxy and circle it. The biggest are the Large and Small Magellanic Clouds that are 180,000 and 210,000 light years away and are 20,000 and 15,000 light years across. All the others are dwarf elliptical galaxies. The closest satellite galaxy to us is 40,000 light years away (the dwarf in the Canis Major), while the furthest is 880,000 light years away (Leo I) (Figure 4.28). The smallest four measure only 500 light years in diameter.

Figure 4.28. Our most distant satellite galaxy is the dwarf Leo I, which is hidden in the light of bright Regulus

Collisions between galaxies are much more frequent than we assumed in the past. Our galaxy has already experienced a number of close encounters with neighbors in the Local Group, such of which caused observable damage. If the Milky Way encounters a small dwarf, it will just suck it in, and its stars will disperse throughout our disk.

In 2005, astronomers discovered a large and elongated group of faint stars that are moving at a right angle to the plane of the disk, and they could find no way of incorporating this into any galaxy model. Further research has shown that it was a dwarf galaxy in the Local Group that was pulled from its orbit by gravitation and will in the future merge with our galaxy. At the moment the stars are approximately 30,000 light years away and can be seen in the direction of the Virgo constellation; that is why it became known as the Virgo stellar stream. So far we have discovered a number of similar stellar streams.

The latest observations show that the globular cluster M 54 in Sagittarius is most probably only the brightest part of a dwarf galaxy that is a member of the Local Group. It seems that this dwarf will also merge with our galaxy sometime in the future. The current distance is 88,000 light years. M 54 is on the visible border in binoculars. Take a look at it when viewing conditions are very good.

A similar situation exists for the globular cluster M 79 in the constellation Lepus, which will (or it might have already begun to) integrate into our halo of globular clusters from the dwarf in the Canis Major. Together with it we should also get the globular clusters NGC 1851 (seen in the Columba constellation, mentioned in this book), NGC 2298 (seen in Puppis), and NGC 2808 (seen in the Carina constellation). M 79 is clearly visible with binoculars, and as you look at the "boring" faint spot of light, remember what's happening there.

A similar fate also awaits the Magellanic Clouds (Figure 4.29). The large one, on its elliptical orbit, has already come so close to our galaxy that gravity has pulled off a part of its matter, which can today be seen across half of the sky, forming what we now call the Magellanic Stream. It is interesting to note that roughly in the same plane as the stream are at least seven other dwarfs that surround the galaxy. Maybe these are the offspring of the same force that caused the emergence of the stream.

Figure 4.29. The Large Magellanic Cloud is our biggest satellite galaxy and can be seen with the naked eye from the southern hemisphere

Our galaxy is not only gravitationally linked with its satellite galaxies but also with the smaller group of galaxies that we call the Local Group. This group currently consists of 47 known members, with 17 still awaiting confirmation of their membership. These are mainly small galaxies, and the measurements of their distances and speeds are not precise.

The largest galaxy in the Local Group is Andromeda (M 31), followed by ours and the spiral galaxy in the Triangulum (M 33). The remaining ones are either elliptical dwarfs or irregular galaxies. The smallest (named GR8, found in the constellation Virgo) is only about 200 light years across.

In their slow gravitational dance around a common center of gravity, the other galaxies in the Local Group also experience close encounters and even collisions. It has been known for a while that Andromeda and our galaxy have been hurtling toward each other with a speed of approximately 500,000 km/h. It looks like they are going to collide in approximately 2 billion years. Computer simulations of the event have shown that there is a 12% chance that Andromeda will take with it our Solar System. But the change of address will only be of a "temporary" nature, for 5 billion years after this event the two galaxies will merge into a single giant elliptical galaxy (Figure 4.30).

Figure 4.30. Collision of galaxies named II Zw 96

Other clusters close to us are the smaller Maffei 1 (5 members, 10 million light years away), a group in Sculptor (14 members, 10 million light years away), the group around M 81 (19 members, 12 million light years away), and the group around M 83 (14 members, 15 million light years away). The most interesting of them all has to be Maffei 1. Its galaxies used to be members of our Local Group, but it seems that when they experienced a close encounter with Andromeda they were catapulted away from us.

Oort's Legacy

Jan Oort (1900–1992) became a significant part of astronomical history in 1927 when he, together with the Swedish astronomer Bertilo Lindblade, determined that our galaxy revolves around its axis. In the 1930s, he concluded from the measurements of the speed of stars close to the Sun that the galaxy has to consist of more mass than we can see in the stars and nebulae. In 1940, he conducted research on the elliptical galaxies NGC 3115 and NGC 4494. He found that these two galaxies also must contain great quantities of unknown matter, which came to be known as dark matter. The nature of dark matter remains one of the greatest puzzles of modern cosmology.

Oort's greatest contribution to twentieth-century science, however, was in radio astronomy, the beginnings of which go back to the 1930s. That was when Karl Jansky first observed radio waves that came from the universe. The visionary Oort quickly realized that radio waves were an ideal tool for studying the structure of our galaxy. In contrast to visual light, radio waves travel undisturbed through interstellar dust and gas.

Oort wondered to which radio wavelength the radio telescopes should be tuned into, so that they will be best suitable for observing the Milky Way. In our Galaxy, hydrogen is the most common element; thus he asked one of his students, Henk van de Hulst, to calculate the energy levels of the hydrogen atoms at which photons would be emitted as radio wavelengths. In 1944, van de Hulst foresaw the existence of the today famous 21 cm spectral line of radiation of the neutral hydrogen. Soon after World War II, the Leiden astronomers reworked the 7.5 m German radar antenna into a radio telescope and started with their observations. In 1950, they finally found the 21 cm spectral line of hydrogen, and a few years later they already mapped the first radio chart of the spiral structure of the Milky Way.

Oort also studied comets and foresaw the existence of a giant cloud of comet cores that circle the Sun at a distance up to 1 light year away. Oort's Cloud, as we call it today, has not been directly seen yet by anyone, yet few doubt its existence.

Oort helped establish the European Southern Observatory, which has gradually developed into one of the most important global centers of optical astronomy. A participant on one of the first expeditions to Chile later said, "Oort would lie an entire night on his back on the wet grass, even though he risked catching pneumonia. He was completely taken over by the Milky Way. The man who has been revealing mysteries of our galaxy through radio waves for the past 25 years could now see it for the very first time in all its glory and beauty under the dark South American sky."

CHAPTER FIVE

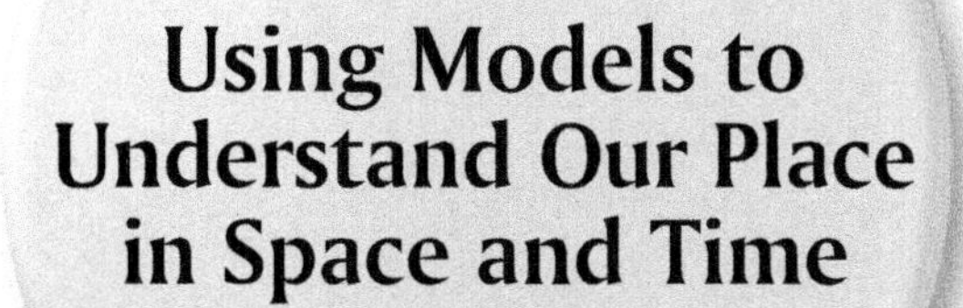

Using Models to Understand Our Place in Space and Time

People have always been very keen on making models. Making scaled-down things that are perfect replicas of the originals is an interesting and entertaining hobby and at the same time we can also learn a lot from it. We make model airplanes, ships, rockets, and cars. But so far nobody has ever made a model of the universe, one in which we could see our Solar System with the Sun and planets, its position within the galaxy, the Local Group, other clusters of galaxies, and further - basically the entire universe as known today.

If you are wondering why, the answer is very simple. It is impossible to make such a model. Of course, we can make smaller scaled-down planets, and they look quite pretty and interesting when lined up next to each other on a table. But if we try to accurately model our Solar System, we quickly realize that it has grown beyond the table, and not only the table but far beyond the borders of our room. For in a true model we have to scale down all dimensions by the same proportion, not just size. Let's take a closer look at this.

Models of Our Solar System

Let us start by making the smallest possible model of our Solar System in which the planets would still remain visible and then we will have a look at what happens to the rest of the universe. We will scale down Mercury, which is the smallest of all planets (with a diameter of 4,880 km), to the size of a grain of sand, half a millimeter across. This means that we have to scale down all of the sizes and distances in the universe by the same factor. In this way we will get a model in the scale of $1{:}10^{10}$. In such a model Earth is a small ball with a diameter of 1.3 mm. The Moon, which is slightly smaller than Mercury, lies 3.8 cm away. The Sun is reduced to a ball with a diameter of 14 cm and is 15 m away from Earth.

B. Kambič, *Viewing the Constellations with Binoculars*, Patrick Moore's Practical Astronomy Series,
DOI 10.1007/978-0-387-85355-0_5, © Springer Science+Business Media, LLC 2010

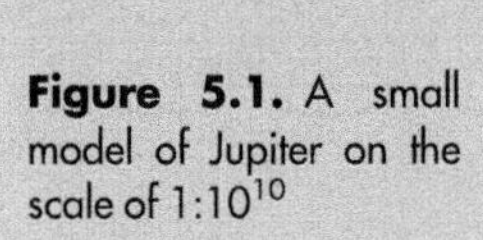

Figure 5.1. A small model of Jupiter on the scale of $1:10^{10}$

It should already be clear that we cannot put such a model in a room, not even in a planetarium. And so far we have only covered the distance between the Sun and Earth! In this model, Jupiter is the size of a marble (diameter of 1.4 cm) (Figure 5.1) and is 78 m away from the Sun. Saturn is slightly smaller and lies 143 m away. In order to reach Neptune, which is the furthest planet in our Solar System, we would need to move 450 m away, and it would only be the size of a small pea (5 mm in diameter).

If we truly made such a model somewhere in nature and looked from Neptune back toward the Sun, we would be astonished at how empty our Solar System is and how strange gravity is that keeps it together (Figure 5.3). Somewhere far away from us, almost half a kilometer away, lies a 14 cm ball, which somehow manages to control Neptune with its gravitational pull and makes the planet circle it. All of us who were watching the Mercury's transit in front of the Sun on May 7, 2003, or Venus's transit on June 8, 2004, had similar feelings (Figure 5.2).

Now let's take a look at the closest star to the Sun. Alpha Centauri is slightly larger than our Sun, but in this scale it is 4,000 km away! In a model in which we can hardly see the smallest planet in our Solar System, the closest star is unimaginably far away.

Figure 5.2. When Venus appeared to cross the surface of the Sun, a number of surprised observers said, "Is Venus really this small compared to the Sun?!" In reality, it is even smaller, for this is not a true model. The Sun in this picture is much further away from us than Venus. In order to obtain the true size relation between the two bodies, we need to enlarge the Sun by 3.6-fold

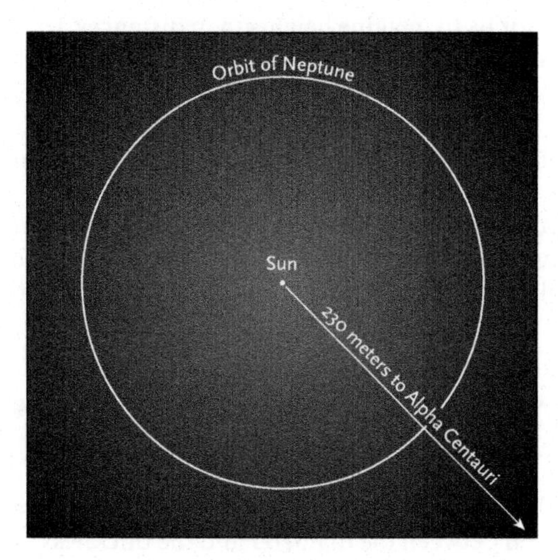

Figure 5.3. If Neptune's orbit is considered the outer boundary of our Solar System and measures 5 cm on the illustration, the closest star is 230 m away! The bright dot in the center represents the Sun's light, for the Sun is too small to be seen (at this scale it is only 0.008 mm in diameter). The stars in a typical galaxy, such as ours, are very far apart when compared to their sizes

Now we can truly be astounded! Our part of the galaxy is even emptier than our Solar System. And in fact, there is no great crowd elsewhere. In this scale it is approximately 25 million km to the center of the galaxy, and the entire galaxy would measure 130 million km! Few can state that they can imagine such distances. A model in the scale of $1{:}10^{10}$ is therefore suitable only for depicting our Solar System.

Models of our Galaxy

In order to make a model of the galaxy that we could relate to, we have to change the scale. Let's imagine that our galaxy is 1 km across. It is possible to imagine a distance over which we can walk in 10 min. With such a model we would use a scale of $1{:}10^{18}$, and in accordance with this we will have to scale down all other sizes and distances. In this case, Neptune's orbit measures only one hundredth of a millimeter in diameter, while the Sun's diameter is already in the kingdom of nanometers!

Where can we expect the first stars to be now? Alpha Centauri is 4.3 cm from the Sun. The other stars can be found scattered around in all directions. The brightest star in the night sky, Sirius, is 8.5 cm away. Vega is 25 cm away, Regulus 76 cm, while Polaris is 4 m away. The stars in Orion's Belt are between 8 and 13 m away from us.

Among the brighter stars in our sky the furthest away are the Delta and Eta Canis Majoris. In our model the first is 17 m away, while the second is 30 m from the Sun. We can see that the brighter stars are also the closest, even though this is not a rule. It is 250 m from the Sun to the center of the galaxy, and the central bulge measures 150 m in diameter. The galaxy disk is on average 20 m thick, but in the place where we live it is only 5 m thick! The halo is spreading around the galaxy. Its denser part, in which one can find most of its globular clusters, would in this model expand in a ball that would be 2 km in diameter, while individual clusters could be found up to 2 km from the center (i.e., in a ball 4 km in diameter). The inner part of the halo has approximately 150 globular clusters that we can imagine as shining meter-size globes. The most beautiful and among northern observers' favorites is M 13 in Hercules. In this model, M 13 is 240 m from the Sun, and its diameter is as large as 1.4 m. Because it lies high above the equatorial plane of the galaxy, it is more than 200 m above the level of the model. (Because it is harder for us to imagine heights than distances, try to think of the known heights of nearby hills, skyscrapers, or houses.)

We should not forget that in this scale even the largest stars are microscopically small. If we would rise above the model of the galaxy, we could see a nice, bright barred spiral galaxy. However, what we see are not the disks of individual stars but merely their light. The distances between the stars are far greater than their sizes.

Let's say that we can somehow imagine our galaxy at this scale, but once we search for the closest big galaxy - M 31 in the constellation of Andromeda, it is nowhere to be seen. We find it as far as 28 km away. Our model in the scale of $1:10^{18}$ fails us at this stage, and we have to scale down everything once more.

Models of the Universe

Now let us imagine that the diameter of our galaxy measures a mere 12 cm, exactly what a CD measures; even the proportion between the diameter and the thickness is about right (Figure 5.4). We have just arrived at the scale of $1:10^{22}$. If we wanted to use a CD to create a model of the galaxy, we have to stick in its center a 1.5 cm ball, for instance from Styrofoam, that will represent its central bulge. Into such a down-sized scaled galaxy there is no use in even discussing where individual stars are located. At this size we can only differentiate between the core and the spiral arms that are wrapped around it. It is a mere 2.6 cm from the Sun to the center of the galaxy. The spheroidal halo of the globular clusters is similar to more than hundred scattered particles of sand in a ball with a diameter of 24 cm. The previously mentioned M 13 in Hercules is, in this model, only one-tenth of a millimeter big and lies 2 cm above the disk.

First, let us construct the Local Group. The Andromeda Galaxy, which measures 24 cm, is 2.9 m away from our galaxy. The spiral galaxy M 33 in the Triangulum lies in the same direction and is a bit further away. Approximately 50 dwarf galaxies, which could be shown by larger cotton ball puffs, are scattered around. In this scale, most of the members would be gathered in a ball with a diameter of 3 m, while the one furthest would be as far as 8 m away. But note that a few of the closest groups to us can be found at a mere 10–12 m away. And this is a distance that is comparable to the size of our Local Group.

Figure 5.4. The model of our galaxy, made from a CD. Here we can see only the distribution of light in the galaxy and not individual stars, which at this scale are already in a subatomic world. The arrow points toward the place where our Solar System is located

As we have seen our galaxy is relatively empty, for the distances between stars are much greater than their sizes. In opposition to this, the distances between the galaxies in the group and the distances between the groups are more comparable to their sizes. This should make it easier to understand why collisions among stars in our galaxy are rare, but between galaxies they are more frequent. The other clusters of galaxies are scattered around in all directions. Their sizes are different – from smaller groups, such as ours, to truly giant clusters with hundreds or even thousands of members.

For an example, let's take a look at the location of one of the closest bigger clusters. Among amateurs the best known is the Virgo Cluster (Figure 5.5), which consists of some 2,000 members. In this model, the cluster is 60 m away from us and is 7 m in diameter.

Clusters and super clusters are not randomly scattered around the universe, as it was once thought. By 1989 it was known, thanks to numerous observations of the sky and the charting of galaxies, that the clusters and super clusters form a sponge-like pattern, with larger concentrations that have been dubbed "Great Walls" and smaller filament-like structures found in between these (Figure 5.6). If we could take a look at the universe from far away, we would see the Great Walls as walls that surround unimaginably large empty spaces or voids. In our model, these empty spaces would be approximately 100 m wide (the large group in Virgo measures a mere 7 m).

Cosmologists estimate that size of the universe today is 93 billion light years. This means that our model must be 93 km across. Can you imagine that?

We had to make three different models to get from being able to picture our Solar System to being able to picture the entire universe. All three of them were static, unmoving. We reduced the space but not also the fourth dimension – time. We know that the entire universe is in motion. The moons circle

Figure 5.5. On images of the galactic clusters it is clearly visible that the distances between the individual galaxies are comparable to their size. The typical distance between them is 10 galaxy diameters. In opposition to this, the stars in the vicinity of the Sun are as much as 100 million average star diameters apart

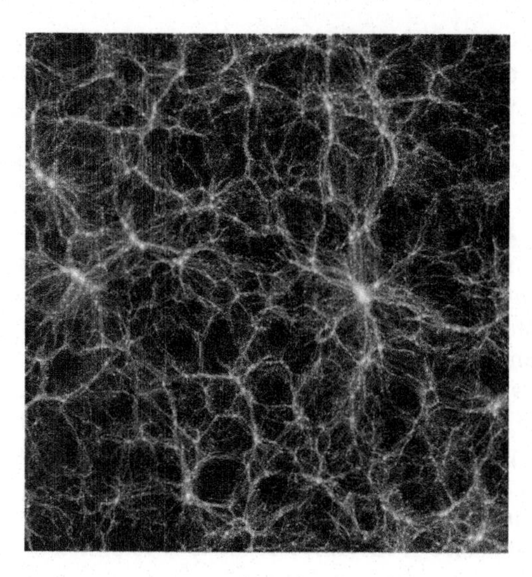

Figure 5.6. If we took a look at the universe from far away, we would see that the galactic clusters (an individual cluster is represented by a dot in this computer graphic) are not randomly scattered throughout the universe but form a sponge-like pattern; they are spread out as walls around vast empty spaces

the planets, planets circle suns, stars go around the centers of galaxies, galaxies circle inside clusters and super clusters, and so on. The entire universe is expanding, and the clusters are moving away from each other, just as if they were placed on an expanding balloon. At whatever galaxy we look, it seems, the other galaxies are racing away from it. And, as the newest observational data show, the expansion of the universe is accelerating!

Timeline of Our Universe

Modern cosmology really began in the twentieth century, when Albert Einstein published the General Theory of Relativity, when Edwin Hubble discovered that the universe is expanding, and when nuclear physics and quantum physics emerged. Only in the last few years have we obtained real observational data that give us the basis for today's picture of the birth and evolution of the universe (Figure 5.7).

Our expanding universe started from a gigantic explosion of space and matter known as the Big Bang, which formed a dense, hot, and nontransparent soup of matter and radiation that filled all of space. However, the young universe quickly expanded and became less and less dense, and at some point it became transparent for its own radiation, which was scattered across space.

The distant echo of this important moment can nowadays be "seen" in the cosmic microwave background radiation (also called the relic radiation) that permeates space and was discovered by Arno Penzias and Robert Wilson in 1965. In the microwave background radiation, we can see light that was created a mere 380,000 years after the Big Bang. This is important because it gives us some key cosmological information. That light has preserved a distribution sample of the unhomogeneous nature of the young universe, from which galaxies and galaxy clusters emerged later on. This is also as far back in time as astronomers can see into the young universe.

In the beginning of the 1990s, cosmologists used the COBE satellite to chart the microwave background radiation and discovered small yet important differences in the temperatures in various directions. COBE could measure extremely precisely the small differences in radiation, but its angular resolution was very poor. Cosmologists were baffled by what might still be hidden in the finer spatial distribution of the microwave background radiation. Was it possible to make conclusions concerning the configuration of the space–time and the conditions in the young universe based on what they were seeing?

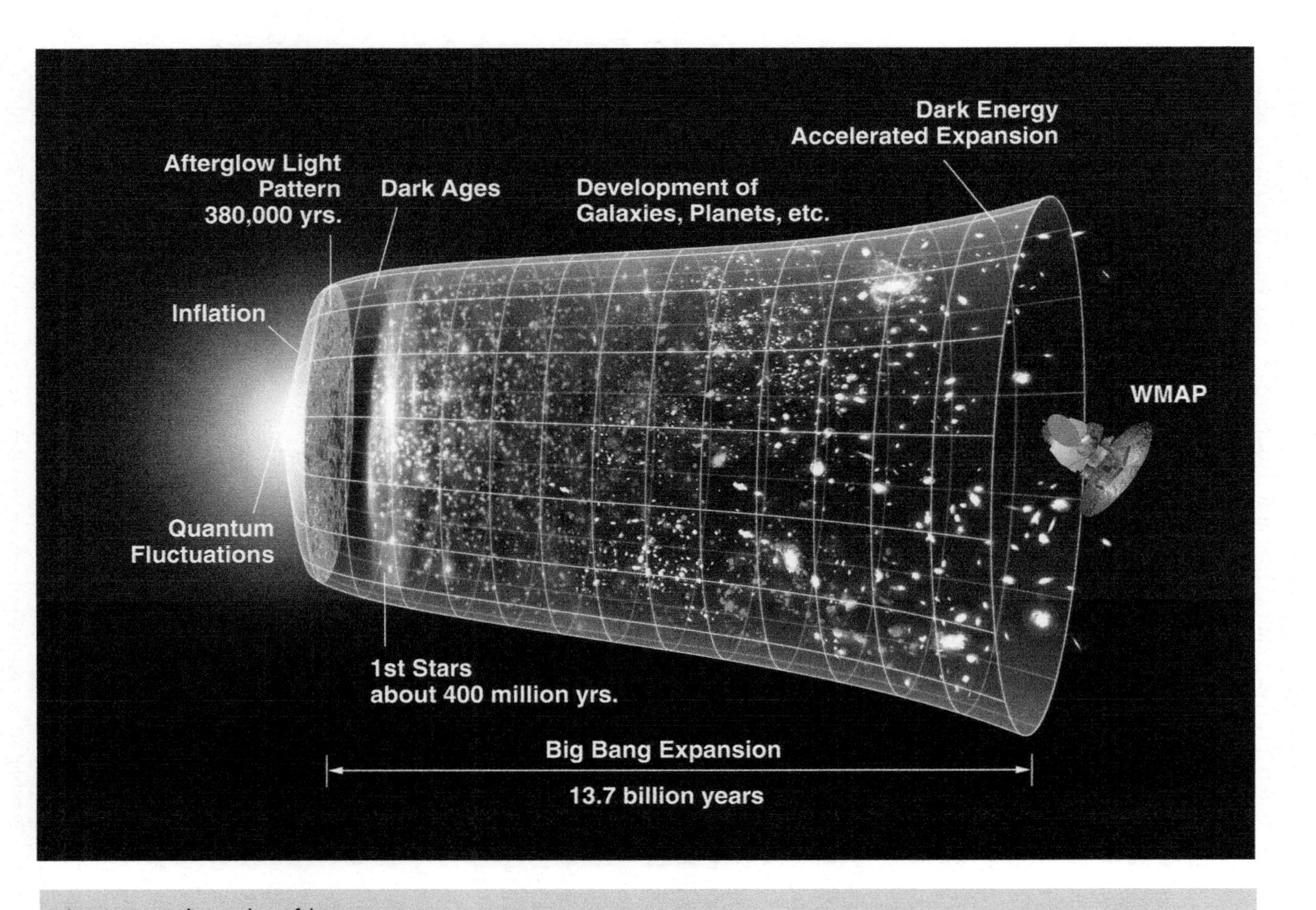

Figure 5.7. The timeline of the universe

The beginning nonhomogeneity that emerged at the birth of the universe should also be apparent in the microwave background radiation. That is why NASA launched the WMAP (Wilkinson Microwave Anisotropy Probe) satellite in mid-2001. The first results of this space observatory were published in early 2003. WMAP measured the spatial discrepancies in the temperatures of microwave background radiation much more precisely than COBE, and cosmologists were able to verify or adjust some very important theories about the universe. They now believe:

- The universe is 13.7 billion years old. The deviation of this measurement is less than 1%.
- The universe is composed of 4.6% ordinary matter, 22.8% an unknown type of dark matter, and 72.6% of mysterious dark energy.
- The model of cosmic inflation of the young universe is almost certainly correct.
- The microwave background radiation was scattered around the universe some 380,000 years after the Big Bang.
- The first stars appeared in the universe some 200 million years after the Big Bang.
- The universe will continue to expand ever faster into eternity.

Now let's have a look at what our dynamic universe would look like if we scaled down the duration of the events by the same factor as the spatial dimension – by a factor of 10^{22}.

All 13.7 billion years of the history of the universe would roll in front of our eyes in a mere 4/100,000 (0.00004) of a second! That is way too fast to catch any events. If the space appears to be incredibly empty, the time compressed into the same scale appears unbelievably fast. In the same way as we had to use different scales for modeling the universe, we have to apply the same solution with time.

So let's change the scale of our model of the evolution of the universe so that the 13.7 billion years lasts 1 year. In this case we need to contract time by the factor of 1.37×10^{10}.

A great number of important things happened during the first moments on January 1 of the Big Bang. Whole books have been written on just the first three minutes (of the real time) of the universe. In our model everything happens in a really, really short period of time. Soon after the Big Bang, the universe spreads out into an unbelievably almost uniform sea of hydrogen and helium. It cools down, and at some moment becomes transparent to its own radiation. This is what we see when we look at it on the images from the COBE and WMAP satellites, when we look at the cosmic microwave background radiation. We are still on day 1, and only 15 min have passed.

The universe keeps expanding and is becoming increasingly cold and less dense. On January 5, the first generation of stars appears. The most massive among these start producing heavy elements in their cores, and within three hours they have spread these across space through giant supernovae explosions. These elements mix with the helium and hydrogen and become a constituent part of the next generation of stars.

The first galaxies, or at least the first clumps that would become galaxies, appear at the end of January. Soon space is filled with them, and the dark matter controls their expansion. The previously mentioned sponge-like structures appear. The universe keeps expanding and evolving. If we were to find ourselves in the region of our galaxy at the end of July, we would not encounter anything familiar. But, in the place where our Solar System would later develop, a huge, massive, anonymous star shining among the gases and dust would be found in the last phases of its life. In August, it would explode as a supernova, and the surrounding gases and dust would be thoroughly mixed and enriched with heavy elements.

At the beginning of September, our Solar System starts developing in the middle of the denser part of the nebula. Already on September 3 a newly born star – our Sun – appears in the center. By September 7, the protoplanets clear their surroundings. This is the day in which we can say that Earth is born. By September 23, it will have cooled down to the extent that it could keep liquid water on its surface.

At the end of September, the first living beings – simple bacteria that do not use oxygen – appear in the ocean. On October 10, we meet for the very first time a recognizable form of life – blue-green algae appear. The great reptiles of the Mesozoic era start crawling on our planet by December 23 and become extinct on December 30 at noon. The first humanoids appear on December 31 at 10 p.m., and the entire written history from Babylon to today rolls through in 14 s. Only 0.9 s ago Galileo looked at the sky through the telescope for the very first time, and this is exactly how old modern astronomy is in this model. When you think about it, you must agree that we have accumulated a vast base of knowledge in a very short period of time!

At exactly at midnight, this book that you are reading is published and within the next second it is already sold out!

How Far Can We Go?

Great adventurers set out on a journey around the world with their motorbike or follow the Silk Road from Europe to China on their bicycles. Astronomers travel much further without really moving far – maybe only to the closest hill. How far we can go really depends on the size of the telescope or binoculars that we use for stargazing, the quality of the optics, observing conditions, and so on. Because we cannot take into account in our book the last two factors, let's assume that we have ideal optics and excellent observing conditions. Our journey into the depths of the universe will therefore depend mostly on the size of the objective of our optical instrument.

In this book, we have chosen celestial objects that can be seen by observers using the naked eye, binoculars, or a small telescope. Of course, we will speak about the visual limits, not the photographic. During direct observation of faraway galaxies the rods in our eye's retina get excited by photons that were created a long time ago in an anonymous star in foreign galaxies. And this is the special joy of observational astronomy.

Let's start with an instrument that we always have with us – our eyes. In most books, we can read that the limit for observing with the naked eye is the large Andromeda Galaxy M 31 (p. 172), which is 2.9 million light years away. This is not always true. It is true that M 31 is so bright that it is clearly visible in places where the sky is dark. However, if you know where to look, the limit is the spiral galaxy M 33 (Figure 5.8) in the Triangulum, which is 3 million light years away. In places with truly dark and clear skies we can easily see M 33 directly; otherwise, we can help ourselves with an observational technique called *averted vision* (the art of looking slightly to the side of a faint object that we want to see). We will certainly be able to see it. The light that is reaching our eye at this moment left the galaxy 3 million years ago. And ever since then the photons have been traveling through the universe with the speed of light. When they set off on their journey, Australopithecus still lived on Earth. We call such old light *fossil light.*

Figure 5.8. The limit for the naked eye: spiral galaxy M 33 in Triangulum

Figure 5.9. The limit for 10×50 binoculars: lenticular galaxy NGC 1316 in Fornax (p. 298)

Even very simple optical aids will significantly extend the limits of visibility. In the table below, we have listed a few bright galaxies according to their increasing distance from us. Most of them have Messier numbers, which means that they were discovered or observed by this famous comet hunter in the eighteenth century. The telescopes of the time, terms of the quality of their optics, were poorer than today's. But it is also true that at the time they did not have electricity, and the night sky even above Paris, from where Messier observed, was very dark.

Observers from the northern geographical latitudes can still see the brightest members in the Virgo Cluster of galaxies, concentrated around M 87, which is 60 million light years away, while the more southern observers can see NGC 1316 in the constellation Fornax, which is also 60 million light years away (Figure 5.9). This is the limit distance that we can still reach with a pair of 10×50 binoculars. Note that these are true astronomical and time distances. Light left the galaxy NGC 1316 or M 87 at a time when dinosaurs became extinct. We could call this *Jurassic light.*

If we want to travel even further, we have to use a telescope and good star charts, such as can be found in the atlases *Uranometria 2000.0* or *The Millennium Star Atlas.* Following NGC 1316, table below presents nine bright galaxies that are further and further away from us. The last on the list is NGC 4889, which is already 320 million light years away and is the brightest member of the cluster in the constellation Coma Berenices. In order to see it we need at least a 15-cm objective and excellent observing conditions.

Galaxy/ Quasar	Type	Size ['x']	Brightness [mag.]	Distrance [millionly]	Constellation	Limit for . . .
M 31	Sb	178 x 63	3.5	2.9	Andromeda	naked eye
M 33	Sc	73 x 45	5.7	3	Triangulum	
NGC 55	SBm	32.4 x 6.5	7.4	7	Sculptor	
NGC 253	Sc	26.4 x 6.0	7.2	10	Sculptor	
M 81	Sb	21 x 10	6.9	12	Ursa Major	
M 83	Sc	11 x 10	7.5	15	Hydra	
M 64	Sb	9.3 x 5.4	8.5	19	Coma Berenices	
NGC 2841	Sb	8.1 x 3.8	9.3	31	Ursa Major	
M 51	SAbc	11 x 7.8	8.4	37	Canes Venatici	
M 104	Sb	8.9 x 4.1	8.3	50	Virgo	
M 109	SBb	7.6 x 4.9	9.8	55	Ursa Major	
M 49	E4	8.9 x 7.4	8.4	60	Virgo	
M 87	E2	7.2 x 6.8	8.6	60	Virgo	
NGC 1316	SABO	12.0 x 8.5	8.9	60	Fornax	binoculars 10 x 50

(continued)

Galaxy/ Quasar	Type	Size ['x']	Brightness [mag.]	Distrance [millionly]	Constellation	Limit for . . .
NGC 5322	E3-E4	5.5 x 3.9	10.0	90	Ursa Major	
NGC 2336	SABbc	6.9 x 4.0	10.3	100	Camelopardalis	
NGC 772	Sb	7.5 x 4.3	10.3	115	Aries	
NGC 7619	E1	2.9 x 2.6	11.0	170	Pisces	
NGC 3646	Sc	3.9 x 2.6	10.8	195	Leo	
NGC1600	E2	2.5 x 1.8	11.0	220	Eridanus	
NGC 1275	E2	2.6 x 1.9	11.6	235	Perseus	
NGC 467	SO	2.4 x 2.3	12.1	250	Pisces	
NGC 4889	E4	3.0 x 2.1	11.5	320	Coma Berenices	15 cm objective
3C 273	Sy1.0	0.1 x 0.1	12.9v	2200	Virgo	20–40 cm obj.

From here onward the galaxies slowly go out of sight, for the other clusters are even further away from us and thus even the biggest and brightest members of these clusters are very weak in our skies. But nature has another good sight for sky watchers - quasars. These extremely compact and bright objects are like lighthouses in the depths of the universe. The brightest in this family - quasar 3C 273 in the constellation Virgo - can, with excellent observing conditions, be viewed with a 20-cm objective, for it shines with a magnitude of 12.9 (Figure 5.10). In order to reach us its light has traveled 2.2 billion years. It is hard to imagine that photons from the quasar, that are at this moment entering our eye, left their starting point at a time when life on Earth was hardly beginning and the seas were populated only by blue-green algae. Could we call this light *primordial*?

Most quasars have a variable brightness; however, the changes are random and cannot be predicted. And thus quasar 3C 273 occasionally "shines" with a magnitude 11.7 and can be glimpsed even with a 15-cm objective!

Today 30- and 40-cm Dobsons are no longer a rarity. Can we see even further with such an instrument? The answer is - unfortunately not! The next brightest quasar on the list is OJ 287, which is 4.2 billion light years away but usually shines with a magnitude of 16. Thus, quasar 3C 273 is the furthest lighthouse that can still be seen by most amateur astronomers.

The quasar 3C 373 lies in Virgo. The chart below depicts the constellation with a rough position of the quasar sketched in. This chart shows stars up to magnitude 5, the same as in the latter part of this book, where the constellations are described.

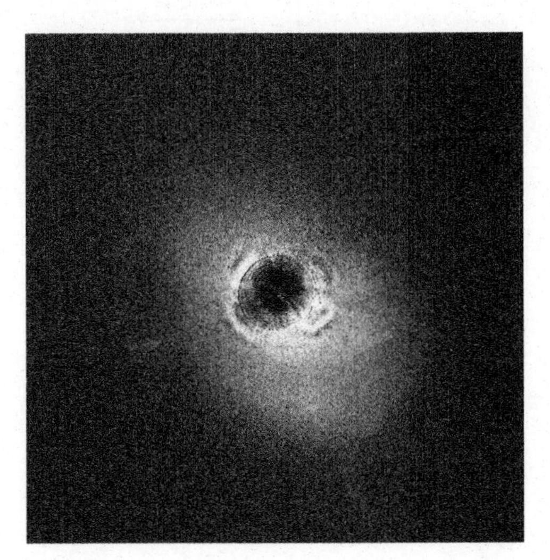

Figure 5.10. A closer look into the heart of the quasar 3C 273 in Virgo

The big chart below depicts stars up to magnitude 9.5, the same as in the detailed search charts that accompany the constellation descriptions in this book. The quasar is much fainter than the faintest stars listed on this chart, but such a chart is useful in leading us past brighter stars to the vicinity of the object you are searching for.

The near area of the quasar, which contains stars up to magnitude 15, can be seen in the inset chart. The width of the field is approximately 50 arcmin. Next to the stars you can also find their magnitudes (written without the decimal point; 115 represents magnitude 11.5) that can help with the exact orientation.

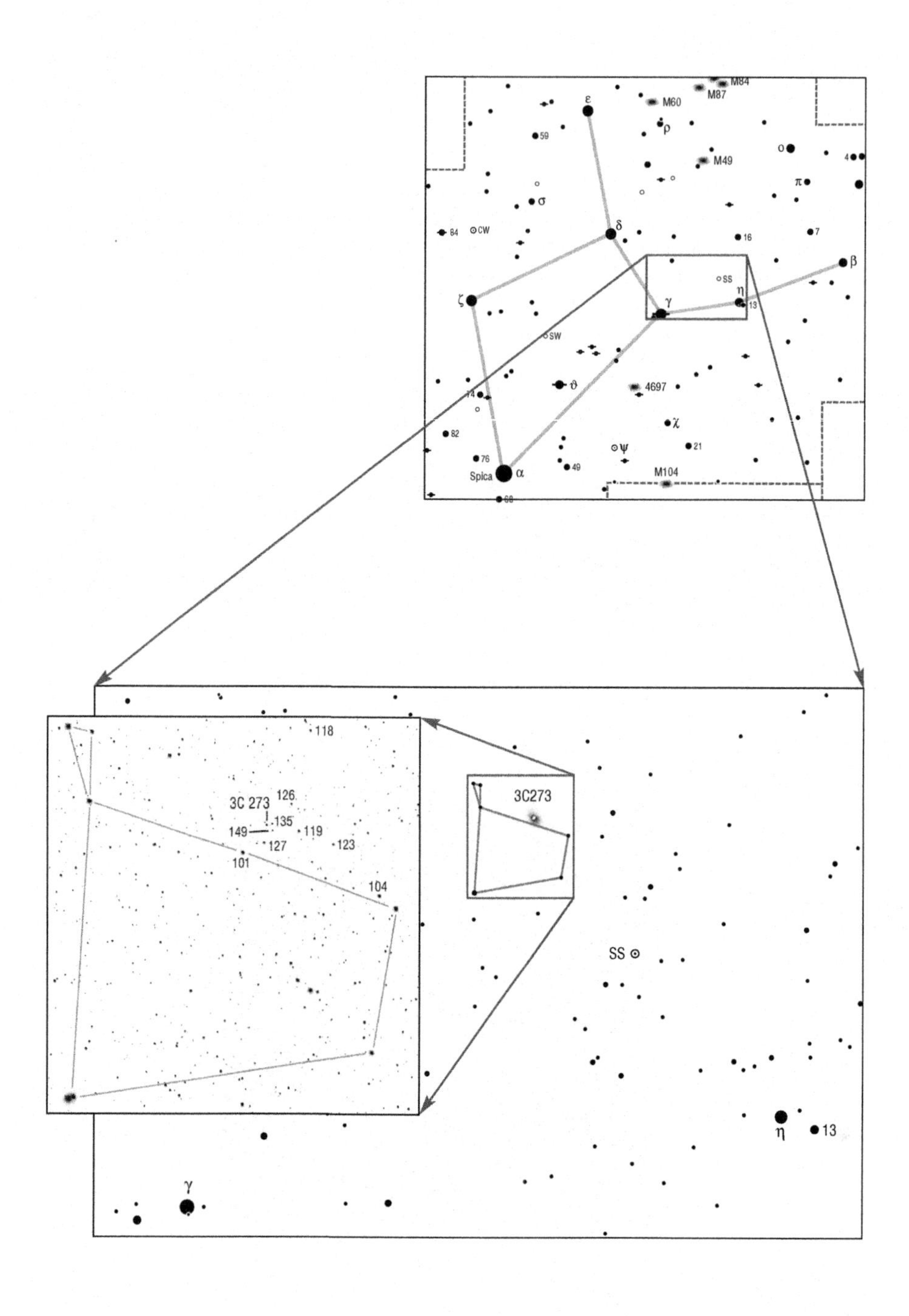

Units of Length In Astronomy

The distances between astronomical bodies (even in our Solar System) are so great that it is inconvenient to measure and express them in standard terms. That is why astronomers use much bigger units.

In our Solar System the basic unit of length is the astronomical unit, which is approximately equal to the distance from Earth to the Sun. It measures 149.598×10^6 km or 149.598×10^9 m. Its symbol is AU (recommended by international standard ISO 31-1) or au (recommended by the International Astronomical Union), but sometimes you might also see ua (recommended by the Bureau International des Poids et Mesures). The use of symbol AU is widespread.

Let's take a look at some examples. Jupiter is 5.2 AU (777.91×10^6 km) from the Sun. This tells us that the planet is about five times more distant from the Sun than Earth is. Neptune is 30.1 AU from the Sun. If its orbit is some kind of measure for the largeness of our Solar System, then its diameter is approximately 60 AU. But Proxima Centauri (the nearest star to Earth) is approximately 268,000 AU from the Sun, and the distance from the Sun to the center of the Milky Way is approximately 1.65×10^9 AU. To express distances between stars or galaxies the AU is not convenient any more. We have to use an even bigger unit. And this is a light year.

A light year is a unit of length approximately equal to the distance that light travels in a vacuum in one year. Its symbol is ly. A light year is equal to about 63,241 AU or 9.46×10^{12} km or 9,460,730,472,580.800 km.

Let's take a look at some examples. Proxima Centauri is approximately 4.2 light years from the Sun, and the distance from the Sun to the center of the Milky Way is approximately 26,000 light years. The nearest large cluster of galaxies in the constellation of Virgo is 60 million light years away, and the nearest quasar is 2.2 billion light years from us.

In professional astronomy, the preferred unit of measurement for such distances is the parsec, which is defined as the distance from which the Earth–Sun distance is seen under the angle of 1 arcsec. This is equal to approximately 3.26 light years.

Defining Distances in Light Years

In astronomy we express distances in light years. In reality we state how old the light is that is coming to us from a certain object. The light from nearby objects on Earth is only a fraction of a microsecond old. The light from the Sun is 8 min and 20 s old. That is how long it needed to travel the distance from the Sun to Earth. And when we are glancing at the Sun, we see it as it was 8 min and 20 s ago. If an eruption takes place on the Sun at this very moment, we will learn about it in 8 min and 20 s and no sooner than that.

When we are watching the universe, we are looking back in time, as if we were sitting in a time machine that could travel only in one direction. Individual stars that we can see with the naked eye are usually a few hundred light years away (and this old), the stars in the Milky Way a few thousand light years away, and the closest galaxies are a few million light years away. At greater cosmological distances, which are measured in billions of light years, we can no longer make a determination of their distance from the time the light needed to travel. We have to take into account the fact that the universe is expanding, and also the geometry of the universe. The age of the light of a galaxy, for instance, of 5 billion light years does not also mean a distance of 5 billion light years. When the object emitted this light that we are currently looking at, it was much closer to us. However, due to the fact that the universe is expanding, it needed longer to travel to us.

In everyday life, we also sometimes use time units to define distances. For instance we say: "The store is a six-minute walk away." This means that the store is some 500 m away. If, on the way to the store, we stop at a bar and drink a beer, we need 36 min for the same distance. In this case we cannot simply deduce the distance from the time one took to get there.

What Is a Quasar?
A quasar (short for QUASistellAR source or radio source) is an extremely bright and distant active galactic nucleus (AGN). In fact, it is a young galaxy that hosts a compact halo of the dust surrounding a super massive black hole at its heart. The black hole consumes large quantities of matter from the host galaxy. The electromagnetic energy output of quasars dwarf every other known astronomical object. The luminosity of quasar 3C 273 in Virgo is about 2 trillion (10^{12}) times that of our Sun. This means that it shines as bright as 100 average giant galaxies such as ours!

Catalogs and Lists of Nonstellar Bodies

A quick glance at the descriptions of the constellations in the latter part of this book will reveal the huge variety of names and labels used for nonstellar celestial bodies. Some of these names belong to galaxies, others to nebulae, still others to globular or open clusters, and so on – to everything that makes up the celestial kingdom. Let's have a quick look at their meanings.

Messier

Celestial objects that differ from pointlike stars intrigued the first astronomers who were observing the sky with their telescopes. There are quite a few such objects in the sky, and in the seventeenth and early eighteenth centuries these astronomers started making the first lists.

At the end of the eighteenth century the French astronomer and comet hunter Charles Messier took a huge step (Figure 5.11). He started systematically enumerating all of the celestial objects that looked like comets and could mislead comet hunters and observers of these celestial travelers. He was the first to observe and list most of these objects; many were discovered by his predecessors and contemporaries, but Messier put them into an organized list.

Many of today's historians of astronomy have asked themselves whether Messier truly only wanted to make a list of objects that were annoying to comet observers or whether he in fact wanted to produce the first complete catalog of celestial objects that live among the stars but are not stars. Note that he also included objects such as the Pleiades (M 45) in his catalog, which had been known to people for millennia, and the great Orion nebula (M 42), which hardly any astronomer would mistake for a comet. On the other hand, the bright and easily seen double cluster in Perseus and some other bright and at the time known objects cannot be found in his catalog.

Figure 5.11. Charles Messier (1730–1817)

Whatever might be the case, Messier gathered and enumerated 103 objects in his list. The first part of the catalog (from M 1 to M 45) was published by the Paris Academy of Science in 1774 in *Memoires de l'Academie.* The addition (to M 68) was published in 1780 in the *Connoissance des Temps.* The final printed version of the catalog was published in 1781 in *Connoissance des Temps* in which all objects up to M 103 were included. Later on various observers added some brighter objects and thus today Messier's catalog consists of 110 objects.

All objects from M 104 to M 110 were known by Messier as well as others; however they somehow did not find their place in the original published catalog. M 104 was added to the list by Camille Flammarion in 1921. M 105, M 106, and M 107 were added by Helen B. Sawyer Hogg in 1947; M 108 and M 109 were added in 1953 by Owen Gingerich, and M 110 by Kenneth Glyn Jones in 1966.

From the 110 objects in the list only 1 found its way there by mistake – M 102. Messier included it just before he handed the manuscript to the publisher. He had copied the information from Pierre Mechain and did not have the time to check it. And, as it was proven later, the coordinates were wrong. Some researchers today are of the opinion that the object was in reality the spiral galaxy M 101, while the others think that Mechain might have had in mind the lenticular galaxy in the constellation Draco that is today known under the name NGC 5866.

Among the remaining 109 objects, M 40 and M 73 are also unusual. The first is a double star, made up of two 9th magnitude stars in Ursa Major. Was this a mistake or did Messier truly see a nebula around the stars? We will never know. M 73 is a group of four faint stars that shine with magnitudes between 10.5 and 12; in the telescope of the great comet hunter they merged into a soft nebula. Today we do not notice any nebulosity surrounding the stars even on long-exposure images. For entirely historical reasons, the "open cluster" M 73 remained in the catalog and can be found in all modern star charts.

Even though the catalog is over 200 years old, it is still the best known and favorite among amateurs and even professionals. The reason for this is that it includes all of the brightest objects and at least one bright, well-known example of every type of object. Most Messier objects can also be seen through binoculars, while all 110 can be seen through a 15-cm telescope in good observing conditions.

The New General Catalogue

The second large catalog that has managed to survive is the NGC – *The New General Catalogue of Nebulae and Clusters of Stars*, which was published in 1888 by Johan L. E. Dreyer (Figure 5.12). The NGC is a listing of all nonstellar objects that were discovered by nineteenth-century observers. Most were contributed by the famous John Herschel (1792–1871), who published his *General Catalogue of Nebulae* in 1864. This is why Dreyer named his work the "new" catalog. NGC contains 7,840 objects of all types that are enumerated in order according to their right ascension (for the year 1860).

Figure 5.12. Johan L. E. Dreyer (1852–1926)

However, astronomy is by no means a dead science, and observers discovered numerous new objects even during the preparation of the NGC catalog for publication. That is why Dreyer published two appendices after the first catalog (in 1895 and 1908). He called these *The Index Catalogue*, or IC. Altogether it encompassed 13,266 objects. The NGC and both IC catalogs can be considered as a single work.

A number of objects from the NGC catalog can be seen through binoculars; most can be seen through 20- or 30-cm telescopes from normally light-polluted observing points, and almost all of them can be seen with a 30-cm telescope during excellent observing conditions. A few hundred objects that include NGC in their name are as bright as the weaker ones in the Messier catalog; thus, we can also observe them with smaller telescopes. A lot of them are also described in this book and are therefore seen through binoculars.

However, Dreyer was not perfect either. Nobody really knows why he failed to include the open cluster M 25 in Sagittarius in his NGC and only later on added it to the IC catalog under the name IC 4725. In addition, the relatively large and easily seen open cluster in Ophiuchus cannot be found in the Messier list or the NGC catalog. In the end, Dreyer finally gave it a name and today it is known as IC 4665 (and is also easily seen through binoculars). Many astronomers find it unforgivable that the open cluster of the Pleiades never got its mark in the NGC or IC catalog.

The period in which Dreyer was preparing and publishing his catalogs was a time of exciting advances in astronomical photography. The photographic discoveries of new, even fainter nonstellar objects were coming so fast that Dreyer discontinued with his catalog work and dedicated his time to his second love – the history of astronomy.

NGC and both IC catalogs were so excellent and systematic that they are still in use even today.

CHAPTER SIX

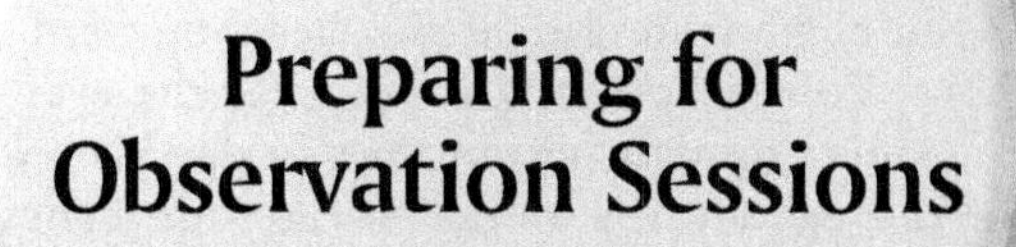

Preparing for Observation Sessions

Before Night Falls

Any serious observer will get ready for evening or nighttime observations during the day. Making a list of objects to search for or observe and that are, at the time of observation, at their highest in the sky is important. Set aside at least half an hour for viewing each object. If you whiz from one object to another you will miss a lot of things that you could otherwise see. Prepare the charts so that you will not need to walk into a bright space often during your observation; doing so impairs night vision. In order to get your eyes readjusted to the dark, you will need at least half an hour each time.

It is advisable that you spend a night with only average viewing conditions for searching and getting acquainted with objects, which can sometimes take many hours. When you know how to find the objects without a problem, then you can make the best use of ideal observation conditions to enjoy a more detailed study of objects.

In order for you not to have to go into a lighted room, have a table and a weak light next to the binoculars, so that you can look at the charts whenever you need to. The light should be white and not – as many manuals suggests – red. Red light can cause a wrongful color adjustment of your eyes and wrong perception of star colors.

B. Kambič, *Viewing the Constellations with Binoculars*, Patrick Moore's Practical Astronomy Series,
DOI 10.1007/978-0-387-85355-0_6, © Springer Science+Business Media, LLC 2010

Adjusting the Eyes to Night Vision

As we already discussed, the lens in the eye creates an image on the retina that houses light-sensitive receptors, also called photoreceptors. When there is plenty of light the cones are active; when light is scarce the color-blind rods are active. During the daytime rods are hidden in the retina. In the lack of light they rise from the retina, and this takes a certain amount of time. To go from daytime vision to night vision (accommodation) can take half an hour or even more. So, every time you come from a lighted space, you need to wait to begin observations for at least half an hour. This is how long it takes for the eyes to adjust to night vision.

The angular resolution is poorer in night vision than in day vision. This is because the rods are not spread as thickly throughout the retina as are the cones (we are not nocturnal creatures). In the small area in the center of the retina (called central fovea) the cones are especially dense; there are even fewer rods than elsewhere on the retina. This is why we can see the weaker celestial objects better when we look slightly past the object instead of stare into it. Knowing this will come in very handy when you are looking at objects that are on the border of visibility. This observing technique is called averted vision, and experienced observers use it routinely.

Observing Conditions

You can stargaze on any clear night. This is true even if the stars in the sky are twinkling merrily, even in populated and brightly lit places, even if the sky is covered by thin and see-through cirrus clouds, and even if the Moon is out in its full shine. But you have to be aware that these are not good observing conditions. You could look only at the brightest celestial objects. It is much better if you use these nights and these viewing conditions to search for those objects that you do not know the location of yet, because binoculars will not do their best in such conditions. It is likely that you will not even come close to exploiting their limits.

If you wish to use your binoculars to their full potential, you have to set off for an observing point with a dark sky and no light pollution and wait for prime observing conditions. But even these are not the same everywhere. When the deep-blue afternoon with a gentle breeze transforms into a dark and clear night, and bright stars twinkle in the sky, you have a night with great transparency. A dark sky and great contrast are ideal conditions for observing galaxies, nebulae, and dim stars. Unfortunately, this is often accompanied by a less than ideal condition - wind turbulence. On such night seeing is often poor.

From average observing sites the seeing is rarely smaller than 1 arcsec, even in excellent conditions. This is most often also the limit for separating close double stars and the limit for numerous details that can be seen on the Moon and planets, even if the theoretical limiting resolution of the telescope is below 1 arcsec. There are only a few places on Earth that seeing drops below this limit on the best nights. This is also one of the reasons why astronomers have sent the Hubble Space Telescope into orbit around Earth high above our atmosphere.

Observers with long experience know that transparency and good seeing usually do not appear at the same time. A humid, breezeless summer day often creates ideal seeing. The night that follows, in which the atmosphere is very still, is excellent for separating close double stars and watching the details on the Moon and planets, even though a slight mist spreads across the sky.

You can also do some things to improve observing conditions. For example, you can select an observing point that is far from a source of warm air, which causes turbulence. Chimneys, rooftops, and asphalt surfaces are all heated by the afternoon Sun. Grass and trees provide the best observing milieu. Also, if keep your binoculars in a warm room, wait for at least 30–60 min for their temperature to cool down and equal the temperature of the surrounding air.

If possible, you should watch the celestial objects when they are at their highest in the sky, when they culminate. The thickness of the atmosphere increases the lower the object is, which means that the lower the object, the greater the disturbances.

Scales of Observation Conditions

Serious observers keep a diary of observations in which they include, apart from the general data (location, date, and hour of observation), the observing conditions and maybe even a sketch or drawing of the observed object.

For a quick and simple evaluation of the observing conditions, appropriate for amateur astronomy, Eugène Antoniadi (1870–1944) came up with a five-grade scale:

I Excellent conditions. The image of the star is totally still.
II Good conditions. The image of the star is occasionally restless, but periods of complete stillness prevail.
III Average conditions. The image of the star moves, and still periods are occasional and short.
IV Poor conditions. The image of the star is constantly moving.
V Very poor conditions. The image of the constantly moving star is smeared.

In this book we add another level to this scale, a level that could be marked with Ia - perfect or exceptional observing conditions. These conditions appear when a totally still atmosphere is accompanied by a very dark sky. Most often you will encounter exceptional conditions in the winter, when there is not a lot of humidity in the air or during temperature inversions when valleys (which are the source of most light pollution) are covered with a thick layer of fog.

Of course, there are a number of other observing condition scales. Some describe the limiting brightness of the stars, and with some you have to count how many stars you can see in predetermined parts of the sky. In yet other ones you need to compare the images of stars with diffraction circles (airy disc), etc.

However, the most shocking scale has to be the scale recently developed by John E. Bortle. It is shocking because it shows clearly how catastrophically we have ruined our dark sky with light pollution.

The Bortle Dark Sky Scale

The limiting brightness of the stars themselves is not the best indicator of the true quality of the sky, which is why the American veteran astronomer John E. Bortle, with over 45 years of observation experience, developed the so-called Bortle Dark Sky Scale. This scale divides the quality of the sky into nine classes, and apart from the limiting brightness of the stars, an important indicator of sky quality is also the visibility of the Milky Way, nonstellar objects, and zodiacal light (Figure 6.1). The scale is shocking to a great many who think they are observing under a dark sky while, according to Bortle's criteria, the sky is only medium dark.

Class 1: Excellent Dark Sky Site (Limiting Magnitude 7.6–8.0)

The zodiacal light, gegenschein, and the zodiac band (these terms are explained at the end of this chapter) are all visible, the zodiacal light is extremely bright, and the zodiac band stretches across the

Figure 6.1 Milky Way and zodiacal light on Mt. Lemon in Arizona in the United States

entire sky. M 33 can be seen with the naked eye. The part of the Milky Way in Sagittarius and Scorpius casts clearly visible shadows. The limiting magnitude of the stars given above (7m6–8m0) is with the naked eye. Venus or Jupiter are so bright as to prevent the adjustment to the night vision. The airglow of the night sky is clearly visible to 15 degrees above the horizon. With a 32-cm telescope we can see stars up to magnitude 17.5, while with a 50-cm one we can see anything up to magnitude 19. If we are observing from a meadow surrounded by trees the telescope can hardly be seen.

Class 2: Typically Truly Dark Site (Limiting Magnitude 7.1–7.5)

The twinkling airglow may be apparent just above the horizon. M 33 is easy to see with averted vision. The summer Milky Way shows a complex structure even with the naked eye, while the brightest parts, when viewed through binoculars, are reminiscent of white marble. The zodiacal light is bright enough to cast weak shadows just before dawn and after dusk, though its color is yellowish when compared to the blue-white of the Milky Way. Clouds appear as dark holes on the starry background. The telescope and your surroundings can be seen only vaguely, except when projected against the starry sky. Numerous Messier objects are clearly visible with the naked eye. The limiting magnitude with the naked eye is between 7.1 and 7.5; through a 32-cm telescope we can see stars up to magnitude 16 or 17.

Class 3: Rural Sky (Limiting Magnitude 6.6–7.0)

Some traces of light pollution are visible low above the horizon. Clouds are faintly illuminated close to the horizon and are dark in the zenith. The Milky Way still appears complex, and globular clusters such as M 3, M 5, M 15, and M 22 are clearly visible with the naked eye. M 33 is easy to see with averted vision. The zodiacal light is striking in spring and autumn, when it extends 60 degrees above the

horizon after dusk and before dawn. The limiting magnitude of the stars with the naked eye is between 6.6 and 7.0; through a 32-cm telescope we can see stars up to magnitude 16.

Class 4: Transition Between a Rural and Suburban Sky (Limiting Magnitude 6.1–6.5)

Light pollution domes are fairly obvious above settlements in a number of directions. The zodiacal light is visible but does not extend even halfway to the zenith. The Milky Way well above the horizon is still

impressive but has lost most of its structure. M 33 is a difficult averted-vision object and is detectable only at an altitude of higher than 50 degrees. Clouds in the direction of light pollution are slightly illuminated but remain dark in the zenith. Through a 32-cm telescope we can see stars up to magnitude 15.5.

Class 5: Suburban Sky (Limiting Magnitude 5.6–6.0)

Only hints of the zodiacal light are seen on the best spring and autumn nights. The Milky Way is very weak or invisible on the horizon and looks rather washed out in the zenith. Light sources can be seen in most directions. Over most of the sky clouds are noticeably brighter than the background sky. Through a 32-cm telescope we can see stars with a magnitude of 14.5 or 15.

Class 6: Bright Suburban Sky (Limiting Magnitude – Approximately 5.5)

Even on the best nights no trace of the zodiacal light can be seen. Traces of the Milky Way are visible only in the zenith. The sky within 35 degrees above the horizon has a grayish glow. Clouds are bright throughout the sky. M 33 can only be seen with binoculars, while M 31 is barely visible with the naked eye. Through a 32-cm telescope used at moderate powers we can see stars of magnitude 14 or 14.5.

Class 7: Transition Between a Suburban and Urban Sky (Limiting Magnitude – Approximately 5.0)

The entire sky has a grayish glow. Strong sources of artificial light can be seen in all directions. The Milky Way is totally invisible or nearly so. M 44 and M 31 are on the border of visibility with the naked eye, while with a medium-sized telescope all Messier objects are only faint reflections of their true images. The naked-eye limiting magnitude of the stars is 5.0 (with great difficulty); through a 32-cm telescope we can barely see stars of magnitude 14.

Class 8: City Sky (Limiting Magnitude – Approximately 4.5)

The sky glows gray or orange and is bright enough to read newspaper headlines without any difficulty. M 31 and M 44 can barely be glimpsed even by an experienced observer on the best of nights. Bright Messier objects can be seen through a medium-sized telescope. Some constellations cannot be recognized. In the best conditions, the limiting naked-eye magnitude of the stars is 4.5; through a 32-cm telescope we can see stars up to magnitude 13.

Class 9: Inner-City Sky (Limiting Magnitude Below 4.0)

The entire sky is lit, even in the zenith. Numerous constellations are unrecognizable, with some such as Cancer and Pisces invisible. Except for the Pleiades almost all Messier objects are invisible. The only objects worth observing through a telescope are the Moon, planets, and the brightest star clusters (if you can find them). The naked eye limiting magnitude is 4.0 or less.

According to the Bortle scale most dark sites in inhabited areas belong to Class 4, with the exception of a few sites that you can find high in the mountains or deep in the deserts.

Maybe even astronomers are not fully aware of the true speed with which our sky is disappearing. The visible differences of the quality of the night sky do not occur on a time scale of generations or decades; they can be noticed from one year to another. Some older readers might still remember the night sky from their youth as being much darker than it is today, a sky in which the Milky Way was clearly visible even from their backyard.

A note by the author of this book: the observation data on the appearance of the celestial objects through binoculars were gathered from 2001 onward on Bloška planota, which at the time belonged to the Class 2 of the Bortle scale. Then light pollution spread across Slovenia so fast that it now is barely Class 4.

Astronomical Seeing
On its way through the atmosphere the light from the stars travels through variously dense and warm layers of air. On the borders, it refracts and changes direction. Because the atmosphere is constantly turbulent, we get a moving image in the eyepiece that dances around and twists and turns, becoming deformed and making it hard to watch. Instead of seeing points of light in your field of view you will see blurred blotches. Astronomical seeing therefore describes how much Earth's atmosphere deforms the images of stars. The size of blurred blotches of otherwise single-point stars is what astronomers call seeing and is expressed in arc seconds.

John E. Bortle
John E. Bortle is an experienced observer of comets and variable stars. He has been observing them since the 1950s, when it was much easier to find naturally dark skies than it is today. Bortle watched all large comets from Arend-Roland and Mrkos in 1957 and numerous lunar and solar eclipses and contributed tens of thousands visual measurements of variable stars to the AAVSO. For over 15 years, he was an editor on comets for the American publication *Sky & Telescope.*

Zodiacal Light and Gegenschein
There are plenty of fine dust particles in interplanetary space and in planetary orbits, including Earth's. These particles reflect the sunlight, which is why we see them in a dark sky in the plane of the ecliptic as a gentle cone-like heap. The zodiacal light is best seen in spring and autumn, when the ecliptic cuts the horizon at a greater angle. In locations with a truly dark sky, the cone spreads into a band of gentle light that runs along the ecliptic across the sky; this light is known as the *zodiacal band.*

Gegenschein, or counter shine, is also reflected sunlight from the dust particles in the ecliptic plane, only in this case the light comes from an area directly opposite the Sun. The dust particles are headed in this direction, with their entire illuminated side turned toward Earth; thus, a bit more reflected light reaches us. In locations with a dark sky we see the counter shine as a gentle circular haze of light on the ecliptic plane, in the middle of the zodiacal band; compared to that haze, it is larger and a bit brighter (see image above). During the night, it travels across the sky in a westerly direction together with the Sun (which is on opposite side).

Gegenshein and the zodiacal band are practically unknown among European astronomers (because of the light-polluted sky). In dark sites, you can still see zodiacal light in the evening and morning, but usually it only reaches a few tens of degrees in altitude.

Searching for Celestial Bodies

Plenty of people who have seen images in astronomy books and gotten all excited about astronomy have bought a small telescope or binoculars, then went under the night sky and, after only a few attempts, gave up stargazing because they could not find anything. Looking for objects and knowing what leads you to them is one of the hardest things in observational astronomy, but one of the easiest if you have proper instructions. This book will provide you with the right instructions. Every object that can be seen through binoculars and is described in the next section of the book is accompanied by a detailed star chart that leads you to it. And this chart is not just a rough map that would get you lost half away across. What you see is a detailed, precise, and reliable guide from the closest bright star all the way to the desired object.

When the object you wish to observe is in close proximity of a truly bright guiding star and they are in the same field of view of the binoculars, you usually will not have any problems. Of course, you need to learn how to direct the binoculars into the desired part of the sky. There are no shortcuts here; only observing experience will help you. If you do not see the desired object in the field of view, you have to first of all make sure that you are looking at the right guiding star. You do this by checking in the field of view and on the chart all the dimmer stars in its proximity, which has to be in equal positions and at the same distances on the chart as in the field of view. If they are not, you are looking in the wrong direction. And this happens quite often at the beginning. You also have to always be careful that you know how the field of view is oriented in the binoculars, i.e., where north, south, east, and west are. On the charts, north is always on the top, south is at the bottom, east is left, and west is right.

It is much harder to find faint objects in an empty part of the sky, far from the brighter stars, but in this book we have prepared detailed charts so even this should not prove to be too difficult. As an example, let us look at the elliptical galaxy M 87 in the Virgo constellation. How can you find it?

In the text you will read that the guiding star to the cluster of galaxies in Virgo, which contains M 87, is the star Epsilon Virginis. You point the binoculars toward it and when you think it is in the field of view, check the surrounding stars, which have to be exactly as depicted on the chart below, which roughly depicts the field of view of the binoculars. You do not have to check all the stars one by one; it is enough if you notice the typical pattern of the stars to the southwest of Epsilon, toward the edge of the field of view, from which the one marked 33 shines with a magnitude 6. The pattern has the shape of the letter V. If you are still not absolutely sure that you have reached the right spot, you can also check the position of the star marked 34, which also shines with magnitude 6 and their brighter neighbors that lie almost in a straight line. Only when you see all of them can you say with certainty: "Yes, this is Epsilon!"

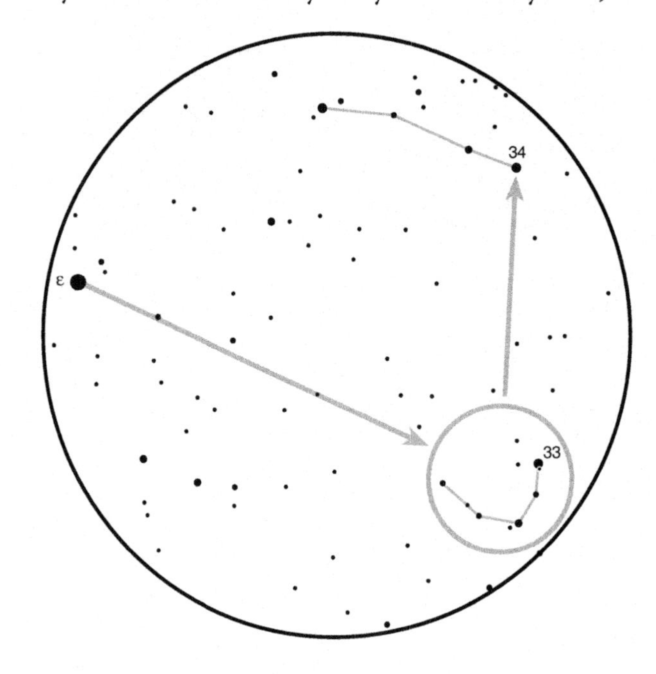

Then you will read in the text that you have to move the stars marked 33 and 34 to the eastern edge of the field of view. What you see then is depicted on the chart below. Roughly in the center you will find a bright pair stars called Rho and 27. The first shines with magnitude 5 and the second with magnitude 6, and both are clearly visible under all observing conditions.

Now there are a number of possibilities to check further, but the most noticeable is the zigzag line of magnitude 7 and 8 stars on the west edge of the field of view. On the way you can also look up galaxy M 60. If you see it, you will probably also see M 87.

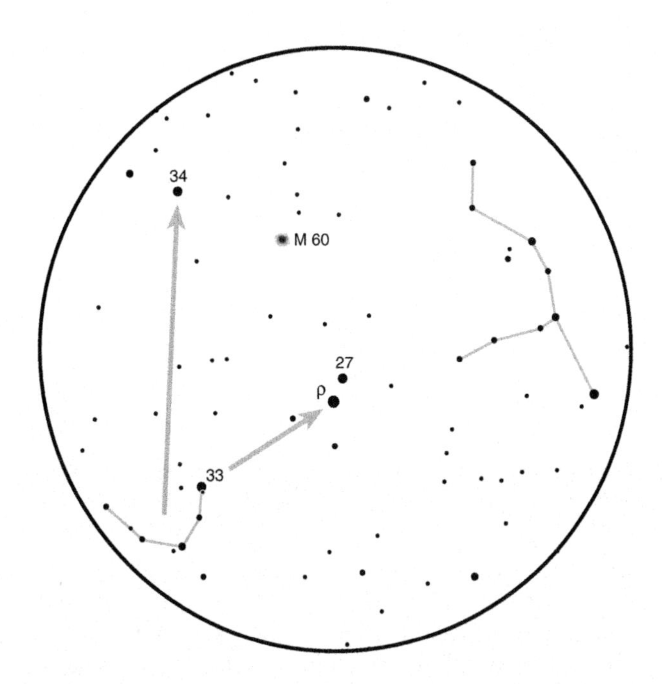

In the next step you move the stars Rho and 27 to the eastern edge of the field of view, and on the northwest edge a small, weak spot of light should be visible. This is M 87. If you do not see it in the field of view but you can see all of the brighter stars from the chart below as well as the circled group of dimmer stars, then you are in the right place, but the observing conditions are so poor at that time that you cannot see the galaxy. Remember the way that leads to it, and you can look for the galaxy another time, in better conditions.

There is no legend alongside the detailed charts that would depict the magnitude of the stars. All of the charts have a common legend (see page 167). However, observers quickly get used to reading star charts and using comparisons, not relying on the magnitudes of the stars.

What we have described here is the best recipe for a successful search for celestial bodies using binoculars, especially those on the border of visibility of the binoculars. When this becomes routine for you any transition to a telescope – which has a smaller field of view – will be much easier.

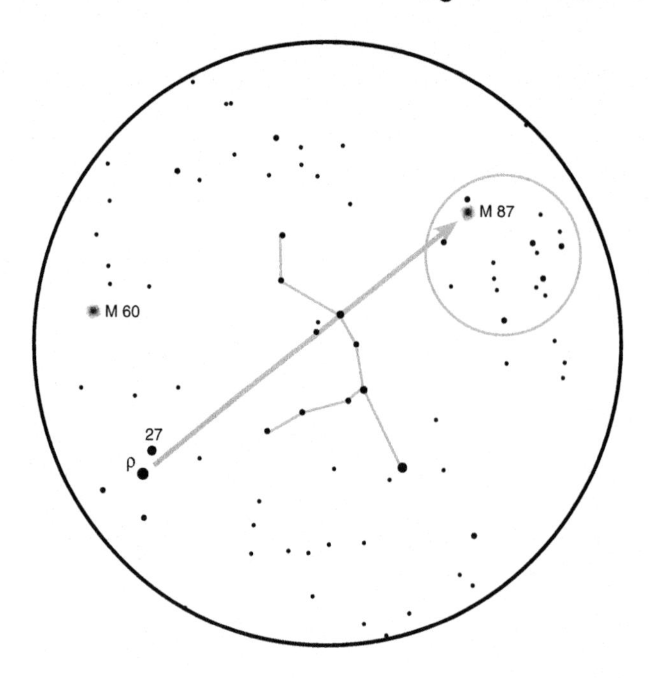

Star Charts, Photographs, and Symbols

On the constellation charts in the next section of this book, all stars of magnitudes 5 are depicted and only those of magnitude 6 that are part of the basic shape of the constellation are shown. The surrounding constellations are marked only with their name, except when the brighter stars in the neighboring constellation help us search for the dimmer objects in the selected constellation. These stars are drawn in a grayish tone. Next to every chart, there is also a 6-degree circle that represents the field of view of the binoculars. On all charts north is at the top, south is at the bottom, east is on the left, and west is on the right. The only nonstellar objects depicted are those mentioned in the text and can be seen in the binoculars. The legend is the same for all charts and is given below.

Magnitudes of Stars	Double Stars	Variable stars	Clusters	Nebulae	Galaxies	Milky Way
–1 0 1 2			Open	Bright		
3 4 5 6		nova	Globular	Planetary	Borders of the Constellations	

Charts to help you search for nonstellar objects are drawn with stars up to magnitude 9.5, i.e., stars that can (in good conditions) still be seen through binoculars. The charts have no symbols for double stars that cannot be seen separately through binoculars, and there are no nebulae or other celestial objects that can only be seen on photographs. However, there are symbols for variable stars, because this kind of star might be near its minimum and thus dimmer or not visible at all in the field of view at the time you are observing it. The charts also do not depict the constellation borders, because these cannot be seen in the field of view of your binoculars. All charts are drawn in the same scale, except when specifically stated otherwise.

Magnitudes of Stars	Variable Stars	Clusters		Bright Nebulae	Dark Nebulae
		Open	Globular		
-1 0 1 2 3	1^m Size of the symbol represents the brightness at maximum.		6^m 7^m		to scale
					Planetary Nebulae
4 5 6 7 8 9	6^m Size of inner black dot represents the brightness at minimum.	Doted circle represents the size of the densest part of the open cluster while the shaded disk represents the whole area of the cluster.	8^m 9^m	>8 to scale and to appearance	just a symbol
					Galaxies
10 11 Stars with magnitudes 10 and 11 are draw n only in open clusters.	Empty circle means, that brightness at minimum is below the magnitude 9.5.		10^m to scale and to brightness	<8 just a symbol	to scale

Detailed charts are included for all objects that are not in the vicinity (in the same field of view) of bright, clearly visible stars that represent the shape of the individual constellation. For all faint and poorly visible objects, such as, for example, galaxies, the book includes detailed charts even in the case where these objects are close to bright stars. Sometimes we really need to know the precise position of such a dim spot of light between the stars in order for us to notice it.

For all charts in which the field of view is marked with a nonshaded circle, this area is 6 degrees wide, except when specifically stated otherwise.

Photographs

Amateur astronomy has undergone many changes in a short time over the past few years, even in Slovenia. When selecting photographs, as author I tried to use as many Slovene photographs made with amateur equipment as I could, even if I could find better foreign ones. It is unbelievable that it is possible to fill such a large book with so many good Slovene photographs. Only a few years ago this would have been impossible.

Photographs of objects made with amateur and large professional telescopes are not in the book merely to liven it up or to drive you crazy because you cannot see a similar image through your binoculars. Images (especially those of galaxies) enliven the imagination. With open clusters they help us seek for the fainter members of the cluster or close double stars within; with planetary nebulae or globular clusters they help us to understand what that weak spot of light that we can see in the field of view truly is. A good image always variegates the observations.

The observer behind the binoculars that we have used to graphically depict object visibility was drawn by Tomaz Lavric. I would like to take this opportunity to thank him.

exceptional celestial object or typical representative of its kind

well seen celestial object

seen celestial object

celestial object on border of visibility

PART TWO

Constellations

CHAPTER SEVEN

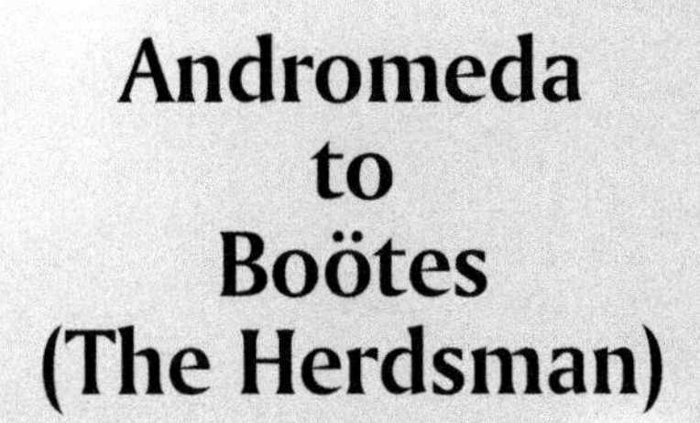

Andromeda to Boötes (The Herdsman)

B. Kambič, *Viewing the Constellations with Binoculars*, Patrick Moore's Practical Astronomy Series,
DOI 10.1007/978-0-387-85355-0_7, © Springer Science+Business Media, LLC 2010

ANDROMEDA (Andromeda)

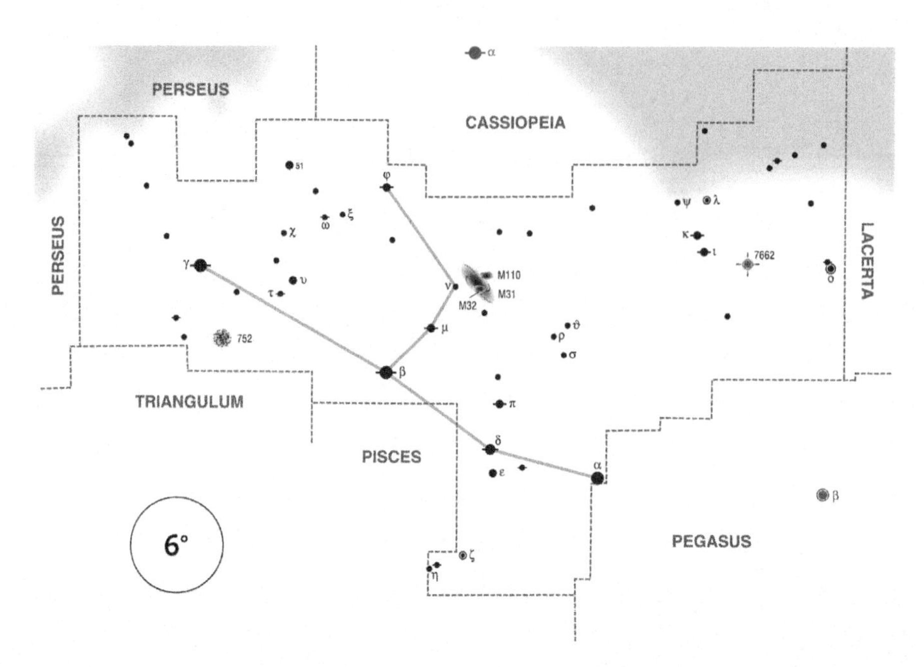

CONSTELLATION CULMINATES		
in begin. September at 2 a.m.	in begin. October at midnight	in begin. November at 10 p.m.

Andromeda is a big and prominent autumn constellation in the northern skies that extends from Pegasus to Capella (Alpha Aurigae). The brightest stars within the constellation are Alpha or Sirrah, orange Beta or Mirach, golden yellow Gamma (all 2m1), and yellow-orange Delta (3m3).

There is no chance of missing Alpha. This star inhabits the northeastern corner of the well-known asterism called the Square of Pegasus. In older star charts and even in some modern tables we might find it under the name Delta Pegasi, which is an older name for this star. However, in modern times it was attached to Andromeda as Alpha Andromedae. Sirrah is the 55th brightest star in the sky. Its distance is 97 light years from Earth. The luminosity of the star is about 100 times that of the Sun.

Beta or Mirach is the 57th brightest star in the sky, a red giant, 200 light years from us. Its luminosity is equal to approximately 400 Suns. It is a binary star. Its companion is magnitude 14.4 and is 27 arcsec (p. a. 202°, 1934) away from Beta. It is a dwarf star with a luminosity 800 times smaller than that of the Sun. The stars are gravitationally bound. The companion star could be separated by binoculars, but it is too faint to be seen.

Gamma is the 61st brightest star in the sky. It is a splendid double star, one of the most beautiful to be seen with small telescopes. The color of the brighter star (2m3) is golden yellow (some reliable observers describe it as soft orange); the fainter (5m0) is greenish-blue. The color contrast is even more emphasized if the eyepiece is not quite in focus but slightly off, so that the images of the stars are a bit blurred. The apparent separation between them is 9.6 arcsec (p. a. 63°, 2004). Unfortunately, the pair cannot be separated by binoculars. Gamma Andromedae is in fact a quadruple star – both stars are doubles. This interesting multiple star group is 356 light years from Earth. The common luminosity of the four stars together surpasses our Sun by over 1,300 times.

The famous Andromeda Galaxy or M 31 (3m5/180′×60′) is undoubtedly the most beautiful celestial object of the constellation. It is the brightest and nearest of all the larger galaxies and the only galaxy that can be clearly seen with the naked eye. It lies approximately 1 degree west from the star Nu Andromedae (4m5), and in a clear, moonless night it can be seen as a small, elongated cloud of faint light. The galaxy appears to be elongated because we see it more or less from its side. It is inclined only by 15 degrees over its equatorial plane toward Earth. We can try to imagine what an outstanding celestial object this would prove to be if we could see it in all its glory, face-on. It would be almost circular, with spiral arms that coil around its bright, denser core. Of course, in our sky it would be much bigger and much brighter.

In binoculars (and in small telescopes) the galaxy will stretch to an approximately 4-degree-long oval spot of light with a well-seen brighter central condensation – a view that undoubtedly disappoints any observer who sees the galaxy for the very first time (Figure 7.1). The reason lies in its disreputableness. When we read about the Andromeda Galaxy, its description is always full of adjectives such as brightest, biggest, nearest, magnificent, spectacular, and so on.

Figure 7.1. In the field of view of an average pair of binoculars the Andromeda Galaxy is visible as a big hazy oval a few degrees in size with a brighter core. Perhaps this does not seem interesting to you. But even a nebulous spot can be interesting if you know what you are looking at!

Even though it is the nearest, it is still very, very far, too far to be seen as much more than a speck of light in our binoculars! On the edge of the galaxy we cannot separate individual stars even with the largest telescopes and observing conditions that are excellent. With larger amateur telescopes some dark lines of dust and cold gases can be seen in the outer regions. Also some of the brightest globular clusters and maybe one or two bright emission nebulae might be visible. Other features of Andromeda Galaxy show up only on long-exposure images. By observing it "live" the Andromeda Galaxy charms us only if we know what we are looking at! When we admire the galaxy in the night sky, we are not merely watching the furthest object in space still visible with the naked eye but the furthest back into the past. The light that is collected in our eyes at that moment left the galaxy 2.93 million years ago, at the time when Earth was inhabited by the first apes! And throughout the entire history of humankind, from the Stone Age to modern times, the photons that were produced in the photospheres of alien stars in an alien galaxy have traveled through space only to end their long journey in our eyes!

The Andromeda Galaxy appeared on star charts a long time before the first telescopes were invented. We know of the existence of this galaxy at least from A.D. 964, when the Persian astronomer Abd-al-Rahman Al Sufi published a book entitled *The Book of Fixed Stars.* The galaxy was mentioned in this book and called the Little Cloud (Figure 7.2).

The Little Cloud was observed for the very first time through a telescope by Simon Marius in 1611 or 1612. Soon after this, the first hypothesis appeared regarding the nature of this cloud. Some astronomers and philosophers proposed that this was a nebula, composed of hot, glowing gases. Others were convinced that this was the place in which a new star and planetary system was in the

Figure 7.2. Al Sufi's chart of the constellations Andromeda and Pisces that includes the Little Cloud, marked with the letter A

process of formation and that the conditions in the nebula were to a great extent similar to those from the beginning of our Solar System. Of course, at that time the distance to the nebula was unknown, as were its dimensions. But, as always in cases like this, intuition was not much help. Astronomers have to wait for the proper technology to be developed. In the case of the Andromeda nebula, as it was called at that time, they had to wait for spectroscopy. The analyses, performed by William Huggins, an English amateur astronomer, in 1864, proved that the light from the Andromeda nebula was produced by myriads of individual stars and not by hot gases.

The increasingly large telescopes from the end of the nineteenth and the beginning of the twentieth century and the development of more and more sophisticated astrophotography confirmed the spectroscopic analyses. Photographic plates showed individual stars on the edge of the Andromeda nebula. However, this was only the first step in lifting the veil from the mysterious haze. At the time astronomers were still of the opinion that this was a cloud of stars within our own galaxy.

In 1917, the world's largest (at the time) telescope, the 2.5-m Hooker reflector at Mount Wilson Observatory was finished. Edwin Hubble used this telescope for photographing the Andromeda nebula. In 1923, he recognized a few Cepheid variable stars on the plates and measured their changes in brightness over a longer period of time. By doing this he established their period of luminosity changes. From the well-known relation between the period and luminosity, and from their apparent magnitude he could derive their distances and therefore the distance to the Andromeda nebula. The first results seemed so unlikely that Hubble checked his calculations over and over again. The result was always the same – the Andromeda nebula was 900,000 light years away! In comparison to the stars this was an incredibly large number. Hubble must have felt similar to Bessel a hundred years before him when he discovered how unimaginably far the closest stars were.

The result astonished the astronomers who gathered in Washington in 1924 at the annual meeting of the American Astronomical Society at which Hubble announced his sensational revelation. The realization that the Little Cloud or Andromeda nebula was not a member of our galaxy but is in fact a vast galaxy, an association of numerous stars similar to our own galaxy, was the end of the guesswork concerning the real nature of the nebula. However, at the same time it was the beginning of a new, even more exciting chapter in the history of astronomy, a period full of astonishing discoveries. Astronomers soon discovered that the Andromeda Galaxy, as it is called now, is one of the nearest galaxies, that it lies – so to speak – in our space backyard, and that other galaxies are much further away. Space became even vaster and emptier.

Today we know that Hubble's estimation of the distance to the Andromeda Galaxy fell way short. Improved measurements of the Cepheids with a 5-m reflector on Mount Palomar, our better understanding of their physical characteristics (they appear in several subtypes), and better knowledge of light absorption in interstellar space show that the galaxy is 2.2 million light years away from us. This is the number found in most twentieth-century books on astronomy. But even this number is not correct. The latest value, derived from the analyses of the measurements by the Hipparcos astrometry satellite is 2.93 million light years!

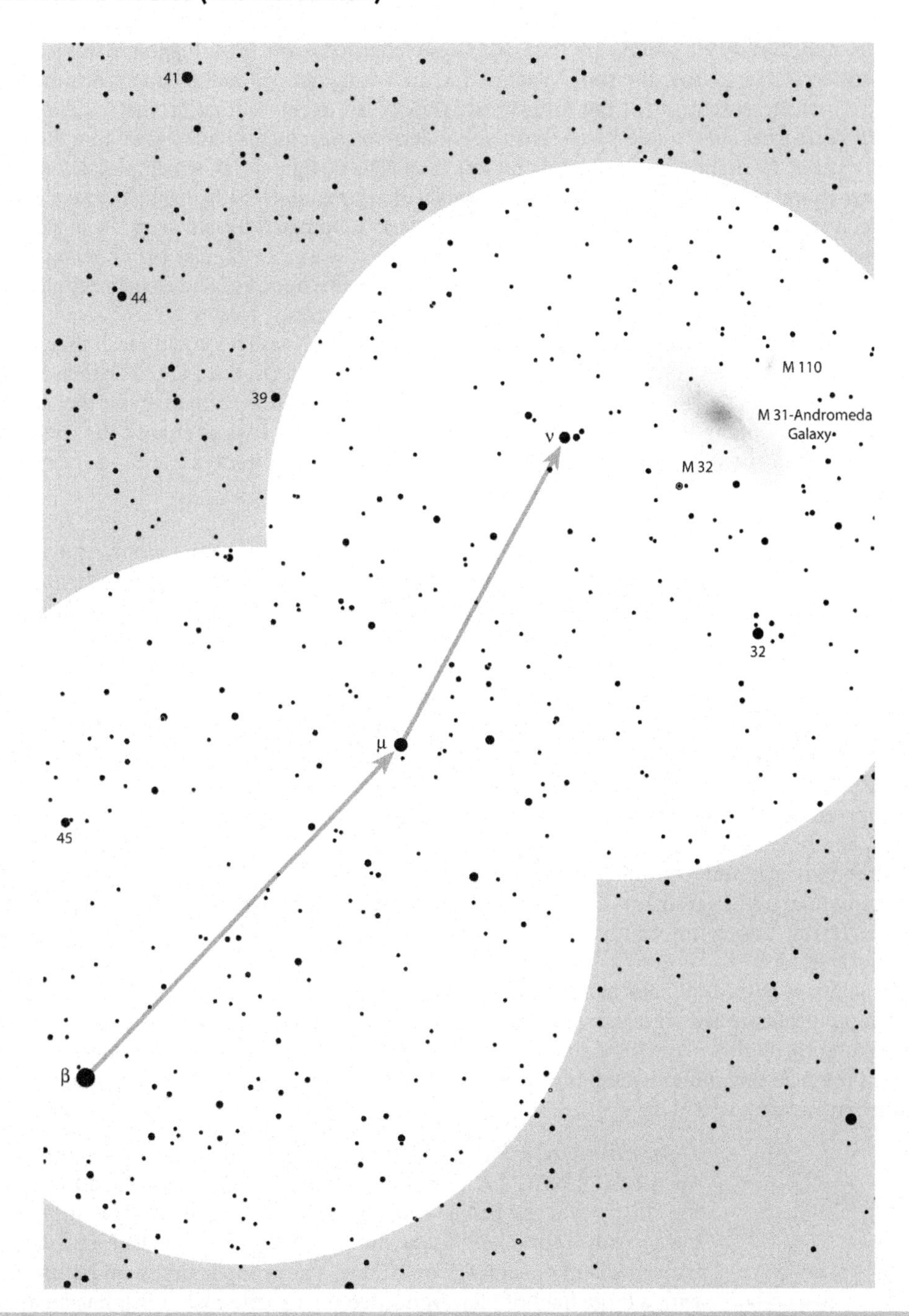

Figure 7.2A. The leading star to the famous galaxy M 31 in Andromeda is bright Beta Andromedae

The Andromeda Galaxy is a member of the Local Group of galaxies and is the biggest in this association, which includes our own galaxy, the spiral galaxy M 33 in Triangulum as well as numerous other dwarf galaxies. The first measurements of the Andromeda Galaxy led to the estimation that its diameter was 110,000 light years from side to side. More sensitive modern measurements have shown that the galaxy is almost twice as big. Its diameter is just a little bit less than 200,000 light years, which places it among the largest known spiral galaxies! Our galaxy has a diameter of approximately 130,000 light years.

Andromeda is a rather normal spiral. The central bulk is dominated by old stars. In it we can find numerous red and yellow giants, but not a lot of interstellar dust and gases. Similar to our galaxy the spiral arms include a lot of dust and gas as well as blue, hot, young stars like our Rigel and Deneb, nebulae like our Orion Nebula, planetary nebulae, as well as globular and open clusters.

Recent observations have shown that the galaxy has a double nucleus. This could mean that it collided with another large galaxy and merged with it at some point in the past. Or, there could be dense clouds of dust and gas in the vicinity of its core that are not transparent to light and so seem to divide the core in two.

The Andromeda Galaxy is relatively large and bright in the sky. It is probably the most photographed object in the celestial sphere. It is suitable for beginners as well as experienced astrophotographers.

The Andromeda Galaxy has at least 14 companions, gravitationally bound to it that circle around the common center of mass and form a subgroup of the Local Group. The brightest of these is the spiral galaxy M 33 in Triangulum (described on page 468). All others are dwarf elliptical or spherical systems. Two that lie in the vicinity of the main galaxy are rather bright and can be seen through binoculars.

M 32 (8m1/$8' \times 6'$) is found approximately 22 arcmin south of the center of the Andromeda Galaxy. M 110 (8m5/$17' \times 10'$) lies approximately 35 arcmin northwest from the center of M 31, but to see it the observing conditions have to be excellent (Figure 7.3). Both of these satellites appear in all photographs of the Andromeda Galaxy and are therefore well known by amateur astronomers.

The other two brighter companions (NGC 147 in NGC 185), which are too faint to be seen through binoculars, lie far from the main galaxy, some 7 degrees north in the constellation Cassiopeia. If the observing conditions are excellent, both can be seen in a telescope with a 10-cm objective lens.

Satellite galaxies with designations Andromeda I, II, III ... X are so dim that they can be captured only with large professional telescopes. Andromeda VIII is of special interest to astronomers, for it is currently going through a really hard time. The Andromeda Galaxy captured it into its gravitational vice, and in the next few million years the small galaxy will be torn to pieces, with its stars dispersed across the main galaxy.

Approximately 5 degrees south and slightly west of Gamma lies the open cluster NGC 752 (5m7/60′) (Figure 7.6). It is not hard to find it, for the cluster and the bright star are in the same field of view of binoculars, but to truly appreciate it you should observe it on a clear and dark night with excellent observing conditions. The group is best seen in instruments with a large field of view, mounted on a stable rack, for it has an apparent diameter of 60 arcmin. The cluster is fully resolved in the binoculars, but how many individual faint stars can be seen depends mostly on observing conditions. On an average night you can see approximately 25 stars. Because Andromeda is at its culmination high in the sky (almost in the zenith in the mid-northern latitudes), you can see approximately 60 stars during excellent observing conditions. On a perfect winter night your binoculars could be pushed to see stars of magnitude 11. In this event, the cluster will appear to be extremely rich (from 90 to 110 stars) and very prominent. The stars in the area of the cluster number almost 600.

Figure 7.3. The Andromeda Galaxy with its elliptical companions M 32 (beneath the galaxy's core) and M 110 (bigger, slightly oval, spot of light above the galaxy)

NGC 752 has another good property for observers: it is very colorful. The brightest and most prominent southwestern couple, labeled 56 Andromedae, consists of two rather bright, conspicuous orange stars with magnitudes 5.8 and 6.1, separated by 3.3 arcmin (p. a. 299°, 2001). The next brightest

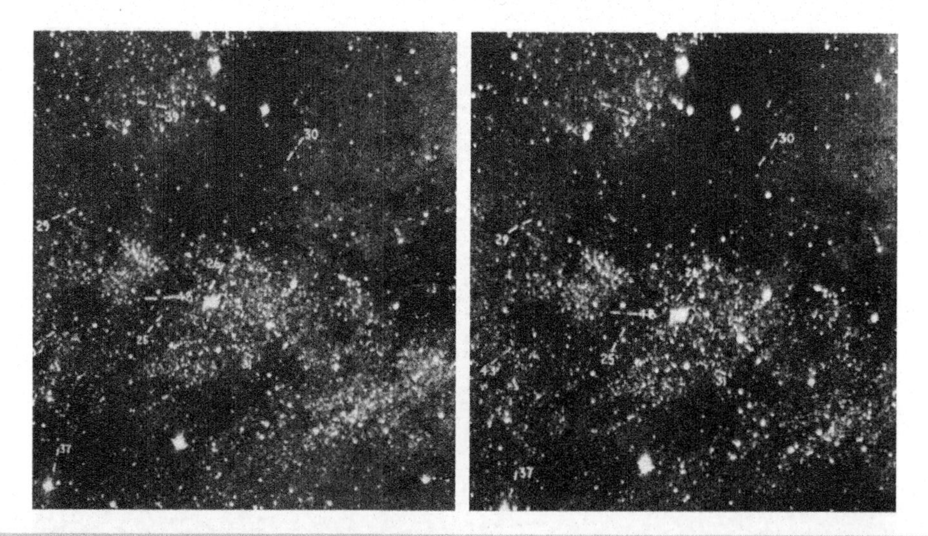

Figure 7.4. On photographic plates, obtained with the 2.5-m reflector on Mount Wilson, individual stars in the Andromeda Galaxy are seen. Within this crowd of stars Hubble looked for Cepheid variables (they are marked on the image). When he found one, he determined its period of pulsation and luminosity. From these data and from the apparent magnitude he could calculate the distance of that Cepheid and therefore the distance of the Andromeda nebula

star is yellow, and some other stars are orange. Although it is true that through a larger telescope we can see more stars, the overall impression of the cluster is completely lost due to the smaller field of view.

As we mentioned earlier, most of the stars are gathered in a field with a 60 arcmin diameter, but some individual members can be found as far as 70 arcmin from the center of the cluster. NGC 752 is approximately 1,500 light years from us, its real diameter is 26 light years, and the most distant members are 30 light years from the center.

Here now is something for those observers who like a challenge. The planetary nebula NGC 7662 (8m3/32″x28″), also known as The Blue Snowball Nebula, can be seen through binoculars as a magnitude 8 object (Figure 7.7). The two stars that lead us to it are both bright – Kappa and Iota Andromedae are both magnitude 4. If Iota is moved to the north-eastern edge of the field of view, you will see four magnitude 6 stars in the middle; three of them are labeled 9, 10, and 13 Andromedae. The planetary nebula lies less than half a degree southwest from 13 Andromedae, but in order to find the exact location it is helpful to first find a little upside down heart, consisting of stars with magnitudes 8 and 9.

Figure 7.5. The southwestern spiral arms of the Andromeda Galaxy. In this image the dark lanes of dust and gas clouds can be clearly seen. The brightest spot is a huge star cloud, so bright that it was given its own number in the NGC catalog – NGC 206. The cloud can also be seen as a faint spot of light in amateur equipment; we just need a telescope with a 20-cm aperture. The brightest stars in the cloud are magnitude 16, but in fact these are all young, hot, and luminous super giants such as our Rigel and Deneb. The stars are still veiled in the nebulosity from which they were born. Just for comparison, our Sun would be a magnitude 29.5 star from this distance and we could scarcely see it through the Hubble Space Telescope

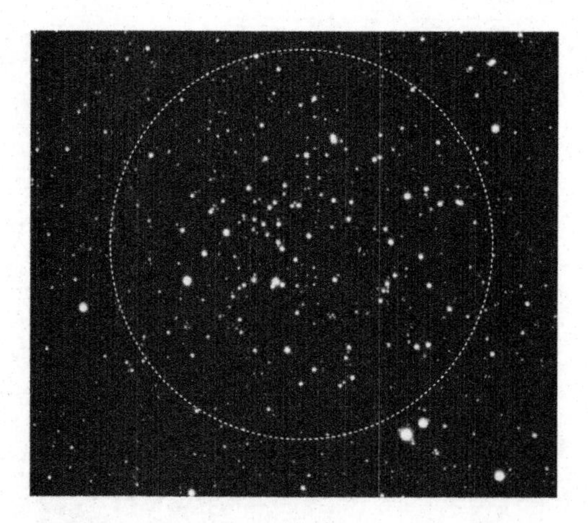

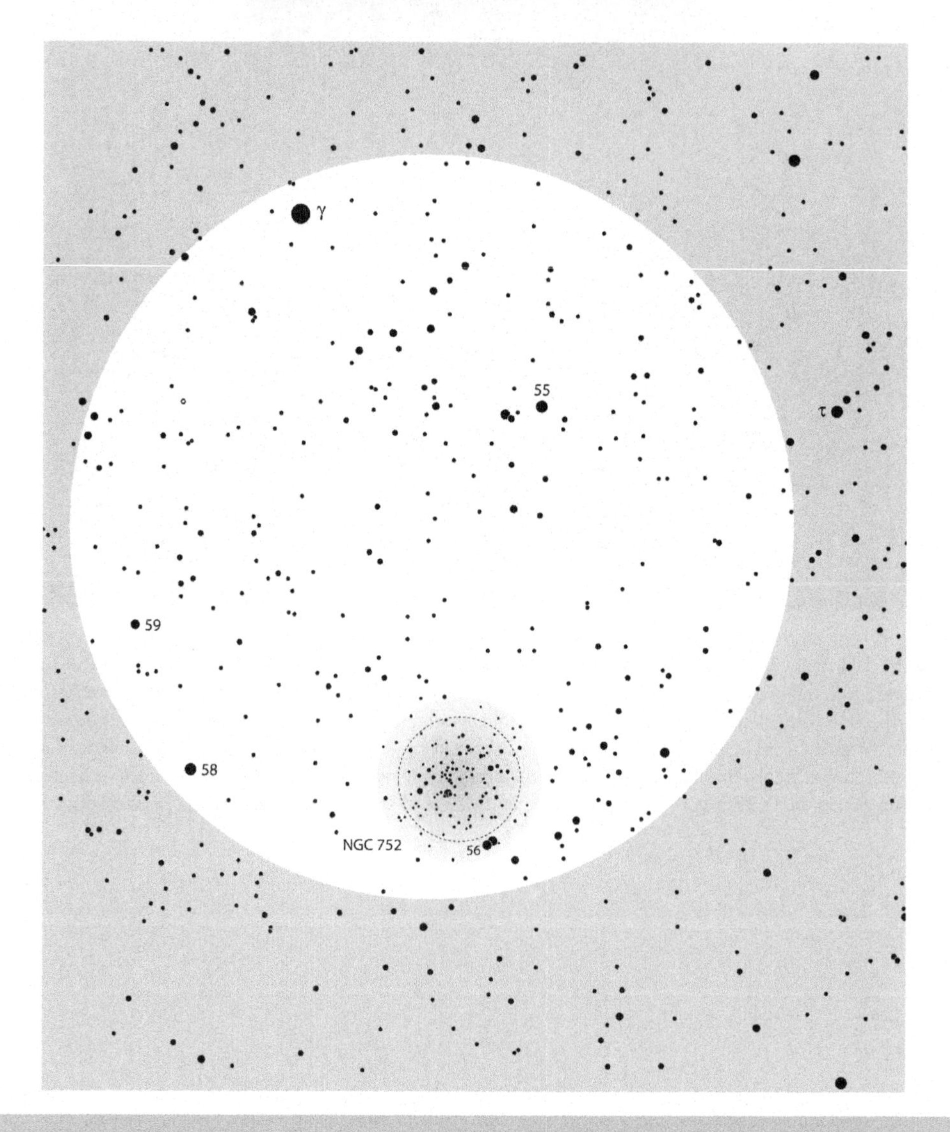

Figure 7.6. The leading star to the open cluster NGC 752 is the bright Gamma Andromedae

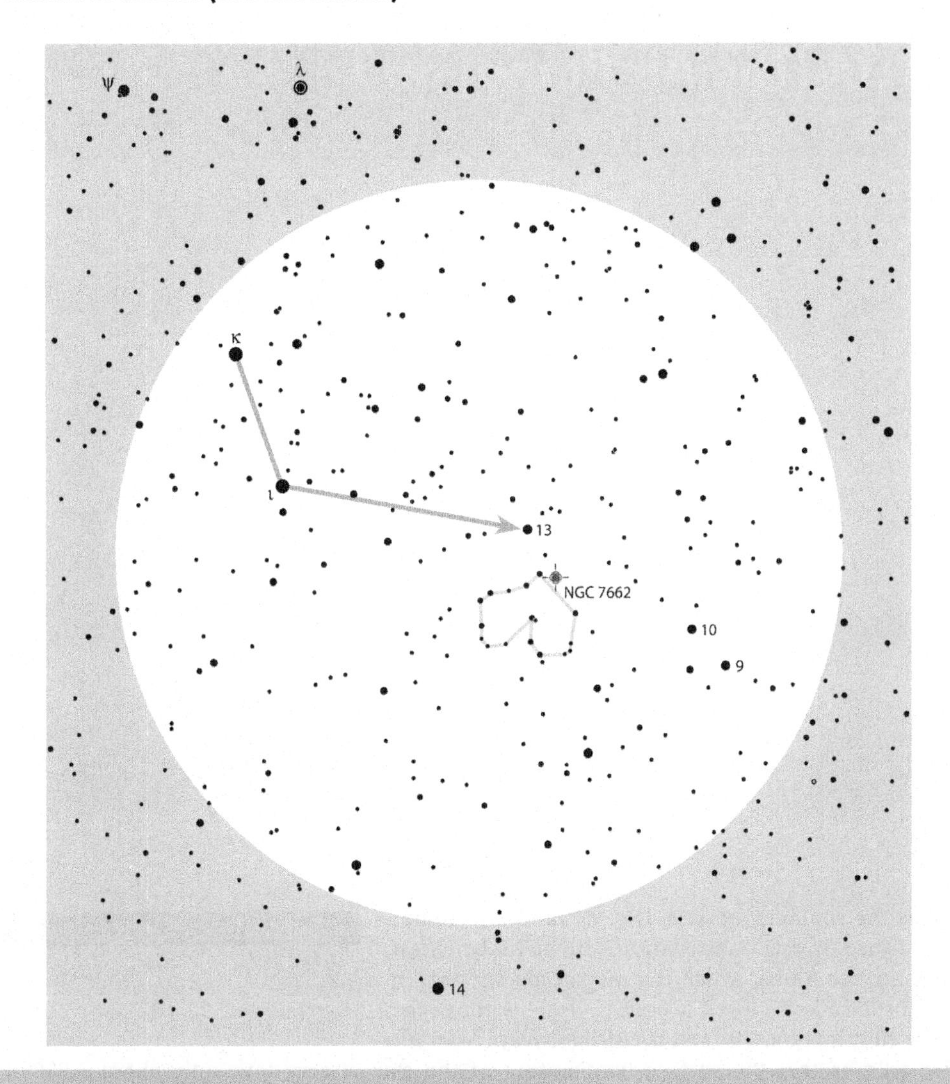

Figure 7.7. The planetary nebula NGC 7662 appears in the field of view of binoculars exactly the same as the surrounding stars. Why should anyone try to find a star that is merely a dot with binoculars? The answer is simple. With binoculars that have a wide field of view the search for faint objects is much easier than with an average amateur telescope. Once we have learned to find the way from the bright stars to the chosen faint celestial object, we can use this knowledge later on when we will own a larger telescope!

ANTLIA (The Air Pump) and PYXIS (The Compass)

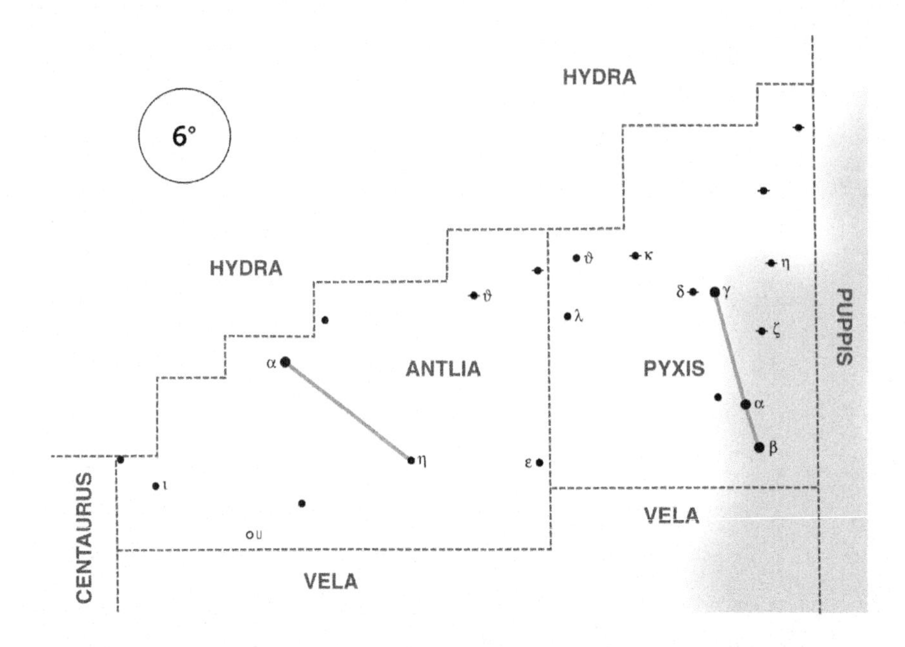

CONSTELLATIONS CULMINATE		
mid January at 2 a.m.	mid February at midnight	mid March at 10 p.m.

Low above the southern horizon lies the faint and rather undistinguished spring constellation Antlia. Its brightest star is the orange Alpha, which is a magnitude 4.3 star. In the mid-northern latitudes, it is only 13 degrees above the horizon at culmination. The star is 365 light years from us. Its color tells us that the surface temperature of the star is very low, only about 3,900 K. The constellation does not contain any interesting objects for observers with binoculars.

Pyxis lies beneath the head of Hydra and east of Puppis. It is a small and undistinguished spring constellation with only two stars brighter than magnitude 4: Alpha (3m7) and Beta (4m0). In mid-northern latitudes, Alpha is only 11 degrees above the horizon at culmination. But even if it were in the zenith, the constellation does not contain any objects that might be of interest to observers with binoculars.

AQUARIUS (The Water Carrier)

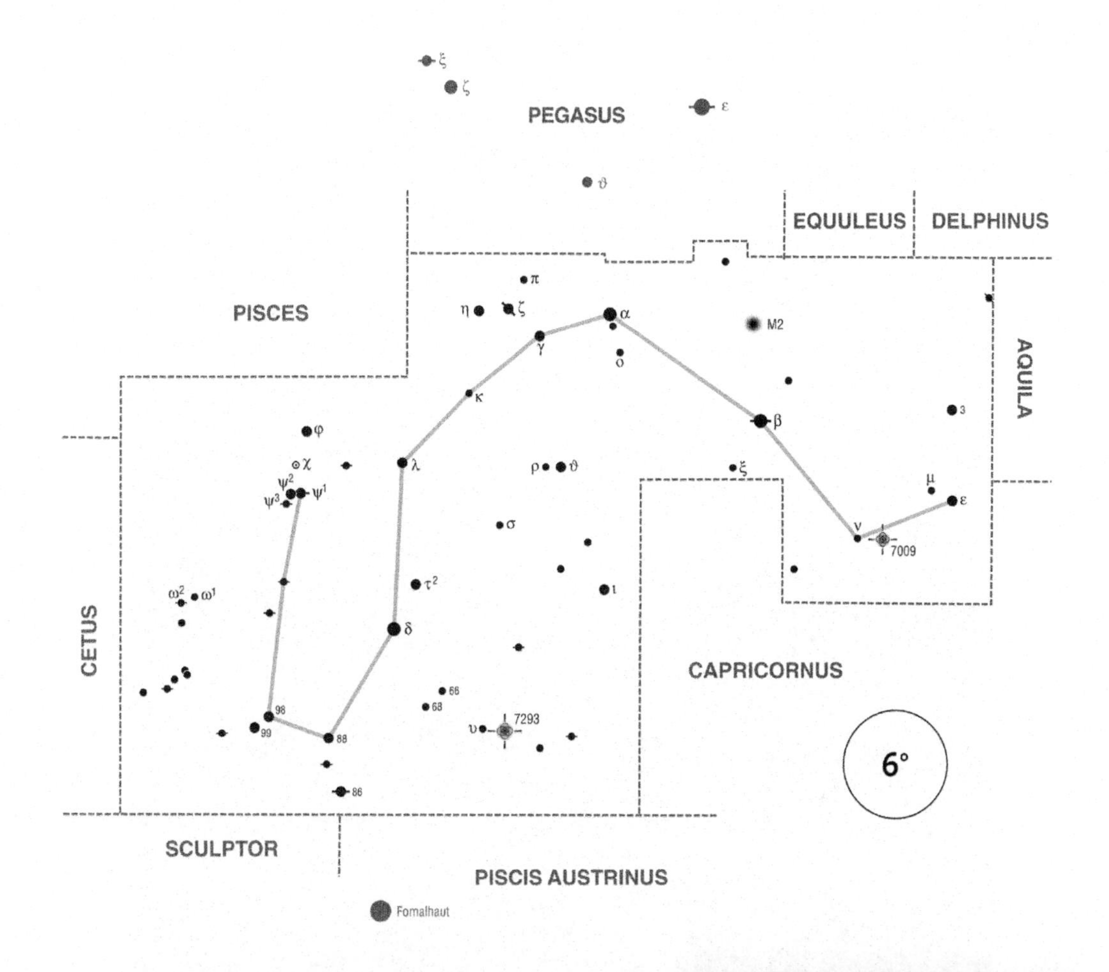

CONSTELLATION CULMINATES		
end July at 2 a.m.	end August at midnight	end September at 10 p.m.

Aquarius is an autumn constellation of the zodiac, situated under Pegasus and north and east of Capricornus. Although it is one of the larger constellations, it is faint. The easiest way to find it is to extend the connecting line between Alpha Andromedae and Alpha Pegasi (both are stars of the asterism called the Square of Pegasus) and follow the line of stars of which the brightest is Alpha Aquarii.

The brightest stars within the constellation are Alpha, Beta (both 2m9), Delta (3m3), and Zeta (3m6).

The yellowish Alpha is a giant, approximately 1,100 light years away from us. Its luminosity is over 6,000 times that of the Sun. The star has a similar surface temperature to our Sun, but since it is larger, its luminosity is proportionately greater.

The yellowish Beta is similar to Alpha. Its distance from us is a little more than 1,000 light years, and its luminosity surpasses our Sun by 5,800 times.

Zeta is a close double with stars of magnitudes 4.3 and 4.5, separated by 1.9 arcsec (p. a. 172°, 2006), so we cannot split them with binoculars. The pair is 75 light years away from Earth. Zeta is an excellent star to test the quality of the optics of smaller amateur telescopes.

Globular cluster M2 (6m5/16′) is shown in the binoculars as a bright, clearly seen, but small spot of light only a few arc minutes in diameter (Figure 7.8). The cluster is easy to find, since it forms a right-angled triangle (with the cluster in the right corner; see chart on next page) with Alpha and Beta). If Beta is caught in the field of view and is moved to the southern edge, the cluster appears on the northern edge.

Figure 7.8. Globular cluster M2

M2 is approximately 37,500 light years from us, and its actual diameter is 175 light years. It contains at least 150,000 stars, the brightest of which are red and yellow giants that shine with magnitude 14 and 15 in our sky. If it seems faint, just remember that our Sun would be a star of magnitude 21 if we saw it from as far away as the cluster!

Planetary nebula NGC 7009 (8m0/44″x23″), popularly named the Saturn Nebula, lies only a degree west from star Nu (4m5) and is easy to find (Figure 7.9). However, through binoculars it is seen merely as a hazy magnitude 8 star. In larger telescopes or on photographs, it has a greenish hue. This light is emitted by double-ionized oxygen atoms that were excited by the ultraviolet light from an extremely hot central star that

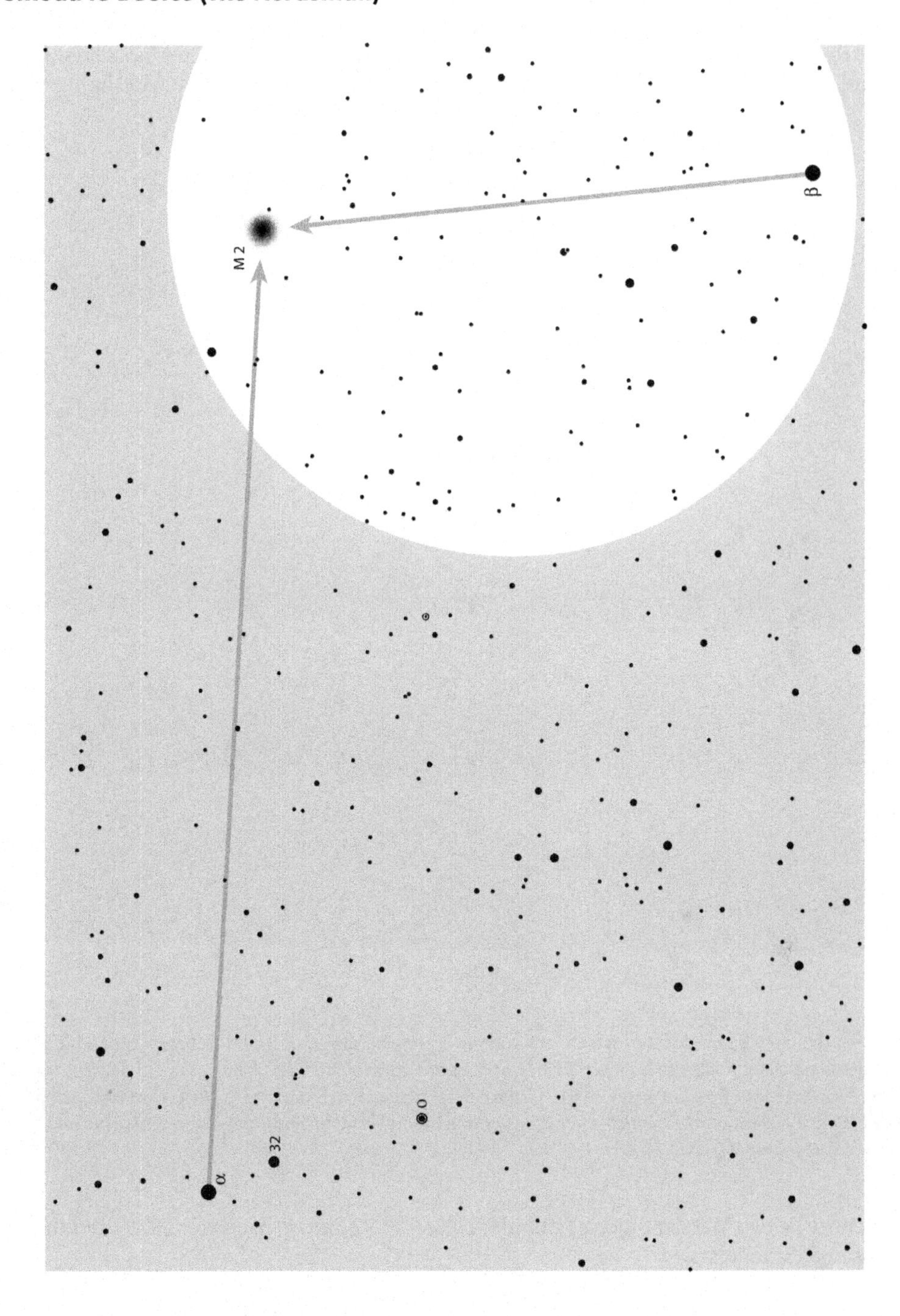
M 2
β
α
32

lights the nebula. The star is a magnitude 12 blue dwarf and cannot be seen in binoculars. Its surface temperature is estimated to be 55,000 K. The planetary nebula is approximately 2,400 light years away.

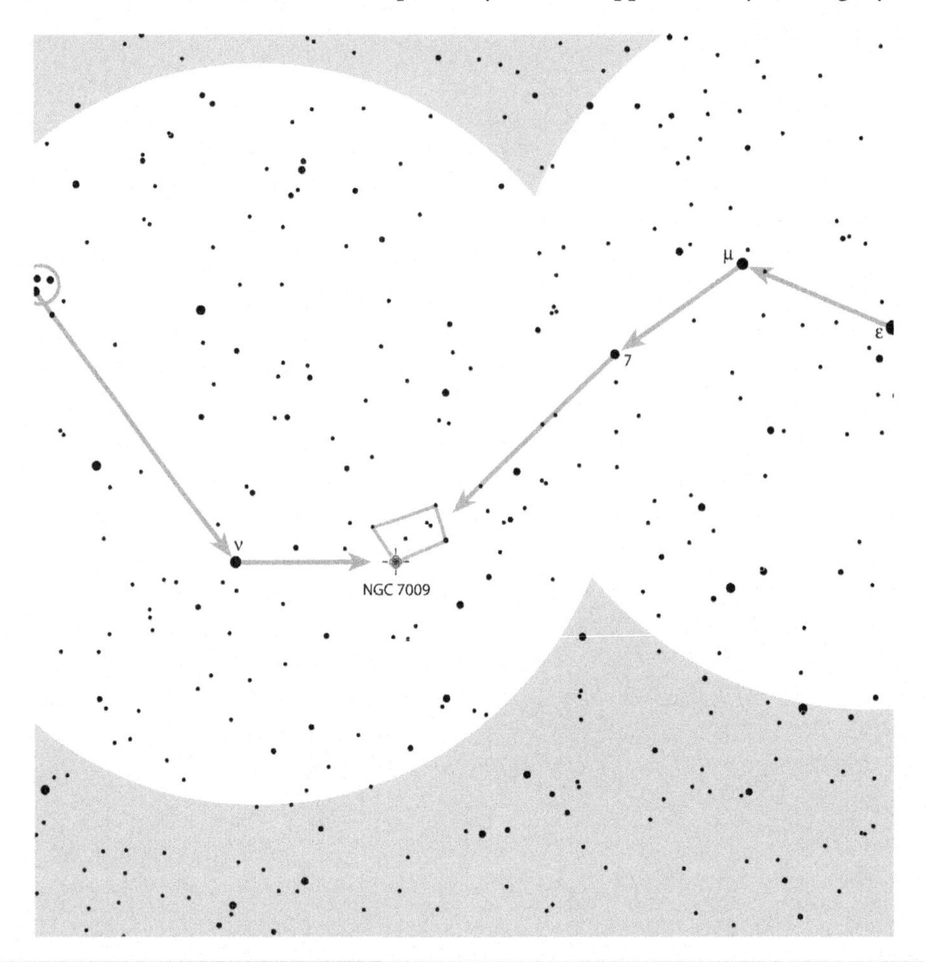

Figure 7.9. The starting point for your search for the planetary nebula should be the bright Beta, which is hidden behind the left edge of the chart. If the star is moved to the northeastern edge of the field of view, some 5 degrees southwest a magnitude 7 triplet of stars (marked on the chart with the circle) can be seen. From here to Nu it is less than 3 degrees. The planetary nebula is in the lower corner of the trapezium, which lies a little more than 1 degree west from Nu and cannot be missed in excellent observing conditions

In its vicinity lie two Messier objects (globular cluster M 72 and open cluster M 73), but they cannot be seen through binoculars.

It is interesting that Charles Messier did not include the planetary nebula NGC 7293 (7m3/16′×28′) in his catalog, although it can be seen even through binoculars. Despite the fact that the nebula is rather bright, its surface brightness is low. In excellent observing conditions, it is seen as a big, faint, and slightly oval spot of light. The star that leads to the nebula is the bright Delta. From there we pass over the stars labeled 66 and 68 to get to Upsilon, which is of magnitude 5. The planetary nebula lies just 1 degree west of it.

As in all planetary nebulae the central star of NGC 7293 is an extremely hot dwarf and is over 50 times smaller than the Sun (i.e., the star is only 2.5 times bigger than Earth!), but its surface temperature is extremely high – over 100,000 K. With its ultraviolet light the star excites gases within the nebula, causing them to shine in wonderful red and green colors, as seen on photographs (Figure 7.10). The red light is emitted by ionized hydrogen atoms, while the green light is emitted

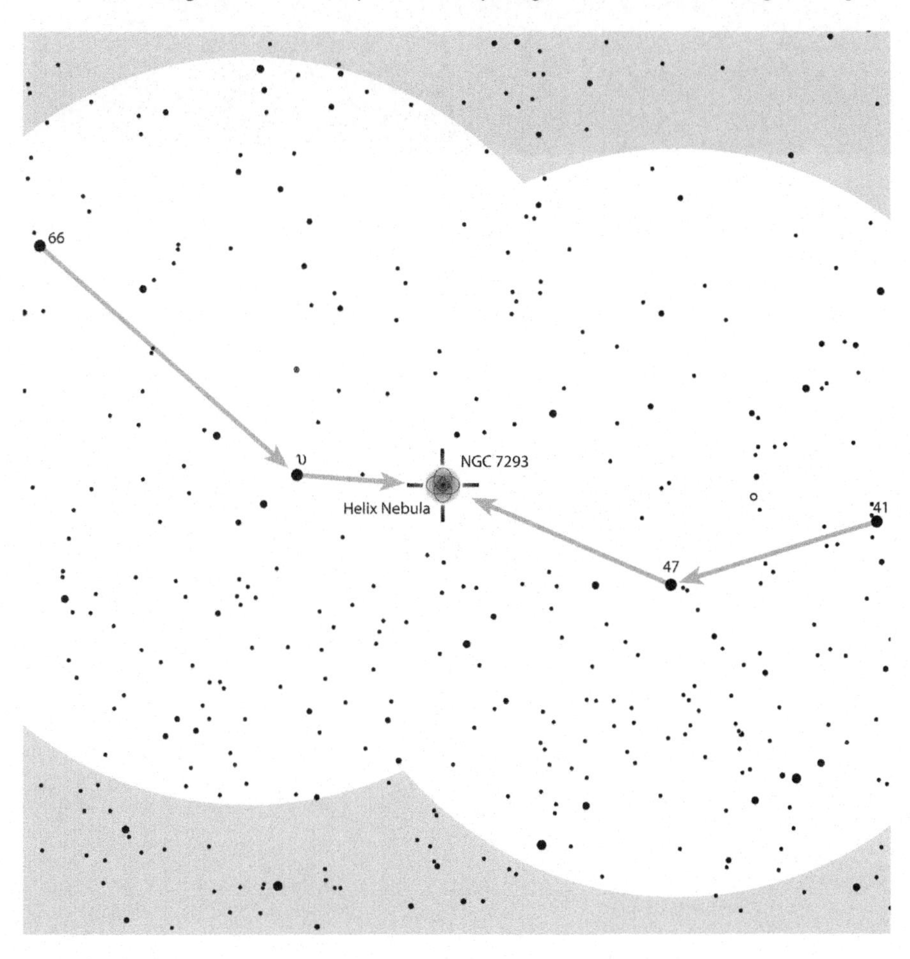

Figure 7.10. Planetary nebula NGC 7293

by double-ionized oxygen atoms. Of course, the nebula in smaller quantities also includes other atoms. The central star is of magnitude 13 and cannot be seen through binoculars.

NGC 7293 is approximately 450 light years from us and is the nearest planetary nebula to Earth.

AQUILA (The Eagle)

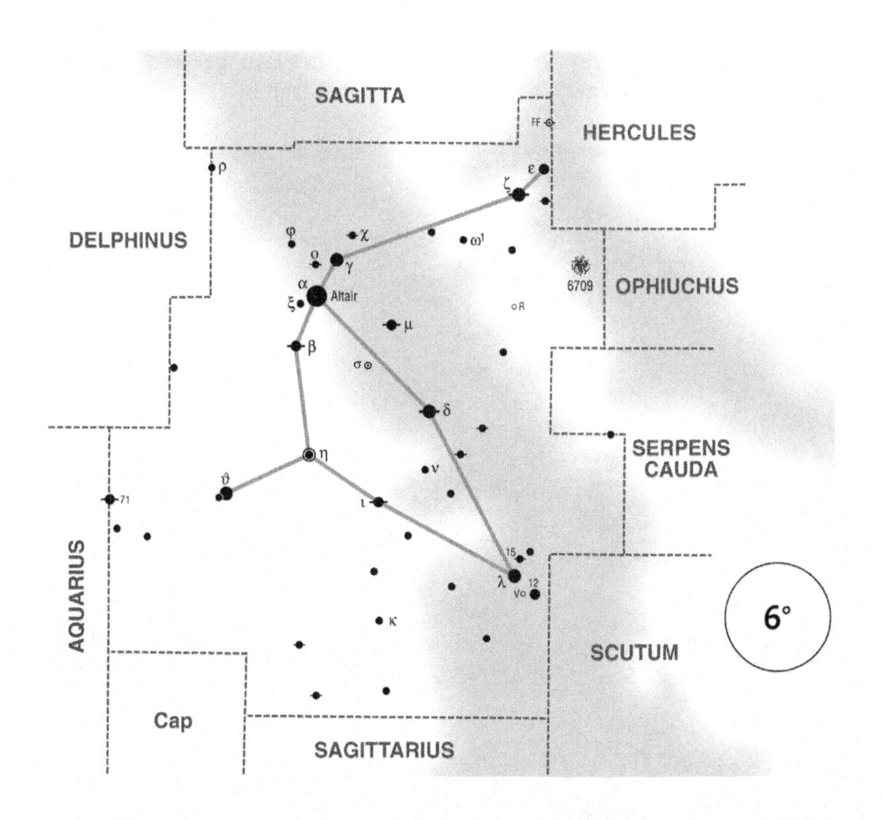

CONSTELLATION CULMINATES		
mid June at 2 a.m.	mid July at midnight	mid August at 10 p.m.

Aquila is a large and splendid constellation in the summer skies. The brightest stars are Alpha or Altair (often Atair; 0m8), Gamma (2m7), and Zeta (3m0).

Altair is the 12th brightest star in the sky. Together with Deneb in Cygnus and Vega in Lyra it forms the famous asterism known as the Summer Triangle. The star is a mere 16.8 light years away, so it is among the closest bright stars. Its diameter is 1.5 times that of the Sun, it has a luminosity of 10 Suns, and it is white in color. The most interesting feature related to this star is the fact that it rotates around its axis extremely quickly, which astronomers concluded from studying its spectrum. The velocity on the star's equator reaches 260 km/s, and the rotational period is only 6.5 h! As a comparison, our Sun needs 25.4 days for one revolution! Astronomers are convinced that the shape of Altair is closer to a flattened ellipsoid than a sphere.

The yellow-orange Gamma is the 116th brightest star in the sky. It is 462 light years from us and is actually a giant with a luminosity of 1200 times that of the Sun!

South of Altair we can notice three stars in an almost straight line. These are Theta (3m2), Eta (variable), and Delta (3m4). Eta is a Cepheid variable star and is a true giant. It changes its brightness from 3m6 to 4m4 as precisely as a "Swiss timepiece," with a period of 7,176641 days. It is 1,200 light years from us, and when it is at its brightest its luminosity surpasses our Sun by 3,000 times. Proper stars to compare the brightness of Eta are Beta (3m7) and Delta.

The constellation lies in the summer Milky Way, thus the entire area is rich in star fields. Beside countless stars and groups of stars we can also see dark nebulae, which are clouds of nontransparent gas and dust in the equatorial plane of our galaxy. Across this constellation runs the so-called Great Rift, a dark lane of nontransparent clouds that apparently divide the Milky Way in two; such dark lanes can be seen in some other spiral galaxies such as the Sombrero in Virgo or NGC 4565 in Coma Berenices.

Figure 7.11. The best known or better described as the darkest (most contrastive) nebulae have the designations B 142 and B 143, and they are situated approximately 1 degree west of Gamma. The nebulae are clearly visible through binoculars; however, they do not show as much contrast as photographs reveal. You should observe them only on really clear and dark nights

To stroll across this part of the Milky Way, it is best to use binoculars with a wide field of view and small magnification (Figure 7.11).

The variable star R Aquilae is a red giant of the Mira type. At maximum, it sometimes reaches naked eye visibility (5m5), but at minimum its brightness slips far below the limit of binoculars (12m). The pulsating period is 284 days; the star reached one of its maximums on March 16, 2008. From this date and from the known period, one can easily calculate the approximate dates of the next maxima.

R Aquilae is one of the coolest stars known. Its surface temperature changes during the pulsating period between 3,500 and 1,900 K. This can be easily deduced from its color, which is a distinct reddish-orange. When the brightness of the star slips toward the minimum, the color becomes increasingly reddish, but unfortunately this effect cannot be seen through binoculars. The star lies in a rather blank area of the sky. The leading star is Zeta, which lies north just a little bit less than one field of view of the binoculars (Figure 7.12A).

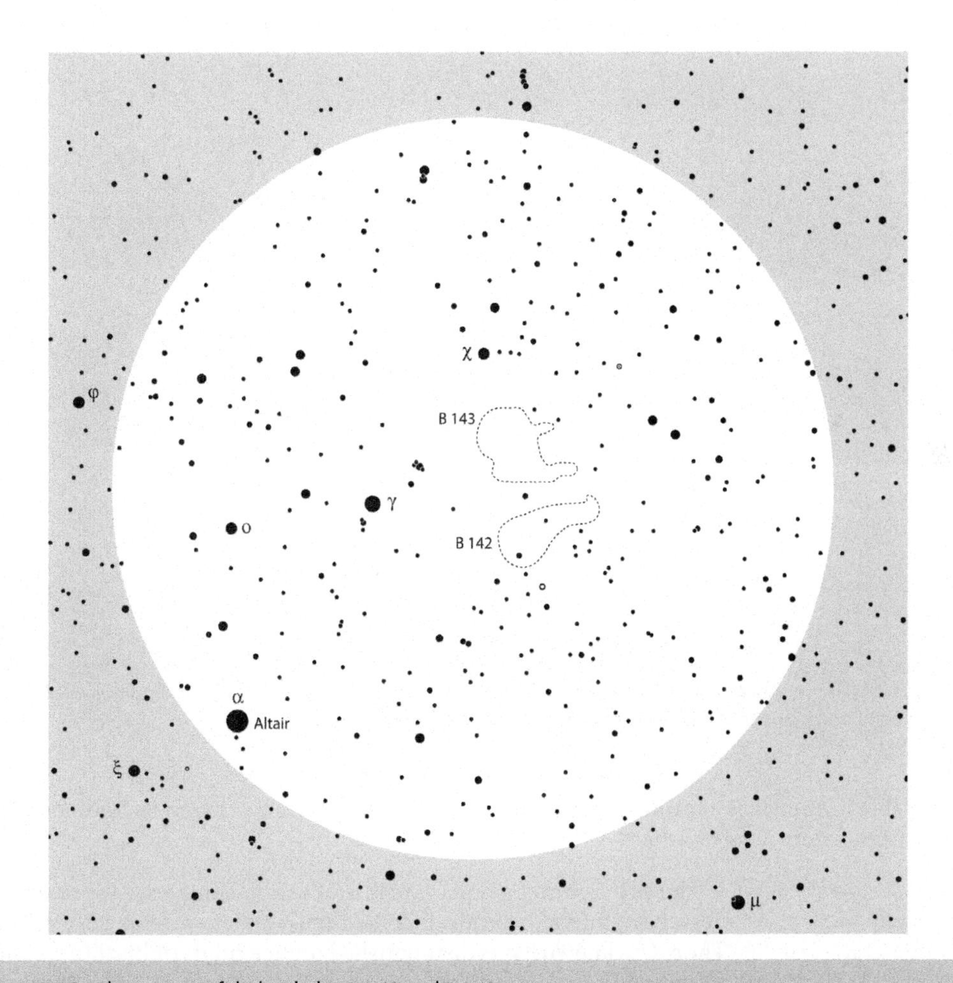

Figure 7.12. The position of dark nebulae B 142 and B 143

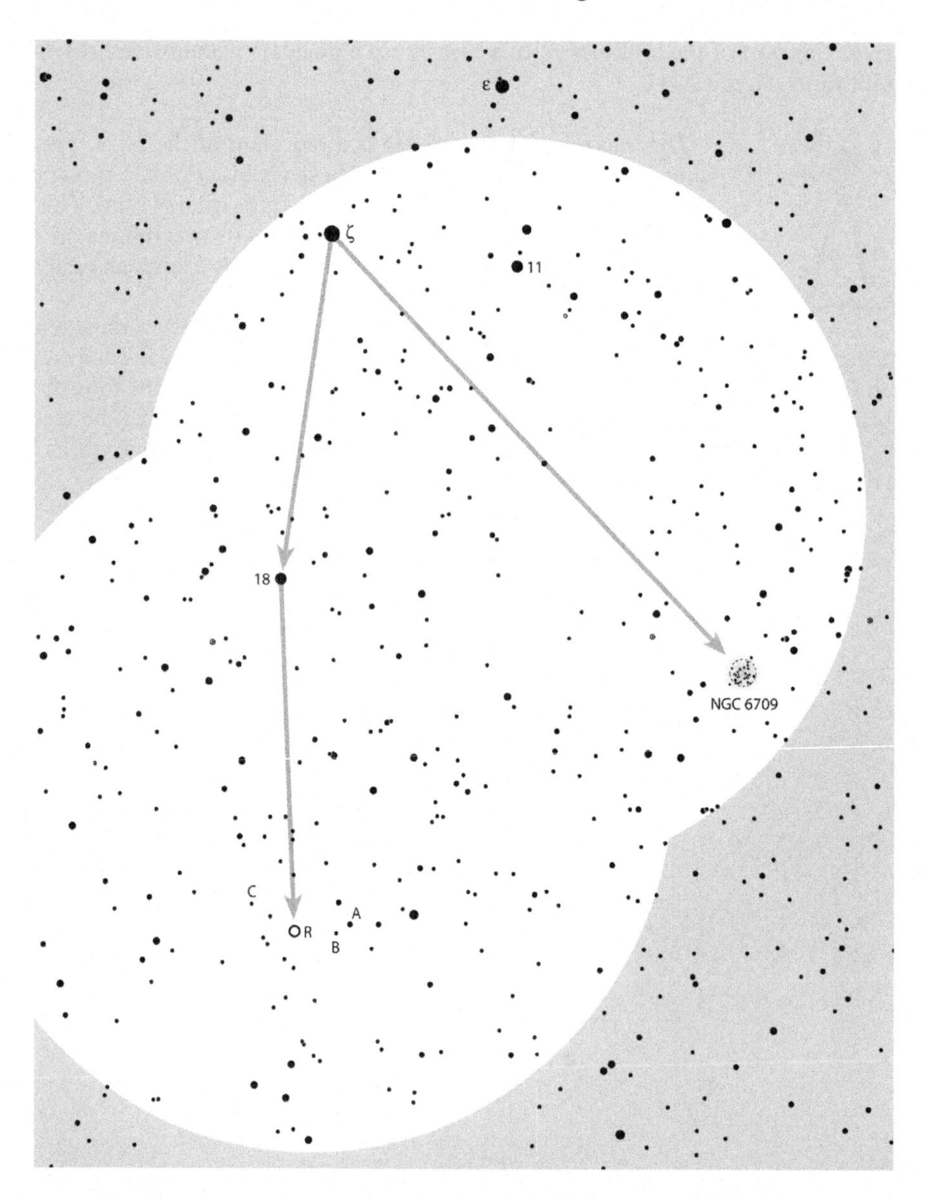

Figure 7.12A. The neighborhood of the variable R with some comparison stars: A (7m8), B (8m5), and C (9m1). The variable star and the open cluster have the same leading star, the bright Zeta

The variable R Aquilae is approximately 700 light years from us. At maximum its luminosity is 200 times that of the Sun.

The sky around Lambda (3m4) is of extreme interest. Approximately 1 degree southwest, we can find the semi-regular variable star V Aquilae. The star is known by its distinguished orange color. Its brightness changes from 6m6 to 8m4 with an average period of 350 days. When identifying the variable star you can make it easier by using the detailed chart that shows its surroundings, including bright Lambda and a magnitude 4 star called

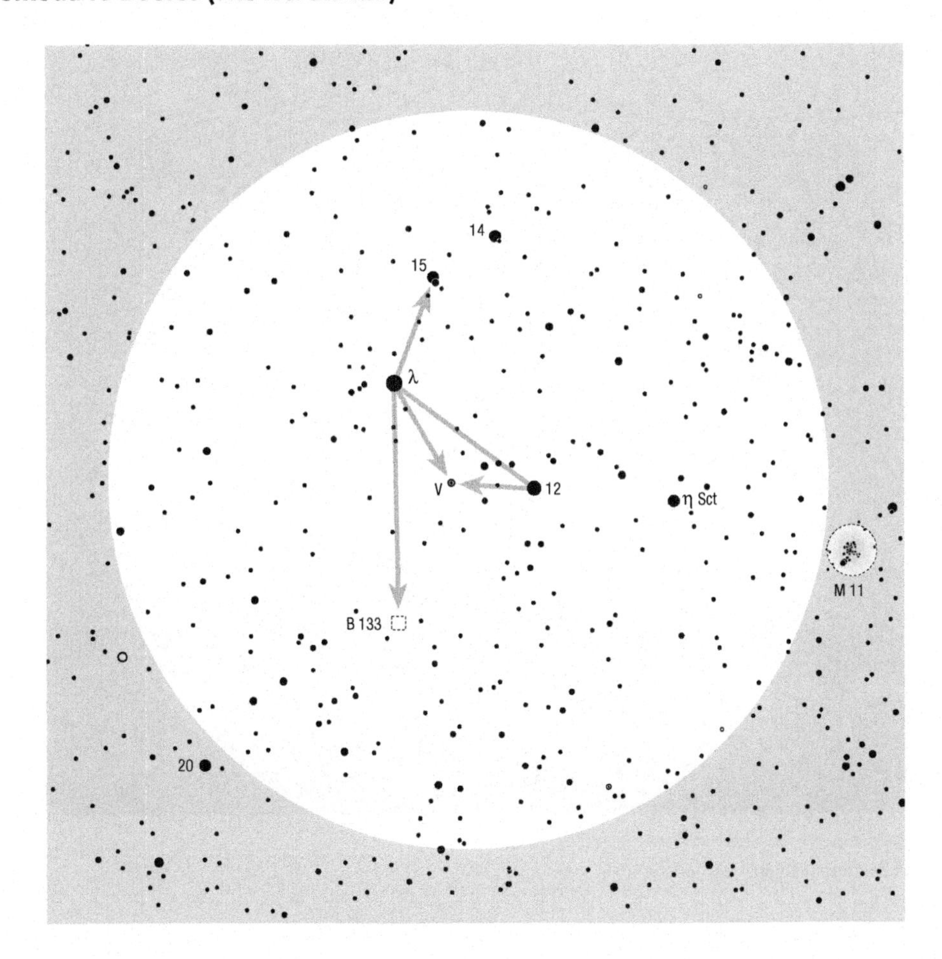

12 Aquilae. This star has a distinguished yellow-orange color and provides a nice contrast for the deeper orange variable. V Aquilae is seen in the binoculars during the entire period. The adjective "semi-regular" tells us that we cannot expect it to behave according to a determined schedule.

In the immediate vicinity we can find a close pair designated as 15 Aquilae. The yellow-orange stars of magnitudes 5.5 and 7.0 are 40.2 arcsec apart (p. a. 210°, 2006).

Some 2 degrees south from Lambda lies the dark nebula B 133 (10′ × 15′), which can be seen through binoculars only with perfect observing conditions and when the sky is particularly dark. You also need plenty of observing experience to notice it.

The open cluster NGC 6709 (6m7/13′) lies in northwestern part of the constellation. This is a group of approximately 40 stars assembled in a rather small part of the sky. The star leading to the cluster is Zeta. If we catch Zeta in the field of view and move it to the northeastern edge, NGC 6709 will appear on the southwestern edge. (The chart showing this can be found alongside variable R on previous page.) The cluster lies in a star-rich field of the Milky Way, which is why it is hard to recognize. In order to see it, you have to choose really clear and dark summer night. The cluster will appear in the field of view as a small and tenuous spot of light. Two stars always shine from the nebula, and with averted vision you can see them even better.

Figure 7.13. Open cluster NGC 6709

ARIES (The Ram)

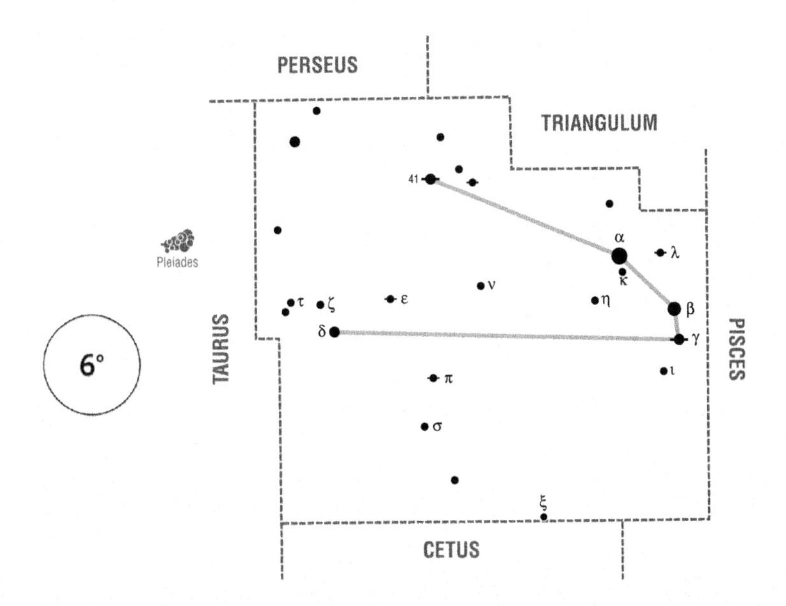

CONSTELLATION CULMINATES		
end September at 2 a.m.	end October at midnight	end November at 10 p.m.

This autumn constellation of the zodiac lies between Pisces and Taurus. It is rather undistinguished, since it contains only three bright stars: Alpha or Hamal (2m0), Beta or Sharatan (2m6), and Gamma (3m9).

The yellow-orange Alpha is the 50th brightest star in the sky. In fact it is a giant 66 light years away. Its luminosity is 49 times that of the Sun in spite of its lower surface brightness of 4,500 K.

Beta is the 104th brightest star in the sky. It is somewhat closer to us (60 light years) than Alpha. Its luminosity is 22 times that of the Sun.

Gamma is a triple star in which the brighter two components are of similar brightness (Gamma A – 4m5/Gamma B – 4m6) and are 7.6 arcsec apart (p. a. 0°, 2006). The pair cannot be separated through binoculars. The third component, Gamma C, is of magnitude 8.6, lies 217 arcsec from Gamma A (p. a. 83°, 2001) and can be seen through binoculars although it is rather faint.

Lambda is a binary star that can be separated through binoculars. Stars with magnitudes 4.8 and 6.6 are 38.2 arcsec apart (p. a. 47°, 2006), so they look like a close pair in the field of view. The brighter star is white, while the fainter one is slightly yellowish, but its color is not clearly distinguished through binoculars due to its faintness. The stars, which are gravitationally linked, circle around a common center of gravity.

AURIGA (The Charioteer)

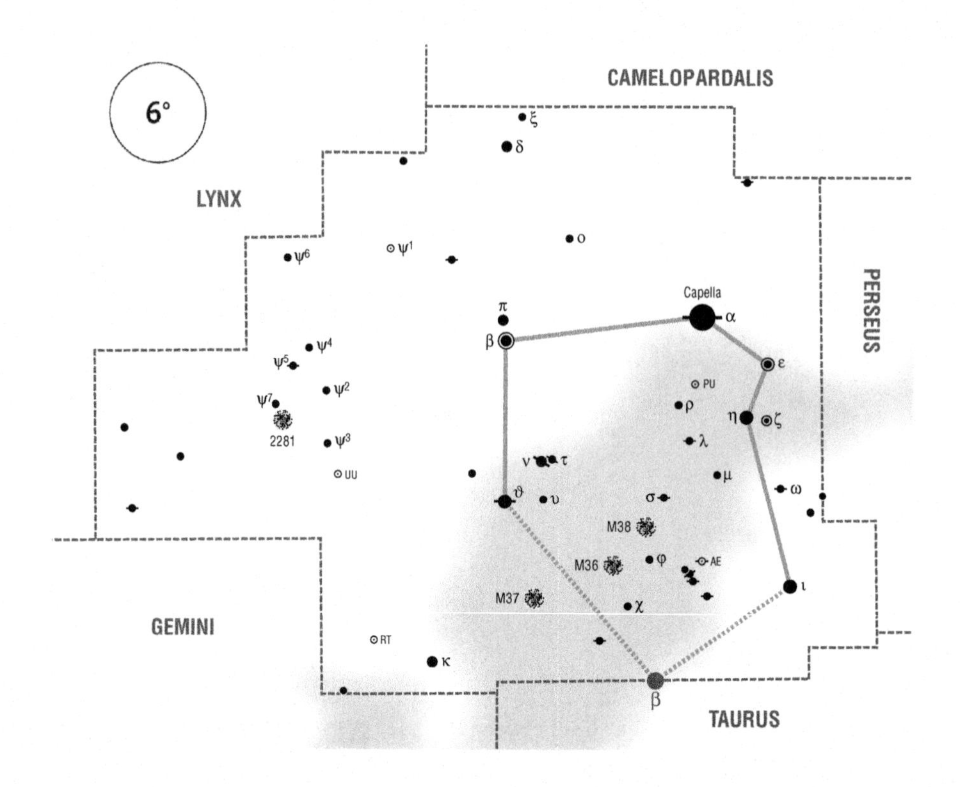

CONSTELLATION CULMINATES		
mid November at 2 a.m.	mid December at midnight	mid January at 10 p.m.

Auriga is one of the brightest winter constellations in the northern celestial hemisphere. It is easy to recognize, since its five leading stars form a clearly visible irregular pentangle. The brightest stars in the constellation are Alpha or Capella (0m08), Beta (1m9), Theta (2m6), Iota (2m7), Epsilon (3m0 at maxima), Eta (3m2), Delta, and Zeta (both 3m7).

Capella is the sixth brightest star in the sky. It is yellowish, like our Sun; its color is even more intense in binoculars. This star is our close neighbor, only 42 light years away from us. Its luminosity surpasses our Sun by 120 times. From its proper motion, astronomers infer that the star might be a distant member of the open cluster Hyades (in the constellation Taurus).

Capella is an interesting multiple star. The components labeled A and B represent a close binary, so close that it was only discovered in 1899 with spectroscopic observations. The pair cannot be separated, even with the largest telescopes in the world. In order to see them as two stars, we need a special technique called long baseline optical interferometry. Astronomers were able to separate the stars with interferometer Mark III at the Mount Wilson Observatory for the very first time in 1994. Capella A and B are only 100 million km apart (the approximate distance between Venus and the Sun), and they move around a common center of gravity with an orbital period of 104 days. The luminosity of Capella A is 70 times that of the Sun; Capella B shines as bright as 45 Suns.

The vicinity of Capella is so rich with faint stars that in the past they were named Capella C, D, E, F, and G, but they are all close to their bright neighbor only when observed from Earth. Physically (gravitational) connected to Capella is a red dwarf, which was given the name Capella H. This is a faint magnitude 10 star (on the border of visibility in binoculars), which lies approximately 12 arcmin southeast from Capella (p. a. 141°, 1895). The actual distance between the Capella A–B system and Capella H is over 9,200 AU, or 0.15 light years. However, this is not all! Capella H is also a binary star with a separation of 3.3 arcsec (p. a. 166°, 1999). Stars with magnitudes 10.5 and 13 are seen and separated through larger amateur telescopes. The Capella system, therefore, includes at least four stars. If we create a model in which Capella A is a 10-cm big ball, Capella B measures 6 cm. The balls are 1 m apart. In this model, the stars in the Capella H system have a diameter of 6 mm each and are 43 m from each other. But the distance from pair A–B is over 11 km! And this wonderful system is gravitationally bound so that all stars move around a common center of mass! Gravity is truly an exceptional force.

There is another interesting fact about Capella. This bright star is circumpolar for mid-northern observers (and those up further north). This means that we can see it all night and every night. In a dark and clear winter sky it glitters almost in the zenith, but on some summer nights, when the sky is clear and dark all the way down to the horizon, you can see Capella low above the northern horizon. Of course, it is not as bright as it is in winter, for its light is weakened by the thick layer of our atmosphere. At least once on a clear spring night, take some time and follow its path along the sky until the morning hours.

Beta is the 41st brightest star in the sky. It is 82 light years from us. Its luminosity is 80 times that of the Sun.

Theta is the 105th brightest star in the sky. Its light travels 174 years just to reach us. The star's luminosity is that of 190 Suns.

Delta and Iota shine in a beautiful orange color that is even more vivid when viewed through binoculars, especially if the image is slightly blurred. Iota is the 113th brightest star in the sky. It is over 510 light years away. The luminosity of the star surpasses our Sun by 1,600 times.

Epsilon is a rather interesting eclipsing binary. One star is an extremely bright super giant, maybe as much as 18,000 times brighter than our Sun and over 100 times larger! Its companion is a mysterious object, so far never seen. We know of its existence, since it periodically passes across the super giant, and its shine is reduced by approximately 1 magnitude. The eclipse period is every 27 years (!), and the fall in brightness lasts for approximately 1 year. The last time this occurred was between January 1983 and January 1984. The next eclipse and fall of brightness is expected to take place around 2010. The nature of the companion star is still unknown. In the past, astronomers believed it was a very young contracting star. Then they assumed it was a black hole with an accretion disc surrounding it. Today, it seems most likely that the companion is a hot bluish star, surrounded by a big and cold cloud of nontransparent matter.

Zeta is also an eclipsing binary star with a period of 972 days. The system is composed of an orange super giant, the color of which is evident through binoculars, and its hot, white blue companion has a luminosity that surpasses our Sun by 400 times. However, the drop in brightness is only 0.15 magnitude; thus it is almost impossible to notice visually.

In the part of the Milky Way that meanders through the constellation, we can notice three splendid and easily found open clusters: M 36, M 37, and M 38. The best way to find them is to start your search at bright Theta. If the star is moved to the extreme northern edge of the field of view, the open cluster M 37 (6m2/24′) will appear on the southern edge. It can be clearly seen through binoculars, but only as a big, rather bright spot of light, with some 20 individual stars of magnitude 9 and 10 shining through the haze. In perfect observing conditions and with averted vision the cluster is even more crumbled, and the number of stars increases to 80! However, if you wish to resolve the cluster completely, you need larger objective lenses and greater magnification.

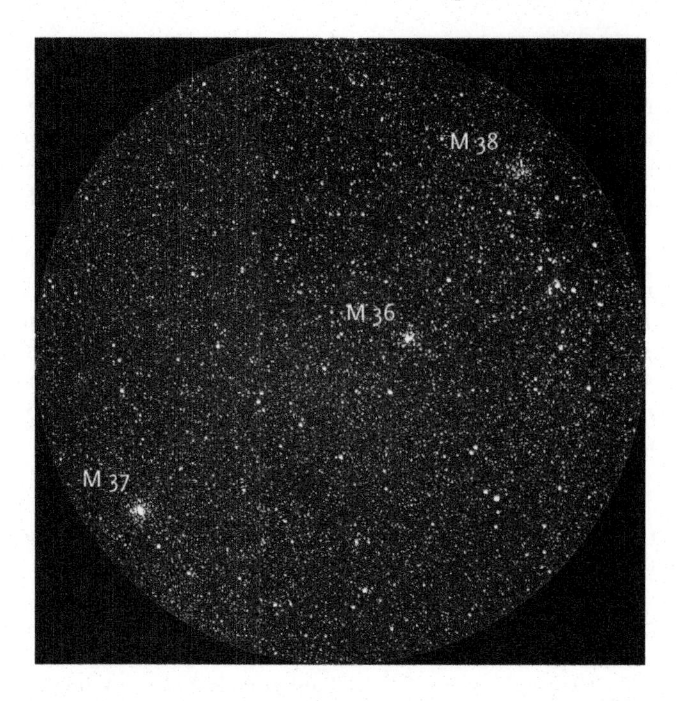

M 37 is a rather rich open cluster with some 300 members. It is 4,400 light years from us, and the actual diameter of its denser part is 25 light years, even though a number of individual members can be found much further away.

M 36 (6m3/12′) lies approximately 4 degrees northwest of M 37. It is clearly seen through binoculars as a bright spot of light, in which we can resolve some 25 individual brightest stars in excellent observing conditions. The cluster is slightly smaller and denser than its neighbors. There are approximately 200 known members. M 36 is one of the youngest open clusters in our galaxy. It consists predominantly of hot and luminous stars. It is 4,100 light years away from us. The actual diameter of the denser part measures 14 light years.

Similar in size to M 37 but slightly closer to us (4,200 light years) is the open cluster M 38 (7m4/21′). Merely 2 degrees north-northwest from M 36, so both can be found within the same field of view in your binoculars, M 38 is also bright and clearly visible (Figure 7.14). In the binoculars the cluster can be separated to a certain degree. We can see approximately 25 individual stars up to magnitude 10 (we can separate even more of them with averted vision). Among approximately 80 stars, the brightest is a yellow giant of magnitude 8.4, with a luminosity of 560 times that of the Sun. Just for comparison: If our Sun was to be observed from such a distance, it would shine like a magnitude 15.3 star and could be seen only through the largest amateur telescopes!

The open cluster NGC 2281 (5m4/15′) is slightly harder to find, because it lies in a rather empty area of the sky with no brighter stars in the vicinity. The stars leading to the cluster are the bright Beta and Theta.

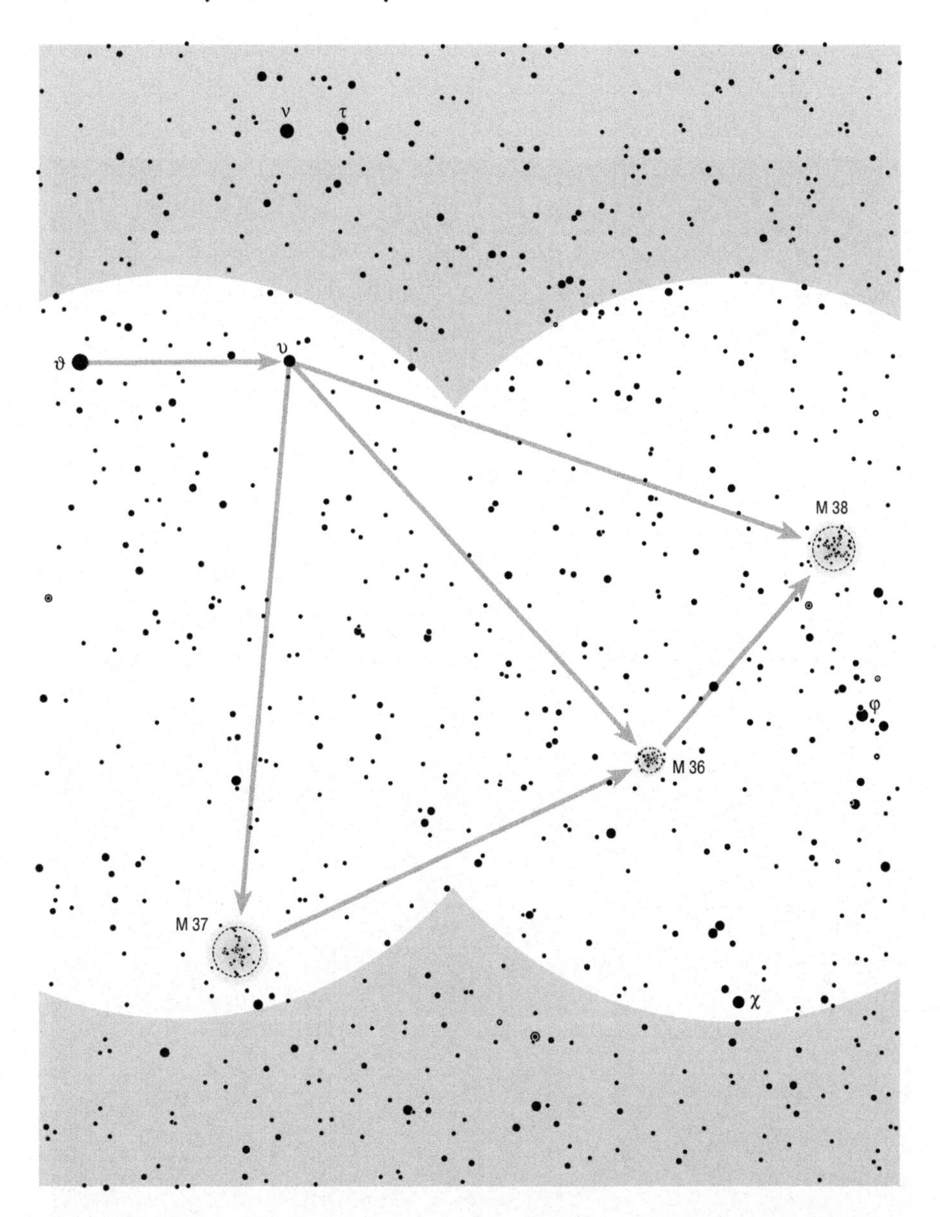

All three form an isosceles triangle with the cluster at the top. Approximately 10 degrees east from the two stars we can find a group of magnitude 5 stars. They form a pattern resembling the upside down letter V that points north (see the detailed chart on next page). The open cluster lies close to the lower, eastern star called Psi-7, which has a distinct orange color and cannot be missed in binoculars.

The open cluster is composed of approximately 100 stars, scattered across a field with a diameter of 20 arcmin. The brightest is of magnitude 7, and some 50 stars have magnitudes up to 11.

Figure 7.14. Wonderful open clusters M 36 and M 38 with their picturesque surroundings

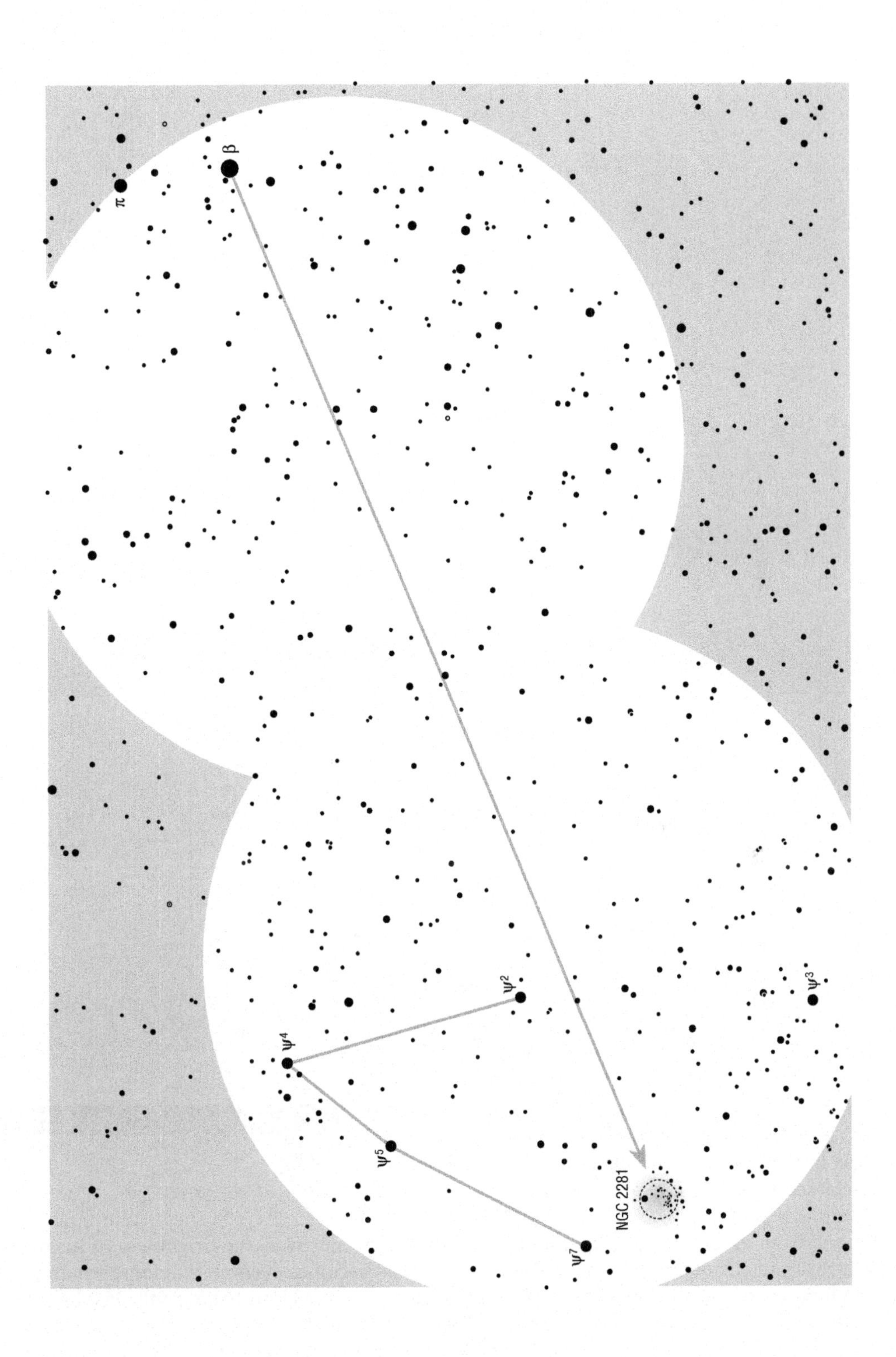
β
π
ψ2
ψ4
ψ5
ψ7
ψ3
NGC 2281

BOÖTES (The Herdsman)

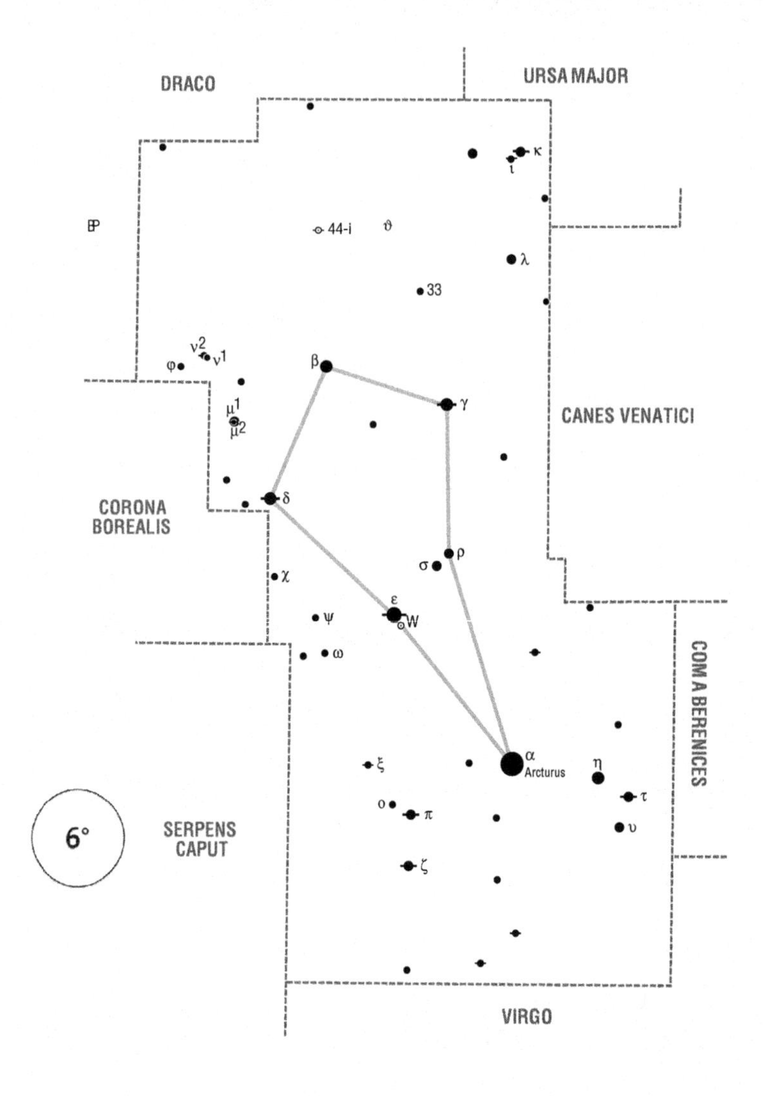

Boötes is a large and prominent spring constellation northeast of Virgo. Its shape resembles a deltoid with the bright Arcturus in the lower corner, which is why some writers like to compare it with a kite or cornet with a scoop of ice cream on the top.

CONSTELLATION CULMINATES		
end March at 2 a.m.	end April at midnight	end May at 10 p.m.

Arcturus is an exceptionally bright and beautiful orange star with the brightness of magnitude –0.05. It really stands out in the sky. The color of the star becomes even more intense when viewed through binoculars if the image is blurred. Arcturus is one of the closest stars, only 36.8 light years away from us. Its luminosity is that of 100 Suns, while its diameter surpasses that of our Sun by 25 times. The surface temperature of the star is only 4,200 K, and its mass is estimated to be that of 4 solar masses. From these data and from its diameter we can estimate its average density, which is a mere 0.0003 of the average density of the Sun. Yes, Arcturus is truly a red giant! Just for fun, astronomers have calculated that this closest red giant heats Earth as much as we are heated by the flame of a candle from a distance of 8 km! It is not a lot, but it is something.

Arcturus is the brightest star in the northern celestial hemisphere (Sirius lies in the southern celestial hemisphere) and one of the brightest in the entire sky. Only Sirius (Alpha Canis Majoris; -1m44) and Canopus (Alpha Carinae; - 0m62) are brighter. With a declination of +19°, it is close enough to the celestial equator to be seen from all inhabited regions of Earth.

Arcturus is also exceptional for its fast proper motion, which reaches over 2.3 arcsec/year. The actual velocity of the star through space is 150 km/s. In our sky, it moves in the direction of the Virgo constellation; thus the deltoid (the shape of the constellation) is prolonging. At the same time, Arcturus is approaching us with a velocity of 5 km/s and will come closer and closer during the next several thousand years. After that it will slowly move away. Its brightness will be diminishing, and in 500,000 years it will no longer be visible to the naked eye.

Two more interesting facts are connected with Arcturus. It was the first star to be observed through a telescope in daylight. The French mathematician Jean Baptist Morin succeeded in doing this in 1635. His "achievement" can nowadays be repeated even with binoculars if the exact position of the star is known to the observer.

In 1933, the light from Arcturus, directed from a telescope to photocells, turned on a switch and opened the great world exhibition Century of Progress in Chicago. Arcturus was selected because at the time it was believed to be 40 light years away from us. The light that opened the exhibition supposedly left the star 40 years previously (in 1893), when Chicago also hosted another world exhibition.

Beside Arcturus the other brighter stars within the constellation are Epsilon (2m3), Eta (2m7), Gamma (3m0), Delta, Beta (both 3m5), and Rho (3m6).

Epsilon is the 81st brightest star in the sky. It is a splendid double star with components of magnitude 2.6 and 4.8, separated by 2.9 arcsec (p. a. 343°, 2005). The brighter star is yellow-orange and the fainter one appears to be bluish or greenish (depending on the observer), even though its spectral type is K (and thus it should be orange). This double cannot be separated through binoculars. The pair is 210 light years from us, and their orbital period has not quite been determined yet. The common luminosity of the stars is that of 360 Suns.

The yellowish Eta is our close neighbor, a mere 37 light years away from us. It is the 109th brightest star in the sky. Its luminosity is eight times that of the Sun.

Xi (4m5) is a wonderful double star, but unfortunately only for observers with telescopes. The two stars with magnitudes 4.8 and 6.9 have an orbital period of approximately 150 years. The pair is only 22 light years away from us. The apparent distance between the stars changes between 1.8 (1912) and 7.3 arcsec (1984). Currently the stars are moving close to each other, and at the moment they are approximately 6.4 arcsec apart (p. a. 311°, 2006). The brighter star is yellowish in color, while the fainter one is reddish with a violet hue; this provides for a splendid color contrast, but unfortunately not for binoculars!

Iota is a magnitude 5 star in the northwestern part of the constellation, just under the brighter Kappa. In fact it is a double star with components of magnitude 4.8 and 7.4, separated by 39.7 arcsec (p. a. 34°, 2004). Through binoculars they are seen as a close couple. The brighter star is white and the fainter is orange.

Delta is an easy double to see through binoculars. Its two stars are magnitude 3.6 and 7.9, and 102.4 arcsec apart (p. a. 79°, 2004). They are approximately 140 light years away from us. The fainter star is fairly similar to our Sun in size and luminosity. The brighter one is larger and hotter. Its luminosity is about 70 times that of the Sun. Both stars are yellowish in color.

The pair Mu-1 and Mu-2 is another easy double to catch through the binoculars. Stars with magnitudes 4.3 and 7.1 are 107 arcsec apart (p. a. 170°, 2002). The brighter star is white and the fainter one is yellowish. They are approximately 95 light years away from us. The two stars are over 3,170 AU from each other. About 53 Solar Systems could be placed in the space between them (when comparing the distances in space with our Solar System, we use 60 AU for its size, which is the diameter of Neptune's orbit). The fainter star is a double itself. Its companion shines with magnitude 7.6 and is 2.3 arcsec away (p. a. 7°, 2006). This pair cannot be separated in binoculars but can be split in a mid-size telescope.

CHAPTER EIGHT

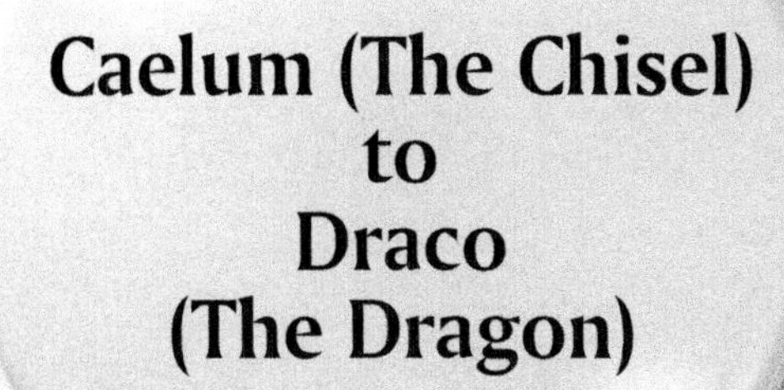

Caelum (The Chisel) to Draco (The Dragon)

B. Kambič, *Viewing the Constellations with Binoculars*, Patrick Moore's Practical Astronomy Series,
DOI 10.1007/978-0-387-85355-0_8,

CAELUM (The Chisel) and HOROLOGIUM (The Pendulum Clock)

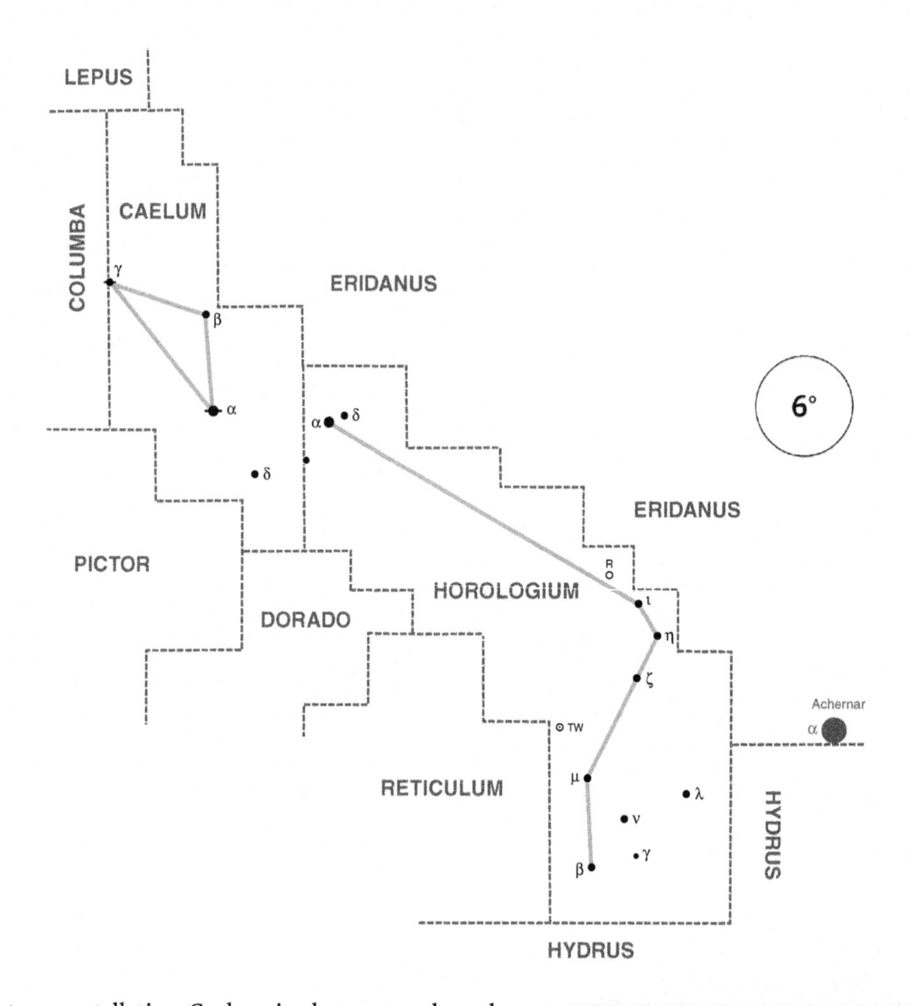

CONSTELLATIONS CULMINATE		
in begin. November at 2 a.m.	in begin. December at midnight	in begin. January at 10 p.m.

The winter constellation Caelum is always very low above the horizon when viewed from mid-northern latitudes. Not that this really matters, for it has hardly any objects of interest for amateur astronomers and absolutely none for observers with binoculars. The constellation can be found west of Columba. The brightest stars are Alpha (4m4), orange Gamma (4m5), and Beta (5m0).

Alpha is similar to our Sun but a little bit hotter. Its light travels 66 years to reach Earth. In mid-northern latitudes, it rises only a few degrees over the horizon during culmination.

A similar statement about lack of interest for us could also be used for Horologium, which is even closer to the horizon. From the mid-northern latitudes we can only see the northernmost part of the constellation, which includes Alpha (3m9). At culmination, the star peeps over the horizon only slightly more than 1 degree, and because of this its light is further dimmed. The rest of the constellation does not contain any interesting objects for observers with binoculars.

CAMELOPARDALIS (The Giraffe)

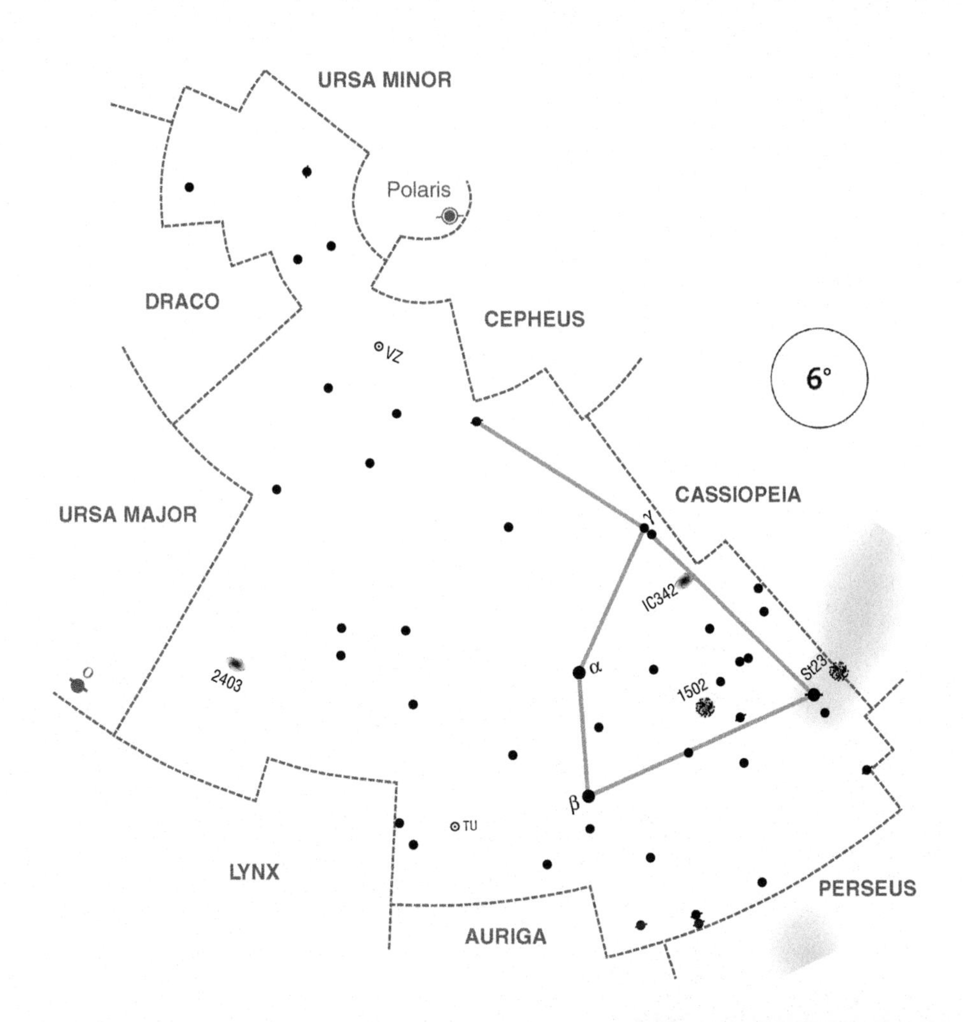

CONSTELLATION CULMINATES		
end October at 2 a.m.	end November at midnight	end December at 10 p.m.

Camelopardalis lies high in the northern skies. In mid-northern latitudes, it is a circumpolar constellation, visible throughout the entire night and every night. If we compare it to its neighbors, Camelopardalis is very sulky. The brightest star is Beta (4m0), which is found between Capella in Auriga and Polaris. In light-polluted areas, the constellation is no longer visible. What a pity!

Open cluster NGC 1502 (5m7/8′) is a small group of approximately 30 stars up to magnitude 11. The interesting thing with this cluster is the double star with the designation Σ485. The two stars with magnitudes 6.9 are 17.7 arcsec apart (p. a. 305°, 2004) and can be seen through binoculars with excellent optics as a touching pair. It is not easy to find the cluster, since it lies in a rather empty part of the sky. With Alpha and Beta it forms an isosceles triangle, with the cluster at the top. Unfortunately, the cluster is roughly 7 degrees away from the two stars, which is a little more than the field of view of the binoculars. The detailed chart given below will be of great help when searching for this cluster.

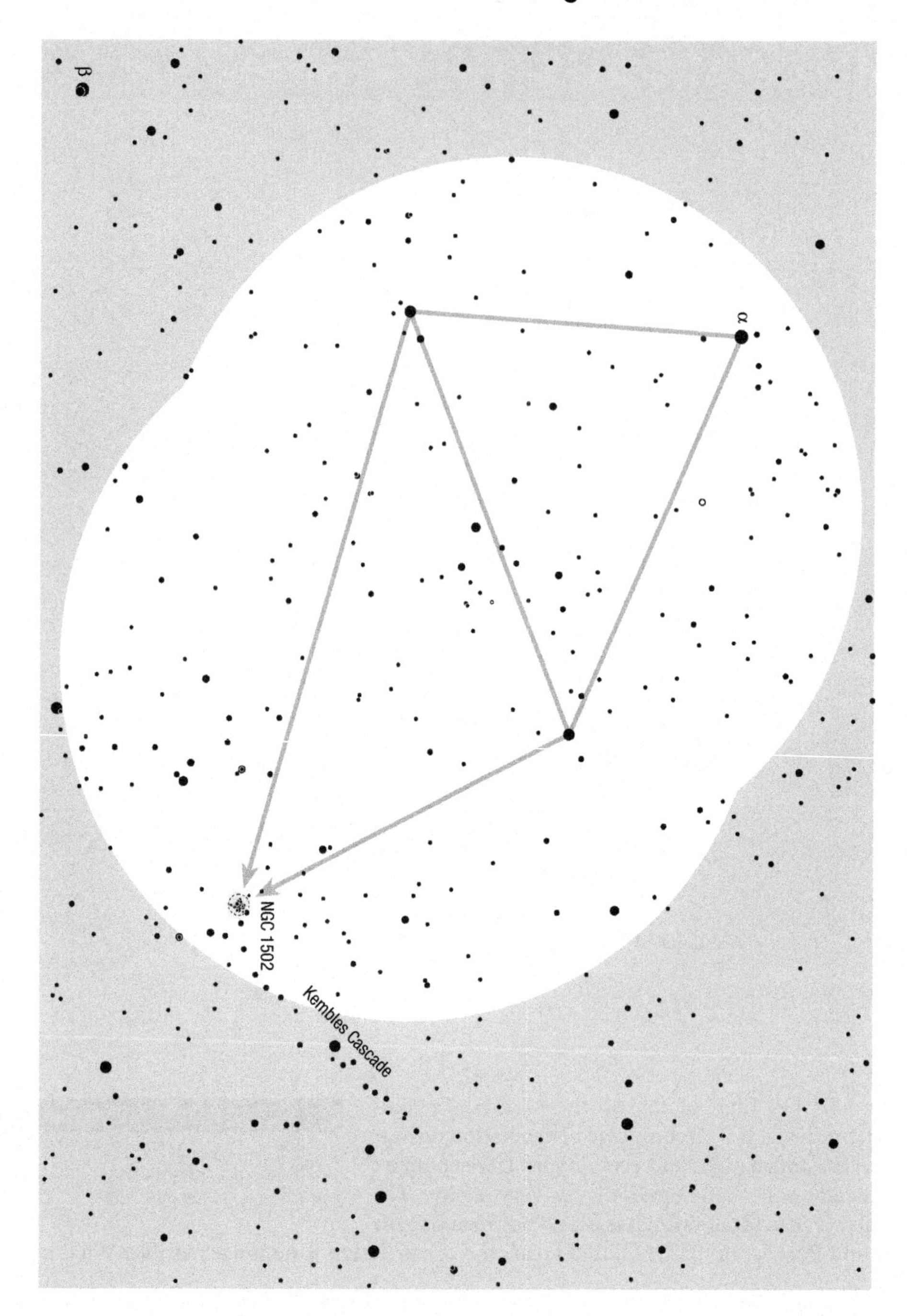

In the eastern part of the constellation, in an almost entirely empty area of the northern sky, we can see the big and bright spiral galaxy NGC 2403 (8m4/18′ × 11′). The star leading to it is Omicron Ursae Majoris. From this star we unfortunately have to move northwest with the aid of magnitude 6 stars. In this case the detailed chart (on next page) is essential.

NGC 2403 is a close neighbor to our Local Group of galaxies (Figure 8.1). It is 12 million light years from us. From these data and from the apparent size one

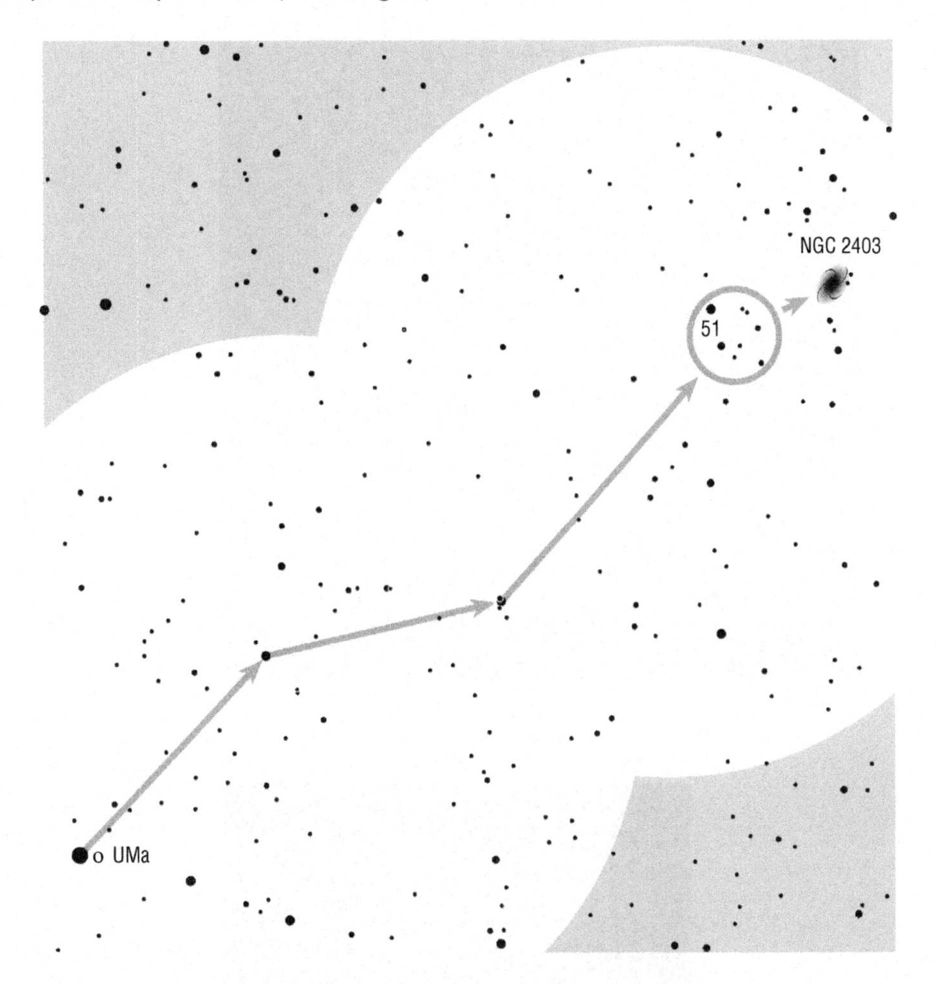

can easily calculate that its actual diameter is approximately 60,000 light years. Most probably, it is a member of a small group that consists of M 81, M 82, and NGC 3077 (all three lie in Ursa Major, 14 degrees east).

The galaxy is seen face on. Through the binoculars it looks like a faint hazy spot of light, but only when there are excellent observing conditions. In perfect observing conditions and using a larger amateur telescope we might notice some irregularities in the otherwise uniform oval of the central nucleus. The spiral arms appear only on long-exposure images. NGC 2403 was the first galaxy outside of our Local Group in which astronomers found Cepheid variables that helped them calculate its distance from us.

IC 342 (8m3/21′) is a spiral galaxy that is seen exactly face on. Despite its promising brightness it can only be glimpsed in perfect observing conditions through binoculars and even then only as a very faint spot of light. Even through larger amateur telescopes we can only see its nucleus, surrounded by a very faint nebula. On long-exposure images this haze breaks into gas and dust clouds, and countless tiny stars that form the spiral arms.

The leading star to the galaxy is Gamma, but this is only a magnitude 5 star. The galaxy lies some 3 degrees south of it. When searching for it, the detailed chart given on next page is of great help. But if you know how to find the open cluster NGC 1502, a clearly seen and distinct line of stars of magnitudes 6, 7, and 8 can lead you to the vicinity of the galaxy. This line of stars is known under the nickname of Kemble's Cascade.

Figure 8.1. Spiral galaxy NGC 2403

Figure 8.1A. Spiral galaxy IC 342

The galaxy lies in the vicinity of the galactic equator, so its light travels through thick layers of gases and dust in our galaxy. Astronomers have estimated that this dims its light by 2.4 magnitudes. What a shame! Otherwise this galaxy would be among the brightest and best seen galaxies in the sky!

IC 342 is approximately 10 million light years away from us. Its real diameter is 50,000 light years. It is a member of a smaller group with the name Maffei 1, which, due to the obscuring clouds, was unknown until 1968.

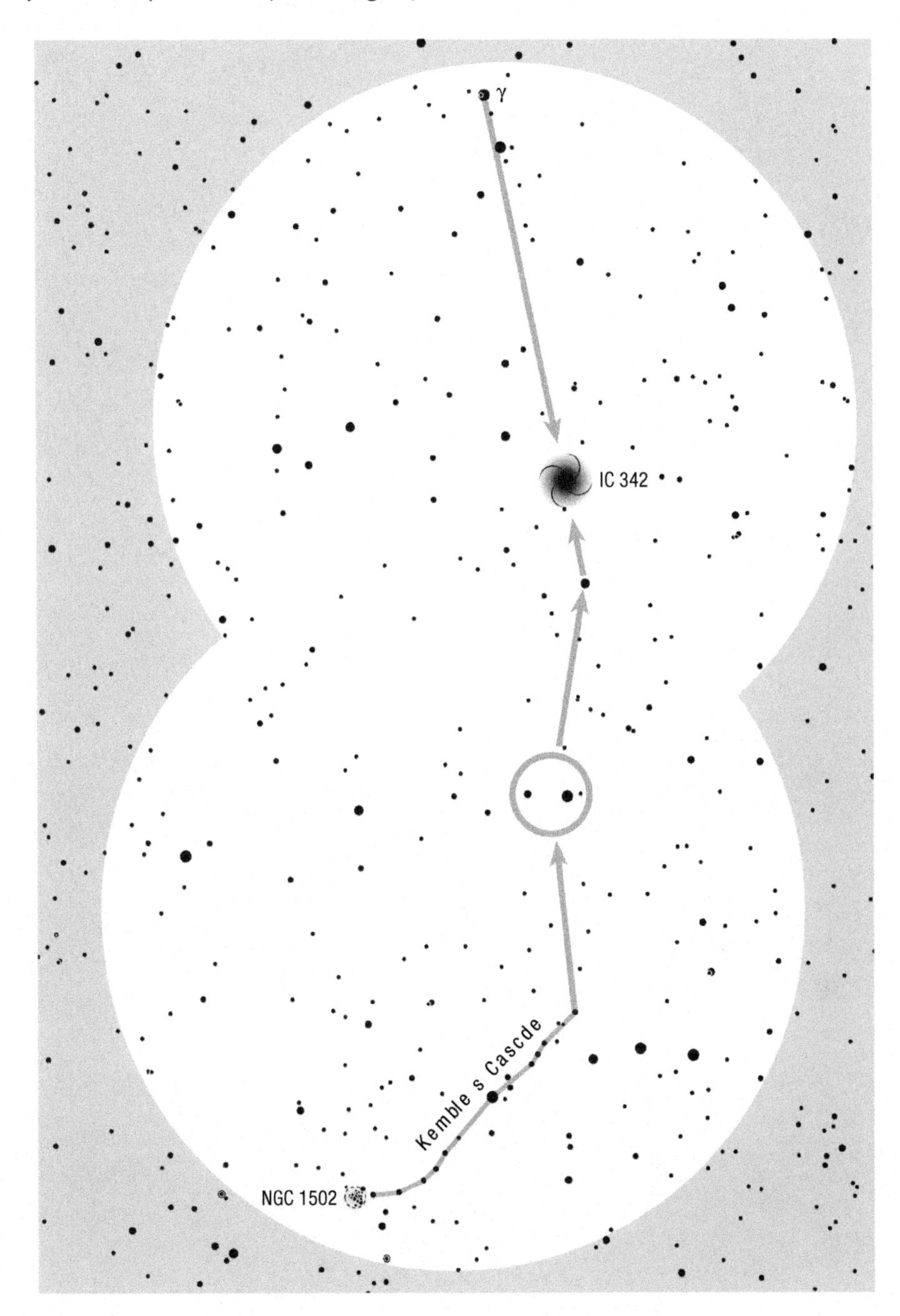

Open cluster Stock 23 (~6m5/18′) is found on the border of constellation Cassiopeia. A magnitude 4 star without a designation leads to it. The cluster consists of one magnitude 7 star, three magnitude 8 stars, two magnitude 9 stars, and a number of fainter ones that melt into the background nebula. Those six stars are always visible, while the fainter haze will only appear when there are excellent observing conditions.

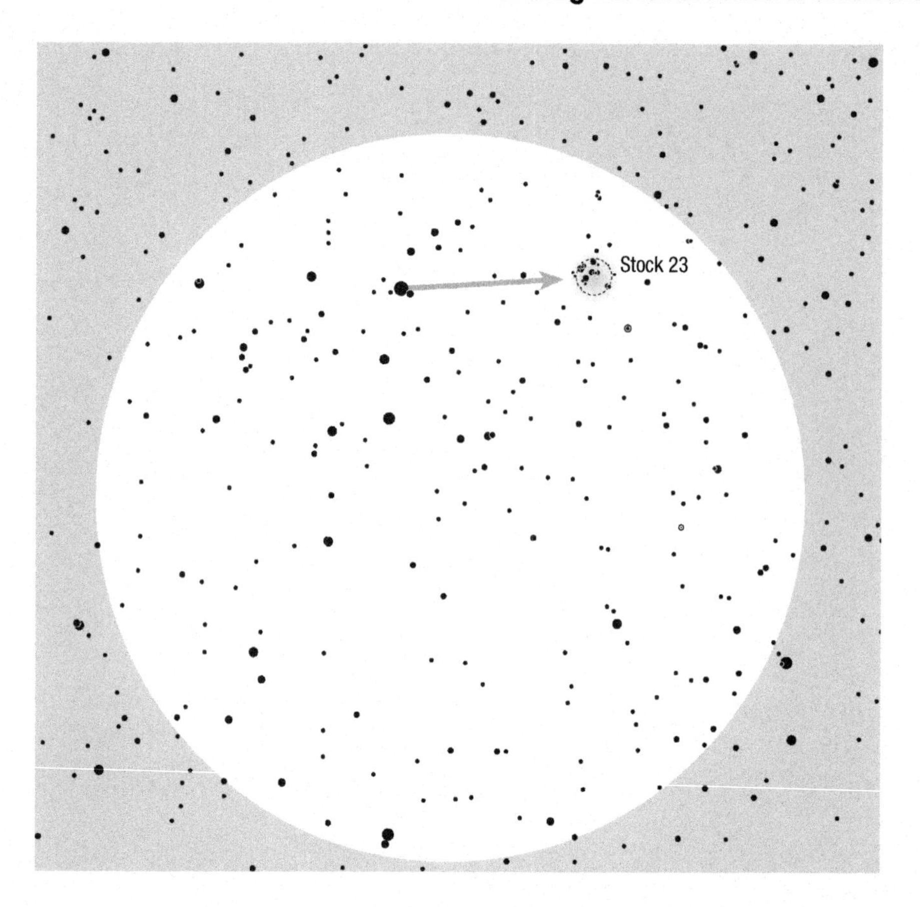
Stock 23

CANCER (The Crab)

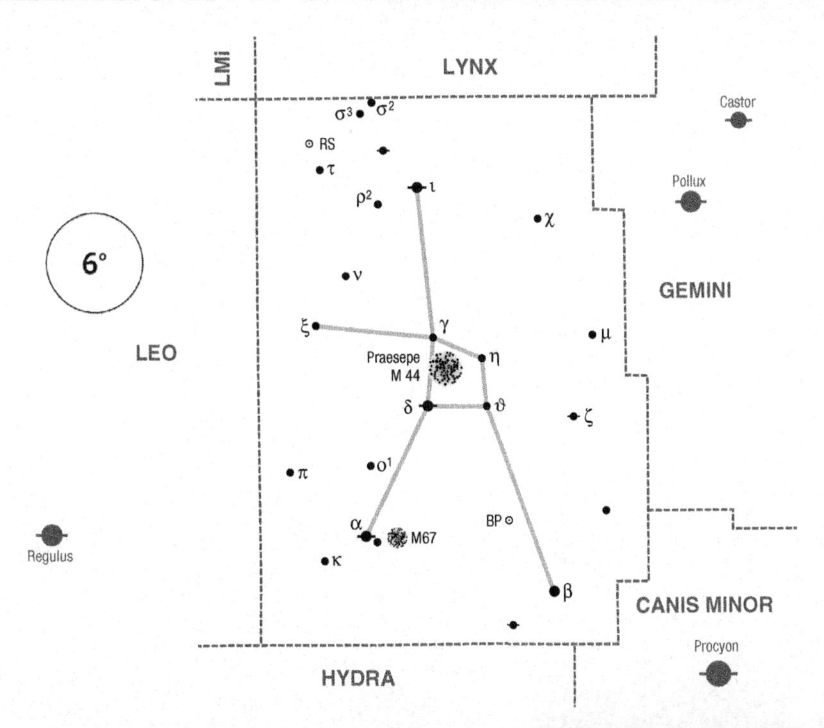

CONSTELLATION CULMINATES		
in begin. January at 2 a.m.	in begin. February at midnight	in begin. March at 10 p.m.

Cancer is one of the smallest and faintest constellations of the zodiac. We would definitely overlook it among the brilliant winter constellations – if it did not contain an open cluster that is clearly visible with the naked eye. Cancer is situated between Regulus (Alpha Leonis) and Pollux (Beta Geminorum). The only stars that are brighter than magnitude 4 are Beta (3m5) and Delta (3m9).

Zeta is one of the most interesting multiple stars in the sky, but unfortunately its components cannot be separated through binoculars. Around Zeta A, which is of magnitude 5.3, moves its magnitude 6.2 companion (Zeta B), which completes one orbit every 59.6 years. The apparent separation varies between 0.6 and 1.2 arcsec. It was at its greatest separation in 1960; in 2005 it was 1.0 arcsec (p. a. 58°), and it will continue to rise until approximately 2020. Both stars are white-yellowish and are 19 AU apart. This is comparable to the distance between Uranus and the Sun. About 5.9 arcsec east-northeast (p. a. 71°, 2006) from the two is another magnitude 6.2 star (Zeta C), which takes approximately 1,150 years to complete one orbit. This star is resolvable in amateur telescopes. The data on its orbit are still relatively unknown. From the irregularities in the proper motion of the star John Herschel (1831) discovered that this is also a binary star. The apparent separation between the magnitude 6.3 and 7.1 components is 0.3 arcsec (p. a. 85°, 2000), and the orbiting period is 17.6 years. And there are more! At least another four fainter stars in the close vicinity could also be true members of the Zeta Cancri system. The measurements are in progress. This interesting multiple star is 83 light years from Earth.

Iota Cancri is a splendid double. Stars with magnitudes 4.1 and 6.0 are 30.7 arcsec apart (p. a. 308°, 2003). In the field of view of the binoculars this is wonderful colorful pair – the brighter star is yellow and the fainter one is bluish. In the immediate vicinity, there are some magnitude 5 and 6 stars that variegate the view.

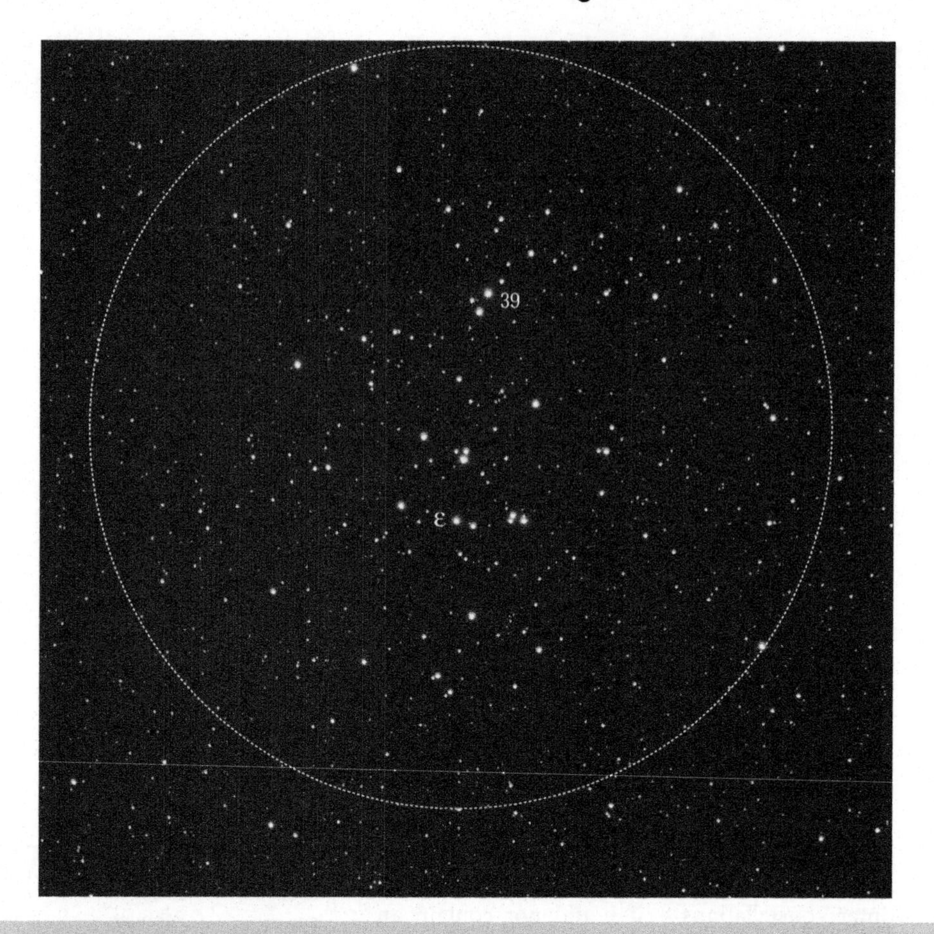

Figure 8.2. Detailed view of the splendid open cluster M 44. The field on the image measures 1.5 × 1.5 degrees

M 44 (3m7/95′), called also Praesepe and Beeheve (what is Latin for "manger"), is one of the largest, closest, and most beautiful open clusters (Figure 8.2). It can be clearly seen with the naked eye as a faint spot of light. Through binoculars, the "nebula" crumbles into numerous individual stars and is truly splendid. The apparent diameter of the cluster is a degree and a half, so it is as big as three full Moons. In this case, the large field of view and the small magnification of the binoculars prove to be an advantage over the larger telescopes, in which the impression of the cluster is completely lost!

Praesepe is one of the rare nonstellar objects mentioned by ancient astronomers. Hipparchus (130 B.C.) called it the Small Cloud. The cluster appeared in modern star charts around 1600, when the German astronomer Johan Bayer entered it under the Latin name Nubilum (the small cloud). The real nature of the "cloud" was discovered by Galileo, who was astonished and intrigued by the fact that Nubilum had broken into numerous individual stars when viewed through his little telescope.

In total, there are over 1,300 stars in the area of the cluster; most of which are true members. In the field of view of the binoculars, you can count approximately 120 stars when there are excellent observing conditions. On a perfect winter night their number could increase up to 170! The 20 brightest have magnitudes 6 and 7. Among the brightest we can find four that shine in a yellow-orange color. The brightest star in the cluster is Epsilon (6m3), which is white. Its luminosity is that of 70 Suns. Our Sun would be a magnitude 11 star if we observed it from the distance of Praesepe; thus, we could see it with binoculars only on a perfect night.

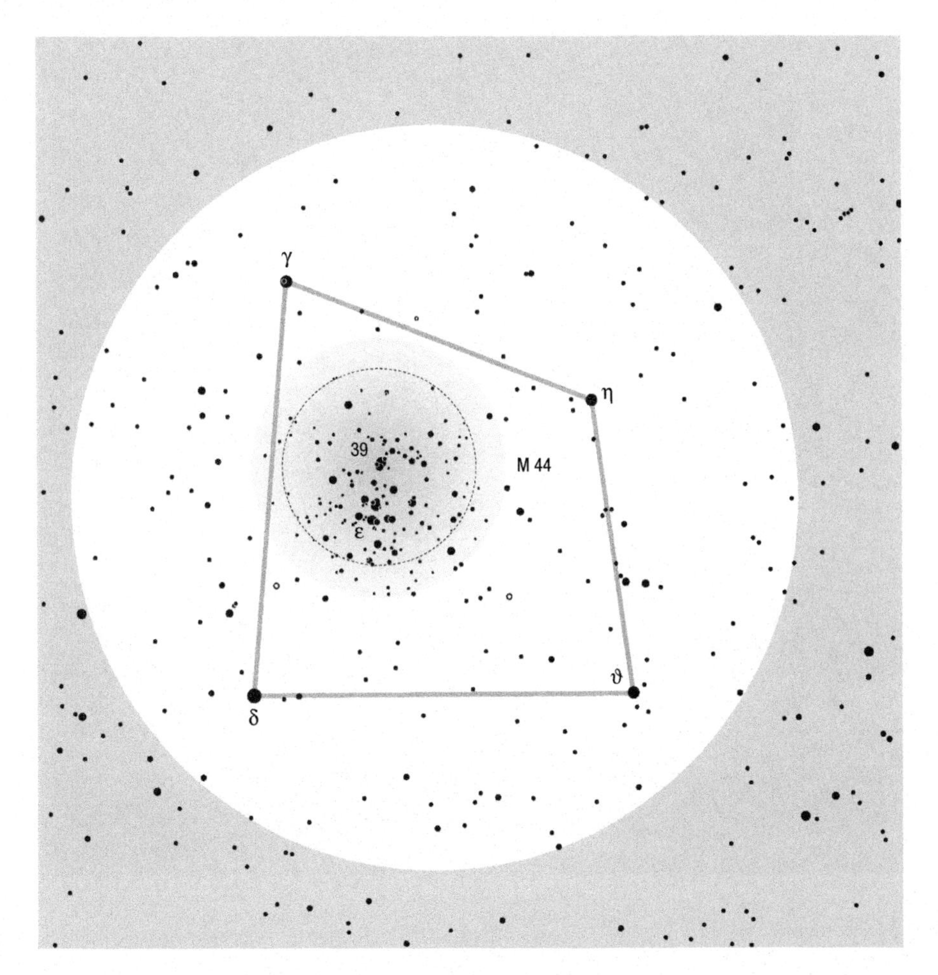

Praesepe is 577 light years away from us. Its actual diameter is 14 light years; however, some individual stars are found as far as 22 light years from its center.

The open cluster M 67 (6m1/30′) lies only 1.8 degree west of Alpha (4m3) and is therefore not hard to find, as the star and the cluster can be found in the same field of view of binoculars (Figure 8.3). M 67 is clearly visible, but only as a 15-arc spot of faint light. During excellent observing conditions, the spot is bigger, and we should see at least one star shining out. With averted vision we can glimpse approximately 10 of the brightest stars distinguished from the haze. In larger amateur telescopes, the cluster is truly splendid! In total this area consists of over 3,000 stars, but unfortunately most of them are faint. Only 50 of them are of magnitude 11 or brighter and can be seen through binoculars in perfect observing conditions. If we observe them through a 20-cm objective lens, the number of stars rises to almost 300!

The open cluster M 67 is 2,700 light years away from us, so its real diameter is only 12 light years. We might feel terribly squeezed if we lived on a planet within that cluster!

Although open clusters are not stable star groups, measurements have shown that this is one of the oldest of its type. The group is probably over 4 billion years old! One of the reasons it has managed to reach this great age is because it is far away from the galactic plane (almost 1,500 light years above), and thus relatively unaffected by the disturbing gravitational pull of the other stars in the galaxy.

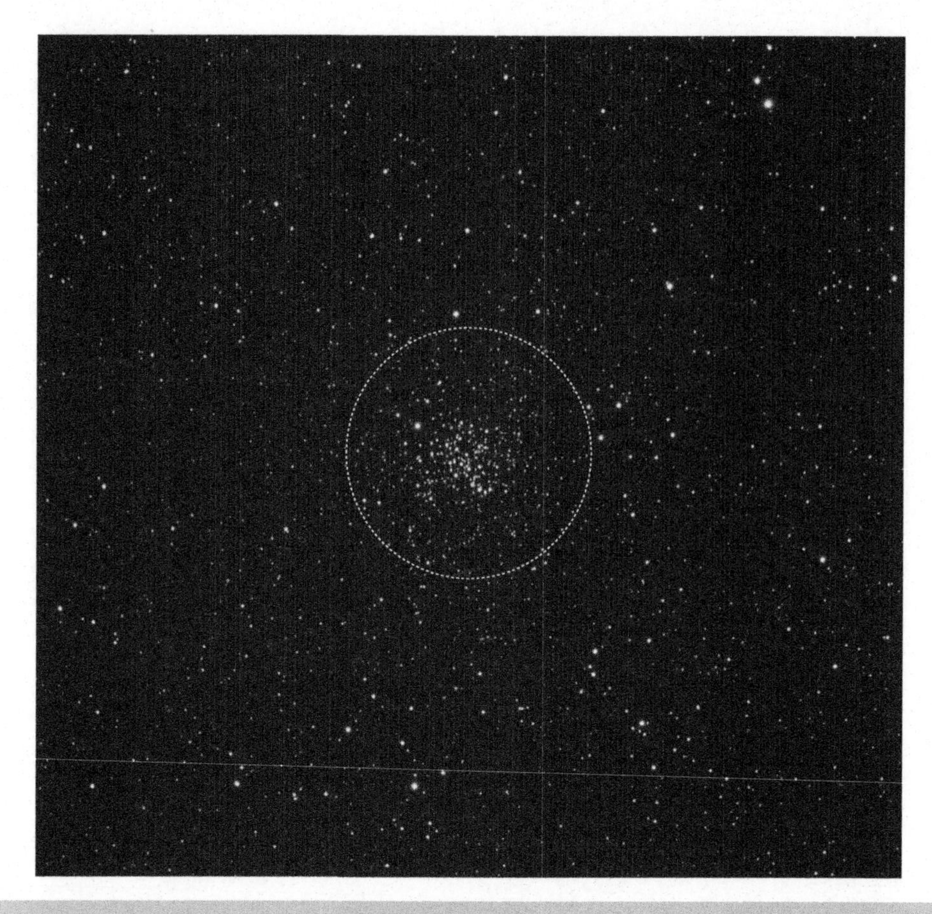

Figure 8.3. Open cluster M 67

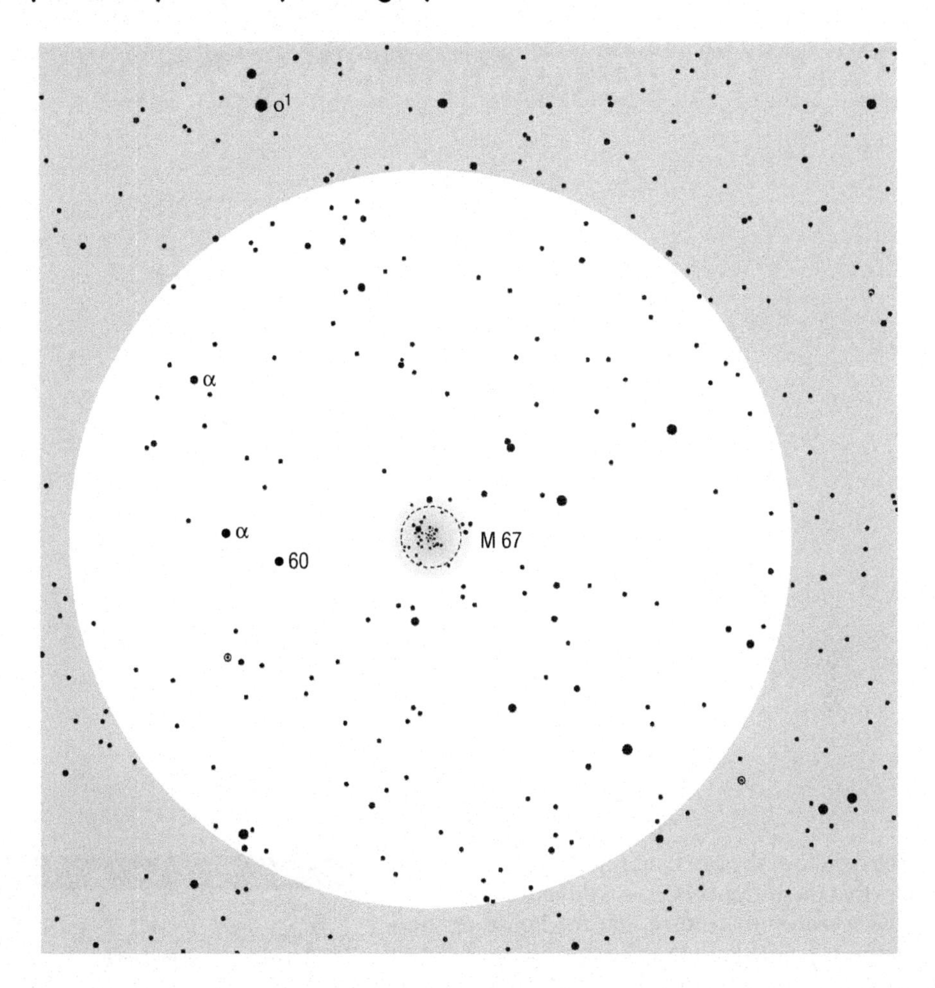
o1
α
α
60
M 67

CANES VENATICI (The Hunting Dogs)

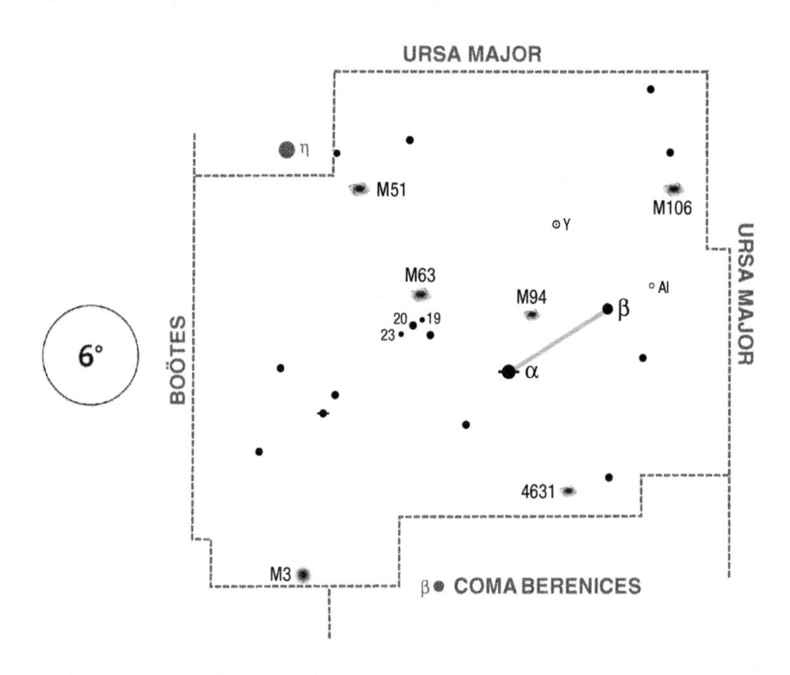

CONSTELLATION CULMINATES		
in begin. March at 2 a.m.	in begin. April at midnight	in begin. May at 10 p.m.

Under the handle of the asterism Big Dipper lies a group of faint stars that the ancient observers did not name. It was not until the seventeenth century that the Polish astronomer Jan Hevelius pointed out the new spring constellation and named it Canes Venatici. The only really bright star in the constellation is Alpha (2m9), called Cor Caroli (Charles' Heart). The next brightest is Beta or Asterion (4m3). Through binoculars, both stars can be seen in the same field of view.

Alpha is one of the most beautiful double stars to be seen through telescopes. These stars, with magnitudes 2.8 and 5.5, are 19.3 arcsec apart (p. a. 229°, 2004). Through binoculars they are on the border of separability. If your instrument has high-quality optics, you should try to split them on a calm night. The stars are gravitationally bound and thus represent a true binary, some 120 light years away from us. The luminosity of the brighter star is 80 times that of the Sun, while the luminosity of the fainter one is that of 7 Suns.

The spiral galaxy M 51 (8m4/11′ × 7′8), popularly named the Whirlpool Galaxy, can be clearly seen through binoculars, but only as a 5-arcmin big spot of light; in excellent observing conditions, it maybe slightly larger and oval in shape. The spiral arms are – of course – not seen, for they are too faint. All that we can see is the brightest, central part of the galaxy's nucleus (Figures 8.5 and 8.6).

M 51 is easy to find. The star leading to it is the bright Eta Ursae Majoris, the end star in the handle of the Big Dipper. If we capture this star and move it to the northeastern edge of the field of view, the galaxy will appear on the southwestern edge. In order to be able to observe it we have to choose a night with at least good conditions; the part of the sky that includes this galaxy should be high in the sky. M 51 is one of those galaxies that the beginner should look at first in order to get the feeling of how "bright and well seen" galaxy looks like when seen through binoculars.

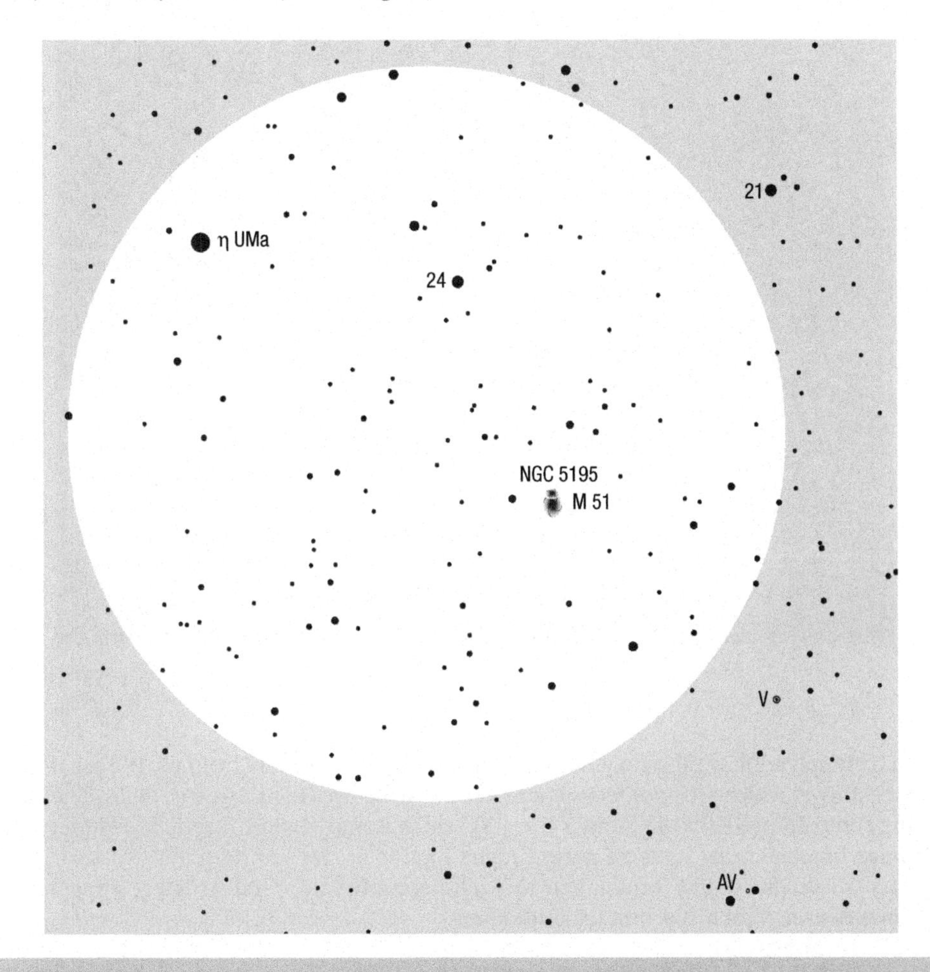

Figure 8.4. The star field around spiral galaxy M 51 in Canes Venatici with the bright Eta Ursae Majoris in its vicinity

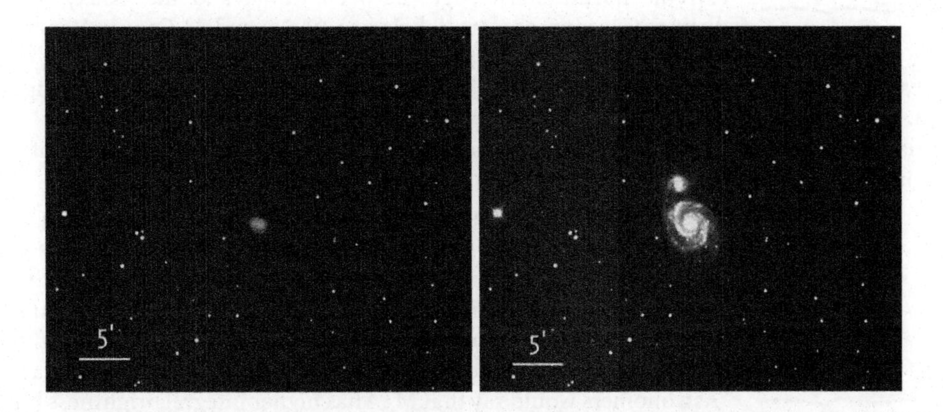

Figure 8.5. Approximate comparison between the appearance of the galaxy in binoculars and on a long-exposure image

Figure 8.6. Spiral galaxy M 51 with its satellite NGC 5195

M 51 is a truly splendid spiral galaxy, some 37 million light years away from us. We see it face on. It was the first galaxy in which astronomers spotted a spiral structure (Lord Rosse in 1845). The galaxy has an adjoining companion NGC 5195 (9m6/5′4 × 4′3), which is more than 1 magnitude fainter and is not visible through binoculars, except on a perfect night, high in the hills or deep in the desert (or both).

M 51 is also a suitable object for amateur astrophotography. Its spiral structure appears on long-exposure images even with a 200-mm telephoto lens.

The spiral galaxy M 63 (8m6/12′3 × 7′6) is easy to find, since it lies in the same field of view as Alpha (Figure 8.7). Through binoculars it is seen as a faint, few arcminutes big, elongated spot of light with a magnitude 8 star beside it. First you have to find Alpha. If the star is moved to the extreme southwestern edge of the field of view, the galaxy will appear on the extreme northeastern edge. Between Alpha and the galaxy there is a distinctive group of stars, labeled as 19, 20, and 23 Canum Venaticorum, which are also visible with the naked eye. The group is approximately 4 degrees northeast of Alpha and represents a good intermediate leading point. The stars form a V-shape pattern pointing toward the north. M 63 lies just 1 degree north.

The galaxy is approximately 37 million light years away from us. Its diameter is estimated to be 90,000 light years.

Only 2 degrees north from the connecting line between Alpha and Beta (both stars are in the same field of view of binoculars, on the extreme edges) lies the rather bright spiral galaxy M 94 (8m2/11′ × 9′1 – chart is on next page). Through binoculars it is seen as a round spot of faint light a few arcminutes in diameter. We see it face on. Although it is brighter than M 63, it is harder to see, since its light is dispersed across a greater area! Astronomers would say that M 94 has higher integral brightness but lower surface brightness than M 63. This galaxy is approximately 14.5 million light years away from us. Its diameter is estimated to be only 33,000 light years.

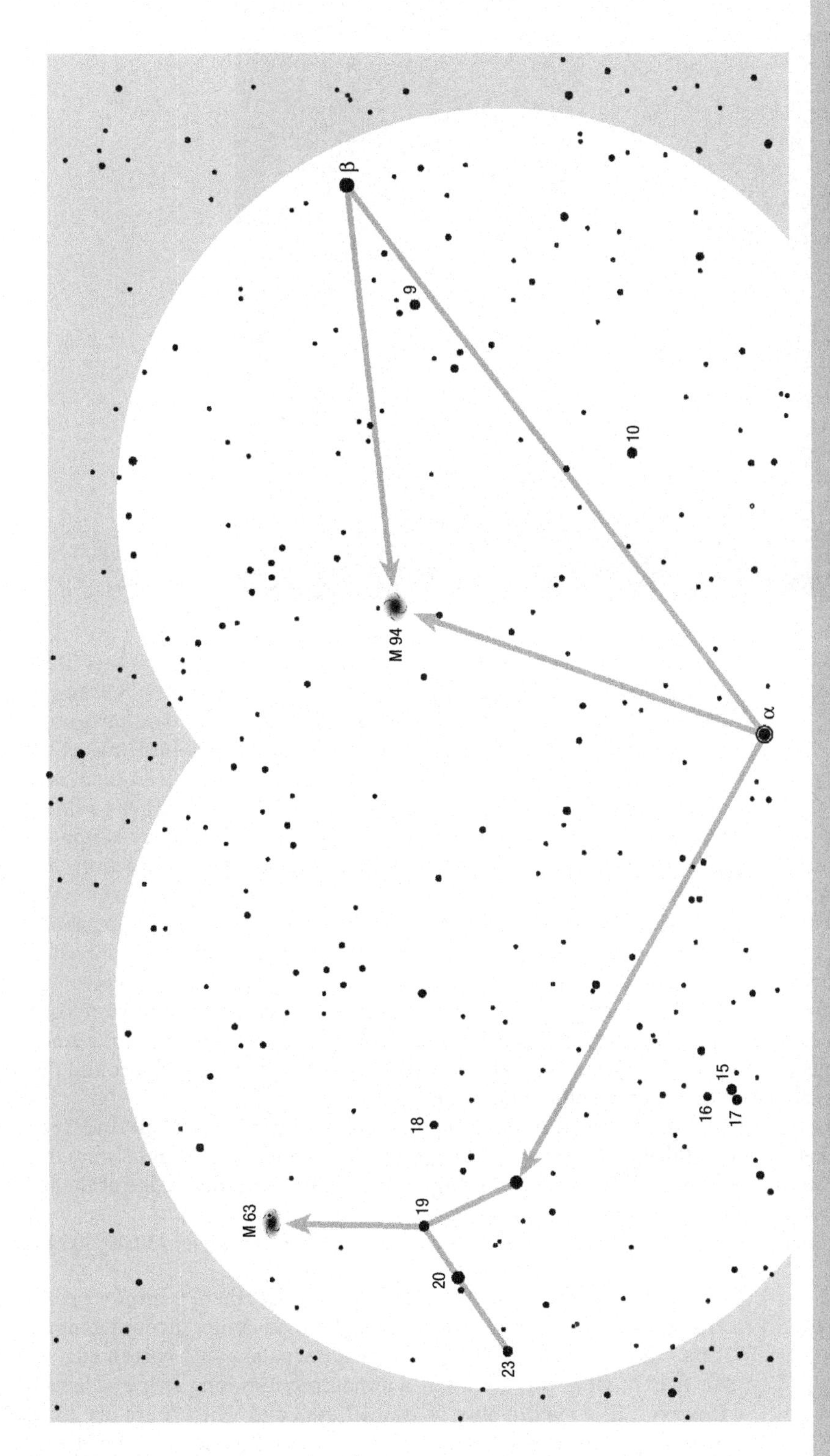

Figure 8.7. Spiral galaxies M 63 and M 94 lie in the vicinity of Alpha and are thus easy to find. If we move Alpha on extreme southwestern edge of the field of view, the rather bright stars 18, 19, 20 and 23, which are arranged in extended V pattern, will lead us directly to M 63. The galaxy has close by magnitude 8 star, which is in binoculars always well seen. When search for the other galaxy – M 94 – Alpha is moved to extreme southeastern edge of the field of view, so that we see Alpha and Beta contemporary. Both stars and galaxy form elongated isosceles triangle with galaxy M 94 on top

Figure 8.8. Spiral galaxy M 63

Approximately 7 degrees north-northwest of Beta lies another Messier's object – the spiral galaxy M 106 (8m4/19′×8′) (Figures 8.9 and 8.10). Despite its promising integral brightness this galaxy is a hard object to observe with binoculars. It is even larger than M 94, which means that its surface brightness is even lower. Through binoculars it is seen as a few arcminutes big, faint, elongated spot of light. First you have to find Beta. The star is moved to the extreme south-southeast edge of the field of view. The galaxy is now only 1 degree from the northern edge of the field of view. If you shift the binoculars in that direction, you will first of all notice a magnitude 6 star, and a degree west of it you should see a group of three stars of magnitudes 7 and 8. The galaxy lies in between. If you decide to try to observe this elusive galaxy, you must wait for a night with excellent observing conditions. When you are certain that the point where the galaxy should be is in the middle of the field of view, you have to wait for your eyes to completely adapt to night vision. Only then should the field of view be carefully examined. The galaxy has to be close to its highest point in the sky! Do not despair if you do not see it. Maybe the observing conditions are not as good as you assumed, or maybe your eyes are tired from searching... Now that you have found the way to the galaxy, you can try observing it at some other time!

Those observers who are not experienced in handling binoculars should first of all find Beta and then, in the same field of view (2 degrees northwest), the group of magnitudes 6, 7, and 8 stars that are gathered around variable AI (6m). If you move this group to the extreme southern edge of the field of view, the galaxy will appear on the northern edge.

M 106 is approximately 25 million light years from us; thus its actual diameter is over 130,000 light years.

For all observers who would like a challenge, there is another galaxy in this constellation, which is on the border of visibility through binoculars. This is NGC 4631 (9m8/17′ × 3′5), a spiral galaxy that is seen edge on. In the field of view, it is shown as a few arcminutes long stripe of faint light. However, it is visible only to trained eyes and when there are excellent observing conditions! In order to see it, you have to use all of the tricks mentioned so far as well as ones we are about to mention: you should be rested, your eyes should be

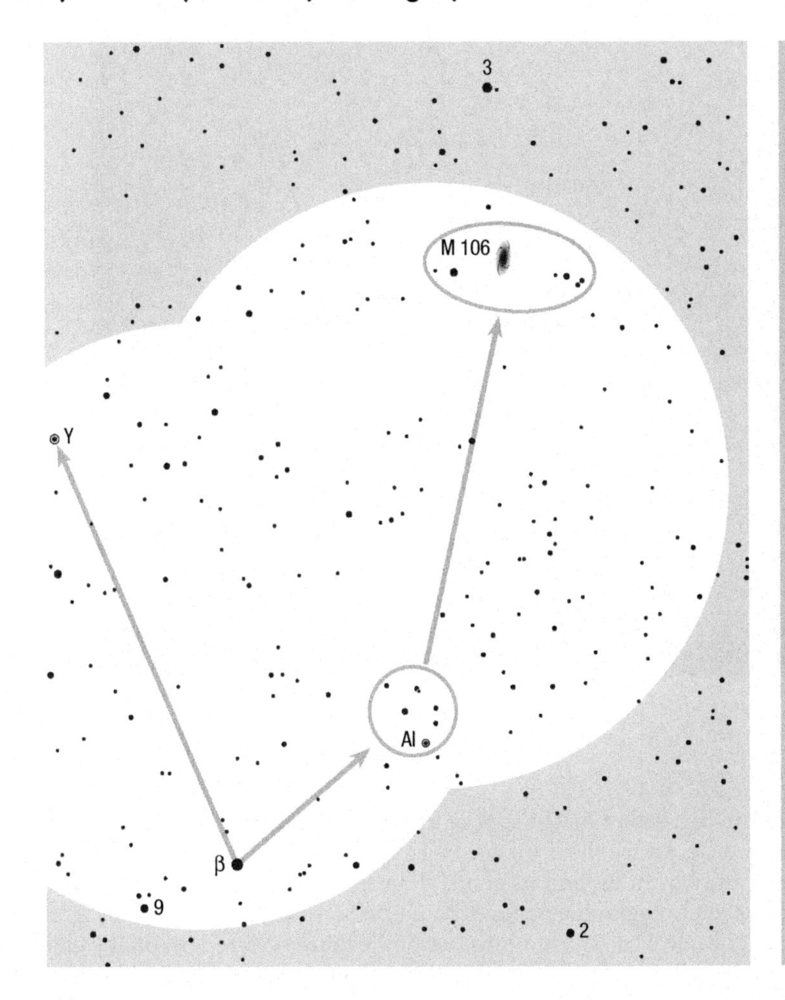

Figure 8.9. The faint galaxy M 106 lies approximately 7 degrees north-northwest of Beta, which makes it slightly more than one diameter of the field of view of binoculars away from it. Beta is also the leading star to Y, one of the most reddish stars, visible with the naked eye. The star is very cold; astronomers have found a lot of carbon in its atmosphere. Y is an irregular variable star. Its brightness oscillates between magnitudes 5.0 and 6.4

totally adjusted to night vision; you have to use averted vision in the examination of the star field where the galaxy should be; you should shake the binoculars delicately... The area with the galaxy also has to be close to its highest position in the sky.

Alpha is the leading star to NGC 4631. If the star is moved to the extreme northeastern edge of the field of view, a group of stars of magnitude 7 and 8 will appear on the western edge. In star charts these are marked as group Up 1. If you shift the binoculars so that now this group is on the extreme northwestern edge of the field of view, the region of the sky with the galaxy will appear on the south (Figure 8.11).

NGC 4631 is approximately 30 million light years away from us. Its real diameter reaches over 130,000 light years, so it belongs among the biggest spirals.

The globular cluster M 3 (6m2/18′) is one of the most beautiful of its kind. It lies on the extreme southeastern edge of the constellation. The easiest way to find it is to use Beta Comae Berenices as the leading star. The cluster lies approximately 7 degrees east, which is half a degree more than the diameter of the field of view of binoculars. If Beta Comae is moved to the extreme western edge of the field of view, and you shift the binoculars a degree east, the cluster will appear in front of your eyes. Although we said that the cluster is splendid and is clearly seen though binoculars, it shows up mostly

Figure 8.10. Spiral galaxy M 106

as an approximately 8 arcmin big, rather bright spot of light. Individual stars on the edge cannot be seen through binoculars. In order to see them we need a larger objective with higher magnification!

M 3 is 33,900 light years away from us, and its actual diameter is 170 light years. At Mt. Palomar Observatory, astronomers used photographic plates to count the stars in the cluster. They managed to count 45,000 stars up to magnitude 22.5. The luminosities of the faintest stars seen on the plates were only 0.6 that of the Sun. Astronomers estimate that the common mass of the cluster is approximately 140,000 Suns and that it contains at least half a million stars (Figure 8.12).

M 3 is situated above the plane of our galaxy, some 40,000 light years away from its center. Now, close your eyes and imagine that we are living on a planet that is circling around a certain star on the extreme edge of the cluster, the edge that is pointing toward the Milky Way. Can you even try to imagine the appearance of the night sky? During one half of the year there are no true nights at all, for the sky is dominated by approximately half a million of the cluster members. Thousands of them surpass our Sirius, even Venus! Most probably night on this planet would resemble our twilight. However, the other half of the year there would be no stars whatsoever. In the night sky of this hypothetical planet one could see only a single celestial object. But what an object! A bright, splendid, spiral beauty – our galaxy in its entire glory, seen fully face on. Its apparent diameter is over 100 degrees! With the naked eye you could clearly see its bright mysterious core, the central bar of stars as well as the gaseous ring around it with myriads of newborn stars and the spiral arms and spurs with countless multicolored stars, magnificent clusters, and nebulae of all kinds! If you stretched out your arms, you could almost touch it. And this beauty of alien nights is our home in space, certainly the true Miss Universe...

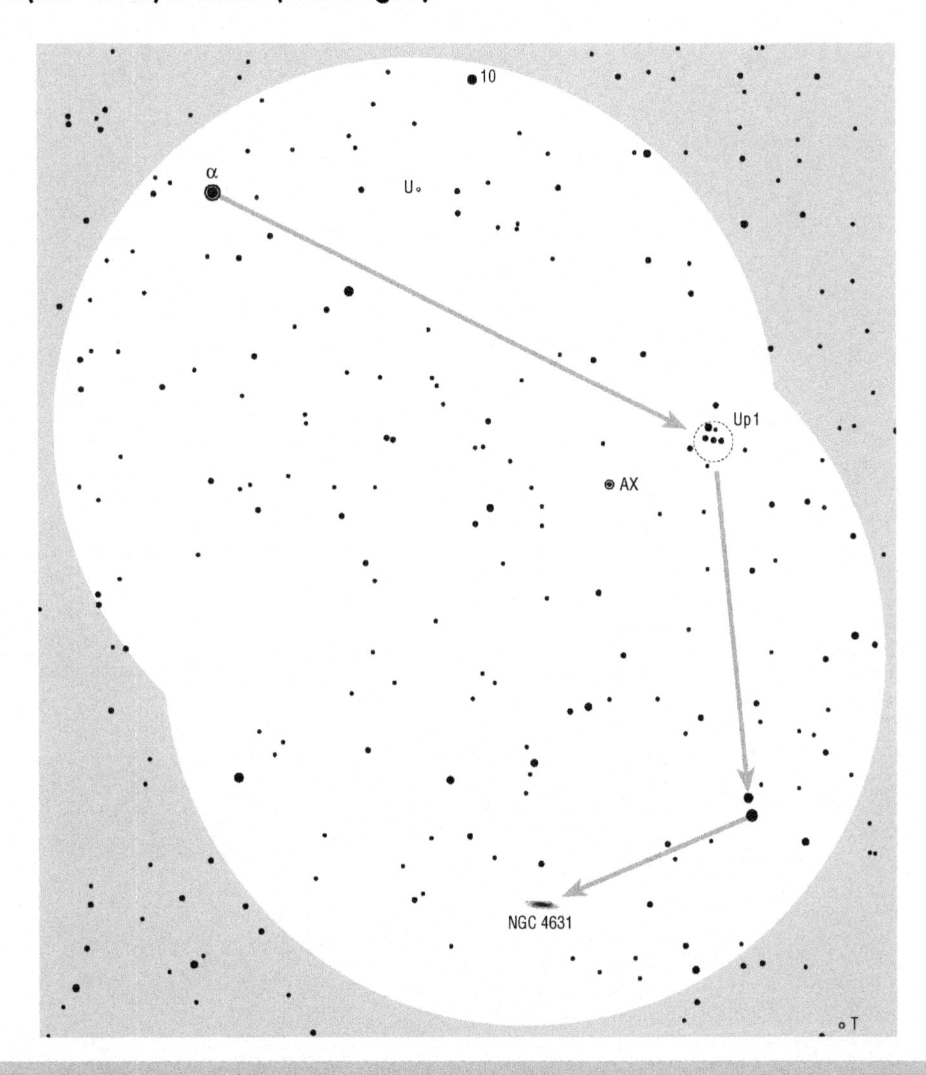

Figure 8.11. The position of the faint galaxy NGC 4631. Alpha and a small group of magnitude 7 and 8 stars called Up 1 lead the way to the galaxy

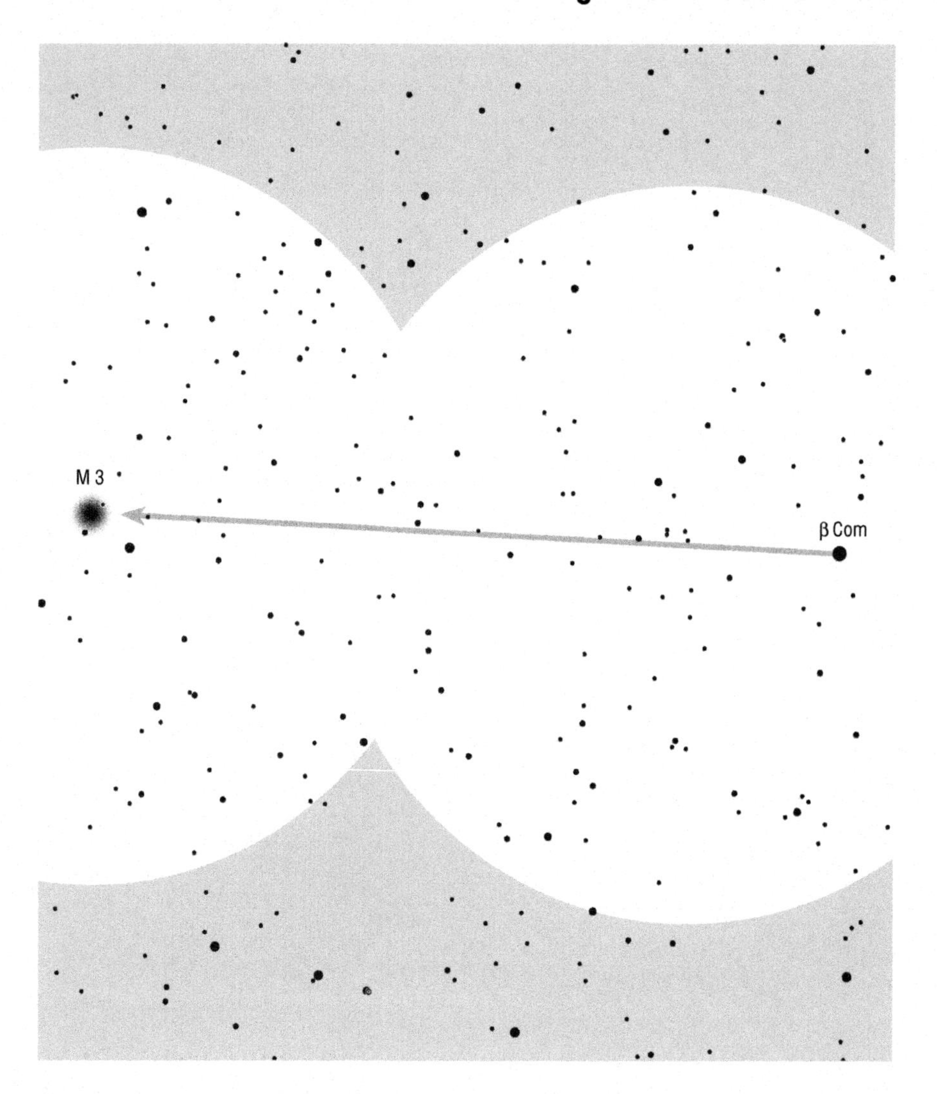
M 3
β Com

Figure 8.12. Globular cluster M 3

CANIS MAJOR (The Great Dog)

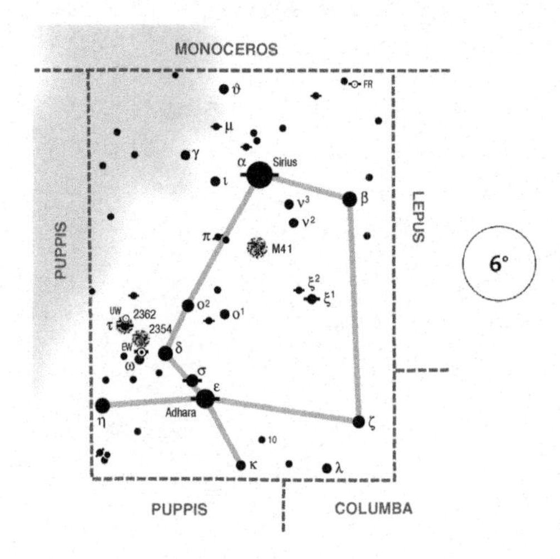

CONSTELLATION CULMINATES		
in begin. December at 2 a.m.	in begin. January at midnight	in begin. February at 10 p.m.

Canis Major is a famous winter constellation, reigned over by Sirius (–1m44), the brightest star in the sky. Sirius is over 0.8 magnitude brighter than Canopus and as much as 1.4 magnitude brighter than Arcturus, which is third on the list of the brightest stars. However, Sirius does not hold the lead position due to its actual size or luminosity, but due to the fact that it is just 8.6 light years from us. On the list of the closest stars it is in fifth place and is the second closest from the stars that are visible with the naked eye (only Alpha Centauri is closer). And if one of the closest stars is 550,000 times further from us than the Sun, our galaxy must be so huge and full of empty space that it is hard to even imagine it.

Sirius is a completely white star. Through binoculars it appears as a glittering diamond; through a telescope it is blinding and can spoil the adaptation of the eyes to night vision! When it is low above the horizon, it flashes in all the colors of the rainbow, as its light travels through thick layers of turbulent atmosphere. The luminosity of Sirius is 20 times that of the Sun, its diameter is 1.7 times bigger, and its mass is 2.03 times greater than that of our star. Its surface temperature is approximately 10,000 K.

Between 1834 and 1844, astronomer and mathematician W. F. Bessel measured irregularities in the proper motion of Sirius and discovered that the star has an unseen companion that circles around the central star every 50 years. Sirius B, as the unseen star has been named, was first seen by Alvan G. Clark in 1862 through a 47-cm refractor. The brightness of Sirius B is of magnitude 8.5, and the separation between the stars varies from 3 arcsec in periastron (in the year 1994) to 11.5 arcsec in adastron (2025). Although the stars could be separated even with a small amateur telescope, it is not quite as simple to see Sirius B, because Sirius A is so extremely bright that it puts it entirely into shade.

The mass of Sirius B, which is approximately the same as the mass of the Sun, has been determined from its orbital motion around Sirius A. Its luminosity, which is just 1/400 that of the Sun, was established from its apparent brightness and known distance. When astronomers combined all of

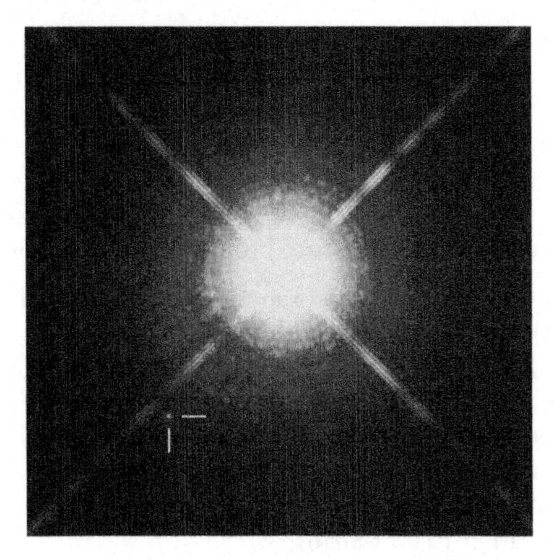

these data, they asked themselves: Is the surface temperature of Sirius B so low as to cause such small luminosity, or is the star itself very small? They received their answer in 1915, when the spectrum of the star was finally obtained. It showed that the surface temperature was high - around 8,800 K. This means that the star has to be extremely small to have such small luminosity. Soon afterward it was unequivocally proved that Sirius B is a white dwarf, the closest one to us. Its diameter is only 0.0084 times that of the Sun, which means that is a mere 11,800 km across (Figure 8.13). Sirius B is even smaller than Earth! Such mass (1 solar mass) compressed into such a small diameter results in an extremely high average density, up to 2,200 tons per cubic decimeter (liter). A sugar cube made off the same mass would weigh about 2.2 tons!

Almost every civilization throughout human history has included Sirius in its myths and legends. Because it is the brightest star in the sky, this in itself is not surprising. However, it is very interesting

Figure 8.13. Comparison between some planets and two of the most famous white dwarfs. In this model, our Sun would measure approximately a meter in diameter. Compare this image with that next to Mira (in the constellation Cetus on page 256)

that all ancient observers described the star as copper red in color and not a brilliant white. How is this possible, as we can be relatively certain that the ancient astronomers wrote down their observations correctly? This is a question that has baffled astronomers for a long time and is as yet unanswered. The companion to Sirius is a young white dwarf. Maybe this star was in its red giant phase 2,000 years ago and contributed to the reddish hue of Sirius. The only problem with this explanation is the fact that astronomers believe the transformation from a red giant to a white dwarf should last approximately 100,000 years, which would mean that in this case it happened exceptionally fast.

Beside Sirius, the constellation is adorned by a number of other bright stars: Epsilon (1m5), Delta (1m8), Beta (2m0), Eta (2m4), Zeta, and Omicron-2 (both 3m0).

Epsilon or Adhara is the 22nd brightest star in the sky. It is 430 light years away from us. Its luminosity is as much as 3,300 times that of the Sun and 160 times that of Sirius.

Delta is the 37th brightest star in the sky. It is very far away, for its light travels 1,800 years to reach us. Its luminosity surpasses our Sun by 43,000 times and Sirius by 2,150 times!

Beta is the 47th brightest star in the sky. It is 500 light years away from us. Its luminosity is 2,900 times that of the Sun and 145 times that of Sirius.

Eta is the 89th brightest star in the sky. It is a super giant with luminosity that surpasses our Sun by 77,000 times and Sirius by 3,850 times! Although it is 3,200 light years away from us, it shines as a magnitude 2 star in our sky!

In the vicinity of Sirius a few additional brighter stars can be found: Iota (4m4), Gamma, and Theta (both 4m1). Theta is an orange star of the spectral type K. Gamma is especially interesting. It is a giant star, some 1,250 light years from us. Its luminosity is 2,700 times that of the Sun and 110 times that of Sirius. However, this star has not revealed all of its secrets yet! It was named Gamma (which should be the third brightest in the constellation) by the seventeenth-century German astronomer Johann Bayer. It is unknown how bright it was at the time, for Bayer did not measure the brightness of the stars himself; instead, he relied on 1,500-year-old data from ancient times. Today this star is fainter than Epsilon, Delta, and Zeta, and even from Omicron Canis Majoris! Supposedly, it disappeared completely in 1670 only to reemerge 23 years later. Since then its brightness has remained stable. This puzzle still needs to be solved.

In Canis Major, one can stumble across the wonderful open cluster M 41 (4m5/38′). On a clear and dark night it can be seen with the naked eye as a tenuous spot of light (Figure 8.14). It lies some 4 degrees south of Sirius, and together with orange Nu-2 (3m9) they form a triangle that can be seen through binoculars in the same field of view. M 41 can be completely resolved through binoculars. In total, it has approximately 300 members that are gathered in an area slightly bigger than a full Moon. Through binoculars we can see approximately 60 stars when there are excellent observing conditions. On a perfect winter night, somewhere high in the mountains or in the middle of the desert this number could increase up to 100! The brightest stars are of magnitude 7. The cluster is very popular among amateur observers for its multicolored stars. The brightest is orange with a reddish hue; there are also numerous yellow, deep orange, and bluish stars.

M 41 has been known since time immemorial. It was mentioned by Aristotle (325 B.C.) as one of the "mysterious clouds" found in the sky.

This open cluster is approximately 2,300 light years away from us; its actual diameter is approximately 25 light years.

Open cluster NGC 2362 (4m1/8′) surrounds the magnitude 4 star called Tau, but an even better leading star is provided by Delta, which lies less than 3 degrees southwest. For observers in mid-northern latitudes, the cluster is rather low in the south, so in order to see it we have to choose a clear winter night in which the sky is dark all the way down to the horizon and when the cluster is close to its culmination. Through binoculars it can be seen as a small and faint spot of light around Tau. Using averted vision

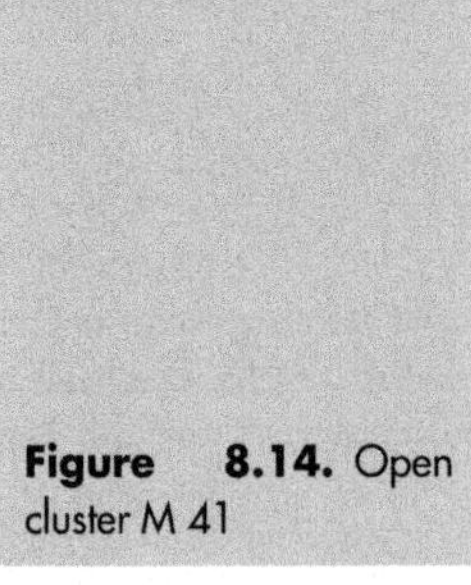

Figure 8.14. Open cluster M 41

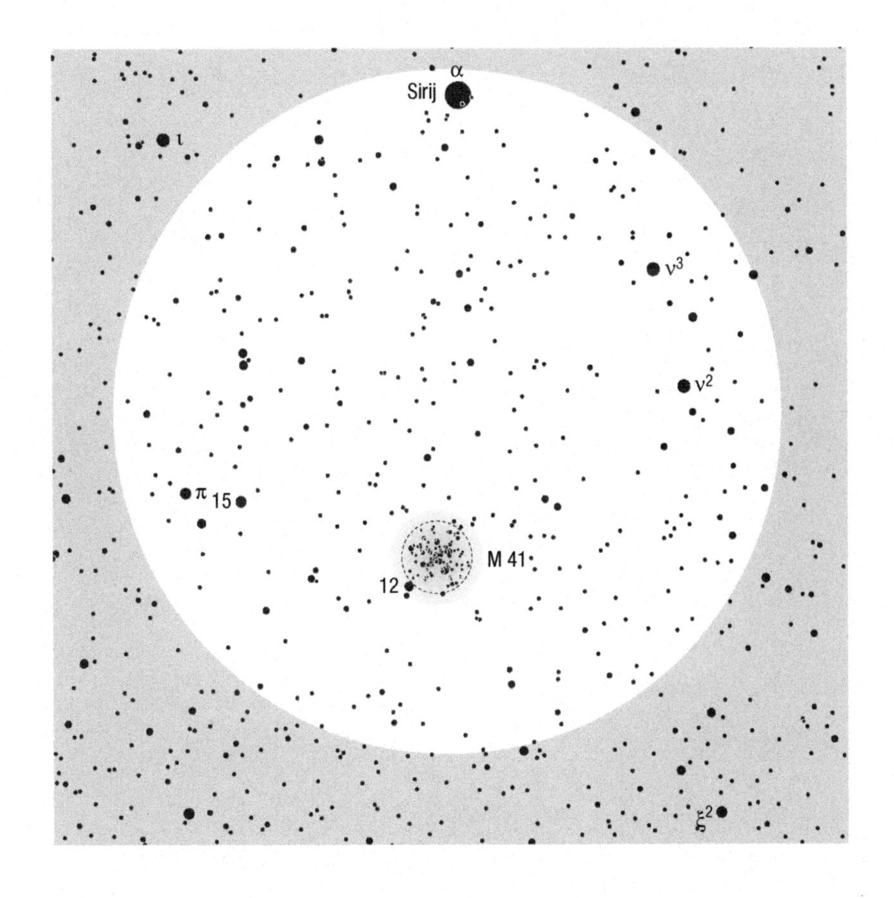

with excellent observing conditions we can glimpse a few of the brightest stars out of the 200 found in the area of the cluster.

NGC 2362 is approximately 5,000 light years away from us; its real diameter is a mere 12 light years. Tau is in fact a double star with a mass estimated to be between 40 and 50 solar masses! The cluster is also interesting to professional astronomers, who have discovered that this is one of the youngest clusters known. Its age is estimated to be only a few million years! It is still surrounded by the cloud of gas and dust from which the stars were born.

If observing conditions are perfect or you are stargazing with a larger pair of binoculars, you can see the open cluster NGC 2354 (6m5/20′) in the same field of view, less than 2 degrees southwest from NGC 2362. Since it is bigger and fainter than its neighbor it is harder to see. Approximately 25 stars are scattered within a 20-arcmin diameter; however, only 15 are true members of this cluster. Because NGC 2354 lies a degree south of NGC 2362, we have to choose a really clear winter night during which the sky is dark all the way down to the horizon to see it, and even then the cluster has to be close to its culmination. Unfortunately, it can still happen that you will watch and observe and yet see nothing.

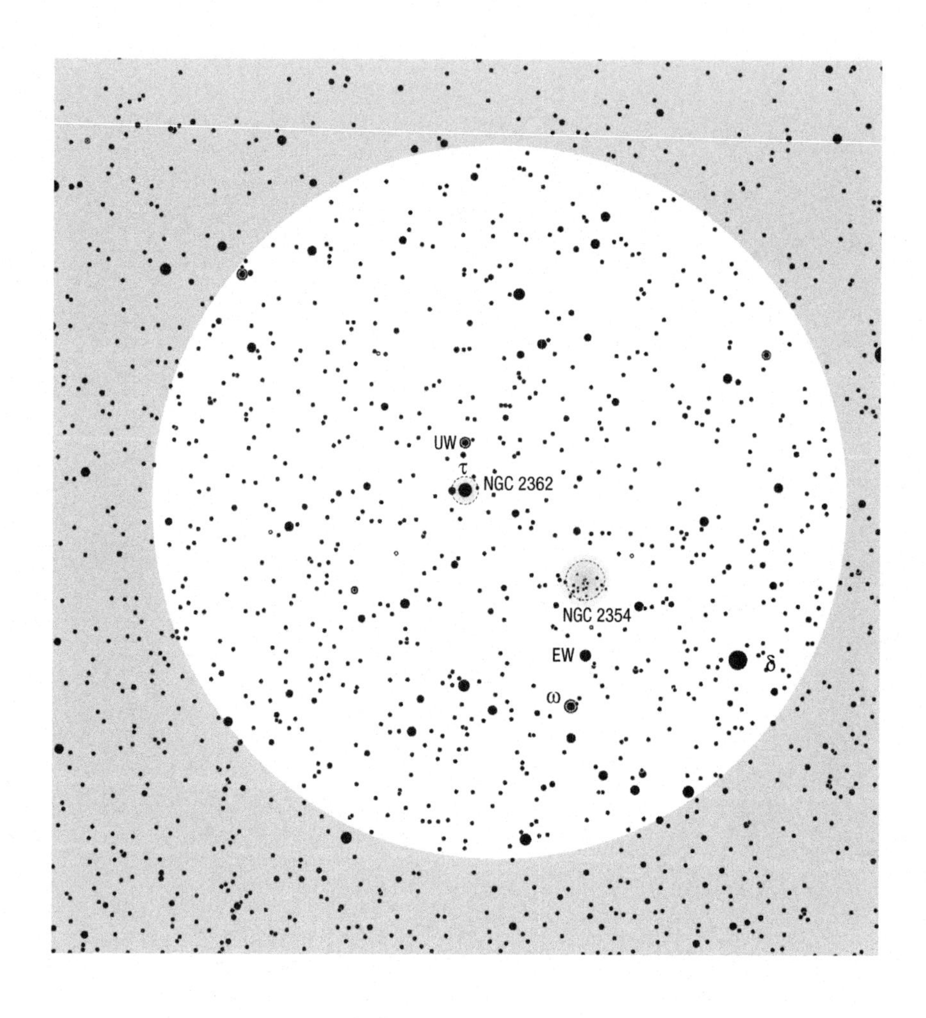

CANIS MINOR (The Little Dog)

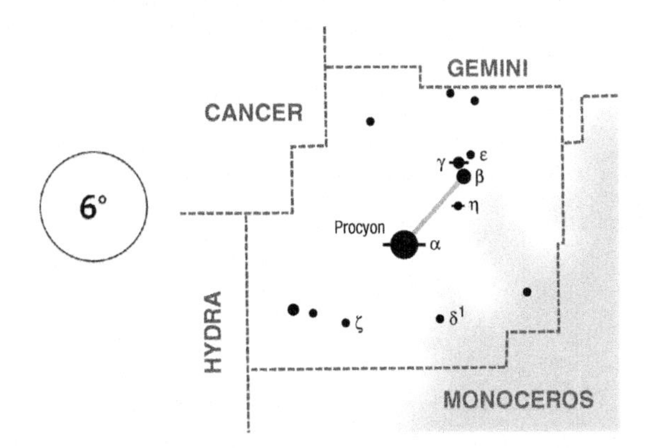

CONSTELLATION CULMINATES		
mid December at 2 a.m.	mid January at midnight	mid February at 10 p.m.

Even though Canis Minor is a small constellation, one cannot mistake it, for it includes one of the closest and brightest stars in the sky – Procyon (0m40). This is the 8th brightest star in ranking, only 11.4 light years from us. Among all stars, it is the 13th closest one and from the ones seen with the naked eye from mid-northern latitudes, only Sirius (Alpha Canis Major) and Epsilon Eridani are closer. The luminosity of Procyon is 6.5 times that of the Sun; the star is twice as big, and its surface temperature is almost 7,000 K.

Actually, Procyon is an interesting double star. While observing the irregularities in the star's movement, A. Auwers (1838–1915) concluded that it has a dim yet massive partner. In 1861, he published that the unseen star circles around Procyon with a period of 40 years. For a long time astronomers searched for the second star in vain. The first to see it was J. M. Schaeberle, who noticed it in 1896 through a 91-cm refractor at the Lick Observatory, which was at the time the largest telescope in the world. The distance between the two stars was 4.6 arcsec. Due to the dimness and proximity of bright Procyon it is extremely difficult to see Procyon B, and it can only be seen in the largest telescopes. Its brightness is approximately of magnitude 11.

The stars circle around each other every 40.82 years, and their average distance is a mere 15 AU (the distance between Saturn and the Sun is 10 AU and the distance between the Sun and Uranus is 19 AU). The interesting part of this double star is that Procyron B is a white dwarf. Its mass is approximately 0.6 that of our Sun and its diameter a mere 28,000 km, which is only twice the diameter of Earth (see picture on page 229). The surface temperature reaches 9,700 K, but because the star is small, its luminosity is only 1/2,000 that of the Sun. Because its mass is relatively big and the star is small, the average density of the matter in the star reaches an exceptional 100 tons per cubic decimeter (liter). Of course, we cannot see all of this through binoculars, but with a little imagination we can – while observing the shining Procyon – imagine these wonderful worlds.

The other interesting star is Beta (2m9), 170 light years away. Beta is surrounded by a pretty group of three stars: Gamma (4m2), Epsilon (5m0), and Eta (5m2). Gamma is orange and has a vivid color in any binoculars.

CAPRICORNUS (The Sea Goat)

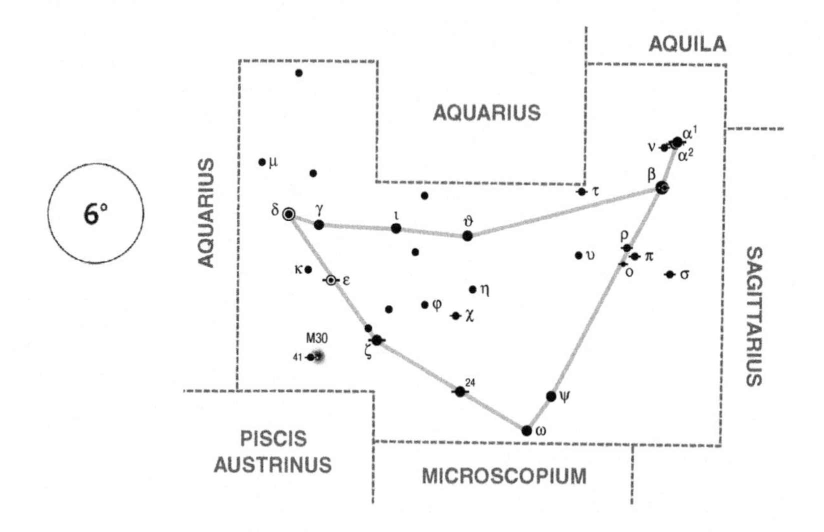

CONSTELLATION CULMINATES		
in begin. July at 2 a.m.	in begin. August at midnight	in begin. September at 10 p.m.

The autumn's Capricornus is one of the less glittery constellations of the zodiac. Its brightest stars are Delta (variable; 2m8–3m1), Beta (3m0), Alpha (3m5), Gamma (3m7), and Zeta (3m8). Despite its faintness the constellation is easy to find. The line consisting of three bright stars in Aquila (Gamma, Alpha, and Theta) points directly to Capricornus.

Delta is an eclipsing variable star. It is only 39 light years away from us. At maximum brightness it is the 143rd brightest star in the sky. The joint luminosity of the two stars is 8 times bigger than that of our Sun.

The double star Alpha, or Al Giedi, can be separated with the naked eye. This pair of yellow stars is approximately 380 arcsec (6.3 arcmin) apart (p. a. 292°, 2002). The brighter star, known as Alpha-2, shines with magnitude 3.6, while the fainter one, Alpha-1, is magnitude 4.3. Alpha-2 is approximately 110 light years away from us, while Alpha-1 is 6 times further.

Beta is also a double star that can be easily separated through binoculars. The companion is magnitude 6.1 and is 206 arcsec away from yellowish Beta (p. a. 267°, 2002).

Omicron is seen in binoculars as a close pair. The stars with magnitudes 5.9 and 6.7 are 22.6 arcsec apart (p. a. 239°, 2006). Unfortunately, this nice couple consists of two white stars without any color contrast. The leading star to Omicron is the bright Beta and a small heap of magnitude 5 stars (Upsilon, Pi, Rho, and Sigma) that lie approximately 4 degrees southeast.

Approximately 3 degrees east-southeast from Zeta lies the globular cluster M 30 (7m2/12′). It is visible through binoculars, but only as a small, faint spot of light with a magnitude 5 star east-southeast from it. Although the apparent diameter of the cluster is 12 arcmin, its extremely dense core measures a mere 1.2-arc min in diameter. M 30 is 26,100 light years from Earth, so its real size is approximately 84 light years, but certain individual members can be found up to 70 light years from its center.

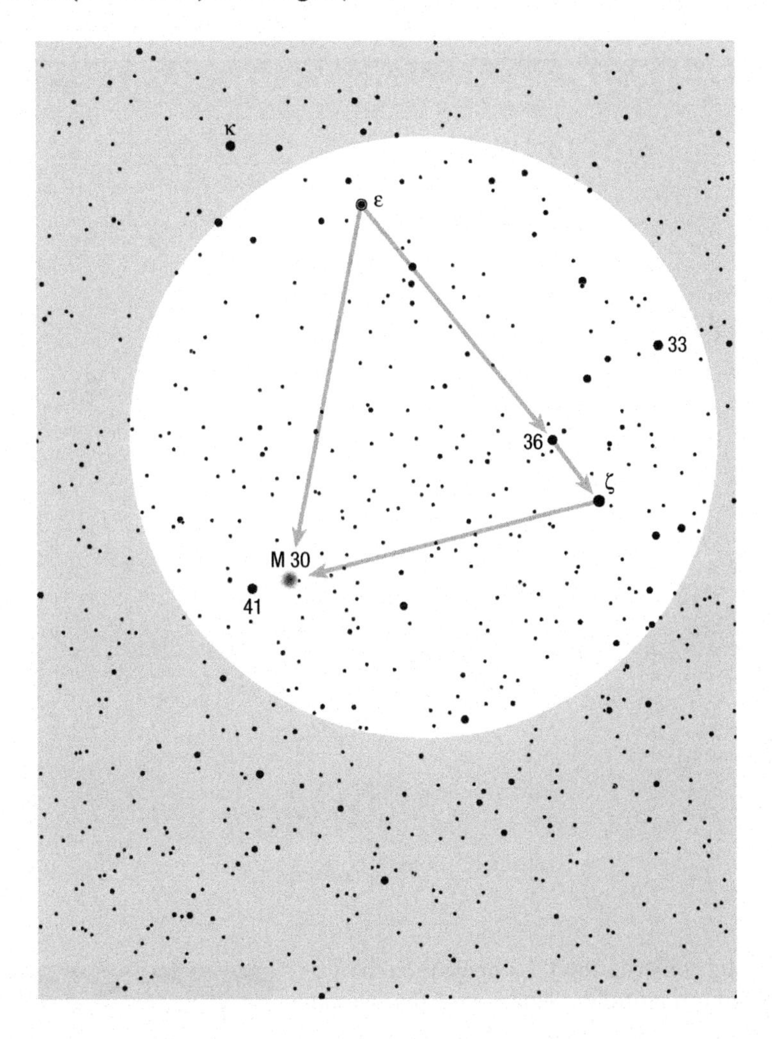
κ
ε
33
36
ζ
M 30
41

CASSIOPEIA (Cassiopeia)

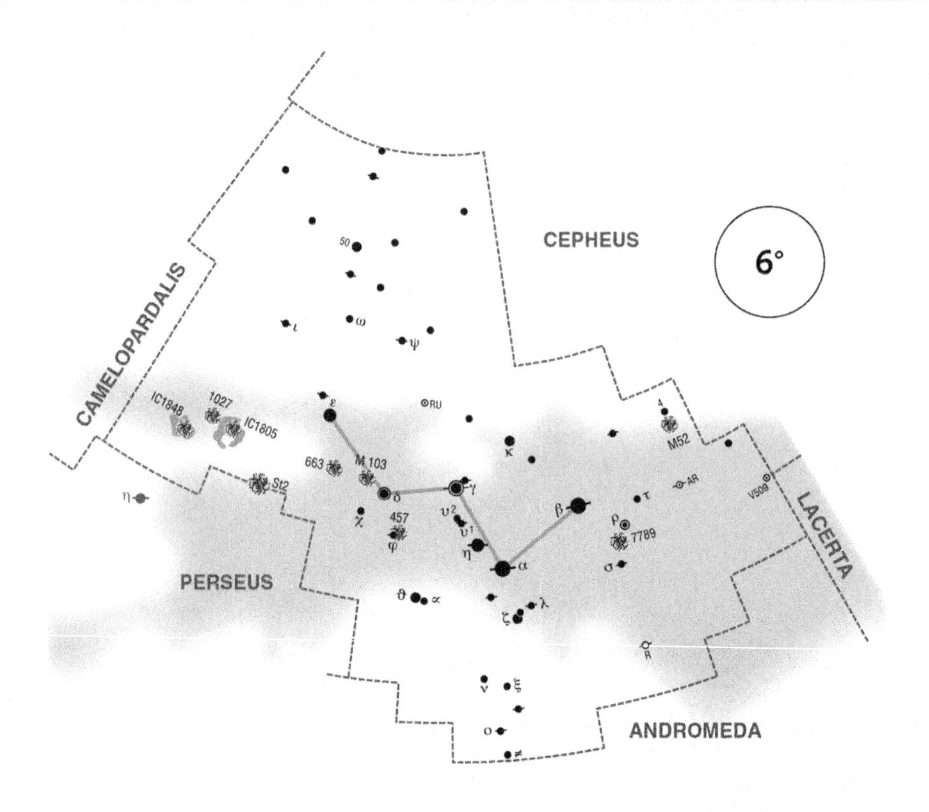

CONSTELLATION CULMINATES		
in begin. September at 2 a.m.	in begin. October at midnight	in begin. November at 10 p.m.

The big and splendid constellation Cassiopeia is circumpolar in the mid-northern latitudes. This means that we can see it every night and for the whole night. It never rises or sets. The arrangement of the five brightest stars resembles a slightly extended letter M (when the constellation is in culmination) or the letter W (when it is low above the northern horizon). If we know its approximate position, there is no way that we can miss it.

The brightest stars are Gamma (2m1 – slightly variable), Alpha or Shedir (2m2 – also variable), Beta (2m3), Delta (2m7), and Epsilon (3m4). The entire area is rich in stars and open clusters as the Milky Way winds it way through the constellation.

Gamma – the central star in the M shape – is the 63rd brightest star in the sky, a subgiant 614 light years away. Its luminosity is as much as 3,700 times that of the Sun. Gamma is an irregular variable star. Its brightness is usually slightly under magnitude 2, but every now and then it shines brighter. The last time this happened was in 1937, when the star reached magnitude 1.6. Because it also has a very uncommon spectrum, astronomers have concluded that it is very unstable. On the basis of spectroscopic data, some have assumed that during its increased brightness in 1937 the star blew off some parts of its outer layers into interstellar space. On long-exposure images a tenuous nebula can be noticed surrounding the star. Its brightest parts are designated as IC 59 and IC 63. Measurements have shown that both nebulae are connected to the star and that similar outbursts have occurred in the past.

The yellow-orange Alpha is the 71st brightest star in the sky, 230 light years away from us. Its luminosity is that of some 480 Suns.

Beta is the 74th brightest star in the sky. Since it is a mere 55 light years away from us, it is considered to be our close neighbor. Its luminosity surpasses our Sun by 26 times.

Delta is the 108th brightest star in the sky. Its light travels 100 years to reach us. Its luminosity is 61 times that of the Sun.

Eta (3m4) is a wonderful double star that can be separated even through small telescopes, but unfortunately not through binoculars. Stars with magnitudes 3.5 and 7.4 are 12.9 arcsec apart (p. a. 319°, 2005). The orbital period of this binary star is approximately 500 years. The pair is known for its distinctive color contrast: the stars are most commonly described as golden and slightly violet. Eta is only 18 light years away from Earth.

Iota (4m5) is one of the most beautiful triple stars in the sky. When there are excellent observing conditions, all three stars can be separated through an 8-cm objective. Two of them form a close double at a distance of 2.9 arcsec (p. a. 230°, 2004) and an orbital period of 840 years. The magnitudes of the stars are 4.6 and 6.9. The third star, which is a magnitude 9.0, is 8.9 arcsec away from the brighter pair (p. a. 99°, 2006). The Iota system also includes a fourth star; however, this one cannot be seen in any amateur telescope. Around the brighter star of the close double, circles a star of magnitude 8.5 with an orbital period of 52 years (in 2002, the separation was 0.4 arcsec; p. a. 66°). This interesting quadruple star is approximately 160 light years from us.

Mu (5m2) is even closer than Beta – it is only 25 light years away from us. This is a dwarf star with a diameter that measures only 0.9 that of the Sun. Its mass reaches only three-quarters the mass of our star, and its luminosity is only 0.4 times that of the Sun. Long ago astronomers concluded from astrometrical measurements that the star had an unseen companion with an orbital period of 18.5 years. The companion was first seen in 1966 through a 2-m reflector at the Kitt Peak observatory. The second star is a red dwarf and is 3 magnitudes fainter. This double star cannot be separated in amateur telescopes.

Open cluster M 52 (7m3/13′) lies on the western edge of the constellation, almost on the border of Cepheus. The easiest way to find it is to continue along the line between Alpha and Beta. The cluster lies slightly further from Beta than the diameter of the field of view of binoculars. If you catch Alpha and Beta in the field of view, move Beta to the extreme southeastern edge and then shift your binoculars another 3 degrees in the same direction. You will see four bright stars. The brightest of these is magnitude 5 and is of a distinct orange color; the remaining three are magnitude 6. The one closest to the brightest one also has an orange tint. M 52 lies less than 1 degree southeast (the chart is on next page). It is clearly seen through binoculars, but only as a small, few arcminutes big spot of light with one brighter and a few fainter stars, none brighter than magnitude 8. It is wise to undertake observations only during excellent conditions; otherwise you will only be able to see two or three faint stars. Of course, open cluster M 52 is much more attractive when viewed through larger amateur telescopes.

The cluster is approximately 5,000 light years from Earth and contains some 100 stars, of which only 12 are brighter than magnitude 11. Its real diameter is around 19 light years.

Open cluster NGC 7789 (6m7/16′) is a clearly visible faint spot of light, some 12 arcmin in diameter. Under excellent observing conditions and with the aid of averted vision you can glimpse a few brighter stars of magnitudes 9 and 10, but they are not truly members of this cluster. The brightest stars in the cluster are magnitude 11 and cannot be seen through binoculars. It is easy to find NGC 7789, since it is together in the same field of view of binoculars with the bright Beta. The hazy spot of light lies slightly more than 3 degrees south-southwest from Beta. The open cluster NGC 7789 is approximately 8,000 light years from us and has a real diameter of approximately 37 light years.

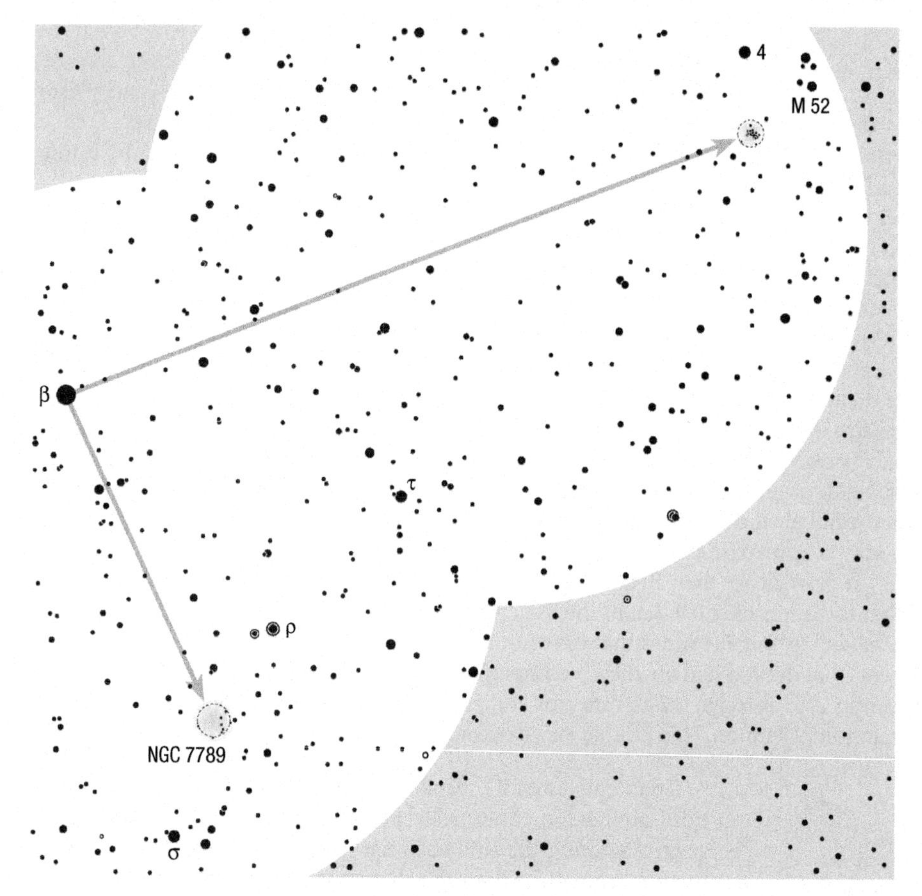

Open cluster NGC 457 (6m4/13′) is easy to spot if you find stars Chi (4m7) and Phi (5m0), which together with Delta (this is one of the stars in the shape of Cassiopeia) form a small triangle. The cluster lies close to Phi, which can be recognized by its close neighboring star of magnitude 7 to the southwest. Through binoculars we can see approximately 10 stars that belong to this cluster. They are scattered across a rather small area only a few arcminutes in diameter northwest of the bright Phi. With excellent observing conditions you can see them clearly, otherwise you must look at them with averted vision, with which you should be able to separate them. The cluster contains slightly over 100 members, but only 30 are stars up to magnitude 11. The brightest members are magnitude 9. NGC 457 is 9,000 light years away. Its real diameter is no larger than 30 light years.

Open cluster NGC 663 (7m1/16′) is easy to find in between Delta and Epsilon, right in the center and slightly southeast of the line connecting the stars (Figure 8.15). This is a small heap of stars that is surrounded by a faint but visible haze – the light from the stars that cannot be separated through binoculars. The brightest stars shine with magnitude 9. In good observing conditions four or five stars can be seen directly, while the others are merged into a tenuous nebulosity. In perfect observing conditions and with the aid of averted vision the haze becomes slightly more crumbled, but in order to see it completely resolved, you need a larger telescope with higher magnification. There are several hundred stars in the cluster; however, only 40 of them are magnitude 11 or brighter. NGC 663 is 6,400 light years from us and has a real diameter of 30 light years.

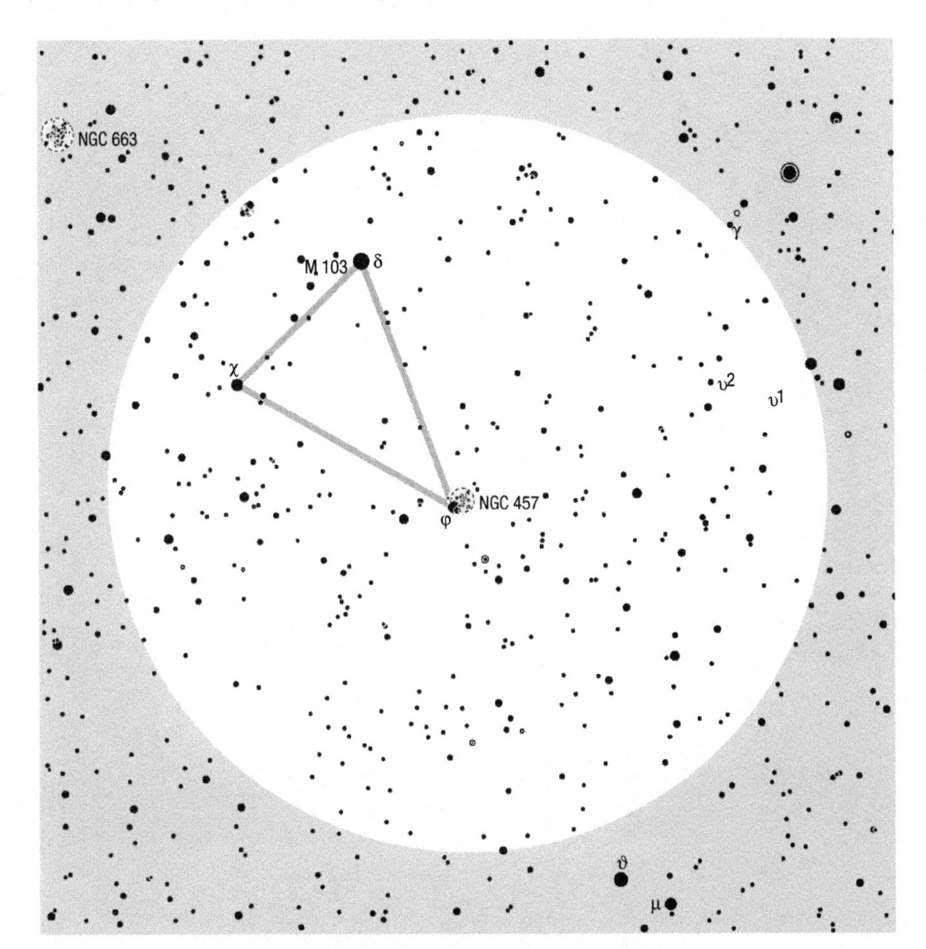

Open cluster M 103 (7m4/6′) is one of those elusive objects that need perfect observing conditions as well as perfect optics (Figure 8.16). The cluster is small and faint and lies close to a rather bright magnitude 7 star, which, when observed through binoculars with not the best optics, completely overshadows the faint nebula. This cluster is only mentioned because it is an entry in Messier's catalog. But any observer who likes a challenge should try to find this one (with a 10 × 50, of course)! The best observing sites are high in the mountains with a clear and dark sky and without any light pollution. Sometimes such perfect conditions can be found during a temperature inversion on winter nights. In this case valleys (together with their light pollution) get covered with a thick layer of fog, but above the skies are as clear and dark as in those ancient times, when electricity had not been invented yet.

Even though M 103 is hard to see, this does not mean it is hard to find. It lies in the same field of view with the bright Delta. This cluster is one of the remotest of its kind. Its light travels 8,500 years to reach us. Its real diameter is approximately 15 light years.

Another open cluster in the constellation is Stock 2 (4m4/60′). This is a big and loose group consisting of approximately 30 brighter stars. The easiest way to find it is to first find the Double Cluster in Perseus (see page 382), which is clearly visible with the naked eye. Stock 2 lies some 3 degrees north-northwest, so they can all be seen in the same field

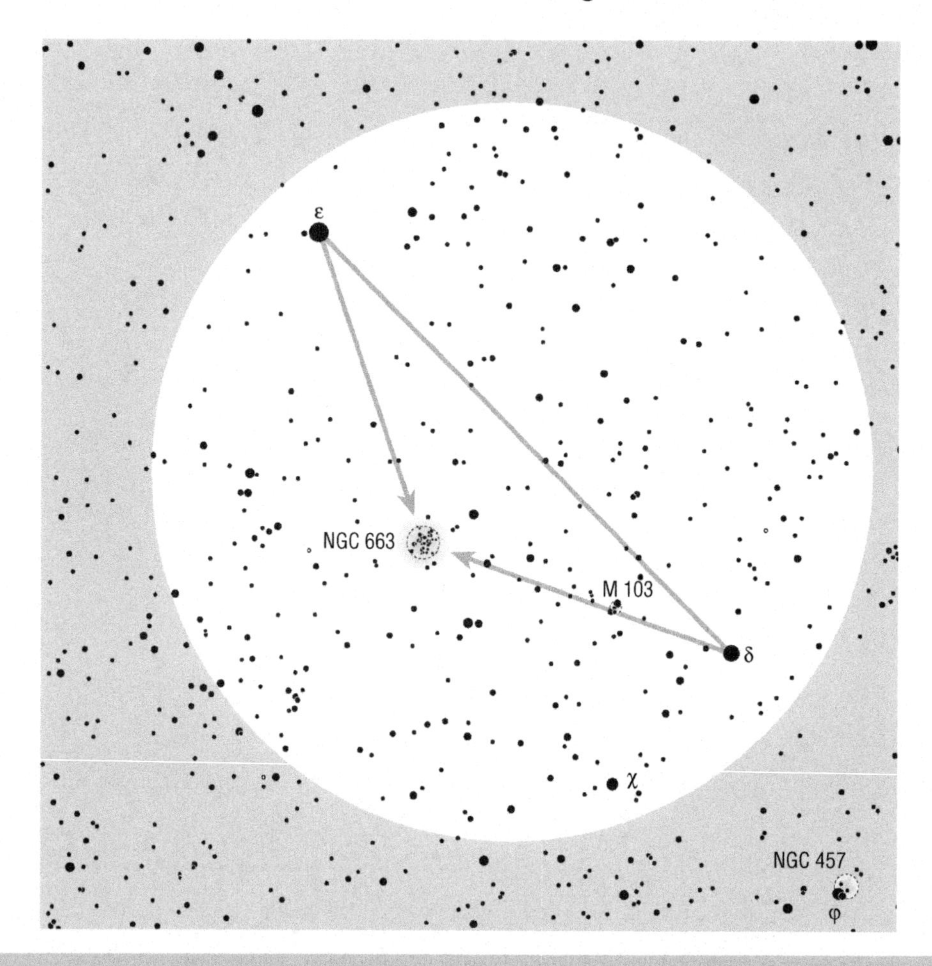

Figure 8.15. The position of the open cluster NGC 663. On the chart we can also see the open cluster M 103, which is visible through binoculars in perfect observing conditions and with excellent optics. The cluster is small and faint and lies close to a bright star, which makes it even harder to see

of view in binoculars. A distinguish arc of stars with magnitudes between 6 and 8 winds from Stock 2 to the Double Cluster (Figure 8.17).

Stock 2 lies in the extreme eastern part of Cassiopeia. This is a very interesting region of the constellation, with a rich part of the Milky Way in the background. It is best seen through wide-field binoculars. On a perfect night, we can see the open cluster Mellote 15 (6m5/22′), which is surrounded by a vast red nebula known as IC 1805 (Figure 8.18). Only a few degrees east lies the open cluster IC 1848 (6m5/12′). This cluster is also surrounded by a red nebula. Although the clusters can be seen through binoculars in perfect observing conditions, the nebulae are reserved for astrophotographers, for they appear only on long-exposure images. The entire area, where we can also find another open cluster NGC 1027 (6m7/20′), is a popular target for amateur astrophotographers.

The clusters are from 1,000 (Stock 2) to 6,500 (Mellote 15 and IC 1848) light years away from us. The distance to NGC 1027 is approximately 2,500 light years. It seems that the nebula surrounding Mellote 15 and IC 1848 are connected and represent just a brighter part of a larger cloud of dust and gases.

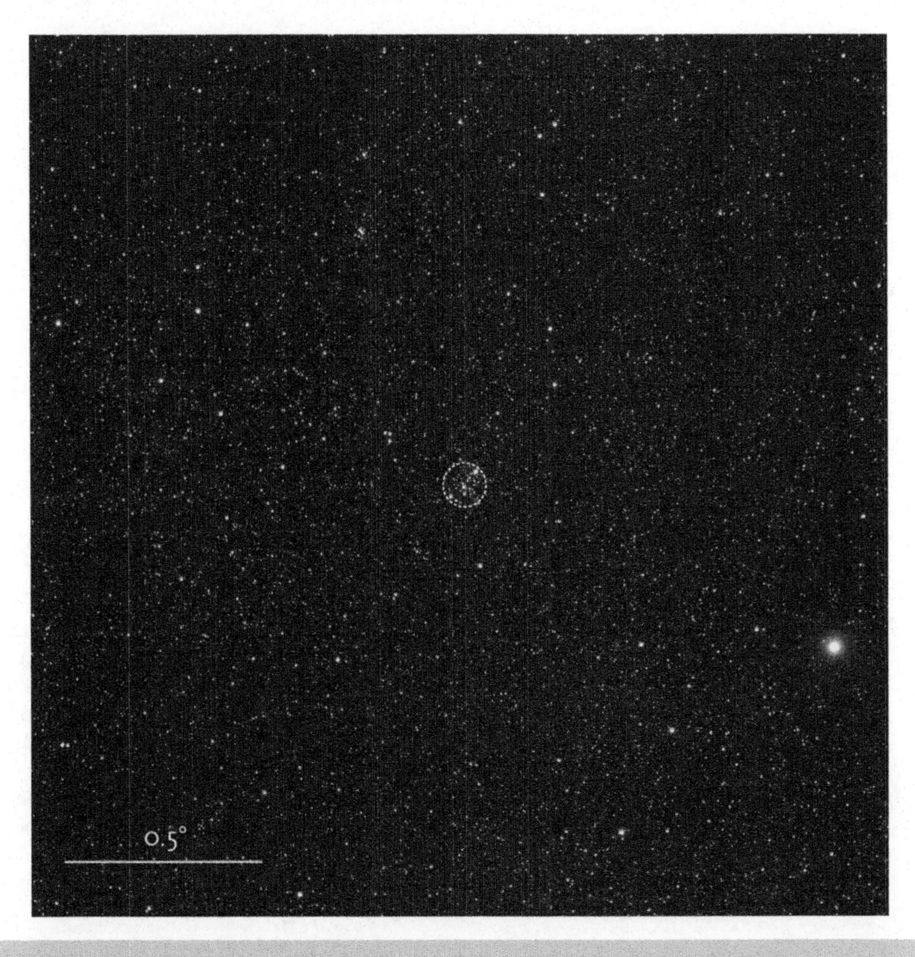

Figure 8.16. Small open cluster M 103. The bright star on the right is Delta

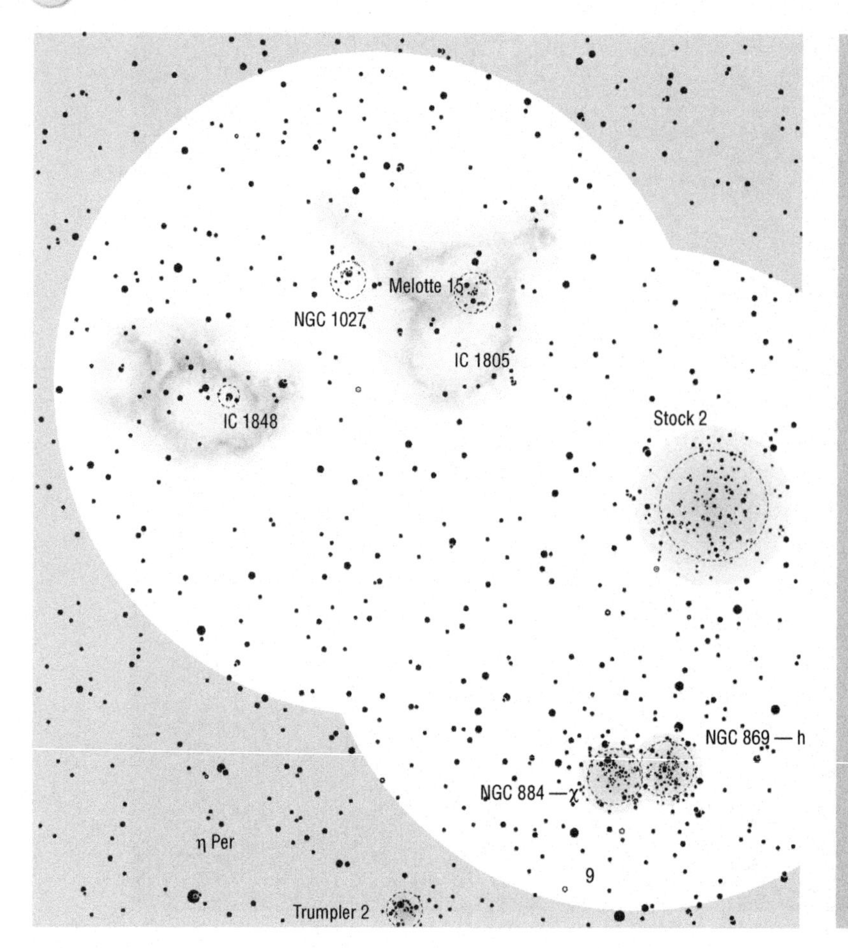

Figure 8.17. The entire area of open clusters is best seen if you aim the binoculars so that you see the star Eta Persei and the Double Cluster in the same field of view. Then you move the binoculars northwest so that the orange Eta disappears from the field

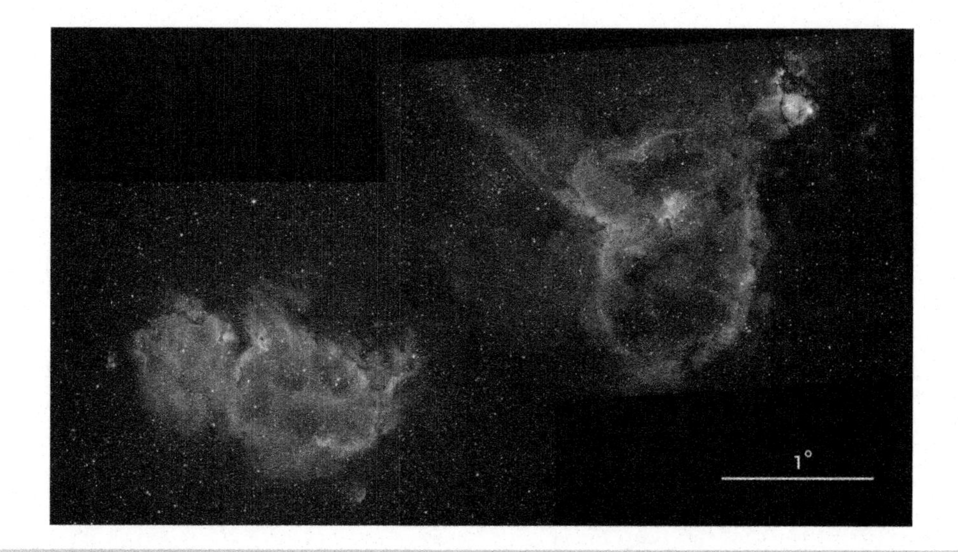

Figure 8.18. Nebulae IC 1805 (*right*), popularly named the Heart, and IC 1848 named the Soul

CENTAURUS (The Centaur)

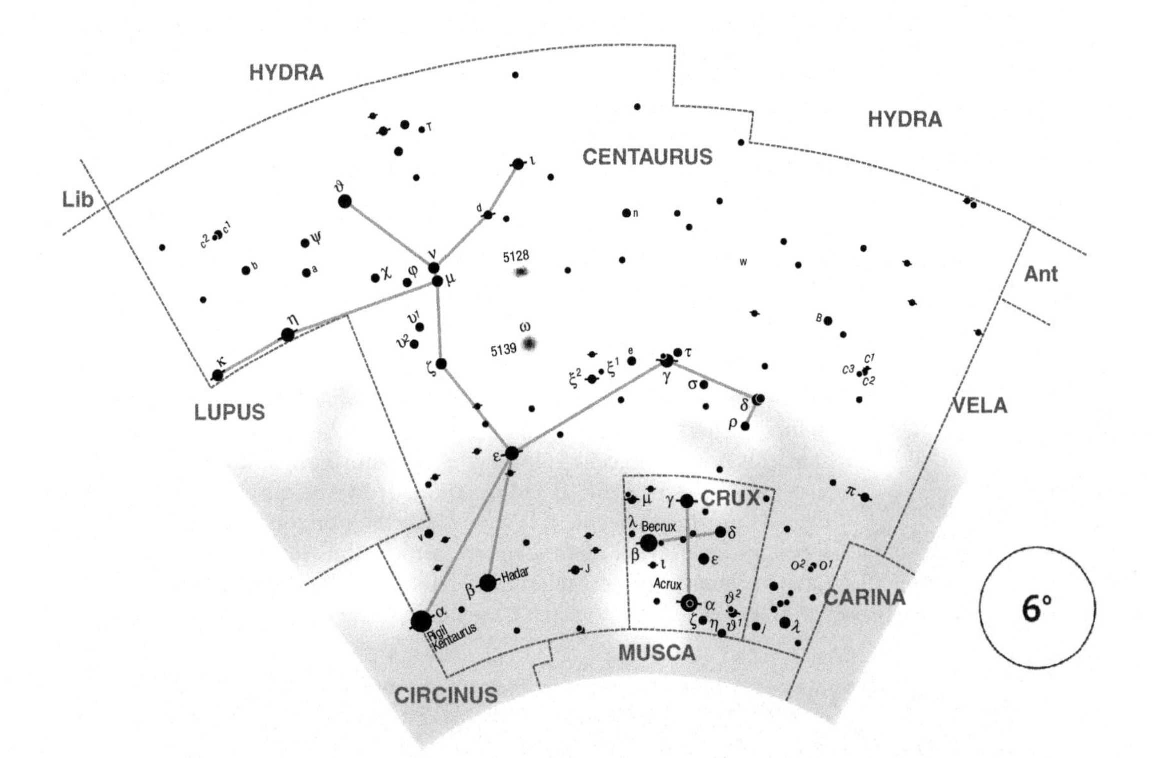

CONSTELLATION CULMINATES		
end March at 2 a.m.	end April at midnight	end May at 10 p.m.

The famous Centaurus constellation is one of the largest in the sky, full of bright stars and other interesting wonders. Unfortunately only its northernmost part is visible from the mid-northern latitudes.

The brightest star Alpha (−0m01) is the 4th brightest star in the sky, as well as the closest bright star. It is only 4.4 light years from us. Its close companion Alpha B is magnitude 1.2 and is orange in color. The pair is separated by 12.5 arcsec (p. a. 224°, 2002). This is too close for binoculars, but they are easily split in any amateur telescope. The Alpha system also includes a third, fainter star in (Alpha C), which has a magnitude of just 13.2. This star is even closer to us than Alpha (4.22 light years), which is why it is called Proxima (the nearest one). This interesting triple star is not seen from the mid-northern latitudes. Our observing site must be lower south than 30 degrees north (at this geographical latitude the star climbs barely to touch the horizon). In the southern latitudes, where the star is high in the sky at culmination, it is a true jewel that shines bright in its yellowish white color.

Theta is magnitude 2.1 and is the 53rd brightest star in the sky. It is 61 light years away from Earth. Its luminosity is approximately 40 times that of the Sun. On mid-northern latitudes the star is only 10 degrees above the horizon at culmination, which is not enough for it to truly shine.

Eta shines with magnitude 2.3, but in the mid-northern latitudes it rises a mere 2 degrees above the horizon. It is the 79th brightest star in the sky and is 309 light years away from us. The star is a giant with luminosity that surpasses our Sun by about 800 times.

Iota (2m7), the 123rd brightest star in the sky, would be quite a notable, if it were higher in the sky. But in mid-northern latitudes it rises only 7 degrees above the horizon, and we always watch it through the thick layer of our atmosphere. The star is 59 light years from us; its luminosity is 20 times that of the Sun.

The following data will persuade anyone that the constellation is truly magnificent when viewed from southern localities. Beta (0m6) is the 11th, Gamma (2m2) is the 64th, Epsilon (2m3) the 77th, Zeta (2m5) the 95th, and Delta (2m6) the 98th brightest star in the sky. For an observer in the southern latitudes, these stars are as bright as the stars in the Big Dipper for northern observers.

Open cluster NGC 3766 (5m3/12′) is composed of approximately 200 stars, the brightest of which are magnitude 8. It is a nice object for southern observers to view through binoculars, but for northern ones it is too low to be seen.

Open cluster NGC 5460 (5m3/25′) has the same brightness and is twice as big as NGC 3766. It can be seen from the southern parts of United States (it is above the horizon), but unfortunately its stars are faint.

But this is just an appetizer. Centaurus also holds two really exceptional deep-sky objects: the brightest globular cluster (Omega Centauri) and the bright galaxy NGC 5128.

Globular cluster Omega Centauri (3m7/36′) is easily seen with the naked eye. It is the brightest among all globular clusters, and it is so bright that before the telescope was invented astronomers thought it was a magnitude 3 star (this is why it is called Omega). In mid-northern latitudes, the cluster is just below the horizon at culmination. However, observers south of 40 degrees north can see it. It is low above the horizon, and its light travels through thick layers of atmosphere, which dims it. In order to spot it we have to choose a perfect night in which the sky is clear and dark all the way down to the horizon. However, if we really want to catch its genuine beauty, we have to travel south.

Omega Centauri is not just the brightest but also the biggest among all globulars. Its mass is estimated to equal 5 million solar masses, which is as big as any other globular cluster. In fact, its mass is the same as the mass of certain dwarf galaxies. Some observational data suggest that it might in fact be the core of a small galaxy that has merged with the ours at some time in the past. Our galaxy's tidal forces might have stripped the stars from the core of a victim, which then dissipated into the Milky Way.

Galaxy NGC 5128 (7m0/18′ × 14′) is the 5th brightest among all galaxies and is clearly visible through binoculars (Figure 8.19). Its declination is -43 degrees. This means that it is theoretically visible from the mid-northern latitudes as well as from more southern places. In the south of Spain it rises 10 degrees above the horizon, in Florida (U.S.) 20 degrees – enough to see it! However it is a true gem for southern observers, where it is high in the sky at culmination. The bright Zeta is the leading star to the galaxy. For those northern observers who cannot see Zeta, the leading star is Mu, whose brightness varies between magnitudes 2.9 and 3.5. Mu and the galaxy are in the same field of view in binoculars. If Zeta is your leading star, you should move the star to the eastern edge of the field of view. You will see the globular cluster Omega Centauri on the western edge. Now you only have to move Omega to the southern edge of the field of view, and the galaxy appears on the northern edge. To have two such prominent sky objects in the same field of view will surely be an unforgettable sight.

Figure 8.19. Splendid galaxy NGC 5128

NGC 5128 is one of the closest galaxies outside the Local Group and is a member of the M 83 group of galaxies. It is 15 million light years away, and its real diameter reaches 70,000 light years. It is one of the most interesting and peculiar galaxies in the sky. It has all the characteristics of a large elliptical galaxy, as well as a pronounced dust belt across the center, which forms a disk plane around it. Beside this, it is also a strong source of radio waves (therefore its designation is Centaurus A), which makes it the closest radio galaxy. All these data suggest that in the past this elliptical galaxy must have "swallowed up" at least one larger spiral galaxy.

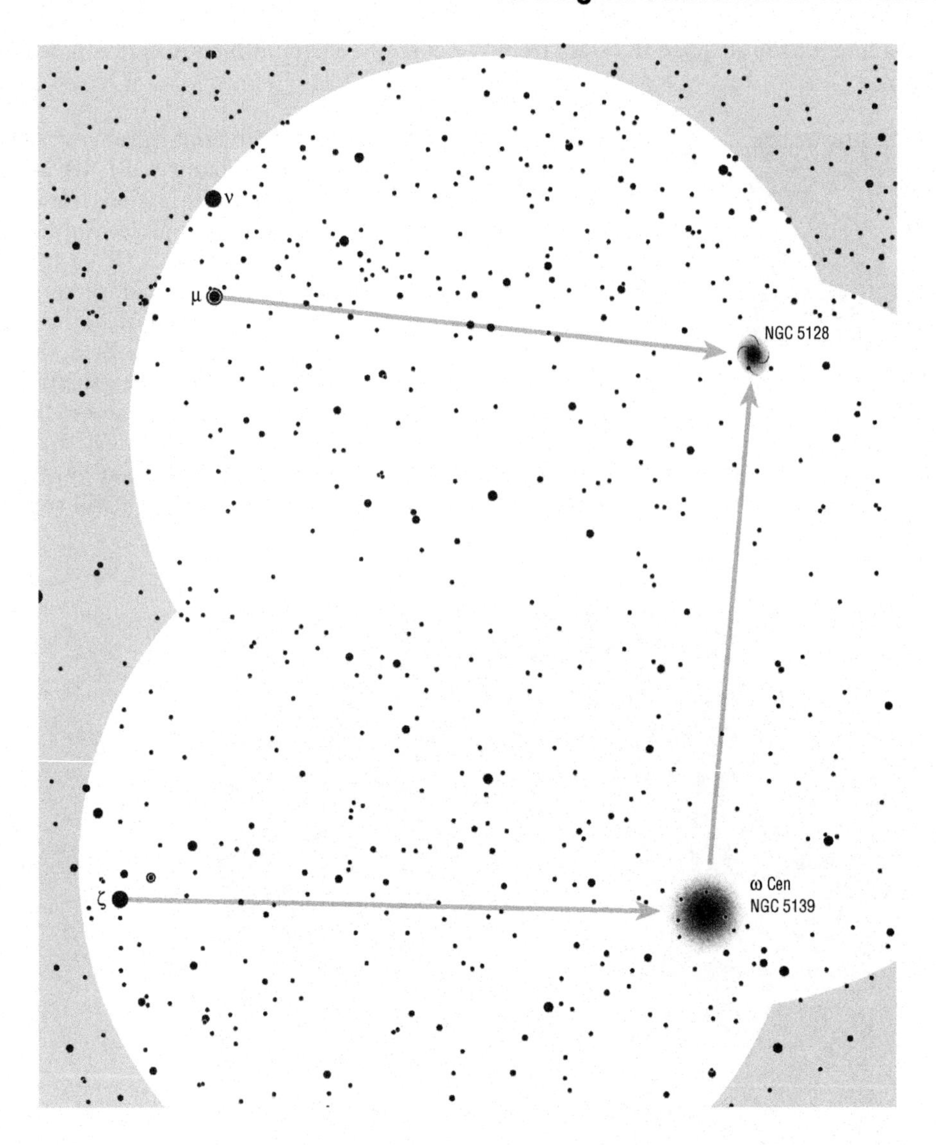
ν
μ
NGC 5128
ζ
ω Cen
NGC 5139

CEPHEUS (Cepheus)

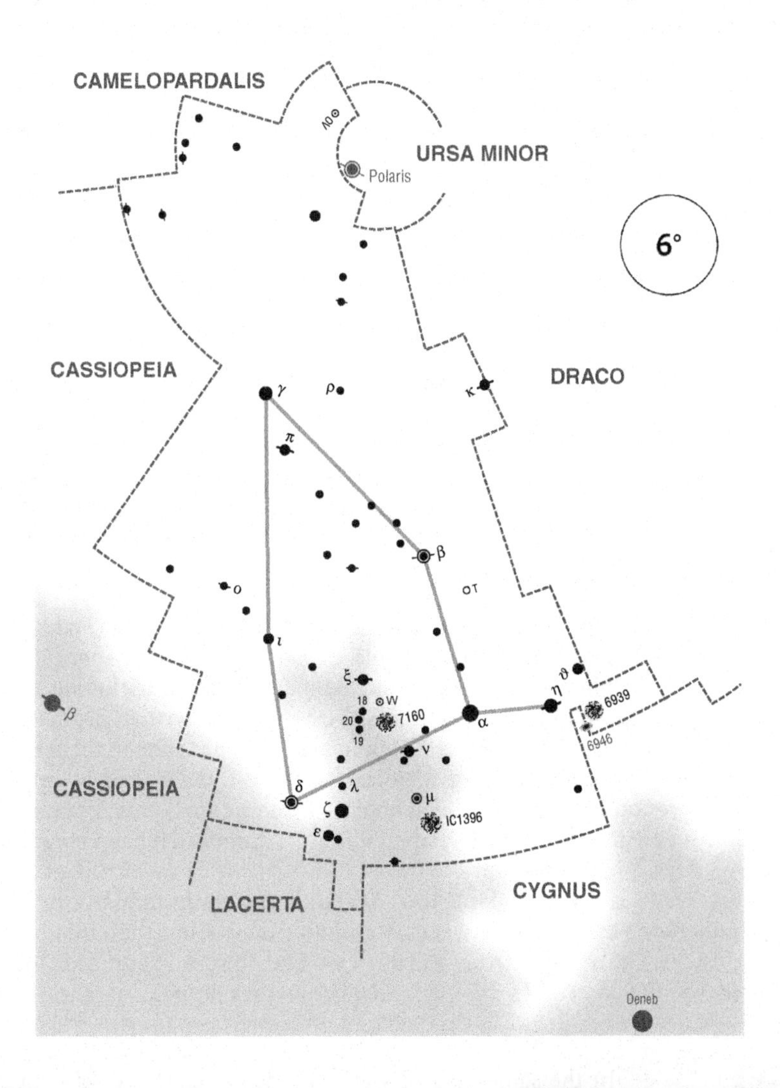

CONSTELLATION CULMINATES		
mid July at 2 a.m.	mid August at midnight	mid September at 10 p.m.

Cepheus is a large but rather faint constellation. For observers in mid-northern latitudes it is circumpolar and dances around the north celestial pole every night and all nights of the year. It never rises or sets. Cepheus is found within the triangle formed by Deneb (Alpha Cygni), Polaris, and Beta Cassiopeiae. On clear evenings, toward the end of the summer, we can find its rather faint stars high in the zenith. The brightest Alpha shines with magnitude 2.4. Other bright stars are Gamma, Beta (both 3m2), and Zeta (3m4).

Alpha is the 88th brightest star in the sky. It lies a mere 49 light years from us. Its luminosity is 18 times that of the Sun. It is interesting for astronomers because of two things that are not visible through binoculars. First, the star rotates around its axis very rapidly. Second, it lies in the part of the sky that is on the path of Earth's axis due to precession. Thus, Alpha Cephei is the Polar star every 25,800 years. The next time this will take place will be around the year 7500.

Beta is a variable star, a blue giant, which pulsates with a period of 4 h and 34 min. However, its brightness changes only by a tenth of a magnitude – too little to notice with average amateur equipment. The star is 1,000 light years from Earth. Its luminosity is as much as 4,000 times that of the Sun.

Eta Cephei is of magnitude 3.4. It is a subgiant, only 45 light years from us. It used to be similar to our Sun, but since it is a few billion years older, it is already on the way off the main sequence of the Hertzsprung–Russell's diagram and is slowly becoming a red giant. Through binoculars, it is a distinct golden yellow in color.

The yellow-orange Zeta is 1,240 light years away from us. It is a super giant with great luminosity – approximately 5,800 times that of the Sun.

Mu is a famous semi-regular variable star with an intense orange-reddish color, perhaps the most intense among all the brighter stars in the northern hemisphere. The Italian astronomer Giuseppe Piazzi named it Garnet Sidus (the Garnet Star). Its brightness changes slowly from magnitude 3.4 to 5.1. The period is not constant; its average value is 730 days. The minimum and maximum brightness is also not always the same. There are no suitable comparable stars in the vicinity of Mu. The closest ones are Zeta (3m4 – lies 4 degrees east), Nu (4m3 – lies 2.5 degree north), and Lambda (5m1 – lies 4 degrees east). The Garnet Star is red super giant, similar to Betelgeuse in Orion. It is very far away; its light needs 5,000 years to reach us. Mu is one of the largest stars we know; with a diameter of 1,420 times that of our Sun, it is the sixth largest! If we put this star into the center of our Solar System, it would extend further than Jupiter's orbit! It is also one of the most luminous stars, with a luminosity that surpasses our Sun by roughly 350,000 times!

Delta Cephei is one of the most famous variable stars; in fact, its name has been given to a whole class of similar short-period variable stars – Cepheids. In 1784, the Englishman John Goodricke was the first to notice its changes in brightness. Today we know that Cepheids are stars that regularly swell and shrink in size, which results in them radiating varying amounts of energy. We can observe Delta's changes in brightness from one night to another if we compare it with its neighbors – Epsilon (4m2) and Zeta (3m4). When Delta is at its brightest, it shines with magnitude 3.5; when it is at its dimmest, it has a magnitude of 4.4. The pulsating period is 5 days, 8 h and 48 min.

Like all Cepheids, Delta is a super giant, too. At maximum its luminosity is 3,300 times that of the Sun. The diameter of the star is between 25 and 30 times larger than that of our Sun. Delta is approximately 1,000 light years from Earth. The star has a magnitude 6.1 companion; they are separated by 40.6 arcsec (p. a. 191°, 2004), so they appear as a close pair through binoculars.

In the same field of view with Mu, merely a degree and a half south-southwest, we can catch a glimpse of the large and very loose open cluster IC 1396 ($\sim$4m/170′ × 140′), which covers as much as 2 degrees of our field of view. In good observing conditions, we see a dozen dispersed stars that do not give the impression of a cluster. On excellent viewing nights, number of stars increases to over a hundred and is a real treat to see. Almost in the middle lies a bright and beautiful triple star, known as Struve 2816. Through binoculars we can resolve only the close pair of stars with magnitudes 5.6 and 7.5, separated by 20.5 arcsec (p. a. 338°, 2006). The third companion is also magnitude 7.5 and is 11.8 arcsec from the main star (p. a. 120°, 2006). It is seen through small telescopes.

The stars in the cluster are surrounded by a faint nebula that can only be seen on photographs. IC 1396 is the brighter and denser part of a much larger association of stars, named Cepheus OB2, which is scattered over a few square degrees to the north. The stars that used to form a classical open

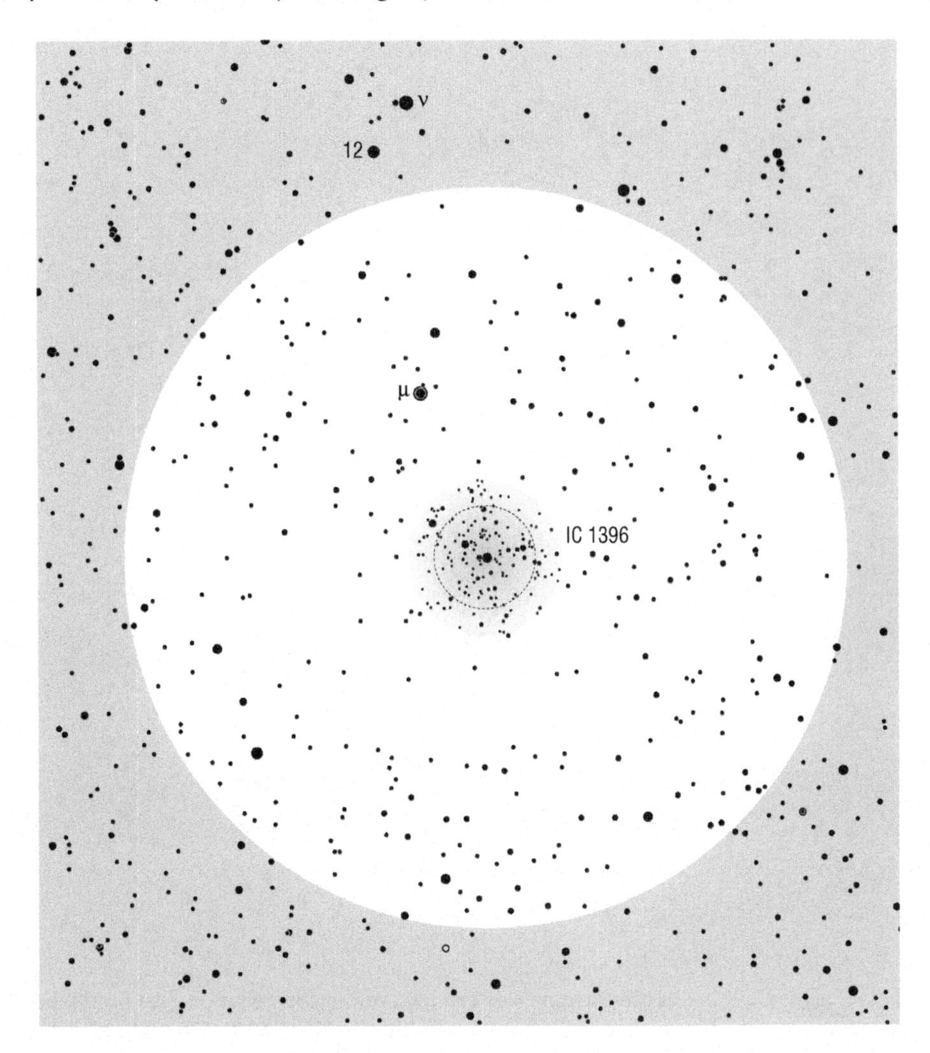

cluster are today only weakly gravitationally connected. They are still moving apart, and in the future they will lose all connection. The stars are approximately 2,500 light years from us.

Two degrees northeast of Nu lies the very small open cluster NGC 7160 (6m1/7′). Within a crowd of faint stars there are half a dozen brighter ones. In good observing conditions, only two magnitude 8 stars can be clearly seen through binoculars. In excellent conditions, you can see another three or four, and you could glimpse a few more with averted vision. This open cluster is therefore not really a celestial jewel. We mention it for those observers who want to see everything that can be seen through binoculars! This tricky cluster can be found with the help of the detailed finder chart. From Nu or a group of stars with designations 18, 19, and 20 you should move the binoculars to the area where the cluster lies. When you are certain that you are seeing the two brightest stars within the cluster (they are always visible), wait for your eyes to completely adapt to night vision. If the observing conditions are excellent, you should see a few more faint stars in the near vicinity, stars that appear as small needle points. When viewed through low-quality optics these stars may be melted into a faint haze, but with averted vision the haze will definitely break into stars.

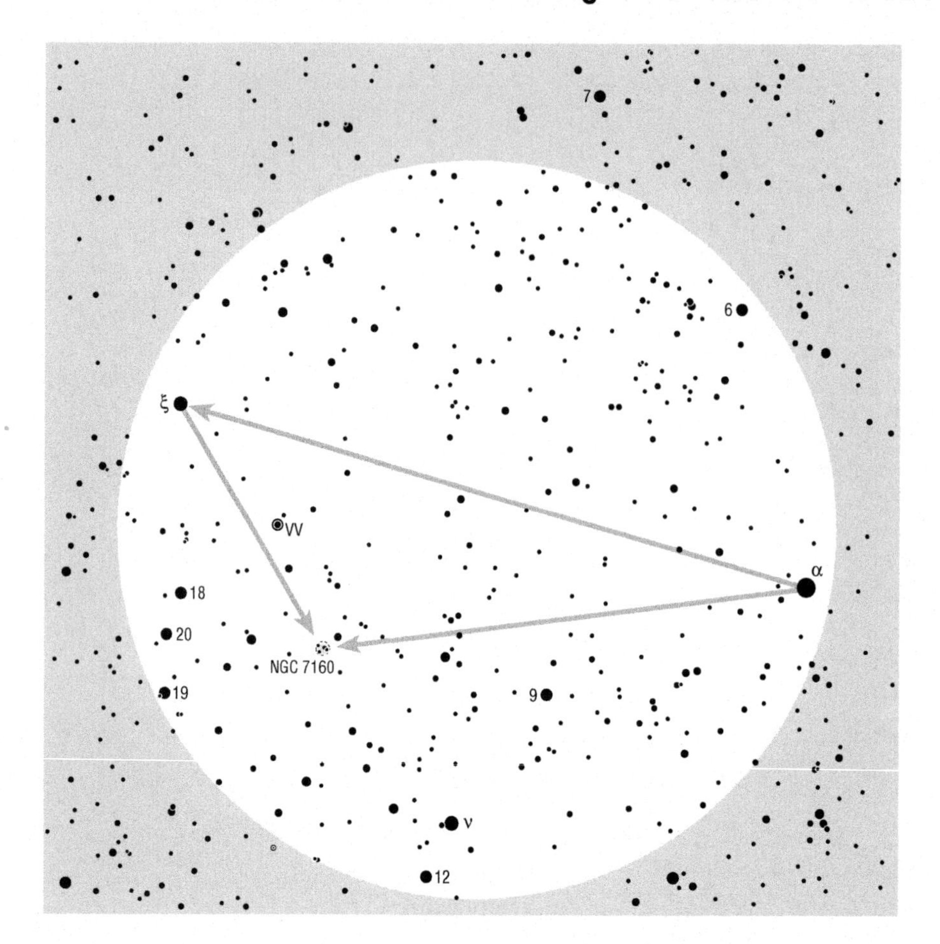

On the extreme western part of the constellation, just on the border with Cygnus, lies the faint open cluster NGC 6939 (7m8/8′), which is on the border of visibility when viewed through binoculars during excellent observing conditions. It looks like a faint patch of light a few arcminutes wide. The cluster is comprised of approximately 100 stars; only one of them is magnitude 11, while all the others are fainter. Less than a degree to the southeast you can find the spiral galaxy NGC 6946; however, this galaxy is already in Cygnus. The leading star to the open cluster is the bright Eta. For accurate positioning, see the detailed star chart below. Searching for and observing such faint objects is suitable only once you have gathered plenty of observing experience. Also, the cluster has to be at its highest position in the sky, or you will not be able to see it.

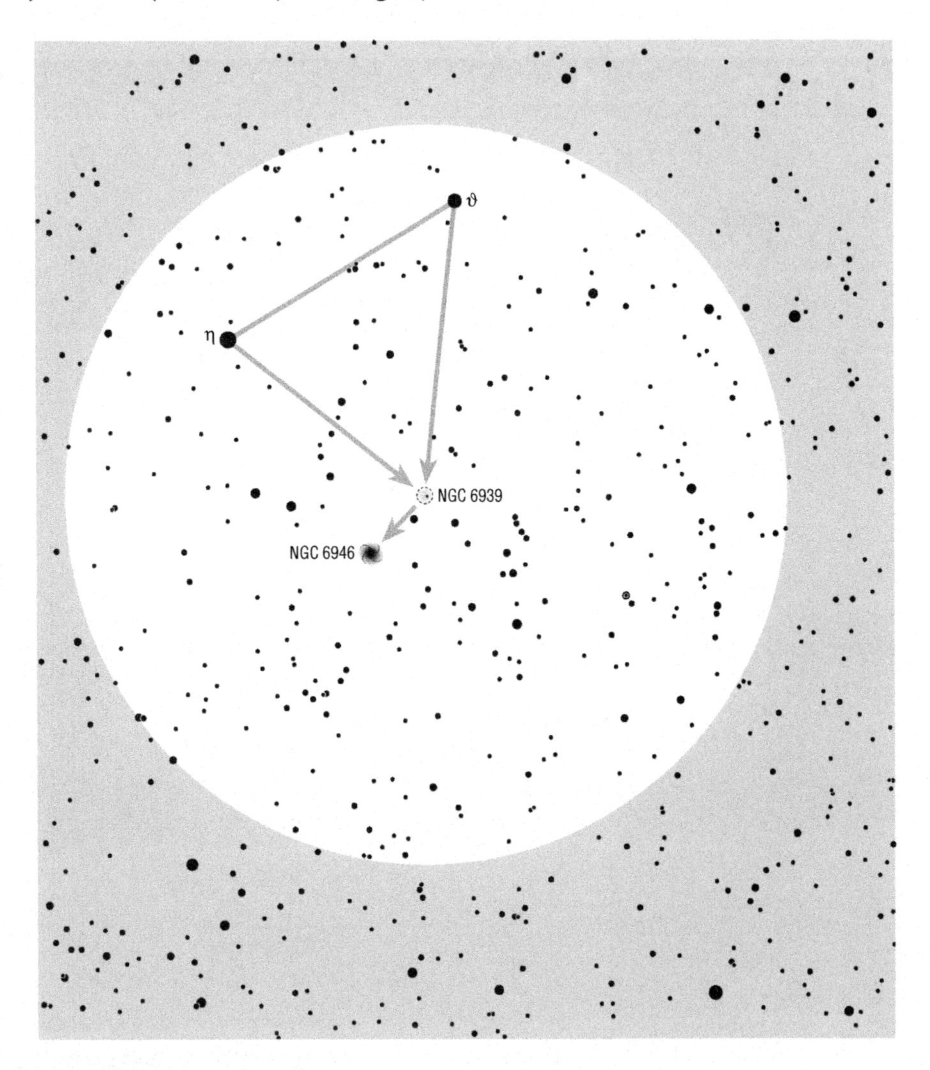
ϑ
η
NGC 6939
NGC 6946

CETUS (The Sea Monster)

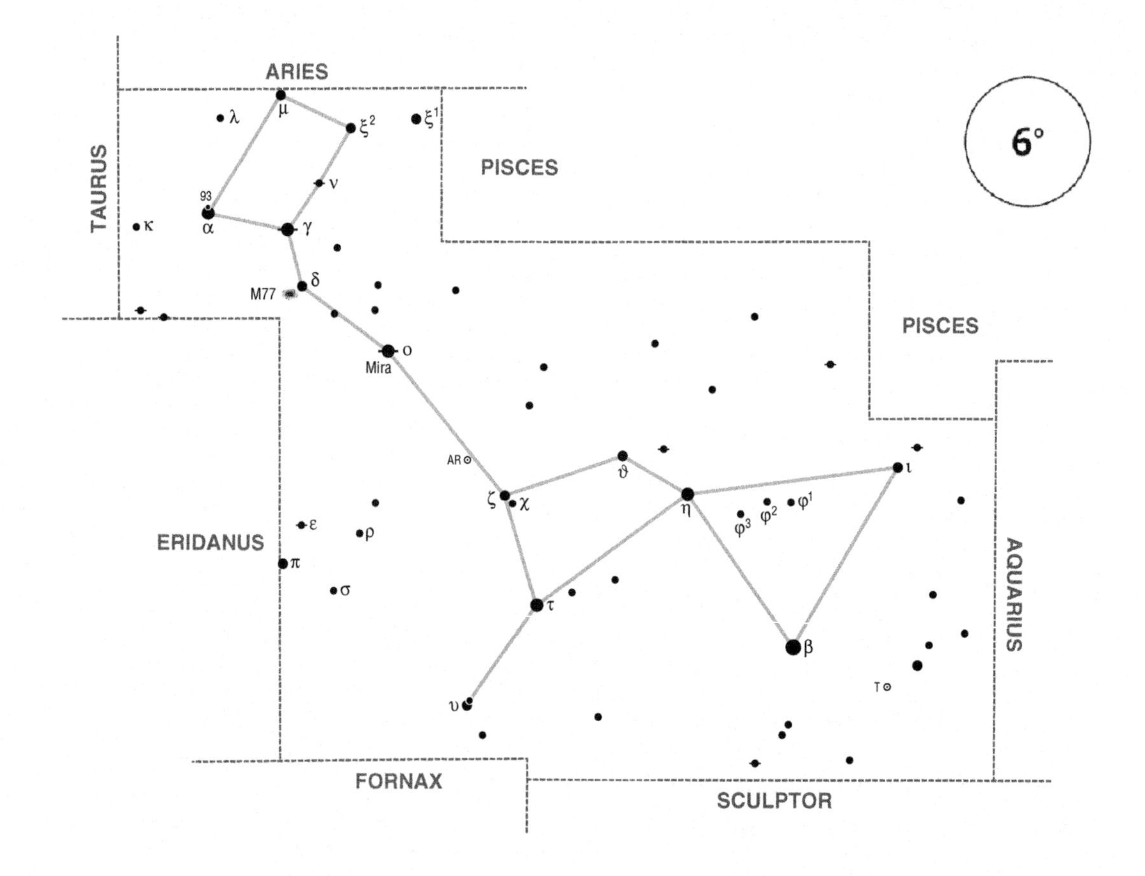

CONSTELLATION CULMINATES		
mid September at 2 a.m.	mid October at midnight	mid November at 10 p.m.

Cetus is one of the biggest constellations in the sky, yet it is rather unremarkable, as only two of its stars are brighter than magnitude 3: Beta or Deneb Kaitos (2m0), and Alpha (2m5). In brightness, they are followed by Eta (3m4), Gamma, and Tau (both 3m5).

Cetus is an autumn constellation. We can find it below Aries and Pisces, east of Aquarius. The biggest part of the constellation lies in the southern hemisphere; only the head of the monster is north of the celestial equator.

The yellowish Beta is the 51st brightest star in the sky; it is 96 light years from us. The luminosity of the star is approximately 100 times that of the Sun.

Gamma is a binary star. Its components, with a brightness of 3m5 and 6m2, are 2.3 arcsec apart (p. a. 298°, 2006). Because the position of the stars has not changed much since their discovery (1836), astronomers assume that their orbital period lasts a few thousands years. In the Gamma system, there is probably another star, a red dwarf with magnitude 10.2, which is 14 arcmin (p. a. 1°, 1923) from Gamma. The real distance between them is over 18,000 AU. These stars cannot be separated or seen through binoculars.

Tau Ceti is our close neighbor, at a distance of merely 11.8 light years. On the list of the closest stars it ranks 19th and is the 5th closest among the stars visible with the naked eye from the mid-northern latitudes. For astronomers it is even more interesting, since it is very similar to our Sun. It is slightly smaller and fainter, but it could have a planetary system similar to ours. In 1960, radio astronomers from Green Bank in West Virginia in the United States started a project known as OZMA. With a large radio telescope they emitted an encoded message to the two closest Sun-like stars, Tau Ceti and Epsilon Eridani. The answer from any eventual intelligent beings could arrive back around 1985. So far we have not received anything.

Alpha is the 94th brightest star in the sky. This is a 220-light-year-distant red giant. Its distinct orange color becomes even more intense when viewed through binoculars. Its luminosity is that of approximately 340 Suns. In the vicinity there is a magnitude 5.6 star known as 93 Ceti, which is a distinct blue in color. The stars are separated by 15.8 arcmin (p. a. 5°) and appear close together in the sky only from our viewpoint. In reality, the fainter star is 500 light years from us, and its luminosity is about 600 times that of the Sun. Thus it has even greater luminosity than Alpha. In the field of view of binoculars they represent an interesting couple, with a strong color contrast, and quite worthy of observation.

The most celebrated star in Cetus is Omicron or Mira (meaning "wonderful, astonishing"), which is a prototype of a long-period variable star. On August 13, 1596, the Dutch amateur astronomer David Fabricus was the first to notice that the star changed its brightness. Since 1638, astronomers have observed and noted every maximum. The average pulsation period is 331 days, but neither the period nor the maximum brightness is steady. The period could be a week shorter or longer than the average value. At its maximum brightness, Mira reaches magnitude 2; in 1779, it even got as high as magnitude 1.7, making it brighter than Polaris. At its minimum brightness, it can fade down to magnitude 10 and is therefore on the border of visibility through binoculars (Figures 8.20, 8.21, 8.22 and 8.23).

Mira reached one of its maximums on November 27, 2008. From the known period, we can easily calculate the approximate dates for the following few maximums: October 23, 2009, September 19, 2010...

The distance to Mira is approximately 400 light years. When it is at its weakest, its luminosity is slightly less than that of our Sun, but when it is at its brightest, it surpasses our star by approximately

Figure 8.20. Long-period variable star Mira at its maximum (*left*) and minimum brightness

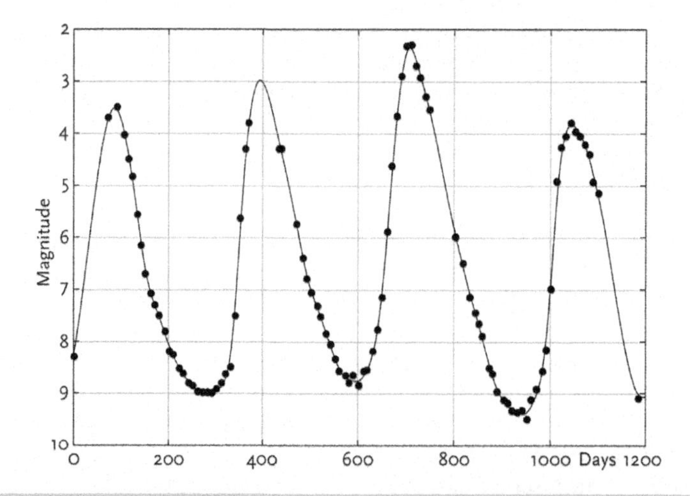

Figure 8.21. The graph of brightness changes of Mira during a period of four cycles. We can clearly see the irregularities in the period, as well as in the minimum and maximum brightness the star reaches within an individual cycle

250 times! Mira is also one of the coldest stars we know. At minimum brightness its surface temperature is around 1,900 K and at maximum it goes up to a mere 2,500 K (the surface temperature of our Sun is 5,800 K).

As are almost all long-period variable stars, Mira is also a red giant. It is one of the largest stars we know; in fact, it is the 23rd biggest. Its diameter is 400 times that of the Sun, or almost 600 million km; however, this changes with its brightness. The star is so big that if we placed it in the center of our Solar System instead of the Sun, it would extend further than the orbit of Mars! Mira has almost entirely exhausted its supplies of nuclear fuel, which makes it unstable (see the chapter in the first section on stellar evolution). The changes in its brightness are a result of periodical expansion and contraction of its outer layers.

In almost all recent books we can read that the Sun is the only star that is seen as a disk; all others are point-like, no matter how large the telescope or the magnification used. This statement no longer holds true. With modern interferometers astronomers can see and measure the true real sizes of some of the biggest stars, such as Betelgeuse in Orion, R in Cassiopeia, and Mira in Cetus (see box at the end). The apparent size of the last one is only 0.029 arcsec. At such a small angle we would see a 1 Euro coin (the diameter is 23 mm) from the distance of 170 km!

If we assume that the mass of Mira is approximately twice as big as the mass of the Sun, we can calculate the average density of the matter in the star at a mere 0.000022 kg/m^3 (average density of the Sun is 1,400 kg/m^3, and the density of air at normal conditions is 1.3 kg/m^3), which is vacuum according to Earthly criteria. Of course, Mira has a very dense and compact core, but its outer layers and its much less denser atmosphere extend far into interstellar space. Jokingly, the star is also known as the "wonderful red vacuum."

All this makes Mira a very interesting star. Even if observing variable stars appears to be a tad boring you should – at least as far as Mira is concerned – observe at least one entire cycle. Through binoculars, the entire cycle can be seen from its maximum to its minimum and back again. It is enough if you observe it once a week and compare its brightness with stars in its neighborhood, marked on the chart (Figure 8.22).

As are many other red giants, Mira is also a double star. Its companion is of magnitude 10.4. The stars are separated by a mere 0.6 arcsec (p. a. 110°, 1998) and cannot be separated through binoculars or even larger amateur telescopes. It is interesting that the companion is almost the exact opposite to Mira. It is

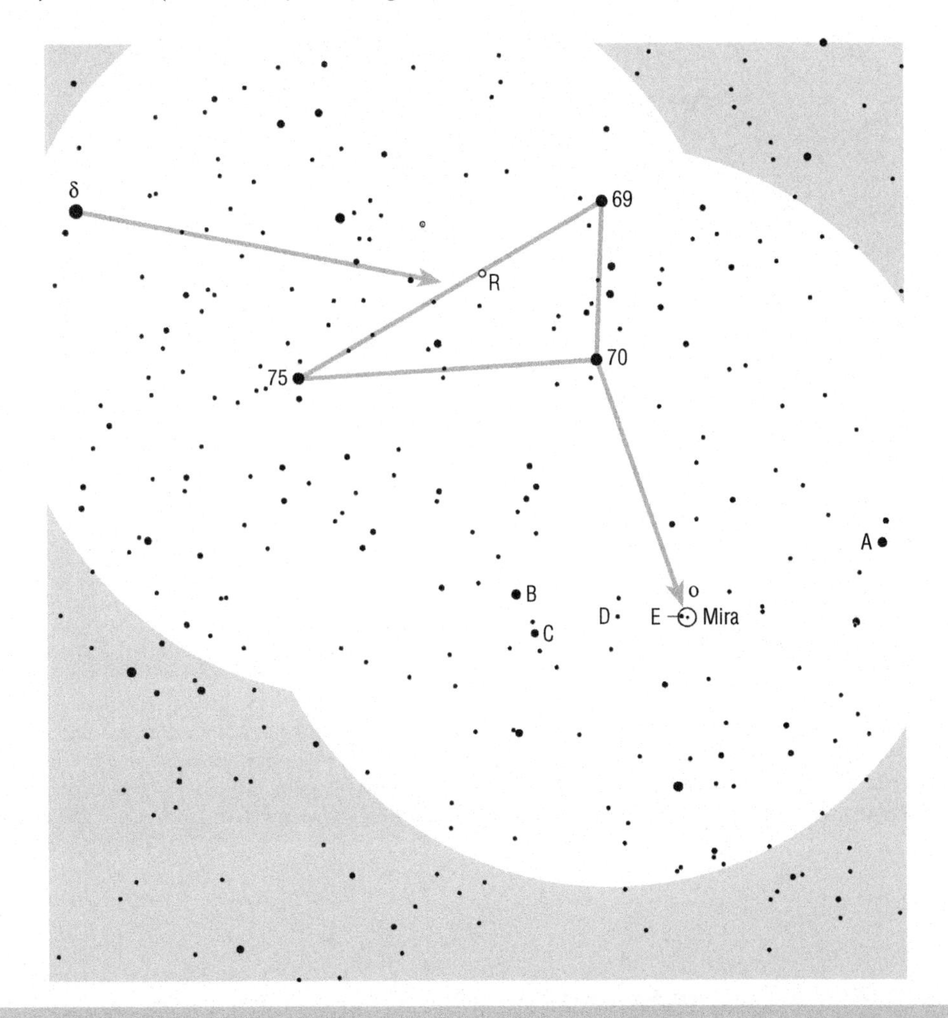

Figure 8.22. The surroundings of Mira. This chart is especially useful when the star is close to its minimum brightness. In the same field of view we can see another variable star known as R Ceti, the brightness of which changes from 7m4 to 14m within a period of 166 days. The star is orange in color. The brightness of the comparison stars are A (5m7), B (6m4), C (7m3), D (8m8), and E (9m2)

very small (its diameter reaches only 1/11 of the diameter of the Sun or 1/4,400 of the diameter of Mira) and very dense. The average density of the star is as much as 3,300 times greater than that of the Sun. It is very likely that Mira's companion is a subdwarf, a star that is in size somewhere between a normal star and a white dwarf. The real separation between the stars is approximately 70 AU, and the orbital period around the common center of mass is approximately 260 years.

The spiral galaxy M 77 (8m9/6′9 × 5′9) is on the border of visibility when viewed through binoculars (Figure 8.24). When there are excellent observing conditions, we can see a small and faint spot of light that appears more like a blurred star than a nucleus of a large and beautiful galaxy. You should have no trouble finding it, since the galaxy lies less than 1 degree southeast from the rather bright star Delta (4m1). During excellent observing conditions, the spot of light is as bright as the magnitude 9 stars in the line that

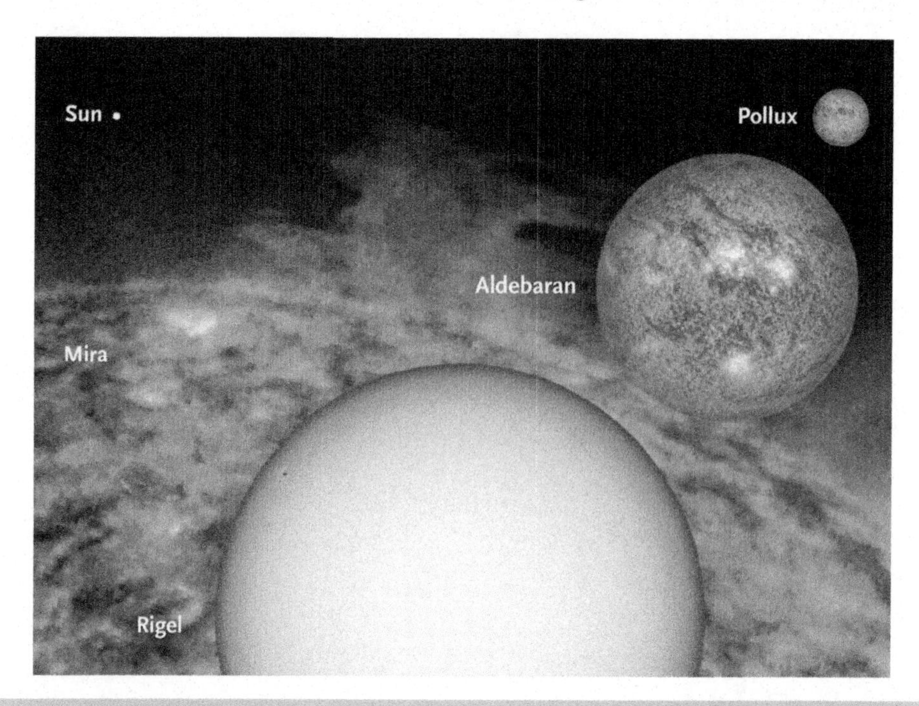

Figure 8.23. A comparison between Mira and some other well-known stars. If the size of Mira is 360 mm and extends far across the edges of the book, the size of the Sun in this model is a mere 1 mm. The diameter of Aldebaran, which is also a red giant, measures "only" 36 mm, and the diameter of Rigel, which is one of the biggest known main sequence stars, measures 66 mm. Earth, which is 110 times smaller than the Sun, would measure only a hundredth of a millimeter in this image!

Figure 8.24. Spiral galaxy M 77. The circle in the center shows the part of the galaxy that is seen through binoculars and small telescopes

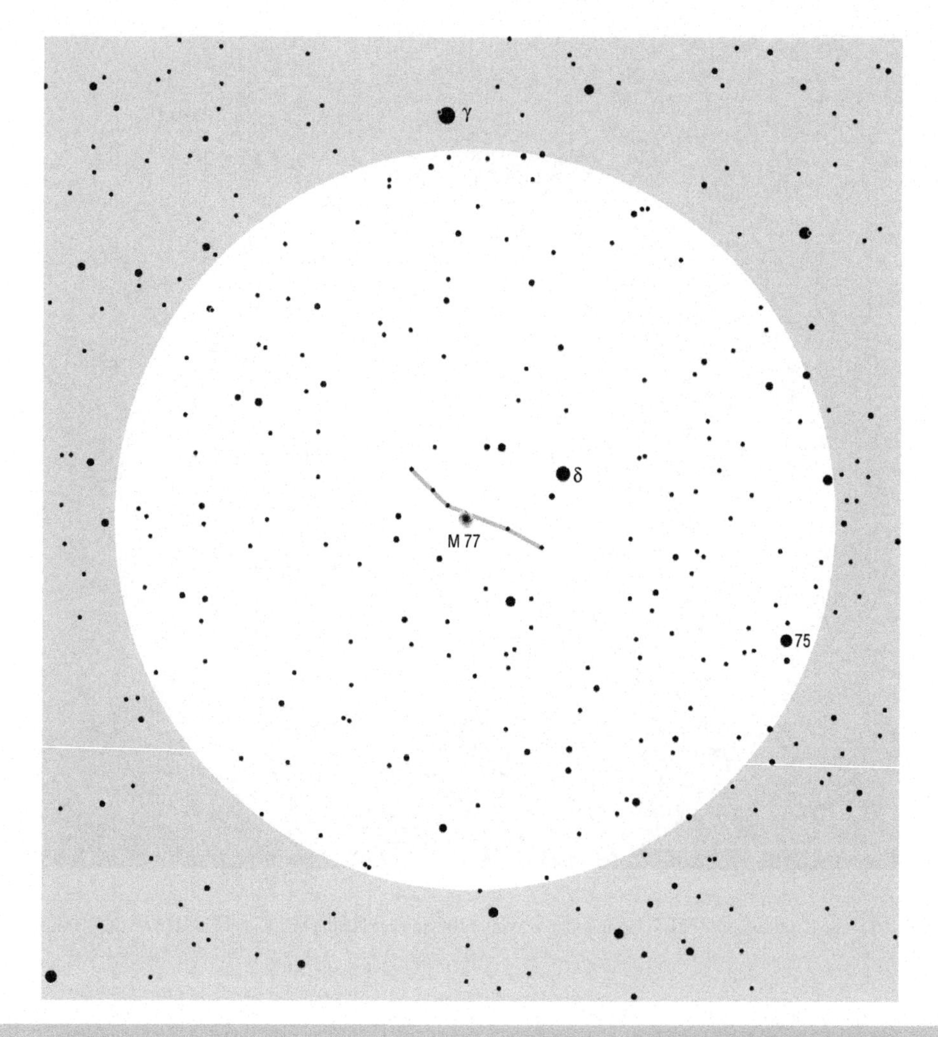

Figure 8.25. The leading star to galaxy M 77 is the rather bright Delta. In the field of view you should pay attention to the line of six magnitude 9 stars. The third star – if you count from the west – is in fact the galaxy's nucleus

spreads in the northeast–southwest direction. In perfect observing conditions, the blurred star extends into a small, faint patch of light. What we see in binoculars is merely the central, brightest part of the galaxy's nucleus. Not even the largest telescope will help us see its spiral arms; for that you would have to photograph the galaxy.

M 77 is approximately 60 million light years from us. Its real diameter reaches almost 100,000 light years. The galaxy is also a strong source of radio waves that radiate from its center, which results in it being classified as a Seyfert galaxy.

Interferometers

An optical interferometer is a combination of two or more telescopes, located far away from each other. By merging the light from all telescopes into a single image, astronomers get the resolution they would get from a single mirror with a diameter as large as the distance between the telescopes (shown here with shaded ellipse). On image is the European Southern Observatory's Very Large Telescope Interferometer (VLTI), which is located on Paranal, a 2,600 m high mountain inside an Atacama desert in Chile.

COLUMBA (The Dove)

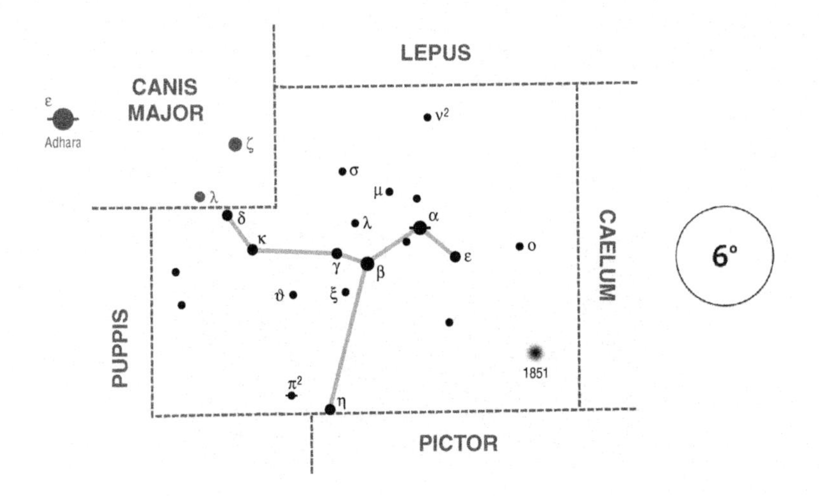

CONSTELLATION CULMINATES		
mid November at 2 a.m.	mid December at midnight	mid January at 10 p.m.

The winter constellation Columba lies south of Lepus, southwest of Canis Major, and from mid-northern latitudes is always low above the southern horizon and therefore not really visible. The southern border of the constellation, where Eta shines, is only a degree above the horizon at culmination.

The brightest stars are Alpha (2m6), Beta (3m1), Delta (3m8), Eta and Epsilon (both 3m9).

Alpha is the 107th brightest star in the sky, 270 light years from us. It is much larger than our Sun. Its luminosity surpasses our star by approximately 450 times.

Beta, Epsilon, and Eta are orange stars, and their color becomes even more intense when viewed through binoculars.

Gamma, which is the magnitude 4 star in our sky, is in fact a giant, approximately 650 light years away. If it was closer to us, let's say 33 light years away, it would appear as bright as Venus.

Columba holds the rather bright globular cluster NGC 1851 (7m1/11′), but from the mid-northern latitudes its visibility is poor due to its low altitude. At culmination, it is only 4 degrees above the horizon. You should attempt to tackle such a difficult observation only once you have gained plenty of observation experience and only in those cold winter nights when the sky is clear and dark all the way down to the horizon. In the field of view of the binoculars the cluster shows up as a faint, few arcminutes large spot of light. Epsilon is the star leading to the cluster. If you move this star to the extreme northeastern edge of the field of view, the cluster will appear on the southwestern edge. Supposedly NGC 1851 emigrated to our galaxy from the Canis Major dwarf galaxy!

COMA BERENICES (Berenice's Hair)

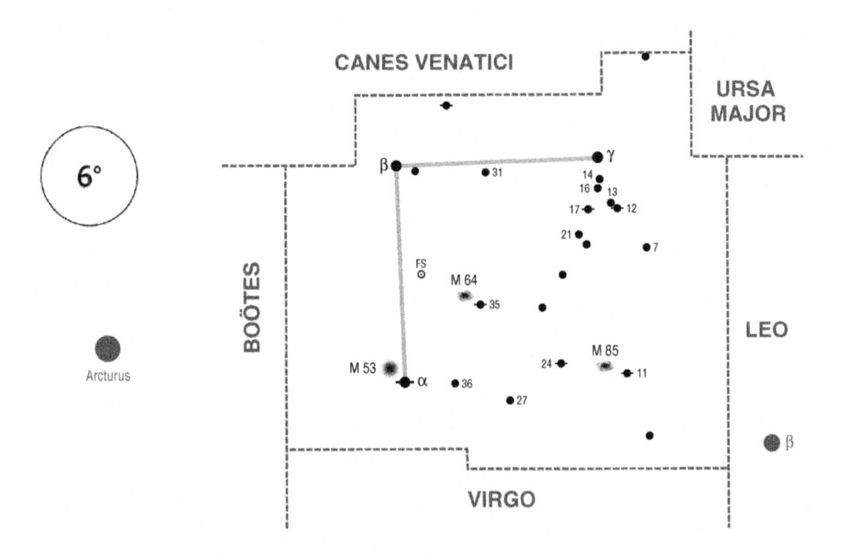

CONSTELLATION CULMINATES		
in begin. March at 2 a.m.	in begin. April at midnight	in begin. May at 10 p.m.

On a clear, moonless night an extended group of faint stars can be seen between Beta Leonis and Arcturus. This is the spring constellation Coma Berenices, which gives the impression of a faint open cluster. The brightest stars are Alpha, Beta, and Gamma, but none of these is above a magnitude 4.

Alpha (4m3) is a close binary. Two stars of almost equal brightness (4m8/5m5) rotate around each other in a period of 25.8 years. The greatest distance between them reaches 0.9 arcsec, and therefore they cannot be separated through binoculars. This is a difficult pair even for the larger amateur telescopes. Alpha is 65 light years from us. The true separation between the stars is therefore a mere 10 AU (Saturn is 9.5 AU from the Sun). The luminosity of each star in the system is approximately 3 times that of the Sun.

The star with the Flamsteed designation of 24 (the way to find it is described in the chart for the elliptical galaxy M 85) is a close double for observers with binoculars. The two stars with magnitudes 5.1 and 6.3 are 20.2 arcsec apart (p. a. 271°, 2006). The pair is very colorful; the brighter star is yellowish and the fainter is bluish. Take a look!

The open cluster (1m8/4°6) mentioned above lies beneath Gamma. In some star atlases it can be found under the name Mellote 111. The brightest stars have designations 12, 13, 14, 16, and 21 Comae Berenices (Figure 8.26). They are all magnitude 5 and can be seen with the naked eye on a clear, moonless night. Other brighter stars such as Gamma, 7, 8, and 17 are not true members of the cluster and are in the same direction merely by chance. With good observing conditions we can see approximately 80 stars through binoculars. But when the night has excellent viewing conditions, their numbers increase to over 150. And this is a sight that only binoculars with a wide field of view can conjure up! In the eyepiece of an average telescope the cluster loses all its charm, as only a few individual members can

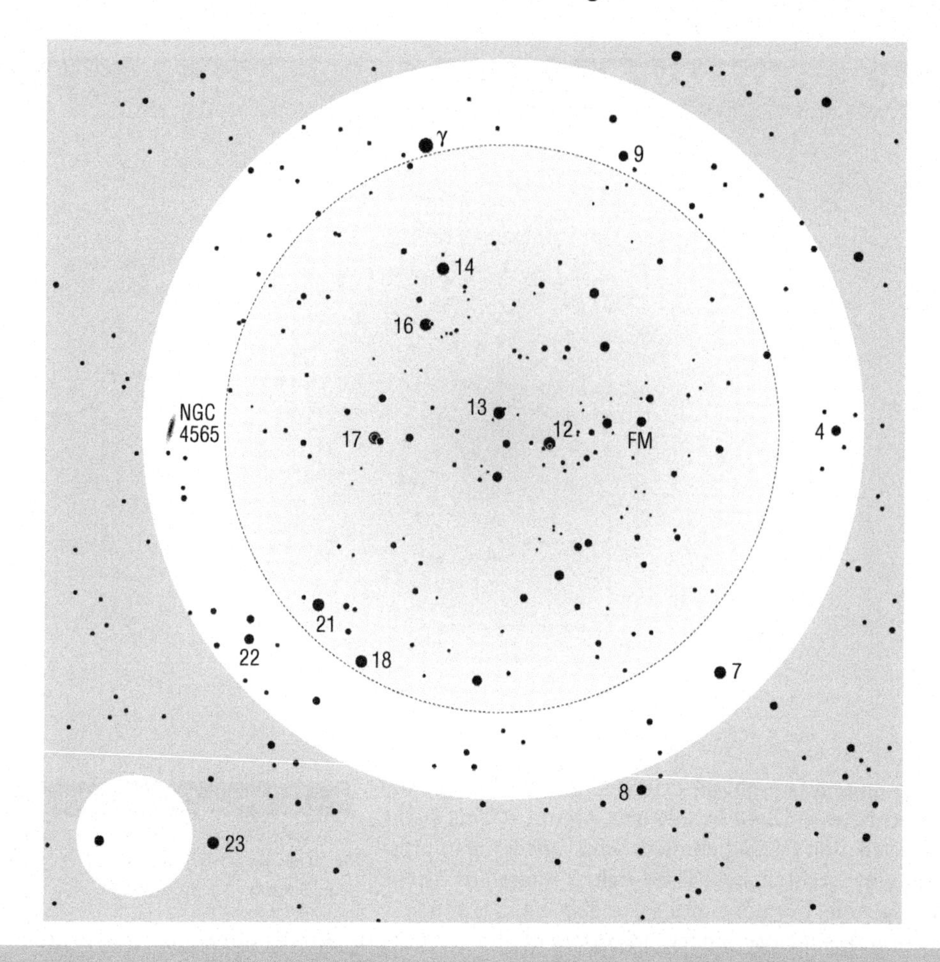

Figure 8.26. A detailed chart of the open cluster in Coma Berenices, with stars up to magnitude 11.5. The larger white circle represents the field of view of binoculars and the smaller the field of view of an average amateur telescope. We can easily see that the cluster is robbed of all of its charm in the smaller field. In the chart, we can also see the position of NGC 4565 (9m6/16′ × 3′), a magnificent spiral galaxy that is seen edge on (see image on opposite page). It is not easy to find with binoculars, and even on a perfect night it appears just as a faint line of light. Of course, the constellation has to be close to its culmination

be seen at any single time. Anyone who will take the time to examine the cluster in detail will find a few close pairs. The brightest among them has a designation 17 Comae. The stars are magnitudes 5.2 and 6.6, separated by 146 arcsec (p. a. 251°, 2002).

The cluster appears so large in our sky because it is one of the closest, at only 290 light years from us. The stars with the largest luminosity are 14 and 16 Comae (50 times that of the Sun), while the luminosity of the faintest members barely reach a third of that of the Sun. For comparison, let's say that our Sun is a magnitude 9.2 star from this distance and would be visible through binoculars.

The rich globular cluster M 53 (7m6/13′) lies only a degree northeast of Alpha and is easy to find, as they both appear in the same field of view through binoculars. Under average observing conditions the cluster is clearly seen, but only as a smaller, few arcminutes big spot of light (Figure 8.28).

M 53 is one of the furthest globular clusters. Its light travels more than 58,000 years to reach us. From that data and from the apparent size one can easily calculate that the real diameter of the cluster is approximately

Figure 8.27. NGC 4565 is a typical spiral galaxy, seen edge on. In its equatorial plane there are numerous dust and gas clouds that are not transparent for light. They show up as *narrow dark lines* that seem to divide the galaxy into two. The equatorial plane of our galaxy includes some quite similar dark regions. If we could see it from an intergalactic distance and edge on, it would be similar to NGC 4565

250 light years! The other interesting fact connected to this cluster is that it lies high above the galactic plane and that it is as much as 60,000 light years from the center of our galaxy. If we could go to M 53 and look back to our home galaxy, we would see it as an 80 degrees big disk of light with clearly seen central bar and spiral arms. This would be definitely the breath-taking sight!

The region of Coma Berenices is otherwise small but rich in all types of galaxies. In some places there are so many of them in the field of a larger amateur telescope that it is impossible to recognize them without a precise stellar chart. Two of them can even be seen through binoculars.

The spiral galaxy M 64 (8m5/9′3 × 5′4), popularly known as the Black Eye Galaxy, lies 48 arcmin east and 26 arcmin north from the magnitude 5 star known as 35 Comae (Figures 8.28 and 8.29). In your search for it, you will find that the detailed chart is of great help! Alpha itself will lead us to the galaxy. If you move the star to the extreme southeastern edge of the field of view, you can recognize star 35 with the pattern of fainter stars in its surroundings. In its vicinity you have to search for a faint, few arcminutes big spot of light. It probably does not need to be emphasized that you have to search for and observe the galaxy only when there are excellent observing conditions, on moonless nights and far away from light-polluted skies and when the galaxy is close to its culmination or even very high in the sky.

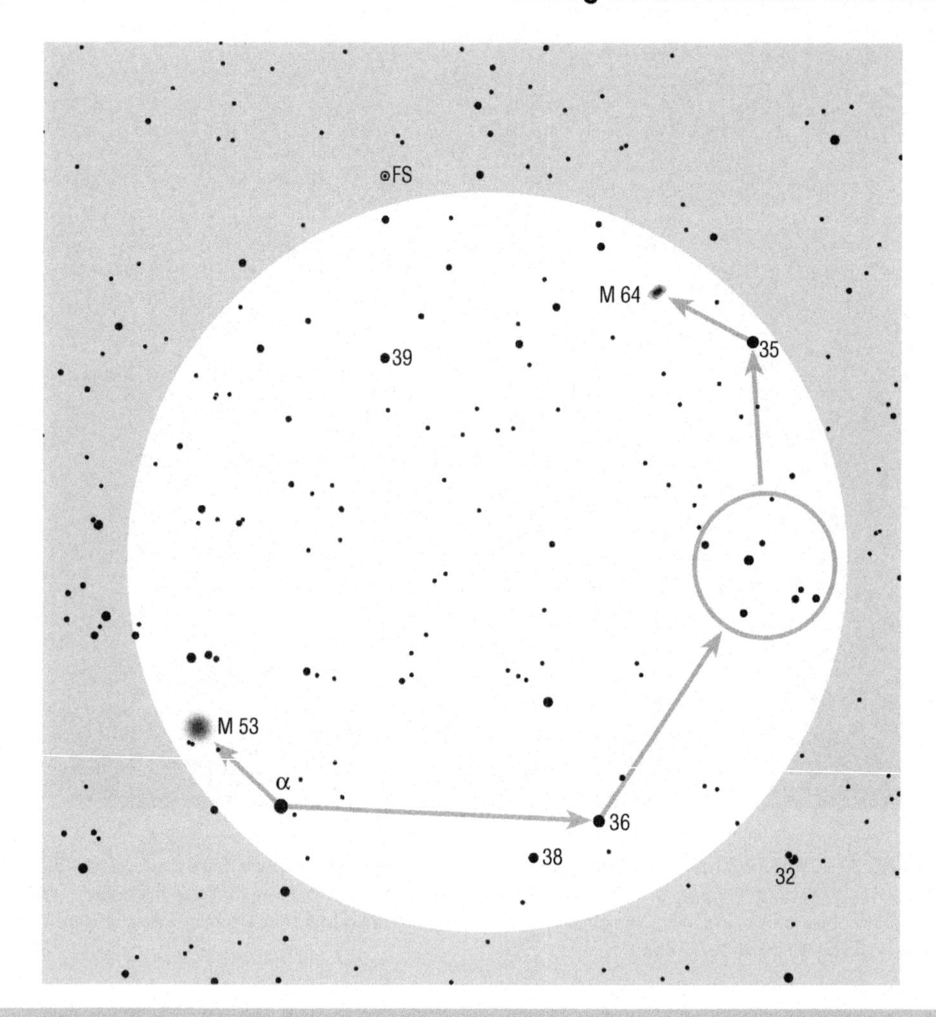

Figure 8.28. The positions of the globular cluster M 53 and the spiral galaxy M 64. The cluster is clearly seen through binoculars, while the galaxy is a slightly tougher object to find. You should attempt to look for it only when there are excellent observing conditions

When you search for such a faint celestial object and move your binoculars from one bright star to another toward the galaxy you are looking for, you often have to switch on the light and glance at the chart. By doing so, you spoil the adaptation of your eyes to night vision. When you are convinced that the point where the galaxy should be (or some other faint object on the border of visibility) is in the middle of the field of view, you have to wait for your eyes to accommodate to night vision once again and then carefully examine the star field. Do not despair if the galaxy is not visible. Maybe the observing conditions are not as good as you thought, or maybe your eyes are tired from the search. You can always also try with averted vision.

M 64 is one of the brightest galaxies in the spring evening sky. It is a true beauty on long-exposure images. Nowadays, when CCD cameras are commonly used even by amateur astronomers, it is possible to make images of distant galaxies of the same quality as professional astronomers made some 30 years ago with the largest telescopes and photographic plates. All you need is average equipment and a lot of knowledge, patience, and experience.

Astronomers are not absolutely certain regarding the distance to galaxy M 64; the best estimate is approximately 19 million light years. Its diameter is thought to be approximately 55,000 light years.

Figure 8.29. Spiral galaxy M 64

A few bright galaxies lie on the border of constellation Virgo. These belong to the well-known Virgo cluster of galaxies. Even Messier included them in his catalog. The brightest are the elliptical M 85 and the spiral M 88, M 91, M 98, M 99, and M 100. All of these galaxies are gathered in a field no larger than a few degrees. Only M 85 (9m1/7′1 × 5′2) can be seen through binoculars, but only as a very faint, few arcminutes big uniform spot of light. You can find the instructions on how to find it in the text beside the chart in Figure 8.30. You should tackle this difficult task only once you have become an experienced observer!

M 85 is approximately 60 million light years from us; its real diameter is approximately 125,000 light years!

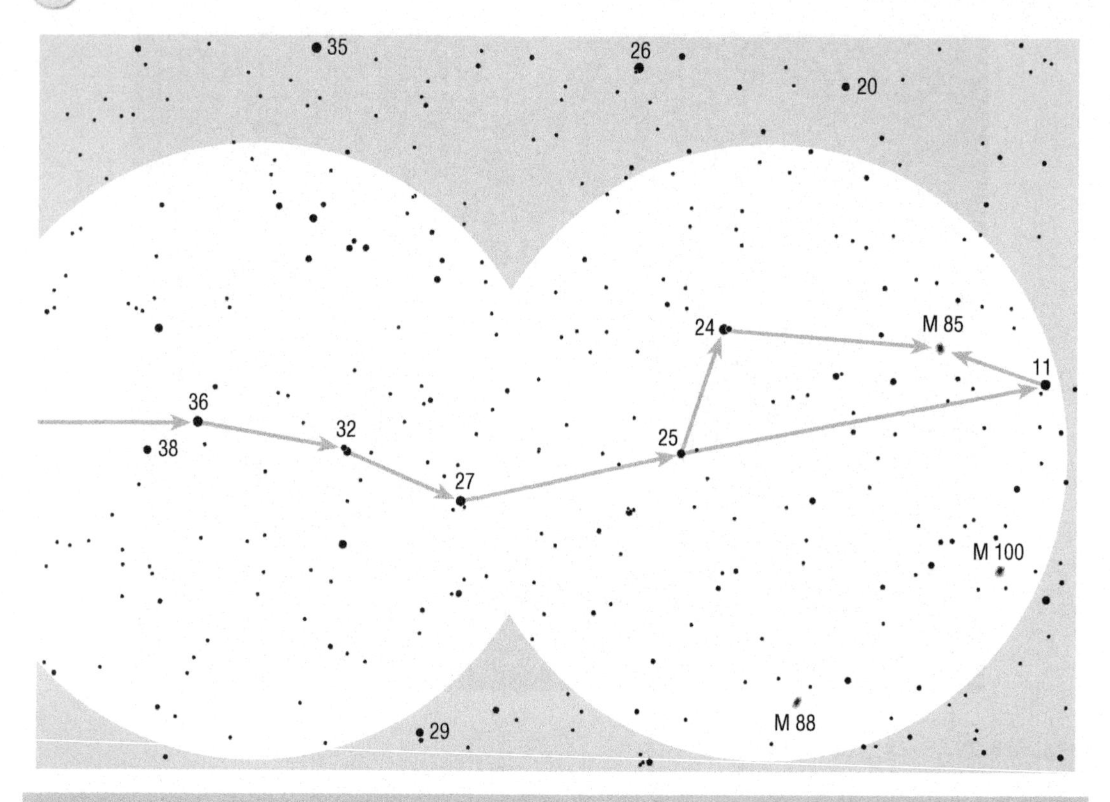

Figure 8.30. The elliptical galaxy M 85 is a tough nut to crack for binoculars. The easiest way to find it is to use Alpha Comae as the leading star. In the chart it is hidden behind the left edge, but this part of the sky is shown on the chart with M 53 and M 64. Use the brighter stars with designations 36, 32, 27, 25, and 24 to slowly move into the vicinity of the galaxy. When you are sure that you have reached the right spot, search for a triangle formed by stars 25, 24, and 11. The galaxy lies only slightly above the connecting line between the magnitude 5 stars designated as 24 and 11. Through binoculars M 85 is seen as a faint, only few arcminutes big flake of light. Its size depends greatly on the observing conditions at the time. On a really perfect night you can notice another two additional galaxies [M 100 (10m1/7′ × 6′) and M 88 (9m6/7′ × 4′)] in the same field of view

Figure 8.31. M 88, which through binoculars appears as a tenuous cloud of light on the border of visibility even when there are perfect observing conditions, is in fact a wonderful spiral galaxy with tightly wound spiral arms and a small but bright nucleus

CORONA AUSTRALIS (The Southern Crown)

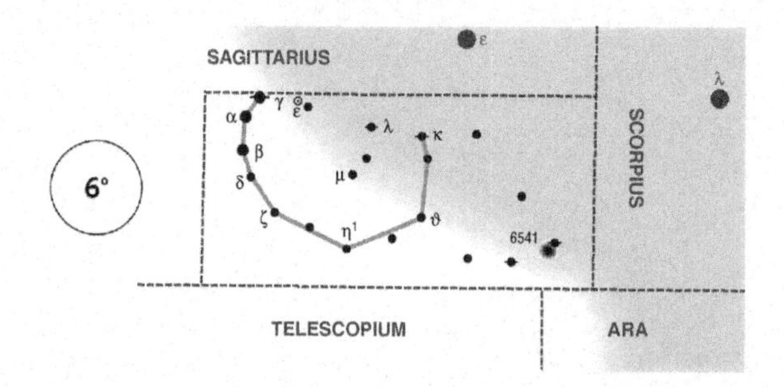

CONSTELLATION CULMINATES		
in begin. June at 2 a.m.	in begin. July at midnight	in begin. August at 10 p.m.

The summer constellation Corona Australis is always low in the sky and most commonly hidden behind a foggy horizon for mid-northern observers. In addition, this constellation does not contain any star brighter than magnitude 4.

The brightest stars are Alpha and the yellowish Beta (both 4m1), Gamma (4m3), and the orange Delta (4m7). For mid-northern observers Alpha is a mere 6 degrees above the horizon at culmination. But from southern Europe or the southern United States, observers have a better view !

Kappa is a double with stars of magnitude 5.6 and 6.2, separated by 20.8 arcsec (p. a. 359°, 2002). It would be possible to separate the pair through binoculars if they were higher in the sky. You could try, but only in excellent observing conditions all the way down to the horizon.

Corona Australis includes the rather bright globular cluster NGC 6541 (6m6/13′ – it is just below the westernmost star with magnitude 5), but it is not seen from the mid-northern latitudes due to its low position in the sky. This is an object for southern observers.

CORONA BOREALIS (The Northern Crown)

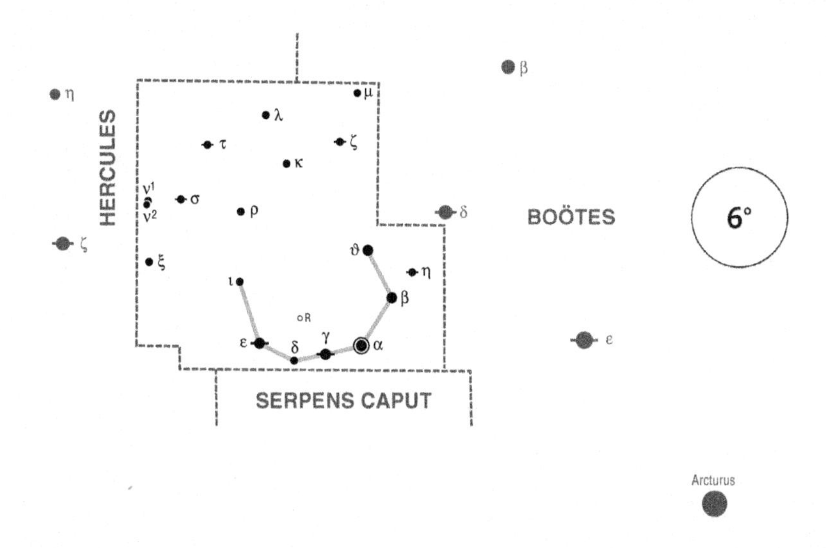

CONSTELLATION CULMINATES		
mid April at 2 a.m.	mid May at midnight	mid June at 10 p.m.

The spring constellation Corona Borealis is small and faint, but due to its characteristic shape of an open semicircle it is easily recognizable. It can be found between Boötes, Hercules, and Serpens Caput. The brightest stars are Alpha (2m2), Beta (3m7), and Gamma (3m8).

Alpha or Gemma (Jewel) is the 67th brightest star in the sky. It is 75 light years from us. Its luminosity is that of 52 Suns. It is a close eclipsing binary with a period of 17.359907 days, but its brightness changes only by one-tenth of a magnitude – not enough to notice with the naked eye or through binoculars.

Corona Borealis is the home to the famous irregular variable star R, which is a prototype of its kind. R Coronae usually shines with a magnitude of 6 and is clearly seen through binoculars. Occasionally and completely unexpectedly its brightness drops by 7 or even 8 magnitudes, and on such occasions it cannot be seen even through larger amateur telescopes. Its diminished brightness usually last a few months, rarely a year or more (Figure 8.32). After this its brightness returns back to magnitude 6. What happens to the star during this?

The R Coronae spectrum is quite unusual at maximum brightness. Strong spectral lines of carbon can be seen within it. Measurements show that its photosphere is composed of over 67% carbon and only 33% of hydrogen and other elements. And it is with the large quantities of carbon that the astrophysicists explained the unusual behavior of the star. Due to the turbulent events in its interior, the surface also occasionally experiences violent outbursts that throw the gases (mainly carbon) into the star's surroundings. In cooler areas, carbon crystallizes and changes into a cloud of graphite particles, which is completely or partially nontransparent for the light emitted by the star. This is when it appears at its minimum brightness. As the cloud expands and becomes increasingly diluted and slowly vanishes into interstellar space, the star begins to shine normally once again (in our sky, of course; the luminosity of the star has not changed, but its light has been blocked by graphite). Variables of this type have probably already survived the red giant phase and are at the very end of their life cycle.

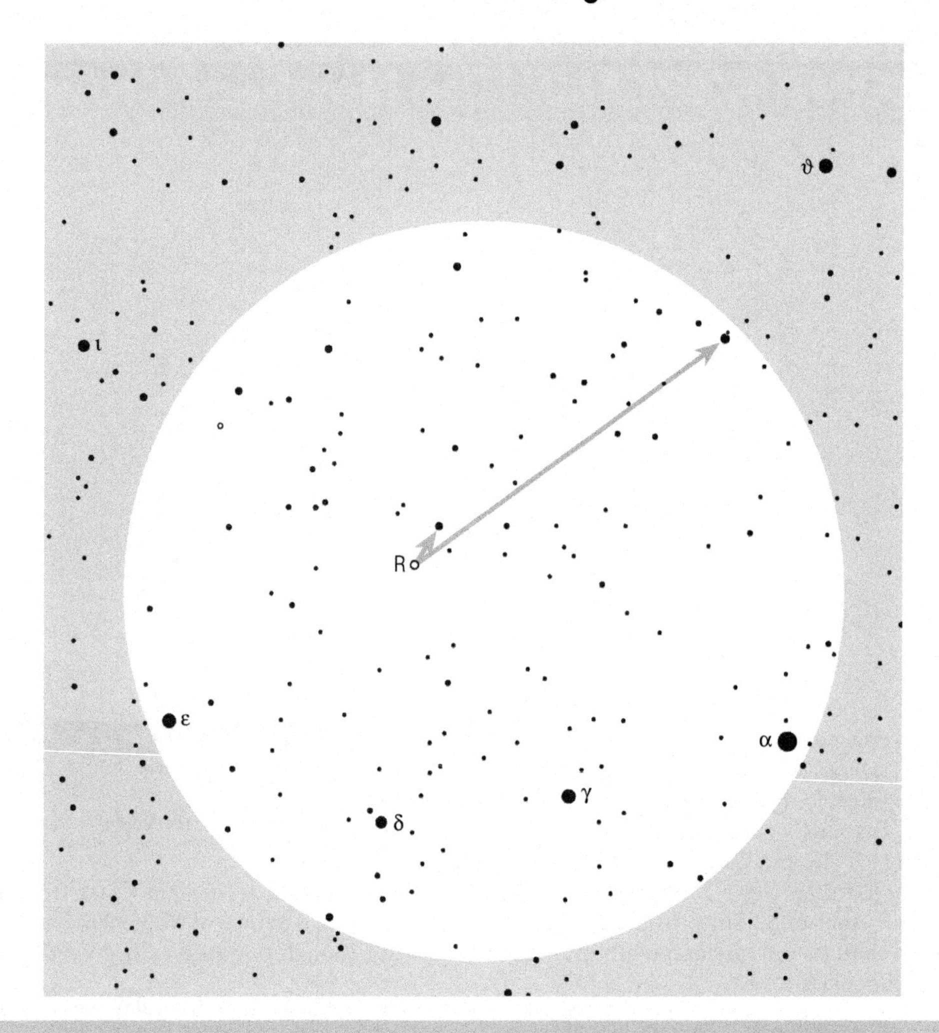

Figure 8.32. The surroundings of the variable star R Coronae. Stars suitable for brightness comparison are its 23 arcmin distant northeastern neighbor, which is of magnitude 7.2, and a star further away, without any label, that is approximately as bright as R

In the field of view of binoculars R Coronae Borealis might well appear merely as one of countless stars, but if we know what are we looking for and what should we expect, it is definitely worth the trouble to find it.

CORVUS (The Crow) and CRATER (The Cup)

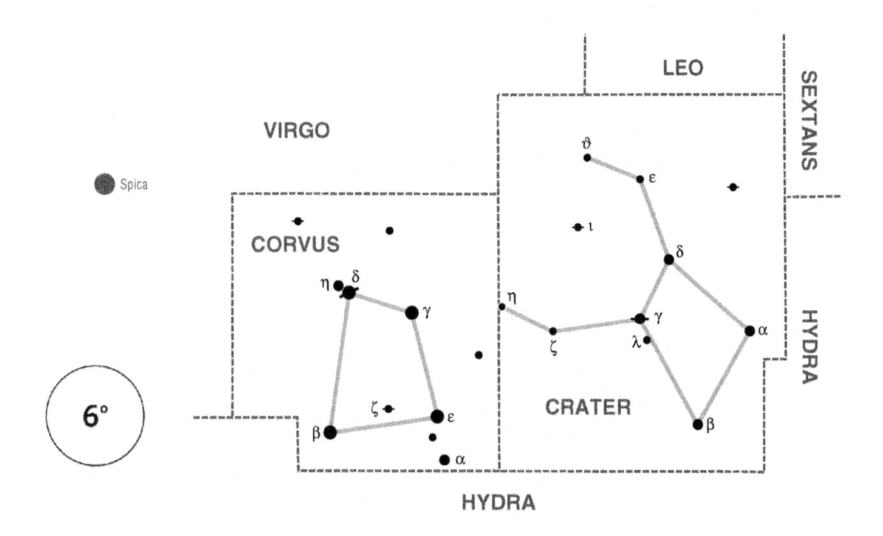

CONSTELLATIONS CULMINATE		
end February at 2 a.m.	end March at midnight	end April at 10 p.m.

The constellation Corvus lies beneath Virgo and is quite easily recognizable, since the four brightest stars form a trapezoid that is clearly defined in a spring night sky. The four stars are Gamma, Beta (both 2m6), Delta (2m9), and the yellowish Epsilon (3m0). The star named Alpha lies under the southwest corner of the trapezoid and shines with magnitude 4.

The yellowish Beta is the 106th brightest star in the sky. It is a giant with the luminosity of approximately 120 Suns. Its light travels 140 years through space in order to reach Earth.

Gamma is similar to Beta. In our sky it appears only slightly brighter. The star is 165 light years from us, and therefore its luminosity is, in fact, slightly greater than that of Beta (180 Suns). Gamma is the 100th brightest star in the sky.

Delta is a double star with an interesting color contrast. Some observers describe the stars as yellowish and pale violet, while others describe them as white and orange. This is a tough pair for observers with binoculars. The two stars with magnitudes 2.9 and 8.5 are 24.3 arcsec apart (p. a. 216°, 2004), so we could theoretically separate them, but in practice it is difficult to locate the fainter star, which is hidden in the light of the brighter one. If you possess binoculars with truly first-class optics, you could attempt to tackle this pair (with small hope of success), but only on a clear and completely calm spring night.

The little and faint constellation Crater lies on the back of Hydra. Among the shining spring constellations, such as Leo and Virgo, the faint stars of the Crater are hardly noticeable. The brightest star is the yellowish Delta, which has a magnitude of 3.6. Together with Gamma and Alpha (both 4m1), it forms a faint triangle. The constellation does not contain any objects that might be of interest to observers with binoculars.

CYGNUS (The Swan)

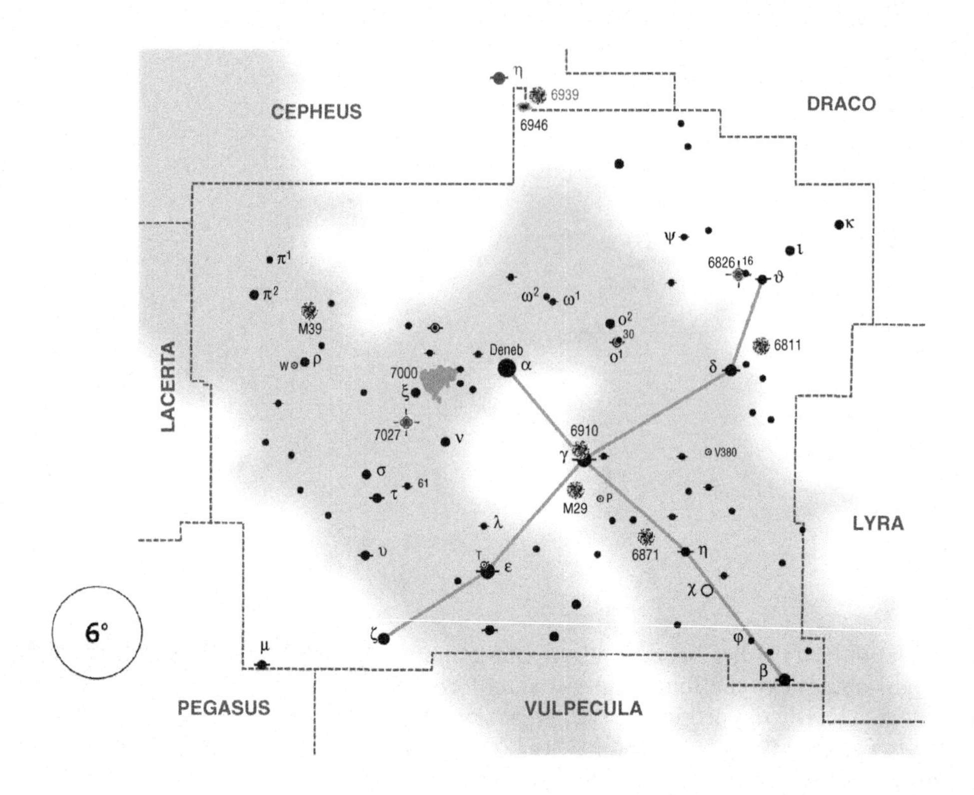

CONSTELLATION CULMINATES		
end June at 2 a.m.	end July at midnight	end August at 10 p.m.

Cygnus is one of the most distinctive constellations in the sky, and we do not need to have much imagination to see in the stars the figure of a bird flying south along the Milky Way. In summer evenings it lies almost in the zenith. The brightest stars are Alpha or Deneb (1m2), Gamma (2m2), Epsilon (2m5), Delta (2m9), and Beta or Albireo (3m1).

Alpha or Deneb is the 19th brightest star in the sky. It is a giant, the luminosity of which is as much as 45,000 times that of the Sun. The star is 1,400 light years from us and is still so bright in our sky. It has a surface temperature of 8,400 K (spectral type A2) and a mass of 25 Suns. The diameter of Deneb is huge, as much as 90 times that of the Sun. If we would place it in the center of our Solar System, it would extend almost to the orbit of Earth (and just to be clear, we are not talking about a bloated red giant but about a regular main sequence star). This is a truly admirable star.

Together with Vega in Lyra and Altair in Aquila, Deneb forms a clearly seen asterism called the Summer Triangle, which can be seen high in the sky on summer evenings. At twilight Vega is the first to appear, but it is soon followed by Altair and Deneb.

Gamma is the 69th brightest star in the sky. It is 1,500 light years from us. This is also a real giant, for its luminosity surpasses our Sun by approximately 21,000 times.

The yellow-orange Epsilon is the 91st brightest star in the sky. Its light travels 72 years in order to reach us. The luminosity of the star is 38 times that of the Sun.

Delta is the 147th brightest star in the sky. It is 170 light years from us. The luminosity of the star surpasses our Sun by 150 times.

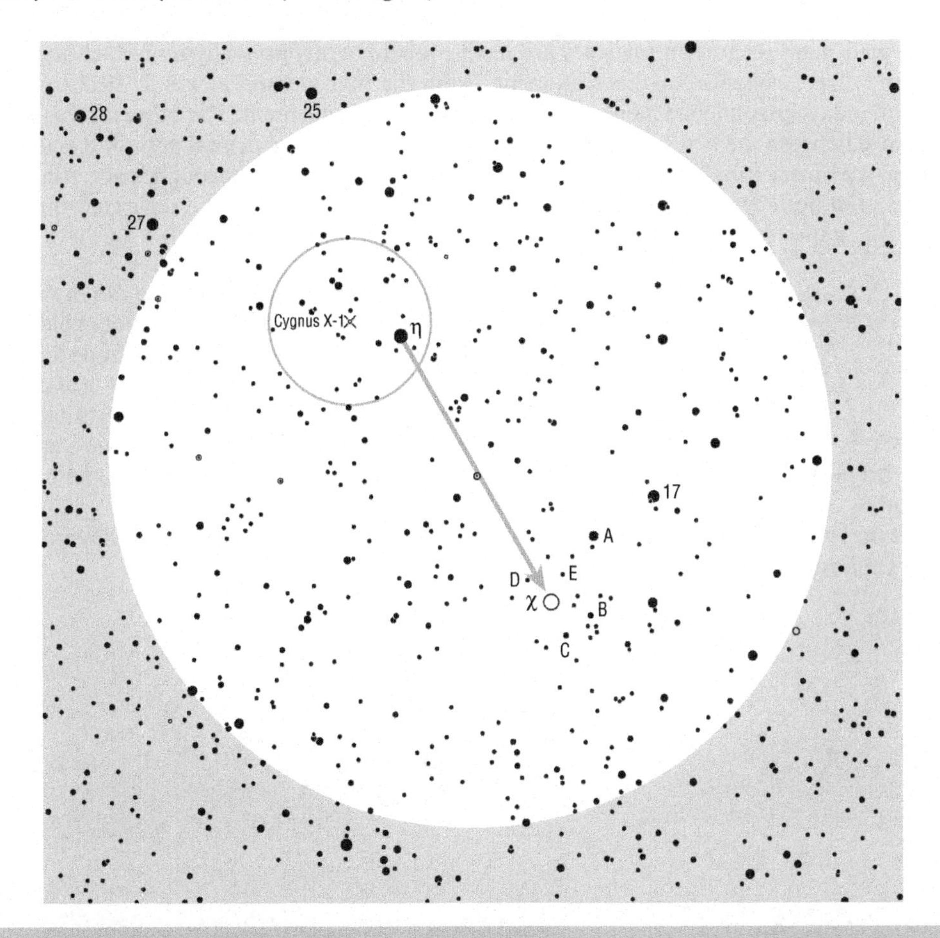

Figure 8.33. Variable star Chi with comparison stars: A (6m4), B (7m5), C (8m1), D (8m7), and E (9m3). On the chart, the area surrounding the X-ray source Cygnus X–1 is marked. This area can be seen on the image on page 286

Chi is a long-period variable star of the Mira type, a red giant that changes its brightness from magnitude 3.3 to 14.2 over an average period of 408 days. At its maximum, it usually shines with magnitude 4 or 5. The star lies 2.5 degrees southwest of Eta (3m9), which is a good comparative star for the maximum (Figure 8.33). When Chi is moving toward its minimum, it is soon hidden among the vast number of faint Milky Way stars in the vicinity. The only way that we can recognize it is by its distinct orange color. The star reached one of its maximums November 19, 2008. From this date and from the known period, one can easily calculate the approximate dates for the next maxima. The distance to Chi is not truly known; however, the estimates range between 300 and 400 light years.

In the opinion of numerous amateur astronomers Albireo (Beta Cygni) is the most wonderful double star in the sky, and luckily the stars can also be separated through binoculars. The main star (3m2) is of a golden yellow color, and its fainter companion (4m7) is of a distinct blue color. The apparent separation is 35.3 arcsec (p. a. 54°, 2006). They are 410 light years from us. The true separation between the stars is 4,400 AU. While observing this double you should think about the fact that you can place up to 73 Solar Systems between them!

Despite their great separation the stars are most probably gravitationally bound, although we do not have any direct evidence for this statement. From the first measurements in 1832 (by F. G. W. Struve) until today, astronomers have not noticed any orbital movement. The cause for this lies in the great distance between the two stars and therefore the extremely long orbital period. It was assumed that Albireo is a binary from the common proper motion of the two stars through space. Albireo A is a binary itself, but both components are so close together that they cannot be separated through any telescope. It is a spectroscopic binary.

Omicron–1 (3m8) is the brightest member of a triple star, known for its wonderful color contrast. Omicron–1 is a red giant vividly golden yellow in color. The star with designation 30 shines with magnitude 5 and is 338 arcsec (5.6 arcmin) away, while the third, even fainter star is of magnitude 7.0 and is 107 arcsec (p. a. 173°, 2000) from Omicron. Both fainter stars are distinctly blue in color. The sight through binoculars supplements the rather bright yellow-orange star Omicron–2 and the numerous fainter stars in the Milky Way background.

Deneb is the star leading to Omicron. If you move it to the extreme eastern edge of the field of view, the bright Omicron–1 appears on the western edge.

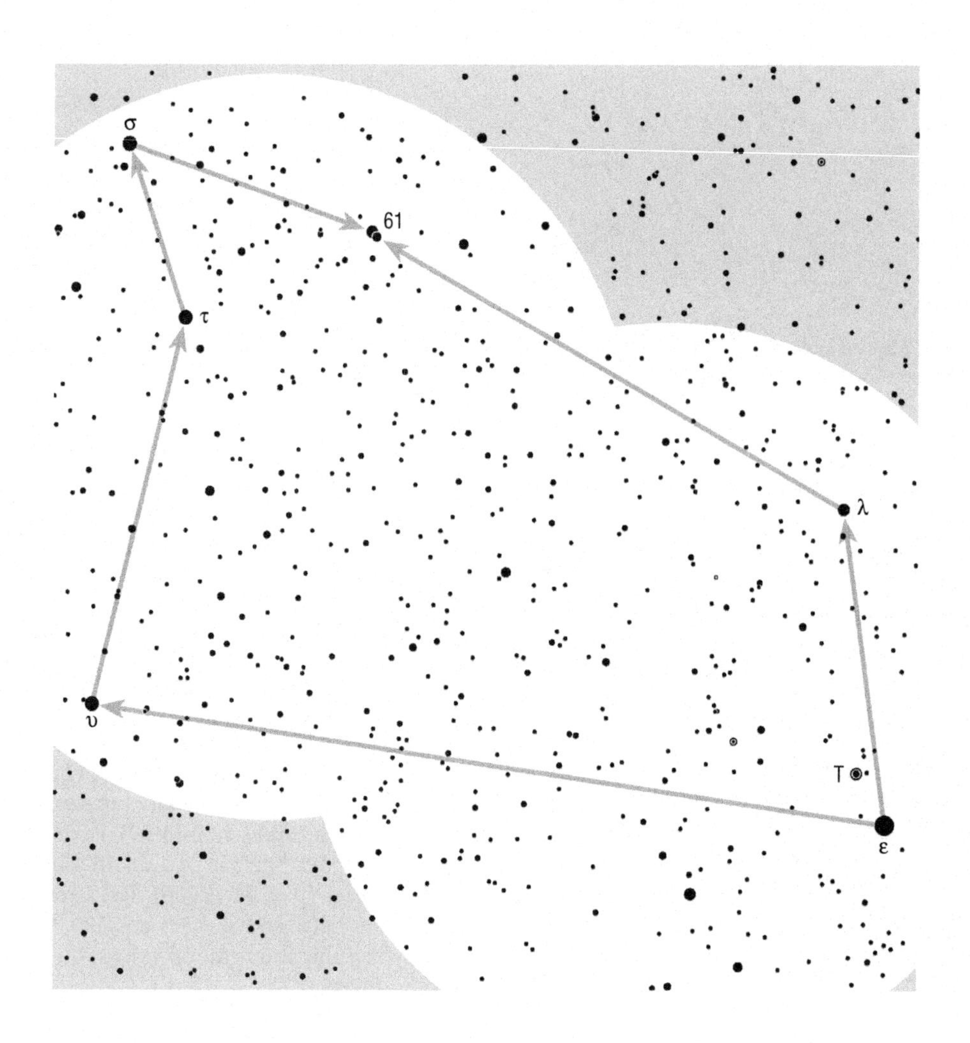

The famous double star 61 Cygni is definitely worth seeing! These distinct orange stars with magnitudes 5.3 and 6.1 are 31.8 arcsec apart (p. a. 151°, 2006), so they are easily separated through binoculars. This double became famous as far back as 1838. That was when F. W. Bessel measured the parallax and determined the distance of the stars to be 10.3 light years away (today's value is 11.4 light years). That was when astronomers could for the very first time start imagining how great the distances between the stars are and how immense (and empty) the universe truly is.

The leading star to this double is the bright Epsilon (shown in the chart above). In the same field of view we can find Lambda (5m), which lies some 2.5 degrees north. If Lambda is moved to the southwestern edge of the field of view, 61 Cygni appears on the northeastern edge.

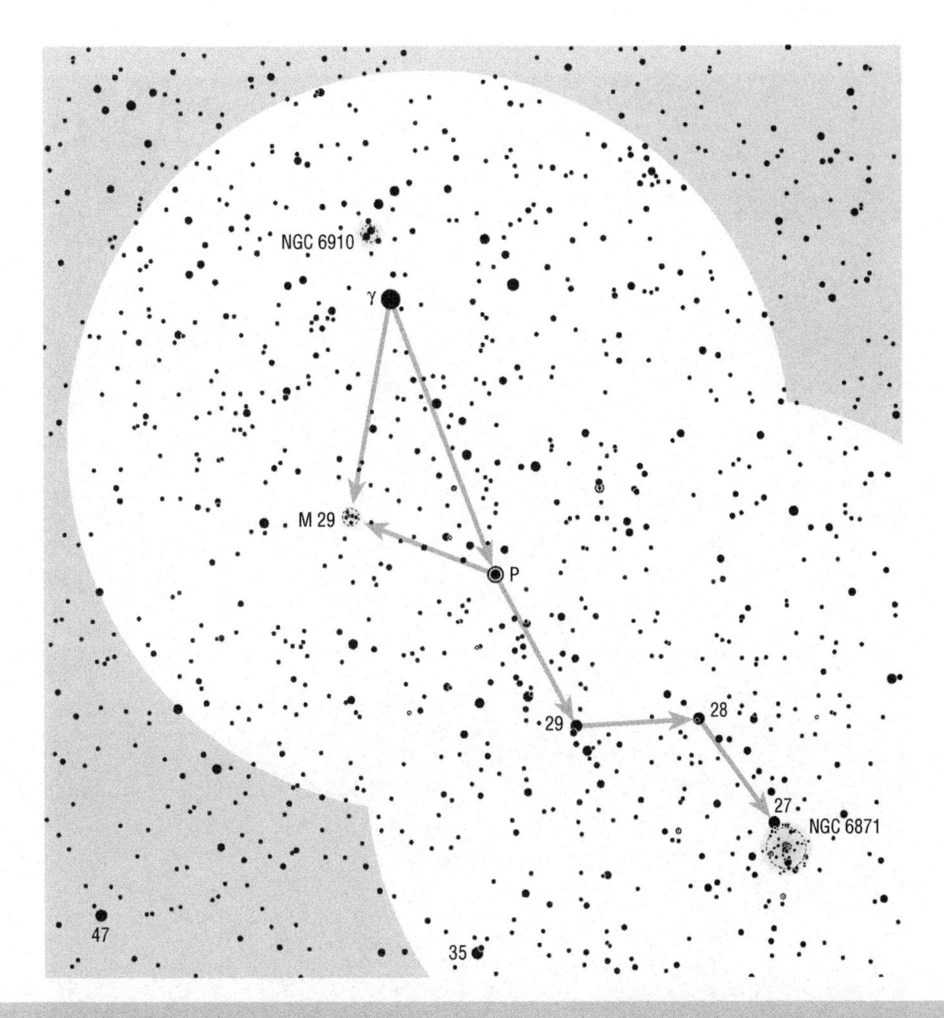

Figure 8.34. The position of open clusters M 29 and NGC 6910. The open cluster NGC 6871 (5m2/20′), which lies close to the magnitude 5 star designated as 27 is even a tougher nut to crack than NGC 6910. The cluster consists of seven clearly seen brighter stars (three are magnitude 7), while the others melt into a tenuous nebula, which can be seen through binoculars only when there are perfect observing conditions. The best time to observe it is on spring mornings or autumn evenings, when the nights are colder and less humid than in the summer

16 Cygni is a close double when viewed through binoculars. It is composed of stars of similar brightness (6m0 and 6m2) that are separated by 39.5 arcsec (p. a. 133°, 2005). The stars are yellowish in color. The fainter member is a double itself; however, we are not able to separate the stars through binoculars. The companion is of magnitude 7.5 and is 15.8 arcsec away (p. a. 76°, 1998). 16 Cygni lies in the northern part of the constellation, close to the planetary nebula NGC 6826, which will be described later. There is also a chart to help you find both objects.

The open cluster M 29 (7m1/7′) is easy to find, since it is in the same field of view with the bright Gamma (Figure 8.34). The trouble is that the cluster is small and that it lies in a very rich part of the Milky Way; thus it is difficult to distinguish it from the countless other stars. There are approximately 150 stars in the area of the cluster, but only around 50 are members

Figure 8.35. Open cluster M 39 lies against a rich background of faint Milky Way stars, which makes it hard to recognize the true members of the cluster. Can you imagine that a certain astronomer had to carefully measure the proper motions of all stars in the area of the cluster for years, that he could recognize its true members, and that we can write down in the books like this "′. . . there are approximately 30 members . . ." Our knowledge of the universe is based on the work of such toilers. The Newtons and Einsteins are extremely rare

of the cluster. Through binoculars we see approximately a dozen stars. The brightest seven are magnitude 9 and are easily separated through binoculars. M 29 is definitely not a very attractive specimen of its kind!

The cluster is some 4,000 light years from us and lies in the neighboring arm of our galaxy. It would be much brighter if it were not obscured by interstellar dust inside it. Astronomers estimate that the dust makes the stars in M 29 fainter by at least 3 magnitudes. What a great shame!

When there are excellent observing conditions we can glimpse a tenuous nebulous spot of light slightly above Gamma. This is the open cluster NGC 6910 (7m4/8′). It is composed of approximately 100 stars; however, the brightest are just magnitude 10. The nearby magnitude 6 stars are not true members of the system, but they prove to be good orientation points for finding the exact location of the cluster. A detailed chart can be seen on Figure 8.34.

The open cluster M 39 (4m6/32′) is relatively large and lies in a rich field against a background of Milky Way stars (Figure 8.35); thus, it is quite suitable for observing it through binoculars with a wide field of view and small magnification. It is rich in stars, but unfortunately most of them are faint. In excellent observing conditions, we can see approximately 35 stars, and as much as 70 can be seen on perfect nights. The 4 brightest stars are of magnitude 7, and 6 of them shine with magnitude 8. The leading star is Rho, which is of magnitude 4 and is clearly seen with the naked eye. Through binoculars they can all be seen in the same field of view. The other way to find this cluster is to find the bright Pi–2. This star and the cluster can also be found in the same field of view.

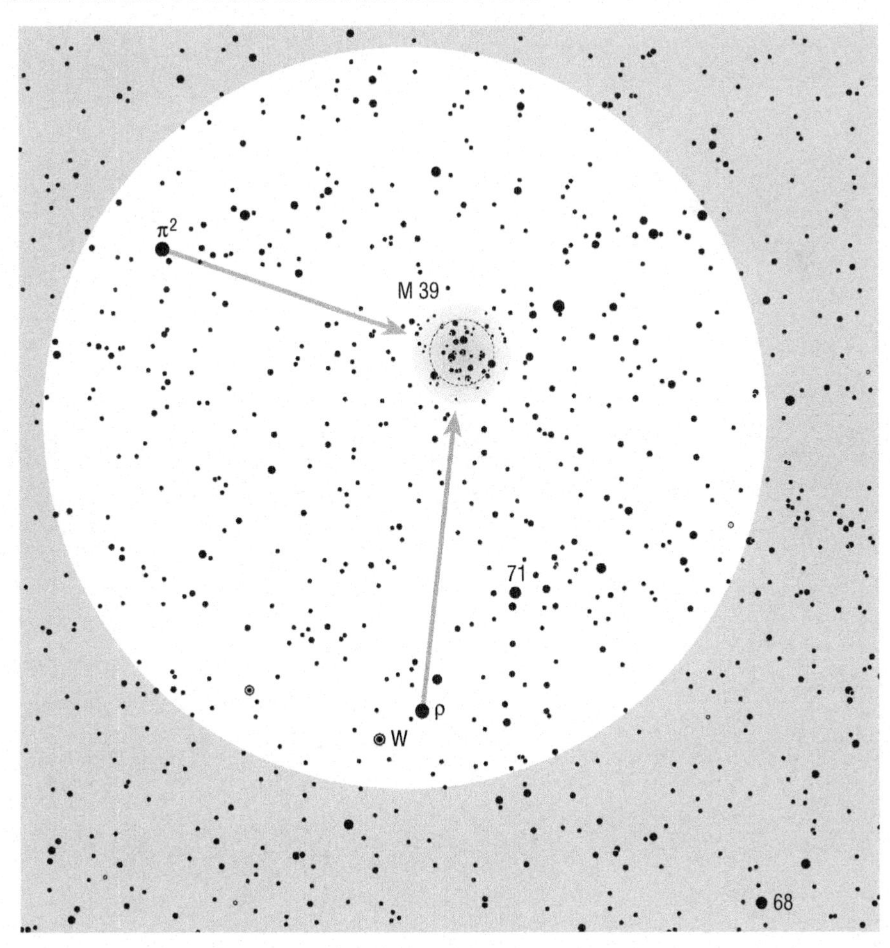

Supposedly M 39 is only 825 light years from us. Its real diameter is therefore only about 8 light years.

Alongside the bright Delta another open cluster awaits us – NGC 6811 (6m8/ 15′). This group of stars is found a mere 2 degrees northwest of the Delta and cannot be missed in the field of view of binoculars. The cluster can clearly be seen, although it appears only as a faint spot of light. A dozen individual stars shine from the nebula on a perfect night, as the brightest stars are of a magnitude 10. In order to separate the cluster into individual stars you need a larger objective.

Cygnus includes some truly remarkable nebulae, which represent an inviting challenge for both amateur and professional astrophotographers. One of these is accessible also to observers relying only on the power of vision. This is the North America Nebula or NGC 7000 ($\sim$4m/120′ × 100′), which lies 3 degrees east of Deneb. If its brightness is exciting to think about, the other data (size 2°×1.7°) quickly sober an experienced observer. The nebula is so large that its surface brightness is, in fact, extremely low; therefore it is seen with the naked eye only when there are perfect observing conditions and even then merely as a slightly brighter patch of light against the Milky Way background. Things do not improve even when viewed through binoculars.

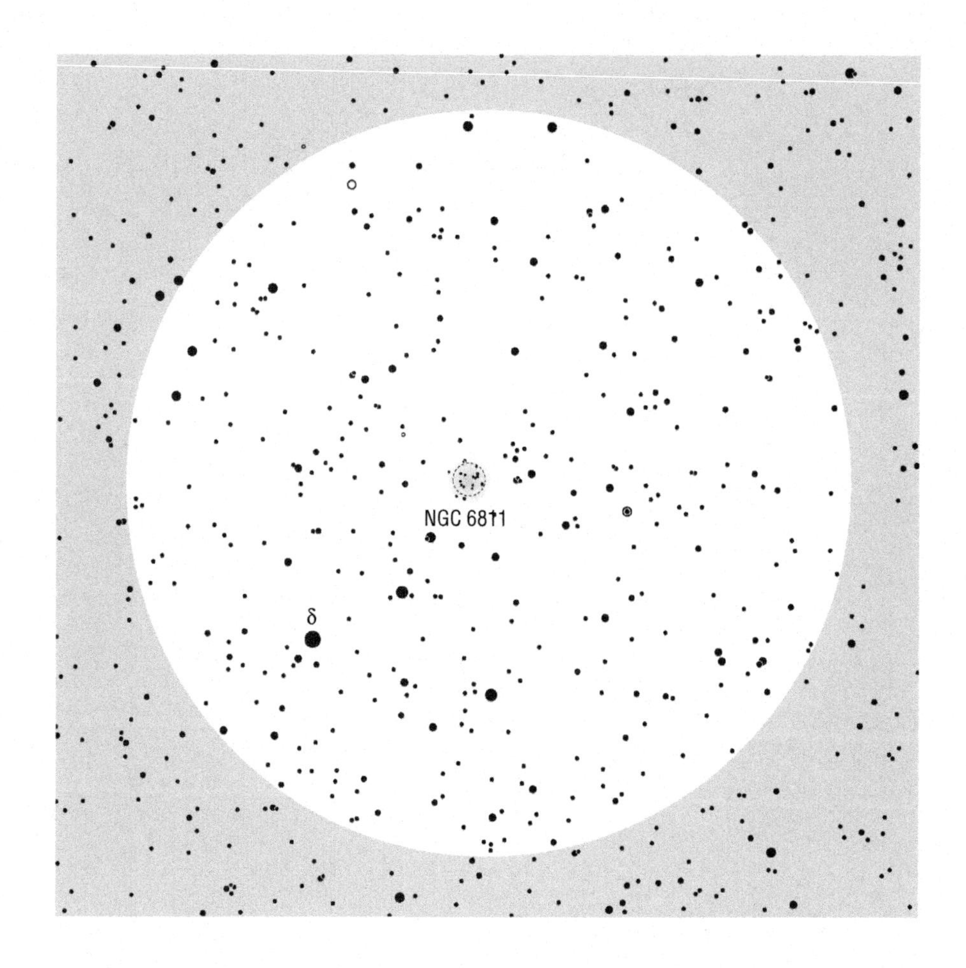

The actual diameter of the nebula is approximately 50 light years. That we can see it at all we probably have to thank the bright and hot Deneb, which excites the atoms in the nebula with its ultraviolet light. The distance between Deneb and the nebula is a few dozen light years. Yes, this truly is an admirable star! The North America Nebula is probably one of the most photographed celestial objects and nearly all astronomy books contain its picture.

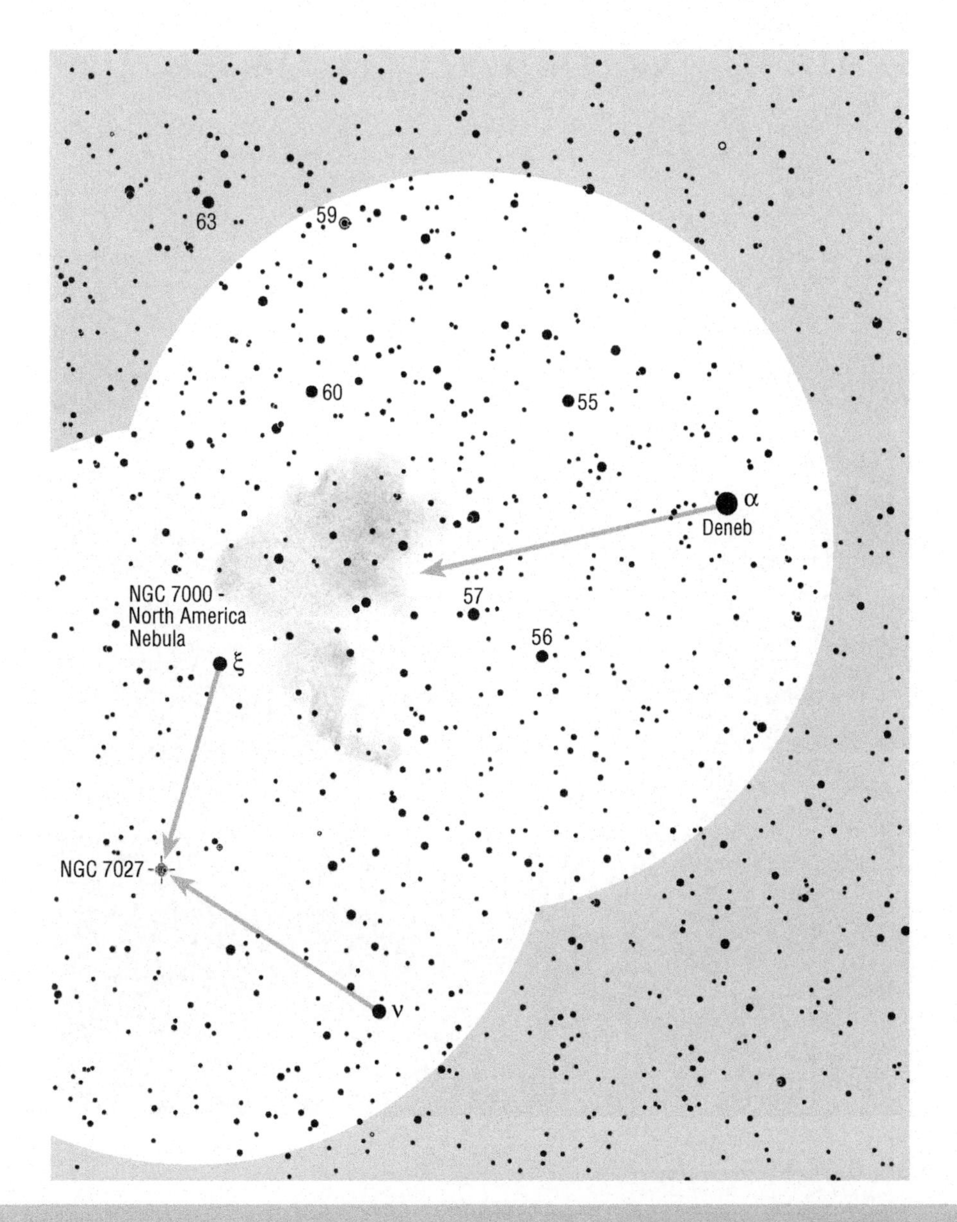

Figure 8.36. Bright Deneb is the star leading to two elusive objects. The North America Nebula demands perfect observing conditions in order to be seen, while the planetary nebula NGC 7027 does not demand such conditions; however, through binoculars it is seen merely as a magnitude 8.5 star and therefore is more or less suitable for gatherers of NGC objects

Figure 8.37. The North America Nebula

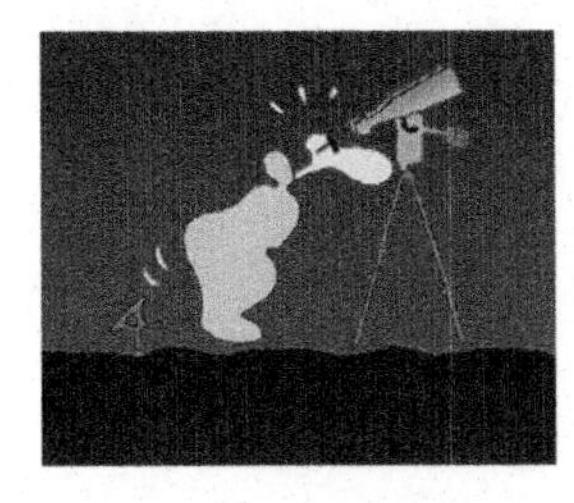

The Cygnus Milky Way is especially rich in star fields, and thus it is worthwhile to take a tour across it with binoculars. The observer is rewarded with breathtaking sights of countless stars, groups, and clusters. Among them we can see darker areas of opaque clouds of gases and dust, which block the light emitted from the numerous stars behind them. Especially dense clouds of stars can be found between Gamma and Beta. Most of the dark clouds can be found between Alpha, Gamma, and Epsilon. In fact we are watching one of the neighboring spiral arms of our Local spur, which is roughly 7,000 light years away. A good pair of binoculars provides an indispensable instrument for panoramic observations of the Milky Way, and in this example they truly outmatch any telescope.

The planetary nebula NGC 7027 (8m5/18" × 10") is an unattractive celestial object for observers through binoculars. However, if you are among those observers who want to see as much as possible with a certain instrument, you can give it a try. The nebula is seen only as a faint point of magnitude 8.5 light. The leading stars are the bright Xi (3m7) and Nu (3m9). The detailed chart is the same as for the North America Nebula. In the vicinity of the planetary

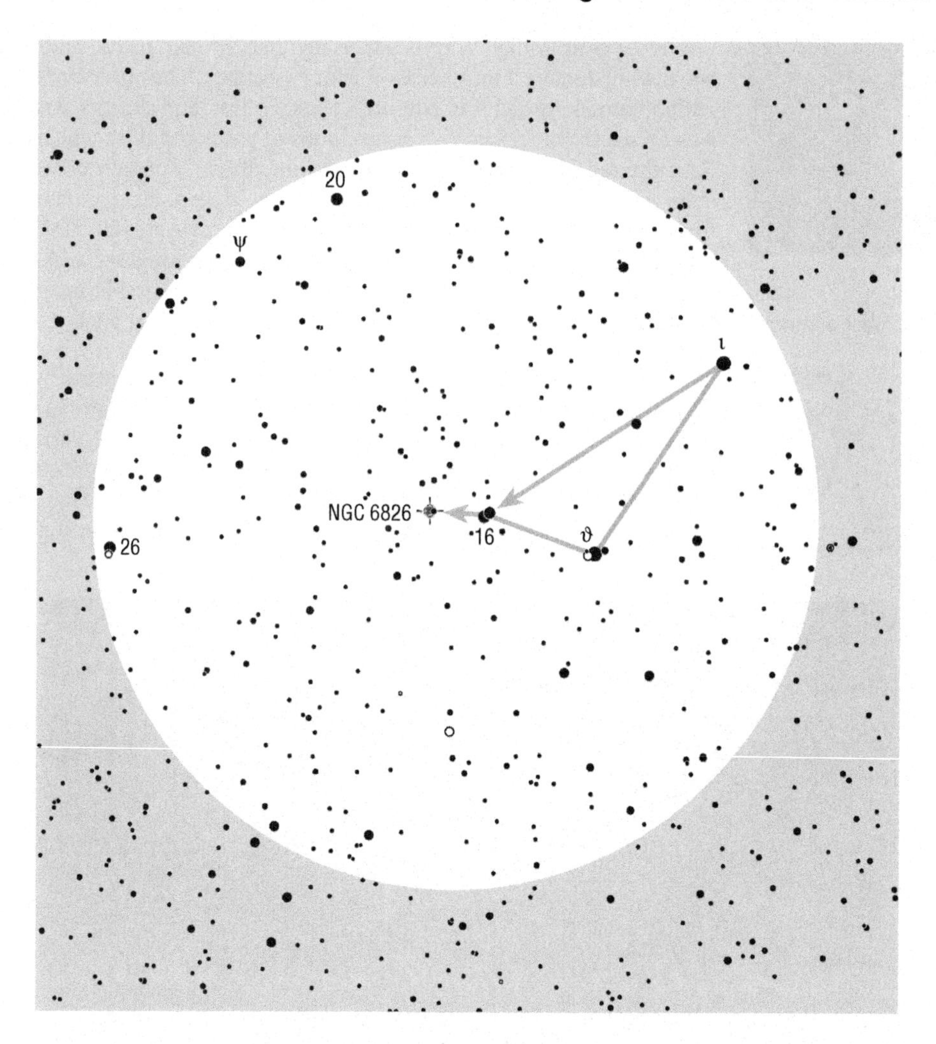

nebula lies a magnitude 9 star, which is slightly fainter than the nebula itself. You should attempt to observe it only when conditions are excellent Figure 8.36.

The same holds true for planetary nebula NGC 6826 (8m8/27" ×24") – it is suitable only for those observers who want to see every last thing they can through their binoculars. The nebula is seen as a faint magnitude 9 star! The stars leading to it are the bright Theta (4m5) and Iota (3m8). Between them and the nebula is the previously mentioned bright double star 16 Cygni. When we are searching for the planetary nebula the detailed chart proves to be of great help. When we go hunting for it, the night has to be excellent and the nebula has to be close to its highest point in the sky.

The summer sky is very poor in galaxies. One of them lies in Cygnus. This is the spiral galaxy NGC 6946 (8m9/11′×9′8) that can be found on the border of Cepheus (Figures 8.38) It is rather bright but big, so it is on the border of visibility in binoculars and therefore suitable for observing only when observing conditions are excellent and only when it is close to its culmination. In the field of view the galaxy is seen as a few arcminutes

Figure 8.38. Spiral galaxy NGC 6946 and open cluster NGC 6939

large and slightly oval spot of light. The star leading to it is the bright Eta Cephei. The star and the galaxy can be found in the same field of view of binoculars. Less than a degree northwest of the galaxy another faint spot of light can be seen – also on the border of visibility. This is open cluster NGC 6939, which lies in Cepheus and is described there (see page 250).

The distance to the galaxy is not known. The best estimates state that it is approximately 10 million light years from us, which makes it one of the closest galaxies to our Local Group.

This is a very special star, well worth visiting at least once and traveling to its surroundings in your imagination. Why? Because this magnitude 9 star is one of the strongest sources of X-rays in the sky! It was the first discovered object with this characteristic in this constellation; thus it was given the name Cygnus X-1. Since 1971, when astronomers made the connection between the star and X-ray source, it has become one of the most serious contenders for containing a stellar black hole! For more on X-ray binaries, look in the first part of this book at the end of the chapter on stellar life.

This strong source of X-rays was discovered in 1962, when astronomers launched the first detectors above Earth's atmosphere. Since 1965, it has been known that this source is a variable. Measurements have shown that the changes in X-ray emission are significant and that they have a period of 50 milliseconds. Fast changes suggest that the source is very small; thus, the object is most likely a highly contracted body due to gravitation. As radio astronomers failed to receive any signal from that area, they excluded the possibility that this was a supernova remnant.

In 1970, the UHURU satellite was launched. This was the first X-ray orbiting observatory, and it collected data on X-ray sources in the sky for over two years. Its detectors were so precise that astronomers were able to accurately determine the position of Cygnus X-1. They discovered that it coincides with a magnitude 9 star designated as HDE 226868 (on the image it is marked with a cross), which lies half a degree east-northeast of the bright Eta. This faint star is actually a very hot and bright blue giant of spectral class B0. Its surface temperature is as high as 30,000 K, and its mass is estimated to be between 10 and 20 solar masses. Spectroscopic research revealed that this is a binary star with an orbital period of 5.6 days. The system is approximately 8,000 light years from us (Figures 8.39).

Shortly afterward Cygnus X-1 was connected with the visual star it attracted the attention of astrophysicists. It had been previously known that a black hole could only be discovered if it was gravitationally bound to another, normal star. The powerful gravitational attraction of the black hole would pull material from its companion. The gases would then spiral toward the black hole. During this process, the matter would be compressed and become so hot that it would emit short-wavelength electromagnetic radiation. Astrophysicists also know that an authentic candidate for a black hole has to have at least 3 solar masses. They immediately started to explore this binary.

Detailed studies revealed that the very hot and dense jet of gas streams toward an unseen companion. It is this hot gas that is the source of X-rays. From the available data, astronomers concluded that the mass of the unseen object is at least 10 times that of the Sun, but it could most probably even reach 16 solar masses. The fast oscillations of the source indicate an extremely small object, with a diameter of less than 150 km. This dimension (in combination with the great mass) excluded the possibility that the unseen object is a white dwarf or a neutron star. It seems like the companion within this binary system is truly a black hole!

However, all of this is merely guesswork. In order to obtain a precise model of the binary system Cygnus X-1 we need to know the inclination of its orbit. We certainly know that the orbit is not seen edge on. In that event eclipses would have occurred, and the blue giant would be known as an eclipsing variable star. Astronomers estimate that the orbit is inclined by 30 degrees over its plane. If this is true, the unseen companion has at least 7 solar masses. The research continues. Cygnus X-1 still remains the most serious candidate for a stellar black hole. Go and visit it!

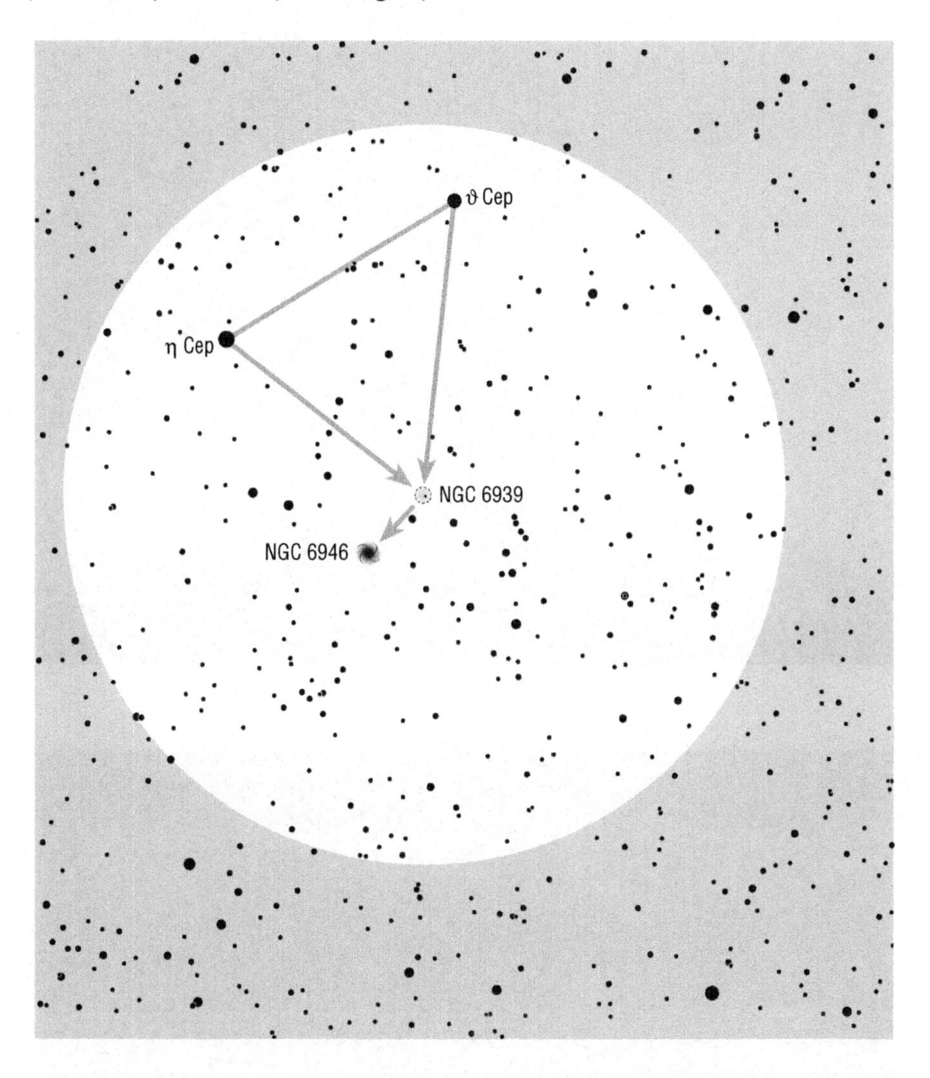
ϑ Cep
η Cep
NGC 6939
NGC 6946

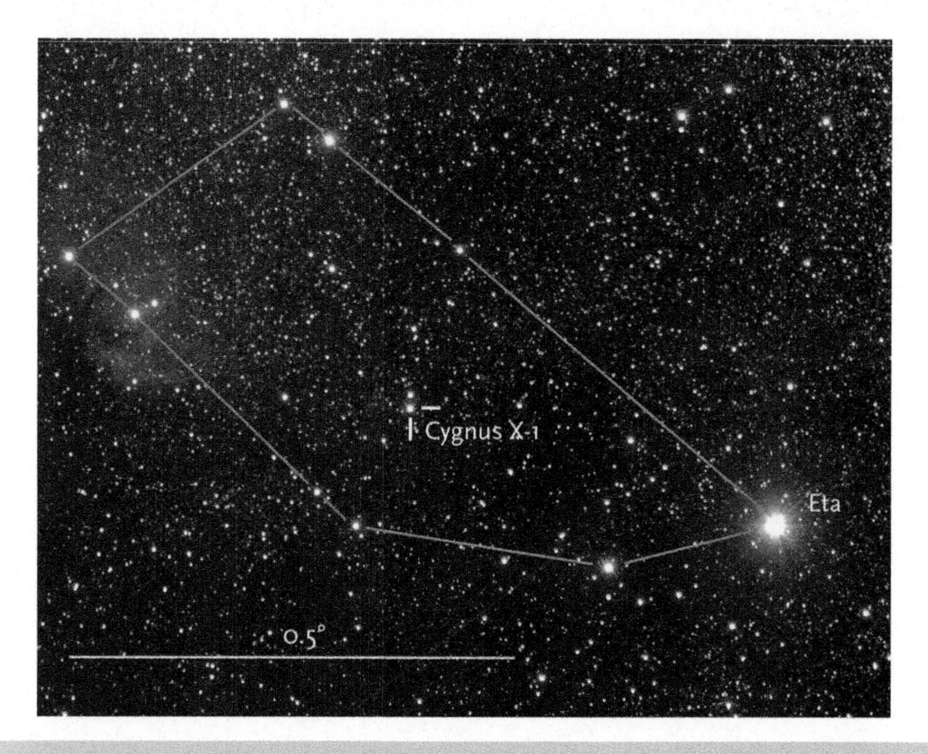

Figure 8.39. The surroundings of the magnitude 9 star (marked with a cross), which most probably hides the black hole. The faint star close above it is 1 magnitude fainter and cannot be found on the detailed chart. Through binoculars it appears only when there are excellent observing conditions. Both stars are separated by approximately 30 arcsec. The brightest star on the right edge of the image is the leading star Eta. A detailed chart can be found next to variable Chi

DELPHINUS (The Dolphin) and EQUULEUS (The Foal)

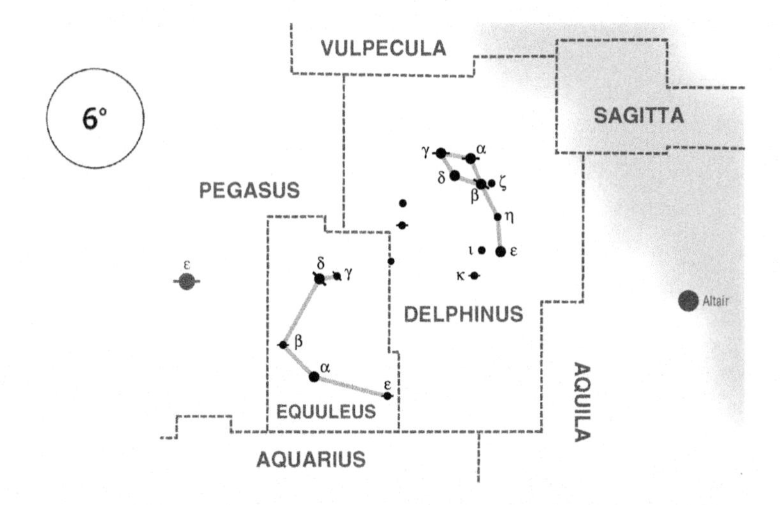

CONSTELLATIONS CULMINATE		
in begin. July at 2 a.m.	in begin. August at midnight	in begin. September at 10 p.m.

Delphinus lies south of Cygnus and Vulpecula. This summer constellation is small but prominent with faint stars just outside of the Milky Way. It looks just like a dolphin leaping from the water during its swim north. A casual observer might mistake it for the Pleiades, which appears over the horizon at least 6 h later.

The brightest stars within the constellation are Beta (3m6), Alpha (3m8), Gamma (3m9), Epsilon (4m0), and Delta (4m4). Through binoculars all of them can be seen in the same field of view.

There are a few interesting double stars that can be seen through telescopes in the Delphinus. Gamma is definitely the most attractive. Stars with magnitude 4.4 and 5.0 are 9.5 arcsec apart (p. a. 266°, 2006). The brighter star is yellow-orange in color, while the fainter is white. However, some observers report different colors for the fainter star, anything ranging from violet to green. Gamma cannot be separated through binoculars.

Equuleus lies east of Aquila, southwest of Pegasus (a single field of view of binoculars west from the bright Epsilon Pegasi) and southeast of Delphinus. It is one of the smallest and faintest original Greek constellations, described in the *Almagest*, the famous book by Ptolemy of Alexandria. The brightest star is the yellowish Alpha, which is of magnitude 3.9. The easiest way to find Equuleus is to use Alpha and Beta Delphini as the leading stars. The constellation does not contain any objects that might be of interest to observers with binoculars.

DRACO (The Dragon)

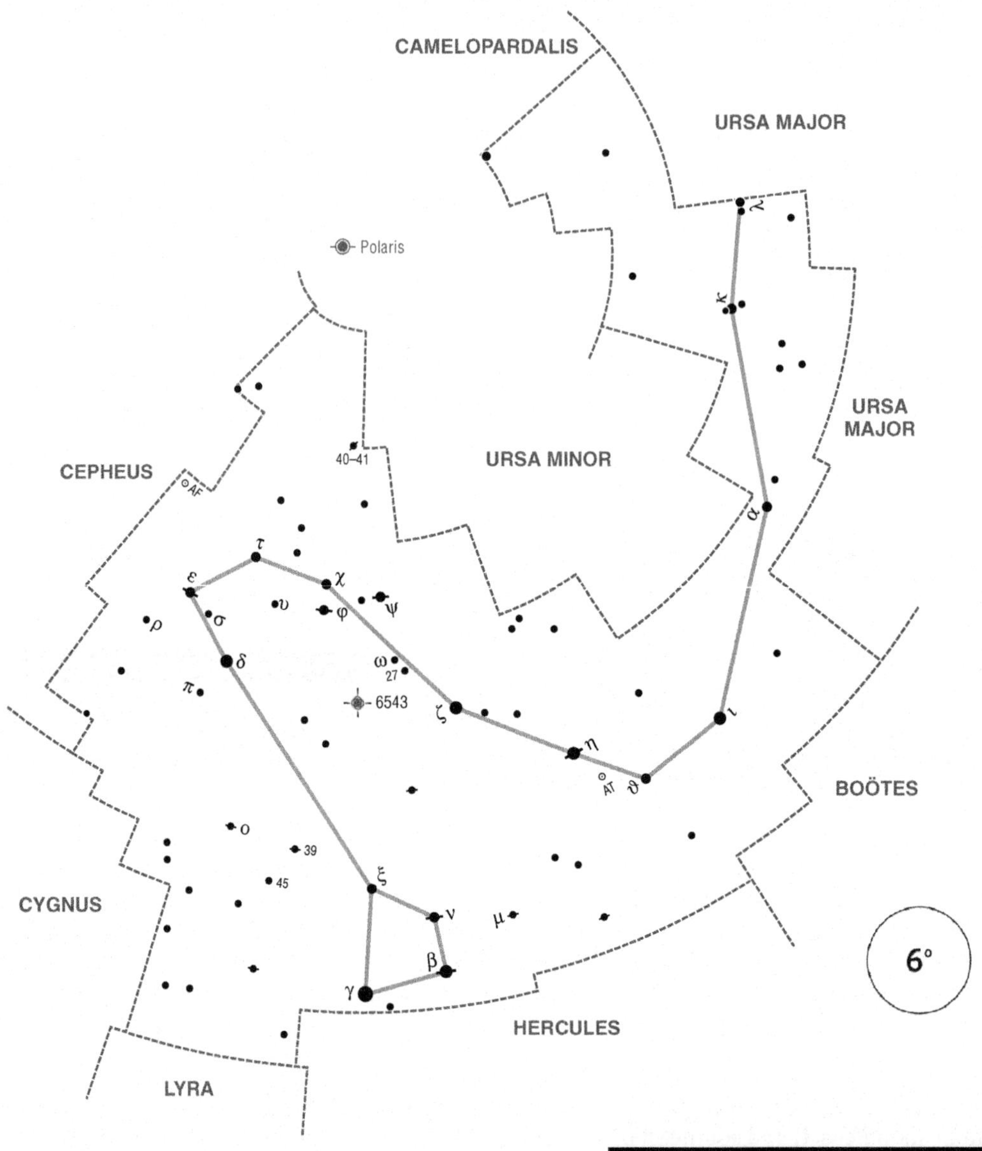

Draco is an extremely elongated constellation. From the mid-northern latitudes it is circumpolar and therefore visible all night every night. The head of the dragon is close to the summer star Vega (Lyra); however, its body extends halfway across the northern skies and divides the constellations of Ursa Major and Ursa Minor. The brightest stars are Gamma (2m2), Eta (2m7), Beta (2m8), Delta (3m1), Zeta (3m2), and Iota (3m3).

Alpha Draconis, or Thuban, lies between Mizar (Zeta Ursae Majoris) and Beta Ursae Minoris. Its brightness is

HEAD AND BODY CULMINATE		
mid May at 2 a.m.	mid June at midnight	mid July at 10 p.m.

TAIL CULMINATES		
end March at 2 a.m.	end April at midnight	end May at 10 p.m.

only of magnitude 3.6. The star is 220 light years from us. In ancient Egypt (around 2800 B.C.), Thuban was the "Polaris," since due to the precession of Earth's axis the north celestial pole was in its vicinity. In 2830 B.C., Thuban was at its closest to the pole, with the separation measuring only 10 arcmin.

The distinct orange Gamma, or Eltanin, is the 72nd brightest star in the sky. It is 148 light years from us. Its luminosity is 200 times that of the Sun. Together with Beta, Xi (3m7), and Nu it forms the head of the dragon.

The yellow Eta is the 118th brightest star in the sky. It is 88 light years from Earth. Its luminosity is that of 45 Suns.

The yellowish Beta is the 130th brightest star in the sky. It is actually a giant, 360 light years from us. Its luminosity surpasses our Sun by 720 times.

Omega (4m7), which lies between Delta and Epsilon, is one of the stars closest to us, since it is only 19 light years away. Arabic astronomers gave it a last name – Al Rakis (means the Dancer) – which is very unusual for such a faint star.

Nu is a double, composed of two stars with an equal brightness of magnitude 4.9. The apparent separation between them is 62.6 arcsec (p. a. 312°, 2005), so they are easily separated through binoculars. The pair of white stars is 120 light years from us. What a pity that there is not any color contrast between the two stars. However, they are helpful in giving a sense of how big an arcminute is in the field of view of binoculars. The actual separation of the stars is approximately 2,300 AU. We can fit up to 38 of our Solar Systems between these two stars!

Psi is a double star. Its components are of magnitudes 4.6 and 5.6 and are separated by 30.7 arcsec (p. a. 16°, 2006). This means that they are seen as a rather close couple when viewed through binoculars. The two stars do not show any color contrast. They are mentioned at this point because of their suitability for testing the quality of your optics and observing conditions.

Omicron is a splendid double star for observers with telescopes. It lies northeast of the dragon's head. This is a pair of stars of magnitude 4.8 and 8.3 that are 36.5 arcsec apart (p. a. 319°, 2003). The brighter star is orange while its companion is blue. We can separate the two stars through binoculars; however, we might have trouble seeing the fainter one. This star can be seen only by a skilled observer, when there are perfect observing conditions, and through binoculars with good optics. Give it a try! The bright Xi is the star leading to Omicron. The two stars, together with two magnitude 5 stars labeled as 39 and 45, form an elongated deltoid, visible with the naked eye.

The double star with designation 41/40 Draconis is composed of stars with magnitudes 5.7 and 6.0, separated by 19.2 arcsec (p. a. 232°, 2006). This pair, which is slightly further away from the main pattern of the constellation, is also suitable for testing the quality of optics in binoculars. The stars in the field of view – if the optics and observing conditions are good – literally touch each other. The double star lies on the extreme northern part of the constellation, a mere 10 degrees from Polaris. The leading stars are Chi, Phi, and Psi, all of which are seen with the naked eye. If these stars are moved to the south edge of the field of view, the magnitude 5 star labeled 35 will show on the northern edge. Three degrees higher to the north lies the double, which is of magnitude 5 and can therefore be seen with the naked eye.

The planetary nebula NGC 6543 (8m1/23" × 17") is seen through binoculars, but only as a magnitude 8 star. It can be found halfway between Zeta and Delta. When we see Zeta, Omega, and 27 in the same field of view and move them to the northwestern edge, as shown in the chart, three faint magnitude 9 stars appear on the eastern edge. These three stars are in line with the north–south direction. Just slightly further north lies the "much" brighter star with a close magnitude 9 companion. This is the planetary nebula we are looking for. The aid of a detailed chart is essential for finding this object. Observing conditions should be at least good. The planetary nebula should be sought only by those observers who want to see absolutely everything that can be seen through binoculars or NGC object hunters.

NGC 6543, popularly named the Cat's Eye Nebula, is considered to be one of the most attractive planetary nebulae in the sky. In images taken with large professional telescopes, its intricate structure can be seen and shows that the central star has ejected its outer layers into interstellar space over 1,500 year intervals. In the best images, taken with the Hubble Space Telescope, we can see 11 rings or gaseous shells around the star. This periodicity of ejecting material by a dying star is not yet fully understood. Some astronomers think that the cause lies in the magnetic activity of the central star, while others are convinced that the central star is actually a double and that the companion star is responsible for the disturbances (Figure 8.40).

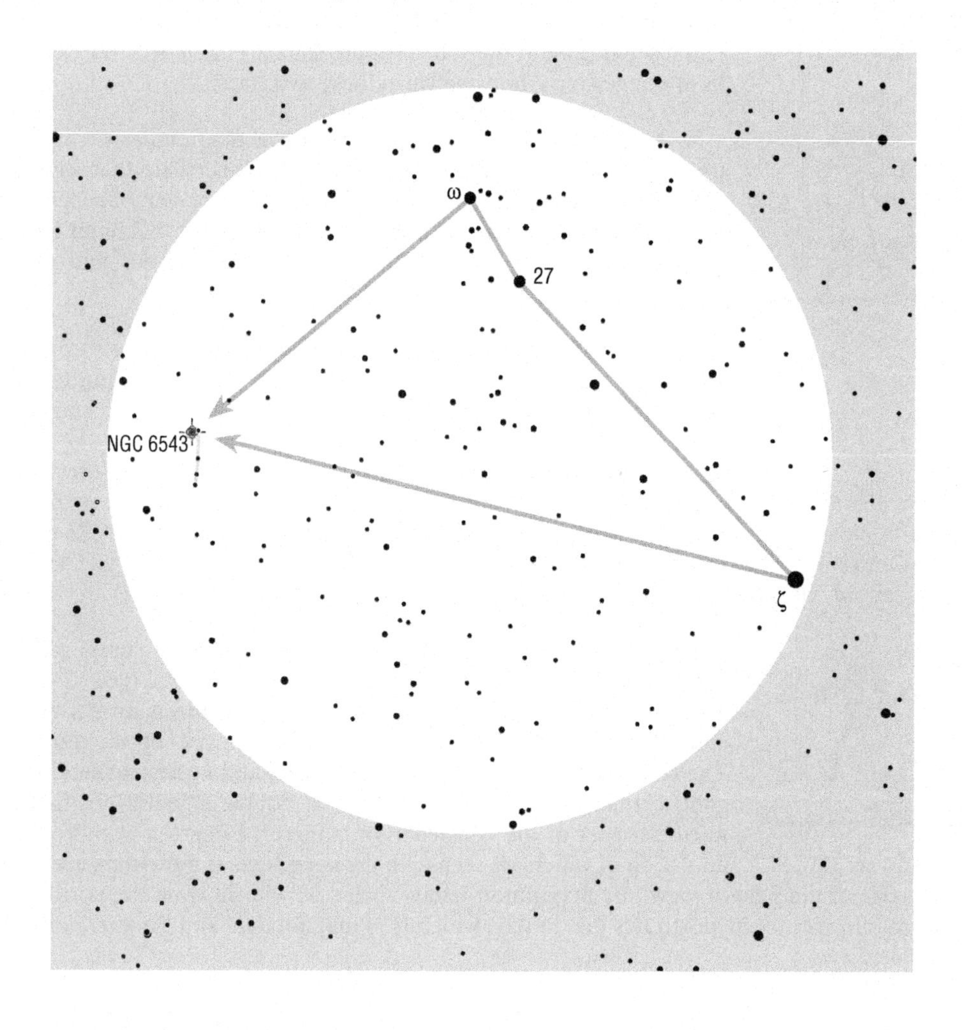

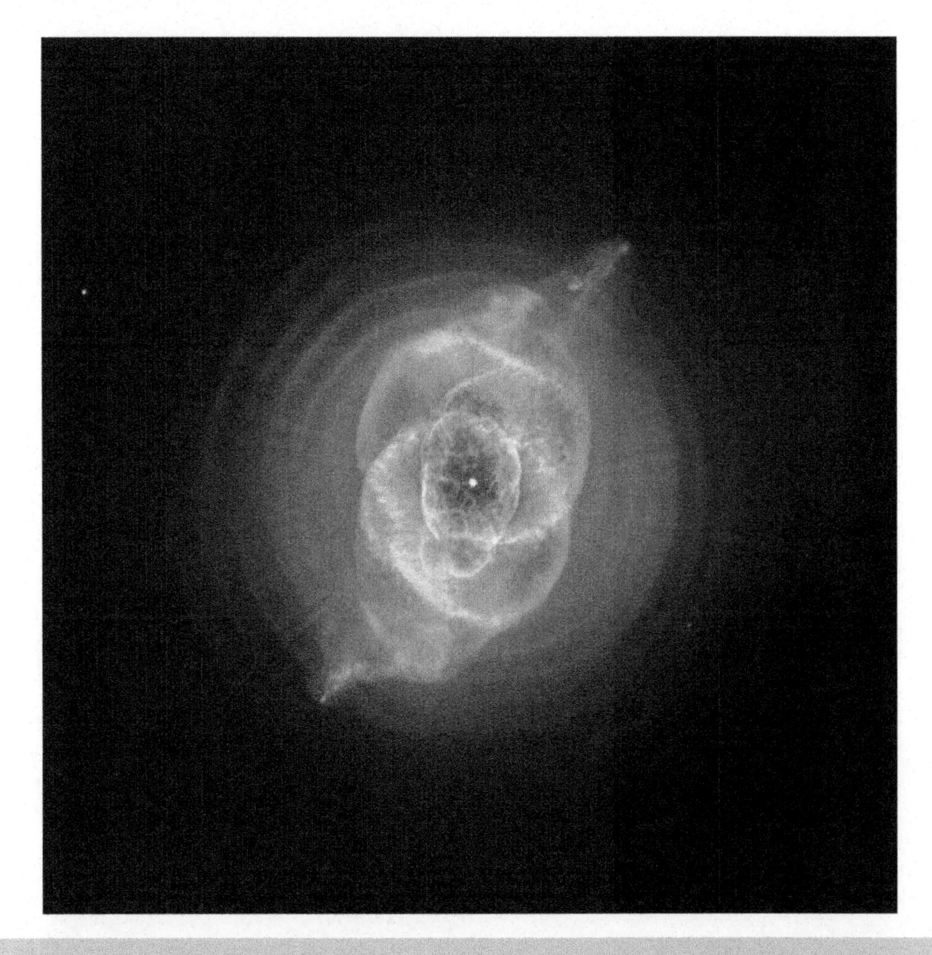

Figure 8.40. Planetary nebula NGC 6543 in Hubble's eye. This image is 1.2 arcmin (1.2 light years) wide

The distance to the planetary nebula has not been determined so far; however, the best estimates place it around 3,600 light years away. At this distance the actual diameter of the nebula is 20,000 AU, or one-third of a light year. The central star that gave birth to the nebula is a hot dwarf with a surface temperature of 35,000 K and a luminosity of 100 Suns.

CHAPTER NINE

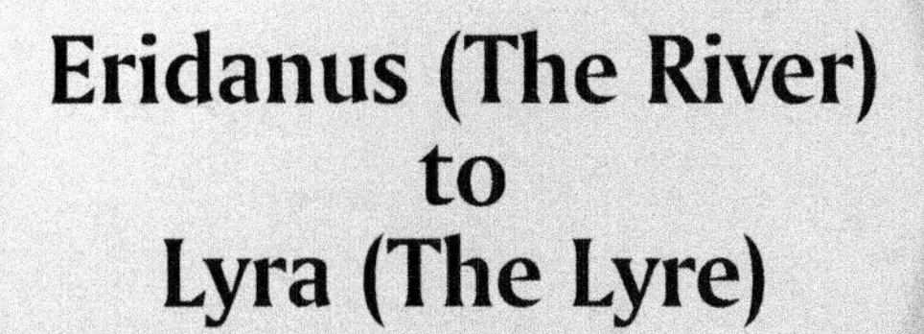

Eridanus (The River) to Lyra (The Lyre)

B. Kambič, *Viewing the Constellations with Binoculars*, Patrick Moore's Practical Astronomy Series,
DOI 10.1007/978-0-387-85355-0_9,

ERIDANUS (The River)

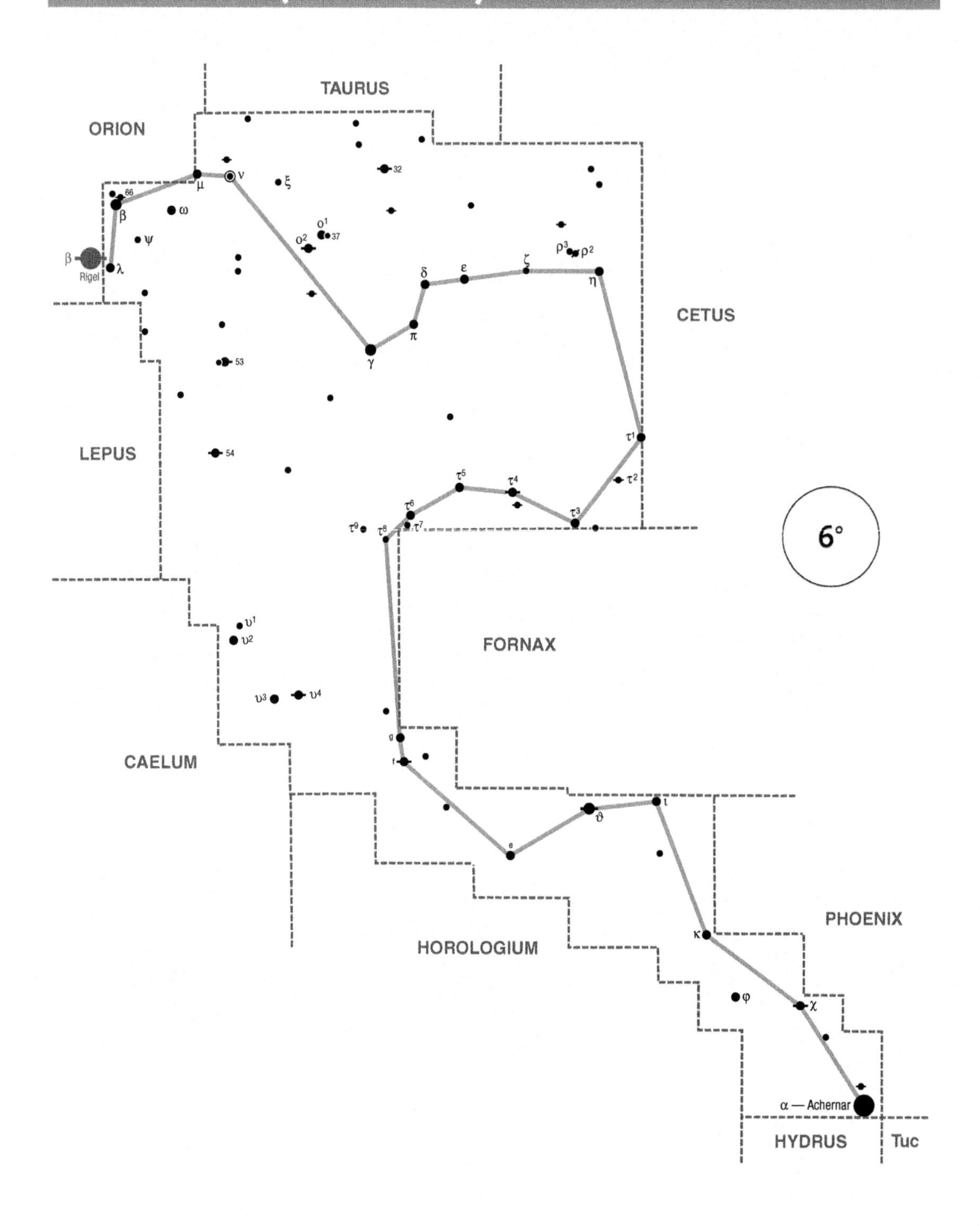

CONSTELLATION CULMINATES		
end October at 2 a.m.	end November at midnight	end December at 10 p.m.

The celestial river Eridaus, which meanders from the celestial equator down to the horizon and ends deep in the southern sky in the vicinity of the south celestial pole, is the 6th largest constellation in the sky. From the mid-northern latitudes only its northern part is visible, i.e., from next to Orion onward.

The brightest star is Alpha, or Achernar (0m45), which cannot be seen from the mid-northern latitudes. It is visible from latitudes such as Cairo or Florida and further down south. Achernar is the 10th brightest star in the sky and is 144 light years away. Its color is bluish white, and it has a luminosity of 1,000 Suns.

By brightness Achernar is followed by Beta (2m8), Theta, orange Gamma (both 2m9), and Delta (3m5).

Beta lies approximately 3 degrees north and slightly to the west of the bright Rigel (Beta Orionis). The star is 89 light years from us and is the 126th brightest star in the sky. Its luminosity is 44 times that of the Sun.

As it is a mere 10.5 light years from us, the yellowish Epsilon (3m7) is one of the closest stars. Among the stars that can be seen with the naked eye, it is the third closest, right after Alpha Centauri and Sirius. Epsilon is slightly smaller and colder than our Sun, but otherwise the two stars are rather similar; there is a chance that planets might orbit around Epsilon. In 1983, the IRAS satellite (Infra-Red Astronomical Satellite) discovered a cloud of cold matter surrounding the star, and this is what eventually gives birth to planets.

32 Eridani is a contrasting pair of stars with magnitudes 4.8 and 5.9. They are separated by 6.9 arcsec (p. a. 348°, 2006). The brighter star is yellow and its companion is bluish. We cannot separate them through binoculars, but we can see them when using even a small telescope.

An interesting triple star, Omicron-2 (4m4), lies northeast of Gamma. At a distance of 16.5 light years it is the 8th closest star visible to the naked eye. The pair A–B is composed of stars with magnitudes 4.4 and 9.7, which are separated by 83 arcsec (p. a. 104°, 2002). They can be separated through binoculars; however, it is extremely hard to see the fainter component. The stars are 400 AU apart, which means that we could put seven Solar Systems between them. The brighter star is yellow-orange in color. The star with designation B is a double in itself and has a very interesting composition – the pair is made up of a white and a red dwarf, which have magnitudes 9.5 and 11.2. They are separated by 8.8 arcsec (p. a. 337°, 1998). To split them you need a larger telescope. At this point they are mentioned because they are something special. Omicron C is one of the least massive stars known to us. Its mass is a mere 0.2 of that of our Sun. Omicron B is one of the rare white dwarfs that can be seen through amateur telescopes. It has a diameter of a mere 28,000 km (this is slightly more than twice the diameter of Earth), but as much mass as half the mass of our Sun. The average density of this star is over 65,000 times higher than that of the Sun. A cubic decimeter (liter) of its matter weighs in at an unbelievable 90 tons! In addition, the gravity at its surface is incredibly high and is 34,000 times greater than gravity on Earth. A man who weighs 60 kg on Earth would – if he could stand on its surface – weigh an unbelievable 2,000 tons on this star!

FORNAX (The Furnace)

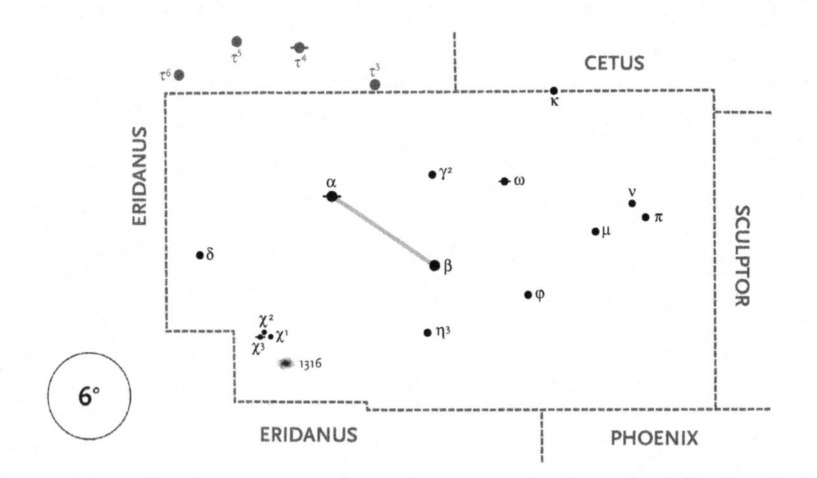

CONSTELLATION CULMINATES		
in begin. October at 2 a.m.	in begin. November at midnight	in begin. December at 10 p.m.

Fornax lies low above the southern horizon for mid-northern observers. The best guide to this small autumn constellation is the meandering line of stars in Eridanus, which are labeled from Tau-3 to Tau-6 on star charts. Fornax has a single star that shines brighter than magnitude 4. This is our close neighbor Alpha (3m8), a mere 40 light years from us.

The constellation is also home to a smaller galaxy cluster, which contains 18 brighter members and numerous fainter ones. This cluster is approximately 60 million light years from us. In places close to the equator its brightest member, the spiral galaxy NGC 1316 (8m9/7′1 × 5′5), can be seen through binoculars, but in mid-northern latitudes it rises only 7 degrees above the horizon, which is not enough. However, if our observing site is lower in the northern latitudes, let's say somewhere in the southern United States or southern Spain, you could try to glimpse it on a perfect winter night. The leading star is Delta and the trio of stars named Xi, all of which appear in the same field of view of binoculars as the galaxy.

GEMINI (The Twins)

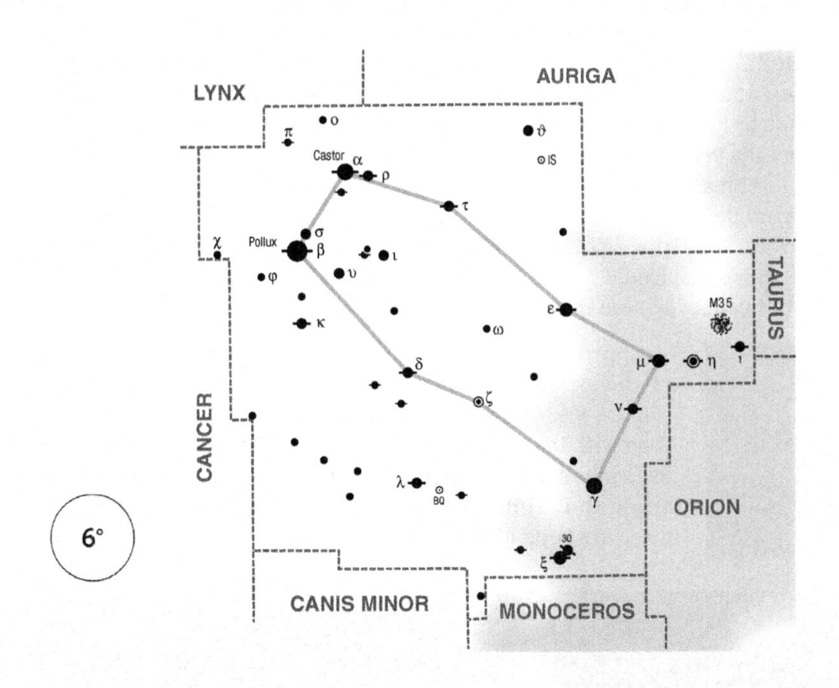

CONSTELLATION CULMINATES		
mid December at 2 a.m.	mid January at midnight	mid February at 10 p.m.

Gemini is a wonderful and prominent winter constellation with celestial twins: Castor (Alpha) and Pollux (Beta). The other brighter stars are Gamma (1m9), Mu (2m9), Eta (3m1 at maximum), Epsilon (3m2), Xi (3m4), and Delta (3m5). The constellation has a characteristic shape that is impossible to miss - parallel lines of brighter stars extend from Castor and Pollux in the direction of Betelgeuse (Alpha Orionis). Pollux is one of the six stars that form the asterism known as the Winter Hexagon. The Milky Way winds across the southwestern region of the constellation, making this area rich in stars and worthy of a panoramic survey through wide-field binoculars!

In spite of their name the twins Castor and Pollux are not similar at all. Pollux (1m2) is an orange star. Its color is well seen even with the naked eye and only intensifies when viewed through binoculars. The surface temperature of the star reaches a mere 4,500 K. Pollux, which is the 17th brightest star in the sky, is known as Beta, even though it is the brightest star in the constellation. It is a mere 33.8 light years from us, and its luminosity is 28 times that of the Sun.

Castor (1m6) is a white, somewhat fainter star - the 23rd brightest in the sky. Its light travels 51.7 years to reach us. In fact, it is a very interesting multiple star, made up of six components. The brightest stars with magnitudes 1.9 (Castor A) and 3.0 (Castor B) are 4.4 arcsec apart (p. a. 60°, 2006), so that they can be separated through amateur telescopes but not through binoculars. The system also has a gravitationally bound third, faint star. It has a magnitude of 9.8 (Castor C), and is separated from the pair A–B by 71 arcsec (p. a. 164°, 2001). The pair A–B has a period of approximately 400 years, while Castor C needs as much as 10,000 years to encircle the bright couple. All three stars, which rotate around a common center of gravity, are spectroscopic doubles!

Gamma is the 44th brightest star in the sky, 105 light years from us. Its luminosity is 130 times that of the Sun.

Mu is the 149th brightest star in the sky, and its light travels 230 light years to reach us. Its luminosity is that of 275 Suns.

The yellowish Epsilon is a super giant; it is about 1,100 light years from us. Its luminosity is nearly 5,700 times that of the Sun. In spite of this gigantic distance, the star shines with magnitude 3 in our sky. Our Sun would be seen only through the largest telescopes if we watched it from a distance as far as Epsilon!

Zeta is among the brightest Cepheid variables – a pulsating giant that changes its brightness in a period of 10.15073 days. At its brightest it shines with a magnitude 3.6, and at its minimum it shines with a magnitude 4.2. The star is some 1,200 light years from us. Its luminosity surpasses our star by about 5,700 times. Stars suitable for brightness comparison are Kappa (3m6) and Upsilon (4m1).

The deeply orange Eta is a red giant. It is a semi-regular variable star with minimal changes in its brightness. Neither the period nor the amplitude is regular with this star. The average value of its pulsating period is 233 days. At maximum brightness the star can reach magnitude 3.1, and at its minimum it is no brighter than magnitude 3.9. Eta is 350 light years from us. The star has a yellowish companion of magnitude 6.1, which is 1.8 arcsec away (p. a. 259°, 2004). The pair cannot be separated through binoculars, but in larger telescopes they represent a splendid couple. Besides, the brighter star is a spectroscopic binary with a period of 8.2 years, but this cannot be seen in any telescope, just in the light of the star.

Approximately 2 degrees northwest from Eta lies M 35 (5m3/28′) one of the most beautiful and brightest open clusters offered by the sky (Figure 9.1). In a clear, dark sky it can be seen as a faint patch of light even with the naked eye. With a pair of binoculars it is not hard to find as it lies in the same field of view with bright Eta. The cluster is seen as a big and bright spot within which some of the brightest members can be seen when there are good observing conditions, and with the aid of averted vision we

Figure 9.1 Underneath M 35 (less than a degree to the southwest) we can see its close neighbor, also an open cluster, NGC 2158 (8m6/5′), which is visible only through amateur telescopes. These clusters appear close in our sky only when observed from Earth; but in fact NGC 2158 is much further away and is one of the remotest open clusters known to us. Its light travels 16,000 years before it reaches us

can see even more of them. The size of the light spot depends mostly on observing conditions. On the best of nights, it extends across at least 20 arcmin, and the number of individual stars that shine out of haze increase! But if this open cluster is observed with a larger telescope and a wide field eyepiece, it turns into a true celestial jewel that will take your breath away, regardless of whether you are a beginner or an experienced observer!

The cluster consists of over 1,000 stars assembled in a cloud with an apparent diameter equal to the full Moon; 90 stars are brighter than magnitude 11, and 50 are brighter than magnitude 10. The brightest star shines with magnitude 7.5, and the next four with magnitude 8. In the detailed chart the stars of magnitudes 10 and 11 are literally overlapping each other (Figure 9.2). In excellent observing conditions we can glimpse a slightly brighter part of the haze in this area, and on a perfect night, when the limiting magnitude of the binoculars almost reaches magnitude 11, this area will be covered in fainter stars.

The distance to the open cluster is estimated at 2,700 light years, which makes its real diameter 22 light years. In the Sun's neighborhood there are only 11 stars within such an area. Can you imagine the crowd we would see in the sky if our planet orbited around a star in this cluster? Most of the brighter stars are blue; however, there are also some yellow and red giants. M 35 is probably not more than 110 million years old, just about middle-aged.

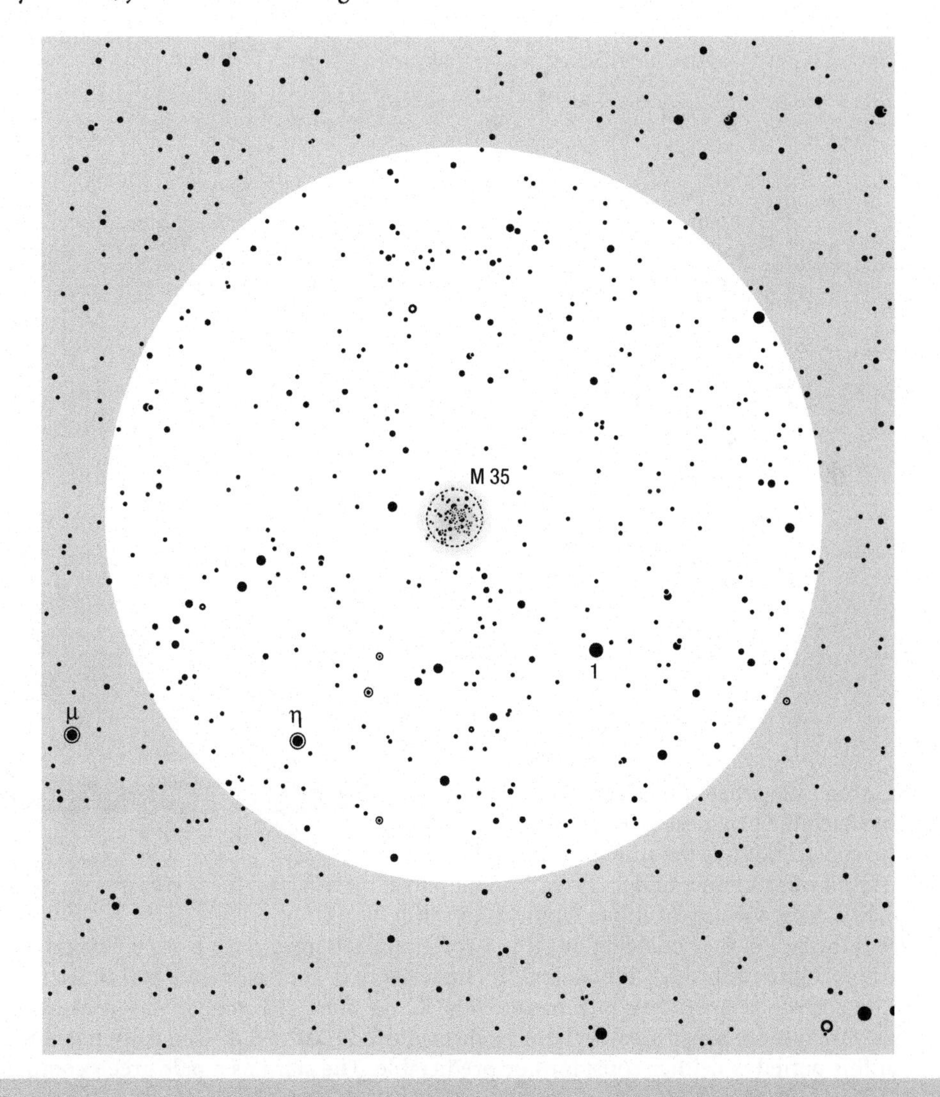

Figure 9.2 The surroundings of the open cluster M 35

HERCULES (Hercules)

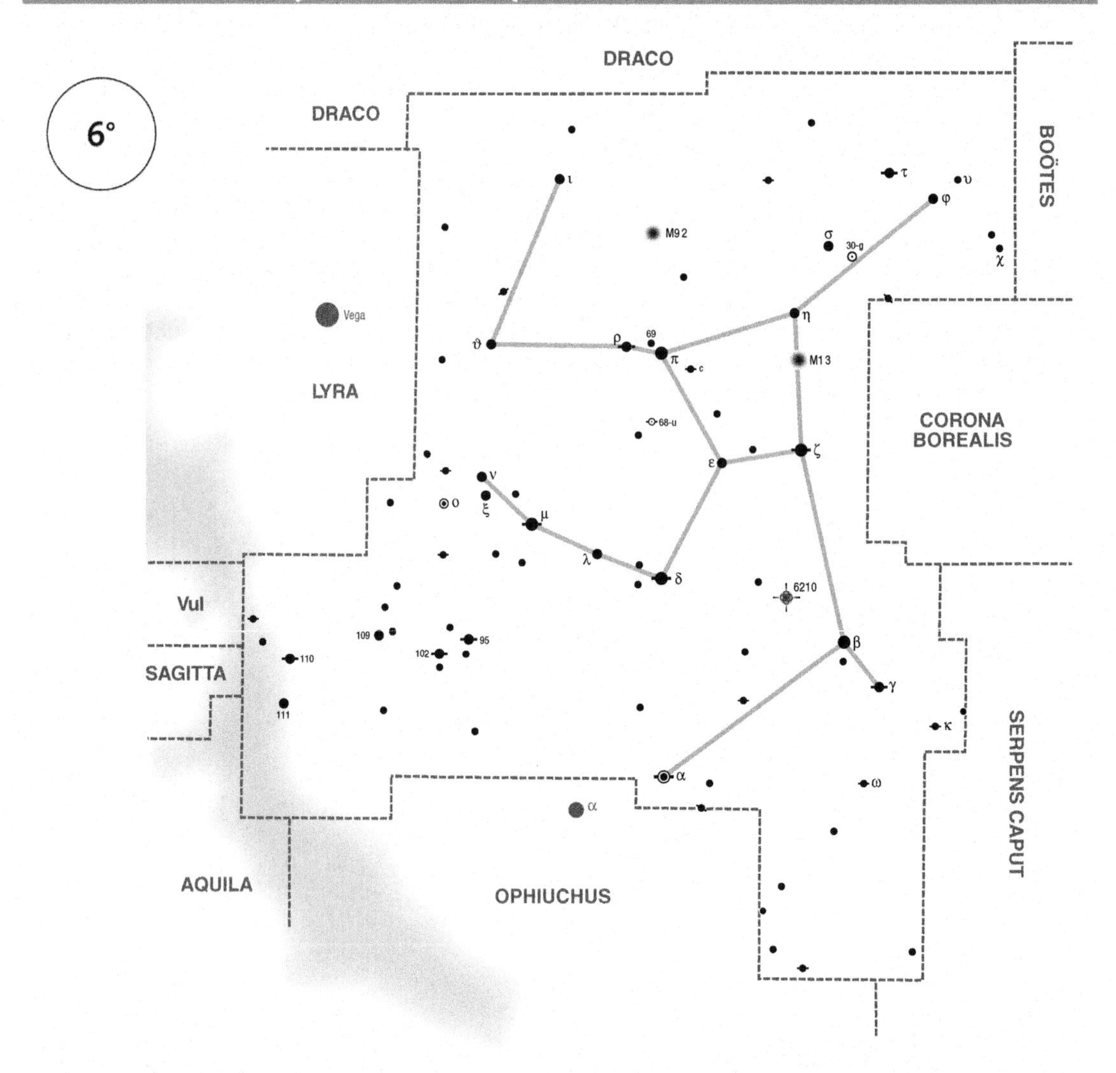

CONSTELLATION CULMINATES		
in begin. May at 2 a.m.	in begin. June at midnight	in begin. July at 10 p.m.

Hercules is a big but rather undistinguished spring constellation, in which the brightest star Beta is only a magnitude 2.8. The constellation is the fifth largest in the sky.

Alpha Herculis, also known under its traditional name Ras Algethi, lies in the vicinity of Alpha Ophiuchi, slightly removed from the rest of the constellation. It is a red giant, 380 light years from us. Its orange color can be clearly recognized through binoculars. Its luminosity is approximately 800 times that of the Sun, while its surface temperature is a mere 3,000 K. As most red giants, Ras Algethi is also a semi regular variable. Its brightness changes from magnitude 3.0 to 3.8. It usually has a period of 90 days, but this period is neither constant nor predictable. The star has a pale green companion of magnitude 5.4, which is seen in small telescopes but not in binoculars. The separation between the stars is 4.7 arcsec (p. a. 105°, 2006). Alpha is 128th brightest star in the sky.

The brightest stars within the constellation are Beta, Zeta (both 2m8), Delta (3m1), Pi (3m2), Mu (3m4), and Eta (3m5).

Zeta, Eta, Epsilon, and Pi form the corners of the asterism known as the Keystone, which is the most recognizable pattern in Hercules and is a sort of trademark of this constellation.

The yellow Beta is the 127th brightest star in the sky, 148 light years from us. Its luminosity is 120 times that of the Sun.

The yellow Zeta is the 132nd brightest star in the sky. For astronomers concerned with stellar evolution, it is a rather interesting star, so it is regularly and carefully observed. It is among the closest stars, a mere 35.3 light years from us. In fact, it is a double star with an orbital period of 34 years. The brighter star is a subgiant, similar to our Sun in mass, but since it is older and the nuclear fuel in its core is slowly running out, it has already started to expand. Currently, its luminosity is seven times that of our Sun.

Delta is an interesting double, composed of stars that are not gravitationally bound; it is only from Earth that they are seen in the same part of the sky. The companion is of magnitude 8.3. The pair is known by its nice color contrast, but the colors are described variously by different observers. Some see them as greenish and violet, others greenish and white, and still others see them as yellowish and blue green. Unfortunately the stars cannot be separated through binoculars (they are 11.8 arcsec apart (p. a. 283°, 2005)), but they represent an easy pair to separate even for small telescopes.

Through binoculars Kappa appears as a close double, with stars of magnitude 5.1 and 6.2. They are 27.7 arcsec apart (p. a. 13°, 2006). The brighter star in the pair is yellowish, and the fainter one has an orange hue. Kappa, which lies some 4 degrees southwest from the bright Gamma (3m5), is easy to find and recognize, once we move Gamma to the north-eastern edge of the field of view.

Globular cluster M 13 (5m8/20′) is definitely the most popular cluster among northern observers. On a clear, dark, moonless night, far from the light-polluted sky, it can be glimpsed even with the naked eye as a magnitude 6 "star," lying between Eta and Zeta (Figures 9.3 and 9.4). The cluster was first mentioned by Edmund Halley in 1715, but he discovered it by mere chance one year before. M 13 is the fourth brightest among the globular clusters. Only Omega Centauri, NGC 104 in Tucana, and M 22 in Sagittarius are brighter. The first two are not visible from mid-northern latitudes.

The cluster is clearly seen through binoculars, but only as an approximately 10 arcmin big spot of light with a clearly differentiated brighter central part. Its size and appearance depend on the observing conditions. However, if we want to separate a few individual stars on the edge of the cluster, we have to have a telescope with a 10-cm objective lenses.

In larger amateur telescopes, the cluster appears truly brilliant – no matter whether you are seeing it for the first or the hundredth time. From its dense center that cannot be separated into individual stars even with the largest telescopes in the world, we can see magnificent arcs of stars that wind around. It is estimated that the cluster includes approximately one million stars. Their common luminosity is over 300,000 times that of the Sun. The brightest members are magnitude 11 red giants, but the actual luminosity of each of them is that of 2,000 Suns. At this distance our Sun would hardly be seen even through the largest telescopes; we would see it only as a magnitude 19 star.

The modern estimate of the cluster's distance is 25,000 light years. It is over 10 billion years old, and it belongs among the oldest clusters in our galaxy. The true diameter of M 13 is approximately 145 light years. This estimate is rough; however, it is hard to determine the exact borders. Most of the stars are gathered in the central part, which is approximately 100 light years in diameter, but a few individual members can be found up to 200 light years away.

Figure 9.3 Magnificent globular cluster M 13. Observers with binoculars will have to be satisfied with a hazy spot of light, as they cannot see it in its full glory

Can we even imagine the density of the stars in the vicinity of the center of such a compact cluster? Images taken with the largest telescopes in the world seem to show that the stars in the center are so neatly and equally arranged next to each other that they almost touch. The reason for this lies in the fact that we are looking at the cluster from a great distance. As previously mentioned, the central part of M 13 measures approximately 100 light years in diameter, or 500,000 cubic light years. If this volume is inhabited by a million stars, the average density is one star per half cubic light year. (For comparison: in our galactic neighborhood the rough estimate of star density is one star per 360 cubic light years!) Of course, the stars are denser toward the center of M 13; nevertheless, it is not such a crowd as it appears at first sight.

It will help us to picture the situation in the cluster if we make a model. Imagine that we have a million grains of sand, which represent the million stars found in the cluster. The size of each of the grains is approximately 0.5 mm, and we have to arrange them inside a sphere 340 km in diameter. If we arrange them in a regular pattern, the distance between two neighboring grains is as far as 2 km. Even if the central part of the cluster is much denser, the distance between two grains will still measure a few hundred meters. This means that by Earth's standards and in comparison to the size of the stars even the densest globular cluster is rather sparsely inhabited; however, this does not hold true for galactic standards. For instance, if we make the same model for our Sun and its vicinity, the closest star, Proxima Centauri, is as far as 15 km away. And by the way, the sand for our model has a volume slightly less than one deciliter and weighs 200 g!

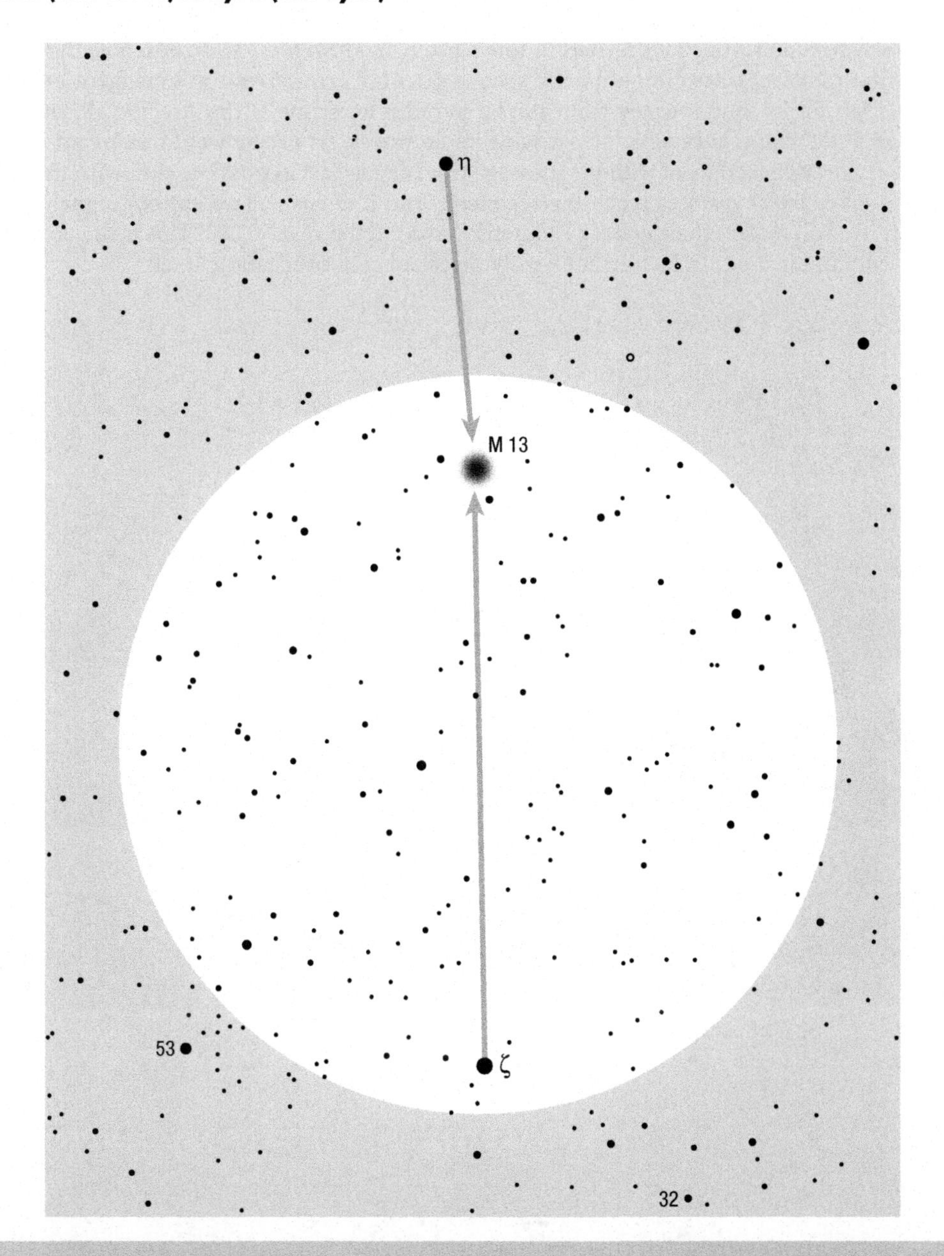

Figure 9.4 Globular cluster M 13 is easy to find. With excellent observing conditions it can be glimpsed with the naked eye. Through binoculars it shows as a big, bright, and highly visible spot of light. The leading stars are Zeta and the 0.7 magnitude fainter Eta. The cluster lies between the two, slightly closer to Eta

It would be quite interesting to live on a planet orbiting a star close to the center of the cluster. The night sky would be full of truly bright stars, so bright that our Sirius, Antares, or Vega, which to us appear distinct, would be seen as faint and insignificant. As much as a few thousand stars would have the brightness of between Venus and the full Moon. There would probably be no night time on the planet, at least not as we know it here on Earth. In the best case scenario, we might experience twilight. The inhabitants of this planet would almost certainly not know anything about our galaxy or the universe, since their sky would be too bright for deep-sky astronomy! For them the entire universe would consist entirely of their domestic globular cluster.

There is another interesting fact related to this cluster. In 1974, M 13 was chosen as the target for one of the first radio messages intentionally sent to possible extraterrestrial intelligent beings. The coded signal left on its long journey from the large radio telescope at the Arecibo Observatory. If somebody lives there and is capable of receiving radio waves, he or she would intercept them. But there is one frustrating fact! Even with the message traveling as fast as possible (i.e., with the speed of light), it will take 25,000 years to reach the recipient. And if aliens will accept our signal and reply immediately, we will receive their message sometime around the year 50,000! This is the down side of interstellar communication. In fact, it can hardly be called communication at all!

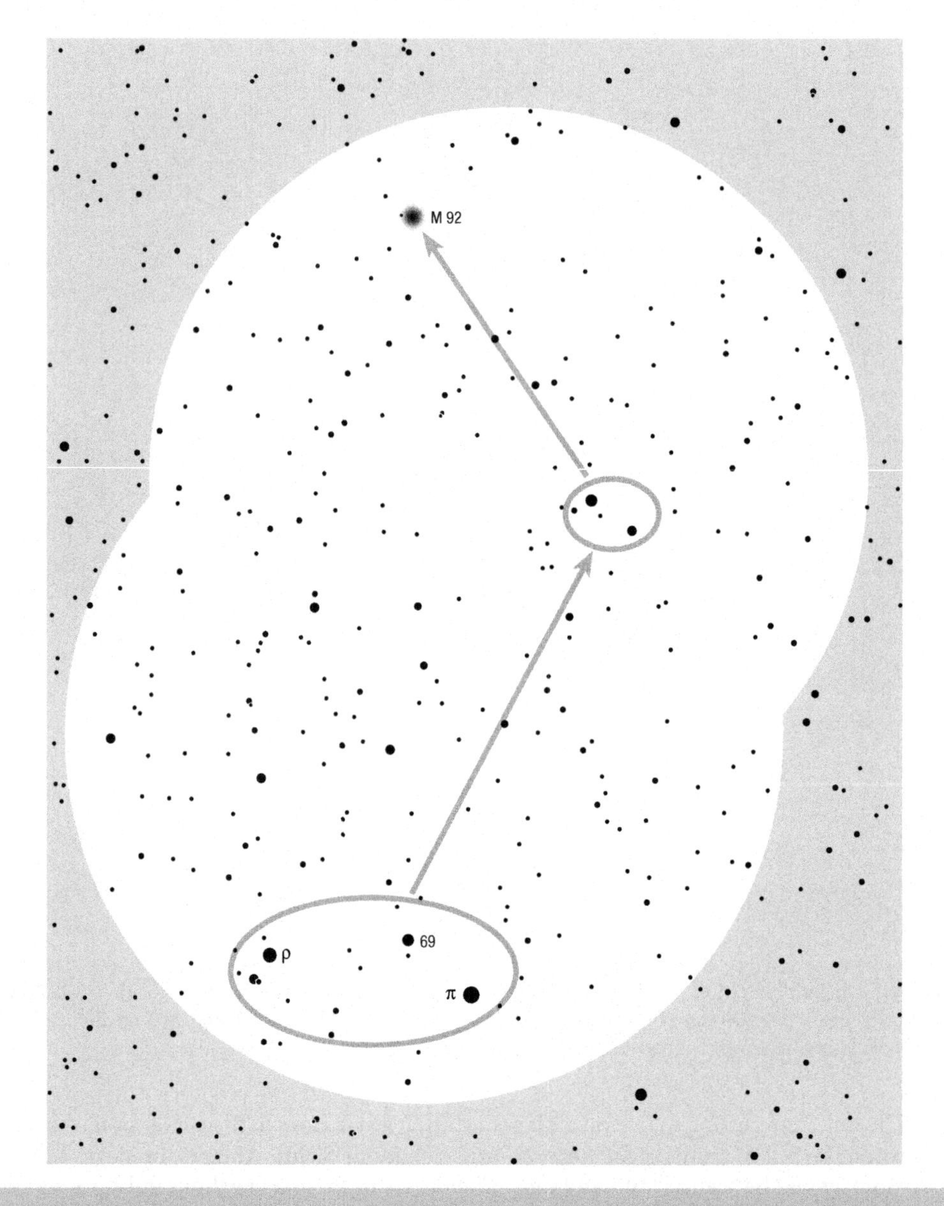

Figure 9.5 The leading star to globular cluster M 92 is Pi, the northeastern star in the Keystone asterism. If the star is moved to the southern edge of the field of view, a magnitude 5 star with no designation, which is considerably brighter than the surrounding stars, appears on the northern edge. Nearby, in the southwest, we can also see a magnitude 6 star. If these stars are moved to the center of the field of view, a hazy spot of light will appear on the northeast; this is globular cluster M 92

Hercules holds another clearly seen globular cluster – M 92 (6m4/14′). It is smaller and slightly fainter than M 13, but it is also on the border of visibility with the naked eye. Unfortunately, it has bad karma, for it is always in the shadow of its bigger and brighter neighbor. If it was in some other constellation, it would most certainly be the prime object.

According to the latest estimates the cluster is 26,700 light years from us and has a diameter of approximately 85 light years. Through binoculars it can be seen as a rather bright spot of light; however, it has a diameter of only a few arc minutes. In your search for it, be sure to use the detailed chart and instructions given in (Figure 9.5).

For all fans of challenges we now present the planetary nebula NGC 6210 (8m8/ 48″ × 8″). It is visible through binoculars, but only as a star-like object. The leading star is the bright Beta. Approximately 4 degrees northeast of it you can spot the magnitude 7 stars, which are clearly seen through binoculars and are in the close vicinity of the nebula. Above them, you should see two faint stars. The one in the southwest is in fact a planetary nebula, which is slightly brighter than the star to the northeast. If you cannot see the stars directly, try using averted vision and they will definitely show. You should proceed with your observations of such faint and elusive objects only once you have accumulated some observation experience.

It is estimated that the planetary nebula is 4,700 light years from us. The central star, the one that generated the nebula, is magnitude 12.5 and cannot be seen through binoculars.

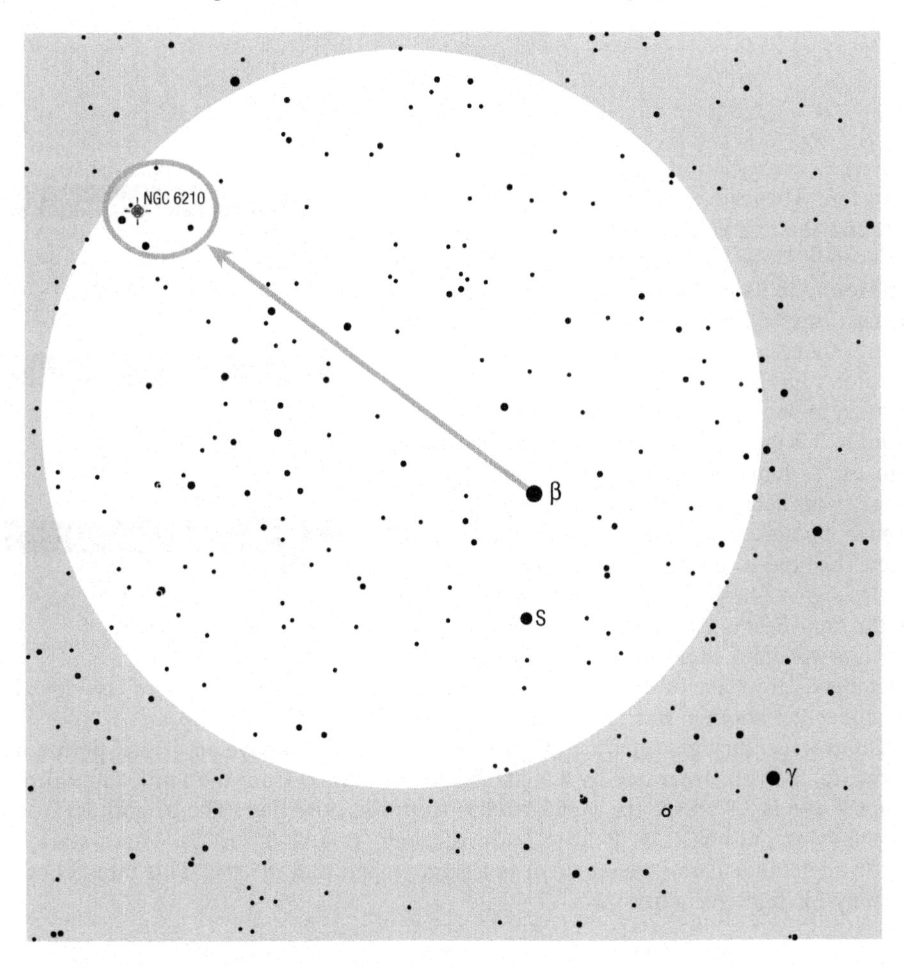

HYDRA (The Water Snake)

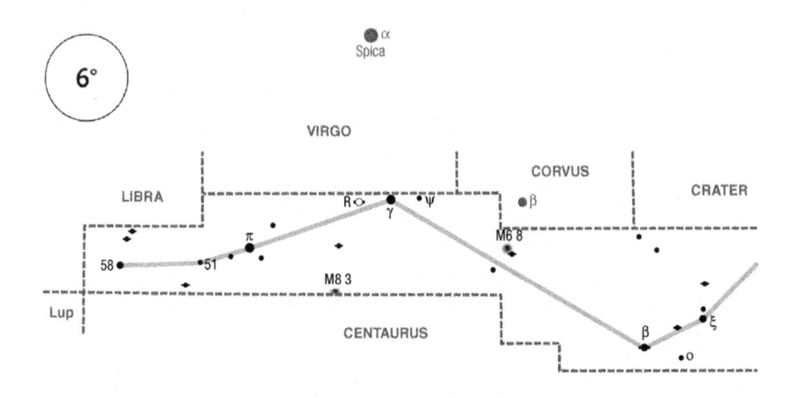

When the largest constellation, Argo Navis, was cut into three, Hydra took its place as largest. Hydra's head can be found beneath Cancer and its tail next to Libra. Despite its size the entire constellation has a single bright star – the orange Alpha (2m0). Because no another bright star can be found in its vicinity, it was given the nickname Alphard, i.e., "the solitary one." In brightness, Alpha is followed by Gamma (3m0), Zeta, Nu (both 3m1), Pi (3m2), Epsilon (3m4), and Xi (3m5).

Alpha is easy to find – the celestial twins (Pollux and Castor) point toward it. It is the 48th brightest star in the sky, 178 light years from us. Its luminosity is 360 times that of the Sun. Through binoculars its beautiful orange color only gets stronger.

The head of the snake lies between Alpha and the constellation Cancer. The head is comprised of the stars Zeta, Epsilon, Delta (4m1), Sigma (4m4), Rho, and Eta (4m3). All stars appear in the same field of view of the binoculars.

TAIL CULMINATES		
mid March at 2 a.m.	mid April at midnight	mid May at 10 p.m.

BODY CULMINATES		
in begin. February at 2 a.m.	in begin. March at midnight	in begin. April at 10 p.m.

HEAD CULMINATES		
in begin. January at 2 a.m.	in begin. February at midnight	in begin. March at 10 p.m.

Epsilon is a splendid multiple star, composed of four visible members. However, on the basis of their movements, astronomers are convinced that the system includes yet another star. The brightest stars with magnitudes 3.8 and 5.3 are the closest together. On average, they are separated by a mere 0.2 arcsec and have an orbital period of only 15 years. In reality, they are separated by 8.5 AU (Saturn is 9.5 AU from the Sun). The third star is of magnitude 6.7 and is 2.8 arcsec (p. a. 303°, 2005) from the close pair. The brightness of the fourth member is only magnitude 12.5. It is 18.1 arcsec away (p. a. 200°, 2005). These stars cannot be separated through binoculars – we see them as a single-magnitude 3.5 star. This interesting system is approximately 135 light years from us.

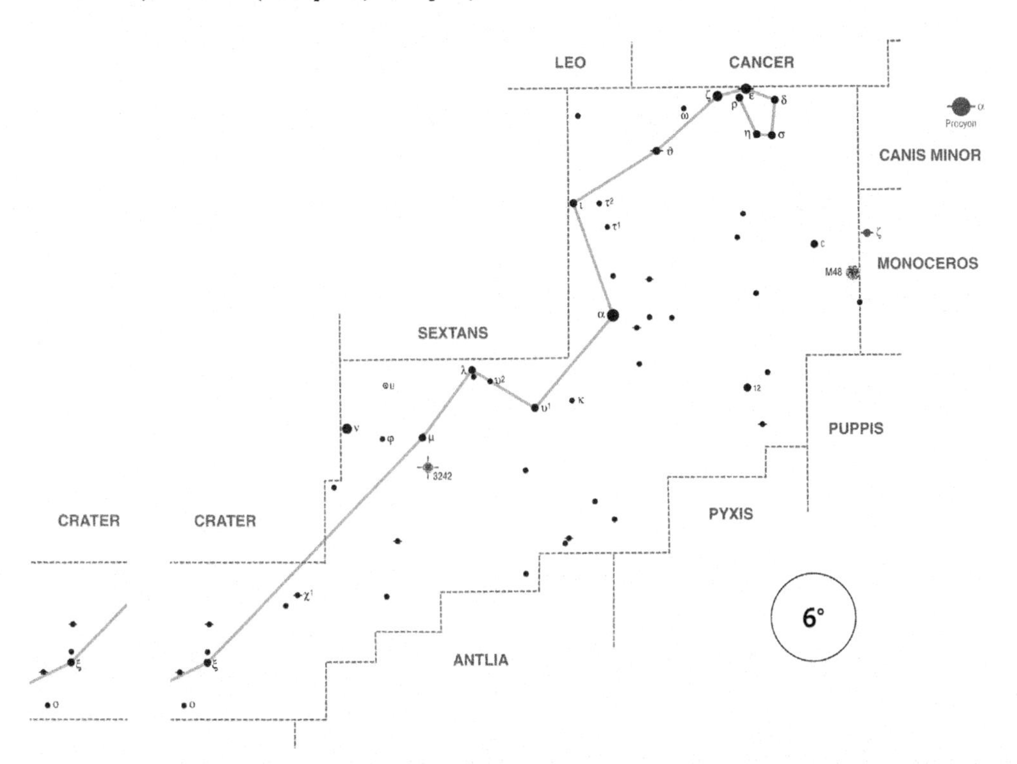

R is a long-period variable star of the Mira type, which changes its brightness between magnitude 3.5 and 10.9 within an average period of 389 days (Figure 9.7). It is found 2.6 degrees from the bright Gamma in the spring part of the constellation. The star can be recognized by its orange color, which becomes increasingly distinctive as the brightness starts to decline. Unfortunately, it is too faint in this part of its period to see its color in binoculars.

The star is especially interesting for astronomers because it exhibits a peculiarity. It is one of the three variable stars (the other two are R Aquilae and R Centauri) with a constantly diminishing period. At the beginning of the eighteenth century it was 500 days, at the beginning of the twentieth century it dropped to 425 days, and nowadays it is 389 days. For us a period of 200 years might seem long, but in the life of a star this is a mere moment. This red giant is most probably in a phase in which its interior is changing extremely fast.

The distance to the star is not truly known. It is thought to be approximately 325 light years. At maximum brightness its luminosity is that of 250 Suns.

The star reached one of its maximums on January 18, 2008. From this date and the known period, one can easily calculate the approximate dates of the next maxima.

Although the constellation Hydra extends over such a vast part of the sky, it houses only a single bright open cluster – M 48 (5m5/54′), which lies in the winter-spring part of the sky, on the border of Monoceros. Charles Messier first saw this cluster in 1771, but he got its position wrong. Later the cluster was given the name NGC 2548. Taking into account its description as well as the fact that there are no similar clusters in the vicinity, astronomers concluded that NGC 2548 is in fact M 48.

With excellent observing conditions the cluster can be seen with the naked eye, but as it lies in an abandoned area, it is rather hard to find with binoculars, let alone with an

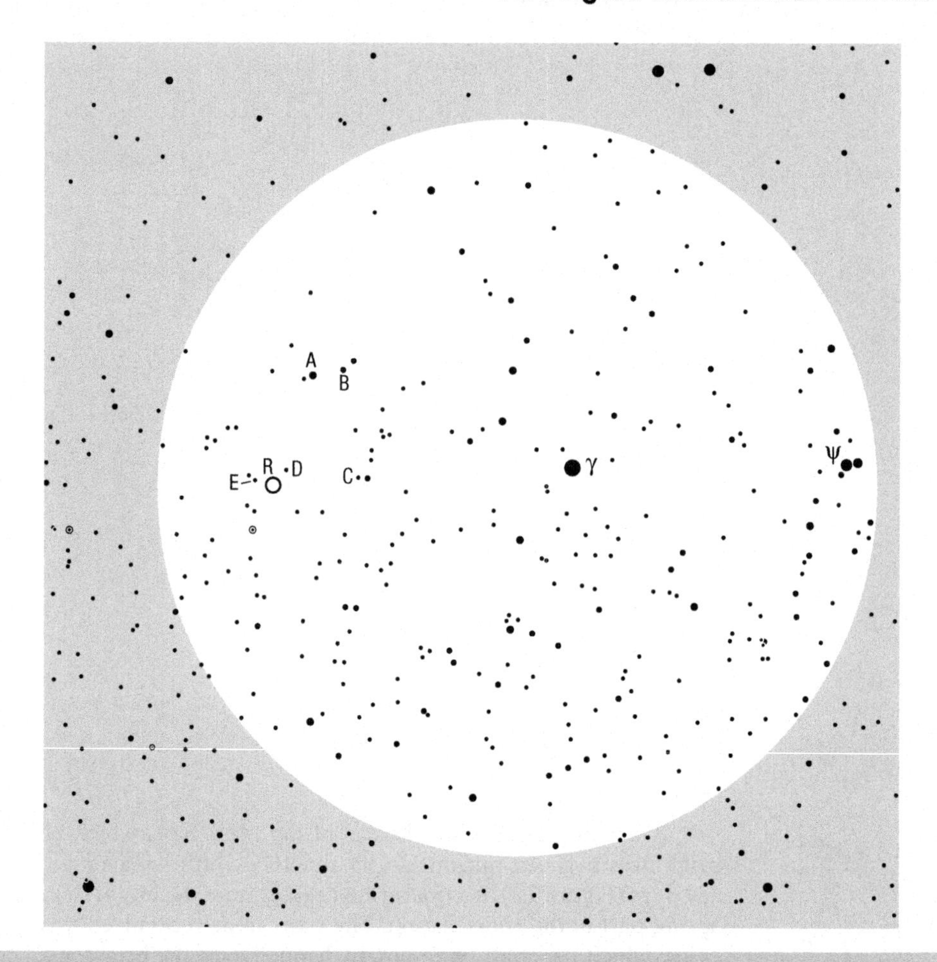

Figure 9.6 The surroundings of the variable R with comparison stars: g (3m0), y (5m0), A (7m3), B (8m0), C (8m5), D (9m0), and E (9m5). The variable star below R is SS Hydrae

average amateur telescope with a small field of view. The stars leading to it are the three stars with designations c, 1, and 2 Hydrae. Star c is of magnitude 3 and is clearly seen with the naked eye. Stars 1 and 2 are of magnitude 6 and can be seen with the naked eye only when there are good observing conditions. All three stars as well as M 48 can be found in the same field of view of binoculars. The cluster is seen as a big and bright spot of light with brighter stars shining from the haze. With averted vision, you can see them more clearly as well as many other ones. The brightest star in the cluster is of magnitude 8.8. At least hundred are brighter than magnitude 11, but we could count as much as 180 if we could see to magnitude 13. In total, there are more than thousand stars in the area.

M 48 is around 1,500 light years from us; thus the real diameter of its denser part is 23 light years.

M 83 (7m6/11′ × 10′) is one of the brightest spiral galaxies in the sky. It is approximately 15 million light years from us. Its real diameter is, therefore, only 45,000 light years, which makes it one of the smallest galaxies of its type. It is situated on the border of Centaurus, approximately 18 degrees south of Spica, so it lies in the spring part of the constellation. When searching for it with binoculars start your hunt at Pi (magnitude 3 star), which is clearly visible with the naked eye. From Pi

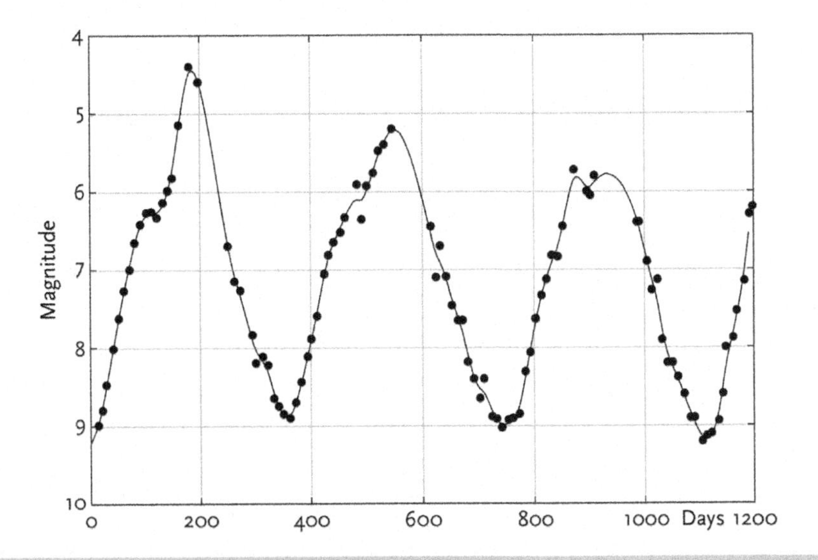

Figure 9.7 Light curve for variable R

follow the line of three stars, the first of which is magnitude 5 (the others are 1 magnitude fainter). Through binoculars the galaxy can be seen as a faint cloud of light, a few arc minutes in diameter. Of course, we can only see the brightest central part of the galaxy. Its spiral arms become visible only on long-exposure images and might be glimpsed in larger amateur telescopes when there are excellent observing conditions.

The globular cluster M 68 (7m8/11′) lies halfway between Beta and Gamma. It has a nearby magnitude 5 double star (seen in binoculars as one). This is also the only star in the vicinity that can be seen with the naked eye. The leading star is Beta Corvi, which appears in the same field of view of binoculars as the cluster. The cluster is seen merely as a faint spot of light. Some stars on the edge of the cluster can be separated

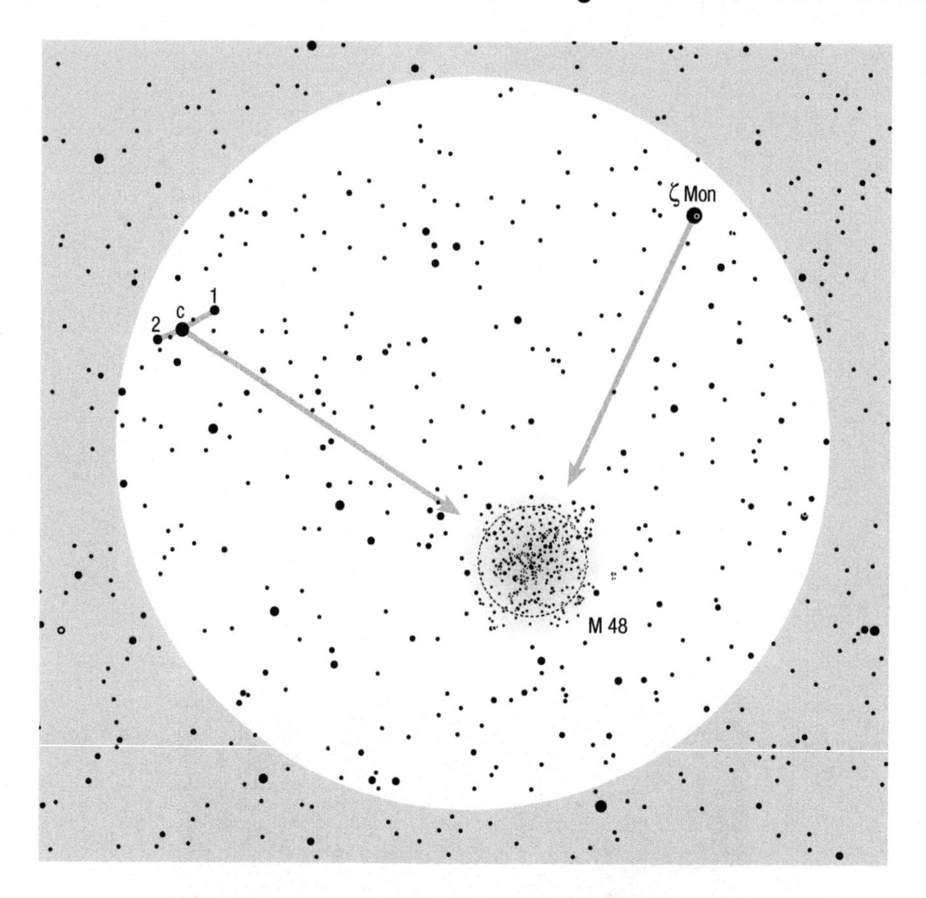

Figure 9.8 You can start your hunt for the open cluster M 48 at Zeta Monocerotis, which is of magnitude 4, or from the triplet of stars with designations c, 1 and 2, as described in the text

through telescopes with at least a 15-cm objective lens, but its real beauty is seen in larger telescopes. With 100,000 stars, M 68 is a very rich cluster. It is approximately 33,000 light years from us, and its real diameter is 106 light years.

The planetary nebula NGC 3242 (7m7/45″ × 36″) is visible through binoculars, but only as a blurred, slightly bluish magnitude 8 star. It lies roughly 2 degrees south of Mu, which is a magnitude 4 star and therefore clearly visible with the naked eye. For its exact position you can look at the detailed chart. The distance to the nebula nicknamed the Ghost of Jupiter is estimated to be approximately 2,500 light years. A simple calculation shows that its actual diameter is approximately half a light year.

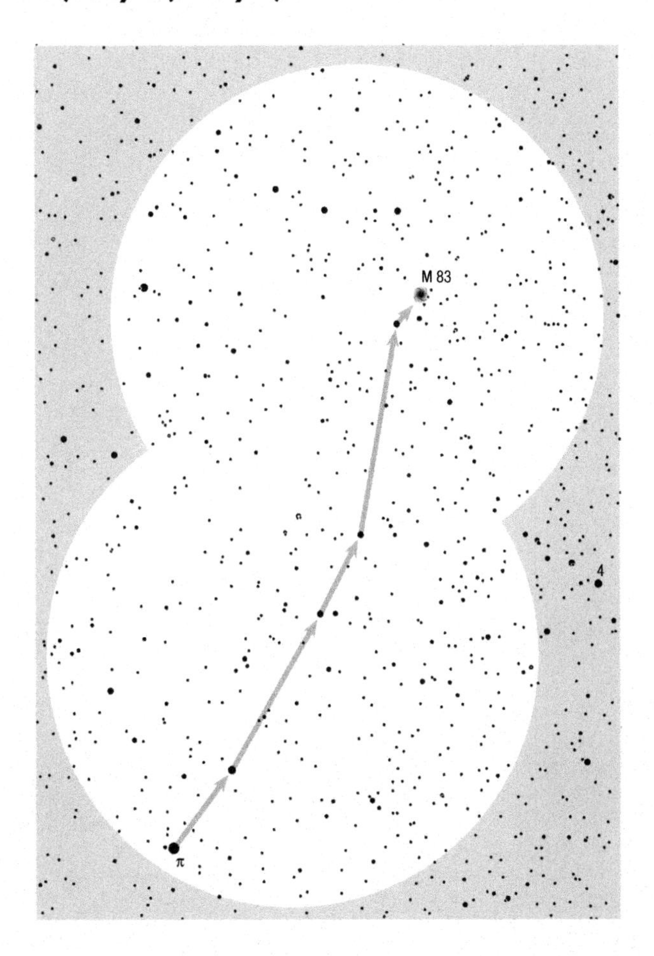
M 83
4
π

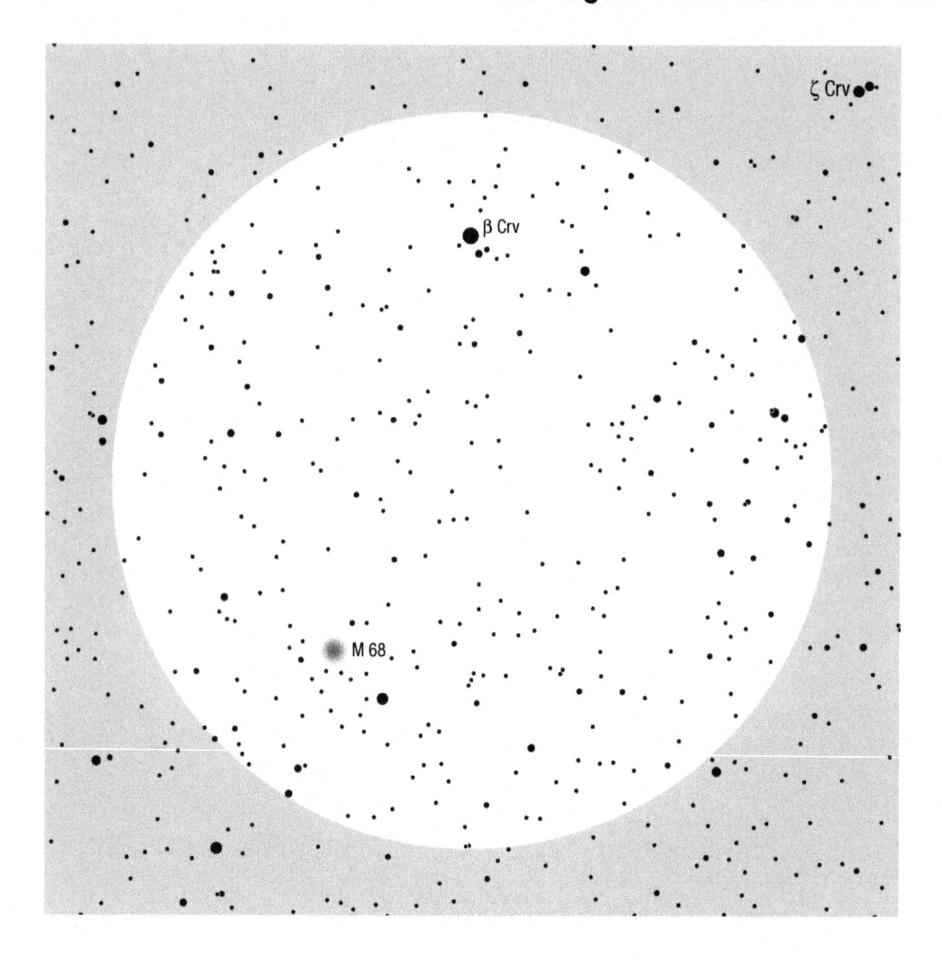
ζ Crv
β Crv
M 68

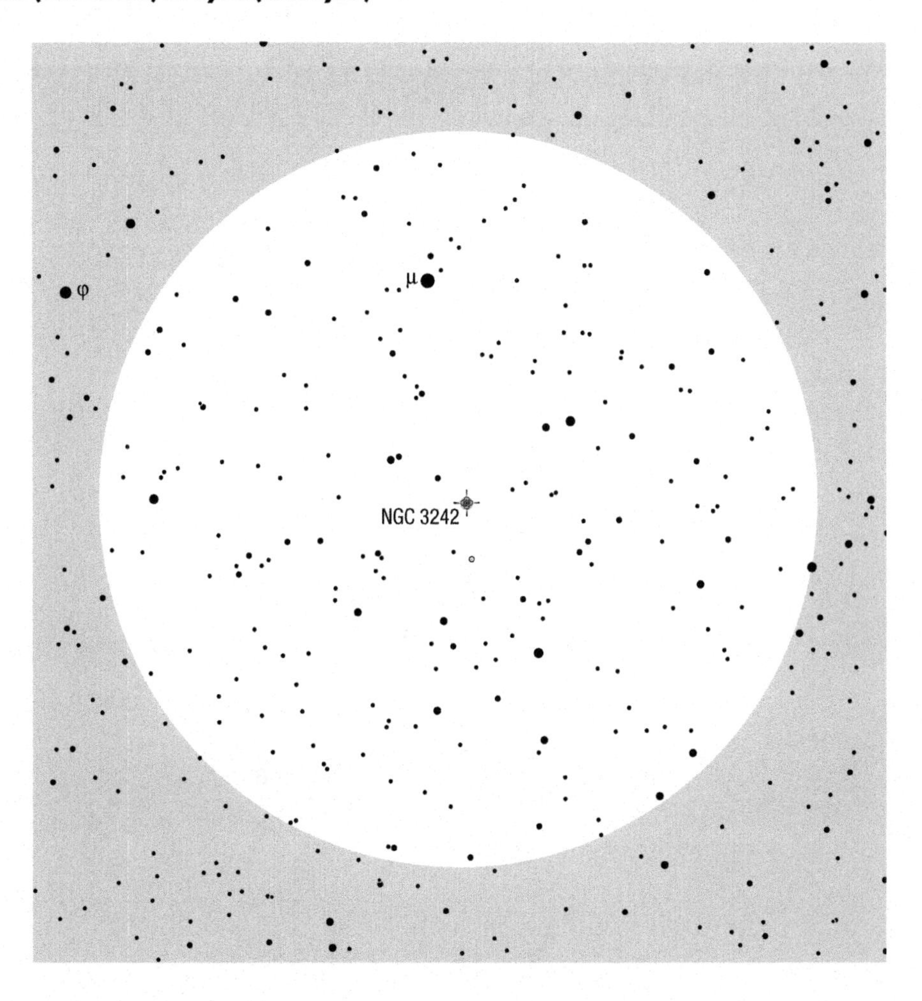
φ
μ
NGC 3242

LACERTA (The Lizard)

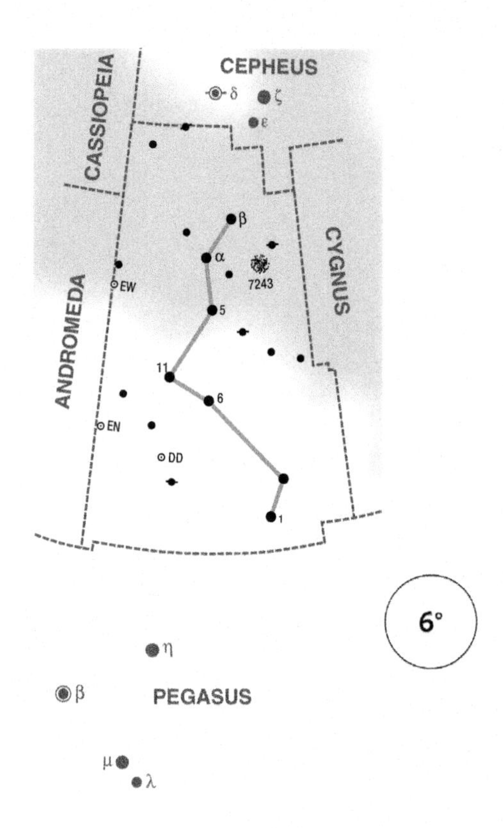

Under Cepheus, east of Cygnus and north of Pegasus, we find the small and faint constellation Lacerta. Its brightest stars shine with only a magnitude of 4 and are hidden by the rich and dense background of the Milky Way stars. You will have to try pretty hard to find this constellation. The easiest way to recognize it is to use Mu and Eta Pegasi as the leading stars.

CONSTELLATION CULMINATES		
end July at 2 a.m.	end August at midnight	end September at 10 p.m.

In the part of the Milky Way that streams through the constellation, there are quite a few open clusters. The brightest among these is NGC 7243 (6m4/21′), which contains approximately 40 stars, but only the brightest five (with magnitudes 8 and 9) can be seen through binoculars. In perfect observing conditions and if the cluster is close to culmination (for mid-northern observers it is almost at the zenith), we can see approximately 15 members. Alpha and Beta are the stars leading to the cluster. But as we mentioned before, you should search for it and observe it only when there are excellent observing conditions. NGC 7243 is approximately 2,600 light years from Earth.

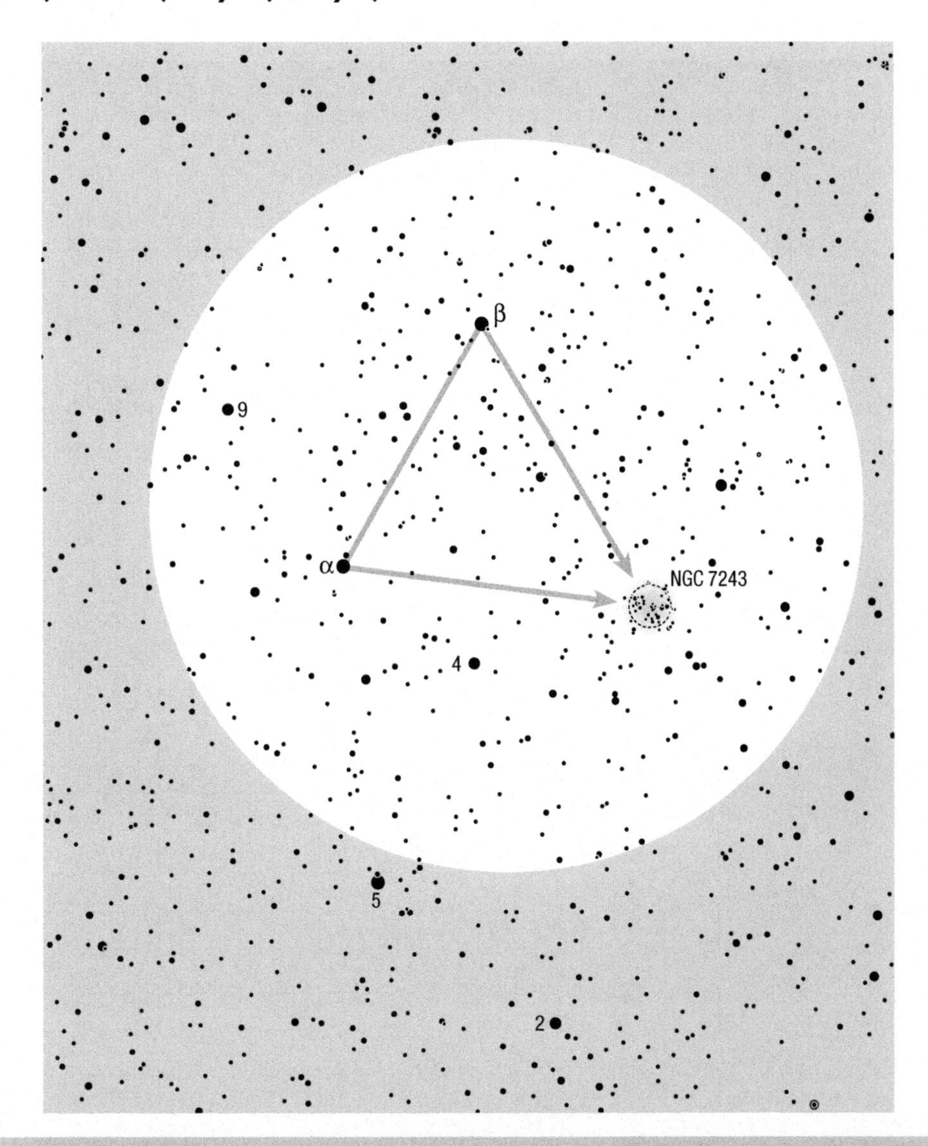

Figure 9.9 The stars leading to the open cluster NGC 7243 are Alpha and Beta. Together with the cluster they form an isosceles triangle. Since this part of the constellation lies on the Milky Way, the entire area is rich with faint stars. With excellent observing conditions, the cluster is easily recognized; however, in slightly poorer conditions the five stars can easily become lost in the crowds of stars in the Milky Way

LEO (The Lion)

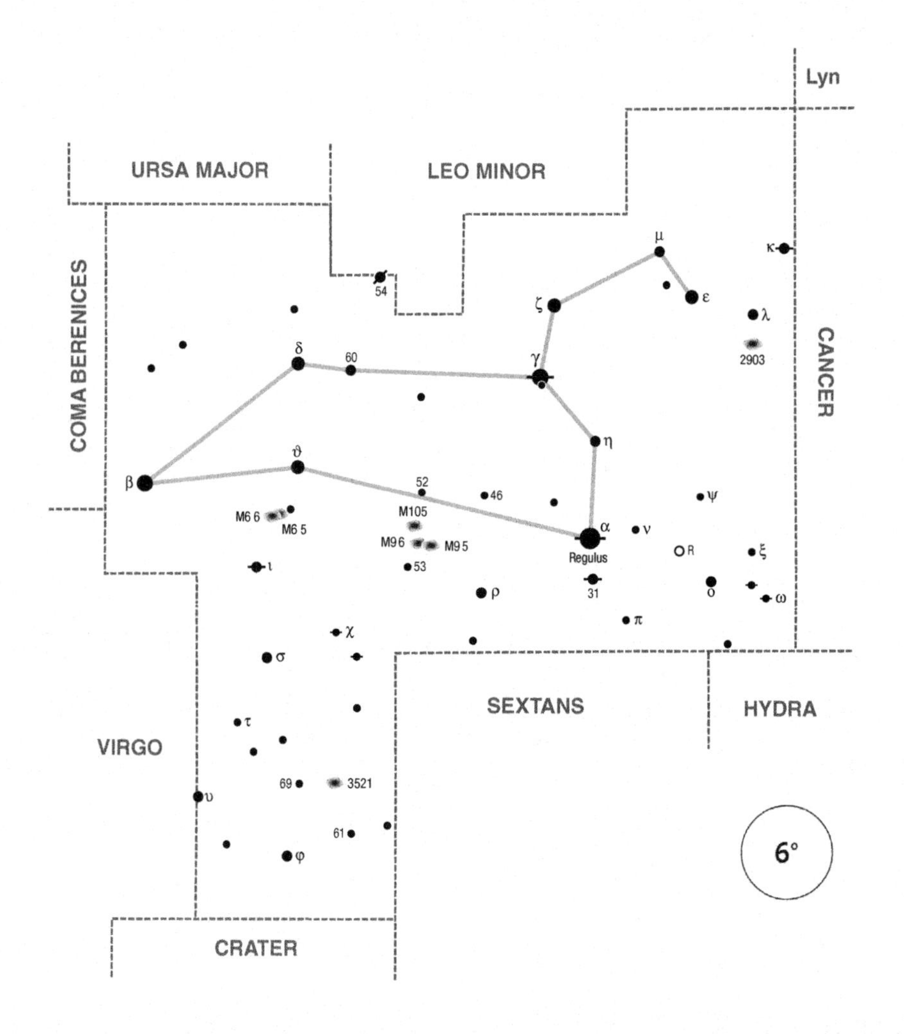

CONSTELLATION CULMINATES		
end January at 2 a.m.	end February at midnight	end March at 10 p.m.

Leo is one of the most beautiful and striking constellations of the zodiac. You really do not need to have much imagination to see the resting king of animals in the pattern of stars. The brightest are Alpha or Regulus (1m4), Gamma (2m0), Beta or Denebola (2m1), Delta (2m6), Epsilon (3m0), Theta (3m3), Zeta (3m4), Eta, and Omicron (both 3m5). Stars Zeta, Mu (4m1), Epsilon, and Lambda (4m3) together with several fainter stars represent the head of Leo. Gama represents its chest, and the rest of the constellation is the clearly visible triangle comprised of Beta, Delta, and Theta, which form the body and hind legs.

Regulus is the 21st brightest star in the sky. It lies close to the ecliptic, so it finds itself in conjunction with the planets or the Moon on occasion. It is a hot star, with a surface temperature of 13,000 K. Its diameter is 5 times that of the Sun, and its luminosity is that of 120 Suns. Regulus is 78 light years from Earth. It is an interesting triple star. The brightest companion is a distinct yellow-orange magnitude 8.2 star, which is 176 arcsec away (p. a. 308°, 2000). The stars are seen and split through binoculars. However, we will find it hard to see the fainter one. The fainter star is a double in itself. The components are of magnitude 8.1 and 13.1, and they are separated by 2.5 arcsec (p. a. 86°, 1943). Because of the low brightness of the fainter star and the proximity of the bright Regulus they are a difficult pair even for larger amateur telescopes.

Beta or Denebola is our close neighbor, a mere 36.2 light years from us. It is the 62nd brightest star in the sky. It is similar to our Sun in its size, but it has a much higher surface temperature (8,500 K), and therefore its luminosity is 13 times higher. If the Sun were to put alongside Denebola, it would be seen as a magnitude 5 star.

Gamma, or Algieba, is the 49th brightest star in the sky. This is truly one of the most beautiful double stars. It consists of two giants – one is orange and the other is yellow. Their brightness is of magnitude 2.4 and 3.6, and they are currently 4.7 arcsec apart (p. a. 127°, 2006). The stars will reach their maximum separation (5 arcsec) in approximately the year 2100. The orbital period is 620 years. They cannot be separated through binoculars. Gamma is 126 light years from us.

Delta is the 97th brightest star in the sky. It is 58 light years away; its luminosity surpasses our Sun by 23 times.

Five degrees west of Regulus we can see two magnitude 6 stars: 18 Leonis (5m6) and 19 Leonis (6m3). Close to the southern star (19) lies R Leonis, a long-time variable star of the Mira type. It has a period of 310 days. The star has a distinct orange-reddish color. At its peak, its brightness reaches a magnitude of 4.4 and can clearly be seen with the naked eye, while at its minimum it slips as far down as magnitude 11.3. The star reached one of its maximums on August 23, 2008. From this date and the known period, one can easily calculate the dates of the next maxima (Figure 9.10).

The star is approximately 600 light years from us, and at its brightest its luminosity is 250 times that of the Sun.

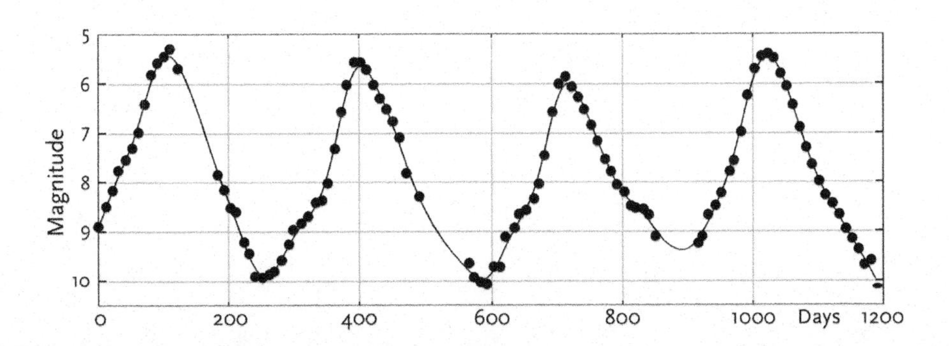

Figure 9.10 The graph of brightness changes for variable R Leonis

This constellation is full of galaxies, some of which can be seen through binoculars. With excellent observing conditions M 65 (9m3/10′ × 3′3) and M 66 (9m0/8′7 × 4′4) are seen as tenuous, small spots of light. They are in the same field of view of binoculars with Theta and Iota (Figure 9.12); the galaxies are in the middle, between the two bright stars. M 66 is brighter and bigger, unmistakably elongated, with a magnitude 8 star nearby in the west. The galaxy is visible in good observing conditions. M 65 is also

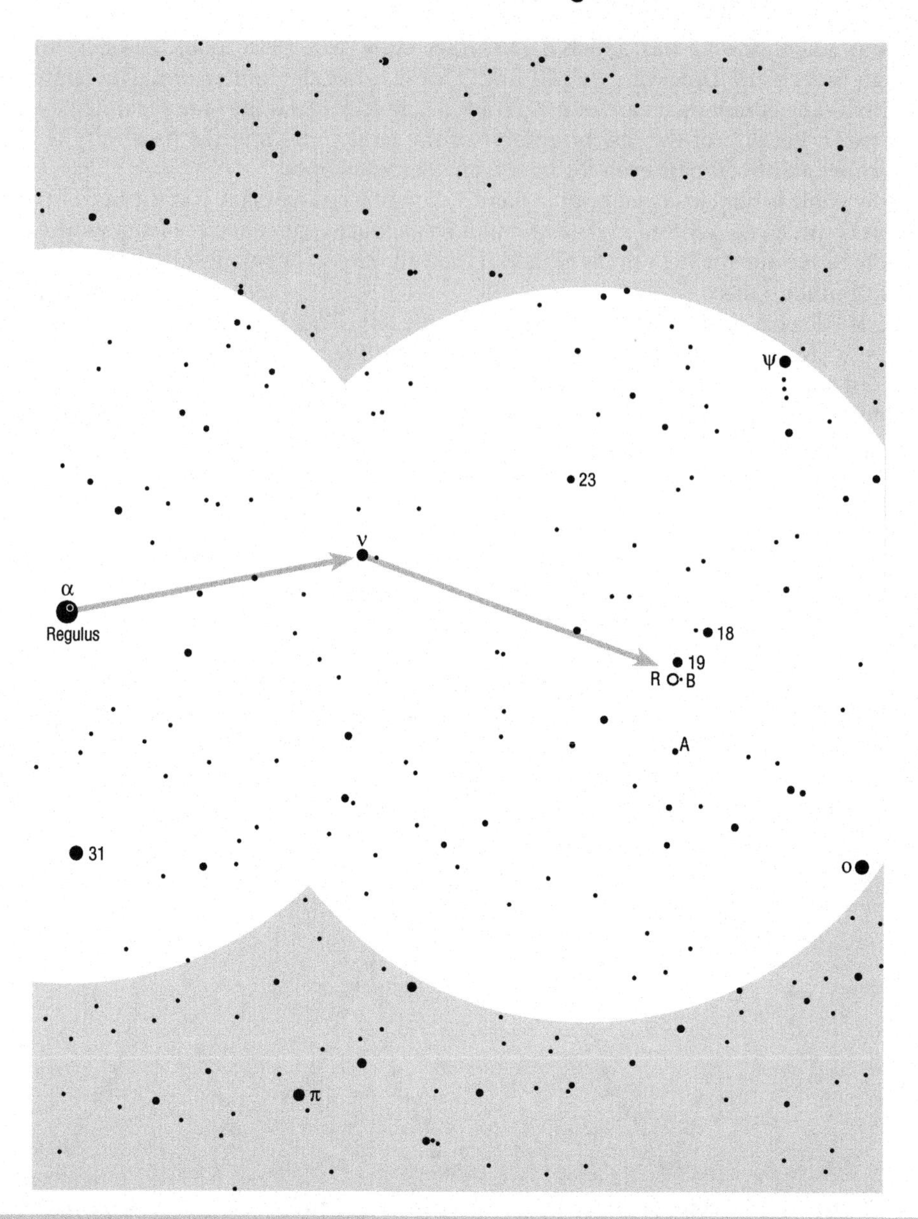

Figure 9.11 The star field between Regulus and variable R. The comparison stars are 18 (5m6), 19 (6m3), A (7m5), and B (9m0)

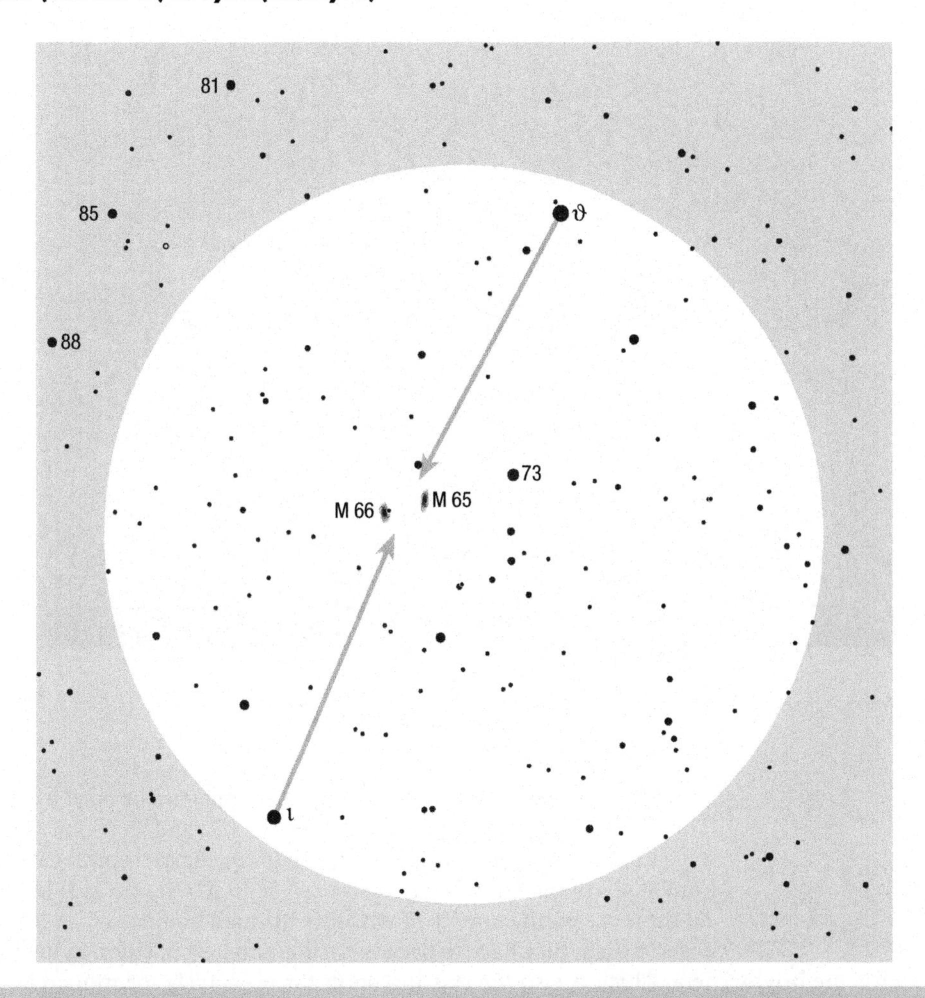

Figure 9.12 Theta and Iota are the stars leading to the galaxies M 66 and M 65. Between both bright stars in close with galaxies lies 73 Leonis, a star of magnitude 5, which is good intermediate checking point

elongated, but it is on the border of visibility through binoculars. In order to observe it you have to choose a really clear and dark night. When observing this pair through binoculars you can experience how the thickness of our atmosphere influences the visibility of celestial objects. When Leo is rather high above the horizon, you can see M 66. When the constellation is approaching culmination, its fainter neighbor M 65 shows up, but it disappears as soon as Leo starts to descend to the west. This is a fact worth remembering.

Both galaxies are spiral, but of course their shape and structure cannot be seen through binoculars or even through larger telescopes. They are 21 arcsec apart and could be called space neighbors. They are 35 million light years from us and 200,000 light years from each other. The actual size of M 65 is 100,000 light years, while M 66 is slightly smaller – 90,000 light years. Looking at the sky from a planet in galaxy M 66 an alien observer might see the neighboring galaxy as a fantastic, 28 degrees big celestial object!

Because of their brightness and apparent size, the galaxies are also suitable for amateur astrophotography (Figure 9.13).

Figure 9.13 Spiral galaxies M 65 (*right*) and M 66

When there are excellent observing conditions, two or maybe even three faint clouds of light can be seen between stars 52 and 53 Leonis. These are the elliptical galaxy M 105 (9m3/4′5 × 4′0), the barred spiral galaxy M 95 (9m7/4′4 × 3′3), and the spiral galaxy M 96 (9m2/7′1 × 5′1) (Figure 9.14). All three are on the border of visibility through binoculars, so in order to observe them they have to be close to their highest position in the sky. The leading star is the bright Theta, which is also the starting point for previously mentioned M 65 and M 66. From Theta you have to move approximately 7 degrees west and slightly to the south. In the field of view, you can see the magnitude 5 star that has been designated 52 Leonis. The area is easily recognized by the trapezoid, formed by the four stars 52, 53, 46, and Rho, which are found at the extreme edges of the field of view of binoculars. The galaxies lie between stars 52 and 53; however, for

Figure 9.14 The elliptical M 105 (*right*), the lenticular NGC 3384, and the spiral galaxy NGC 3389. The last two are not seen in binoculars

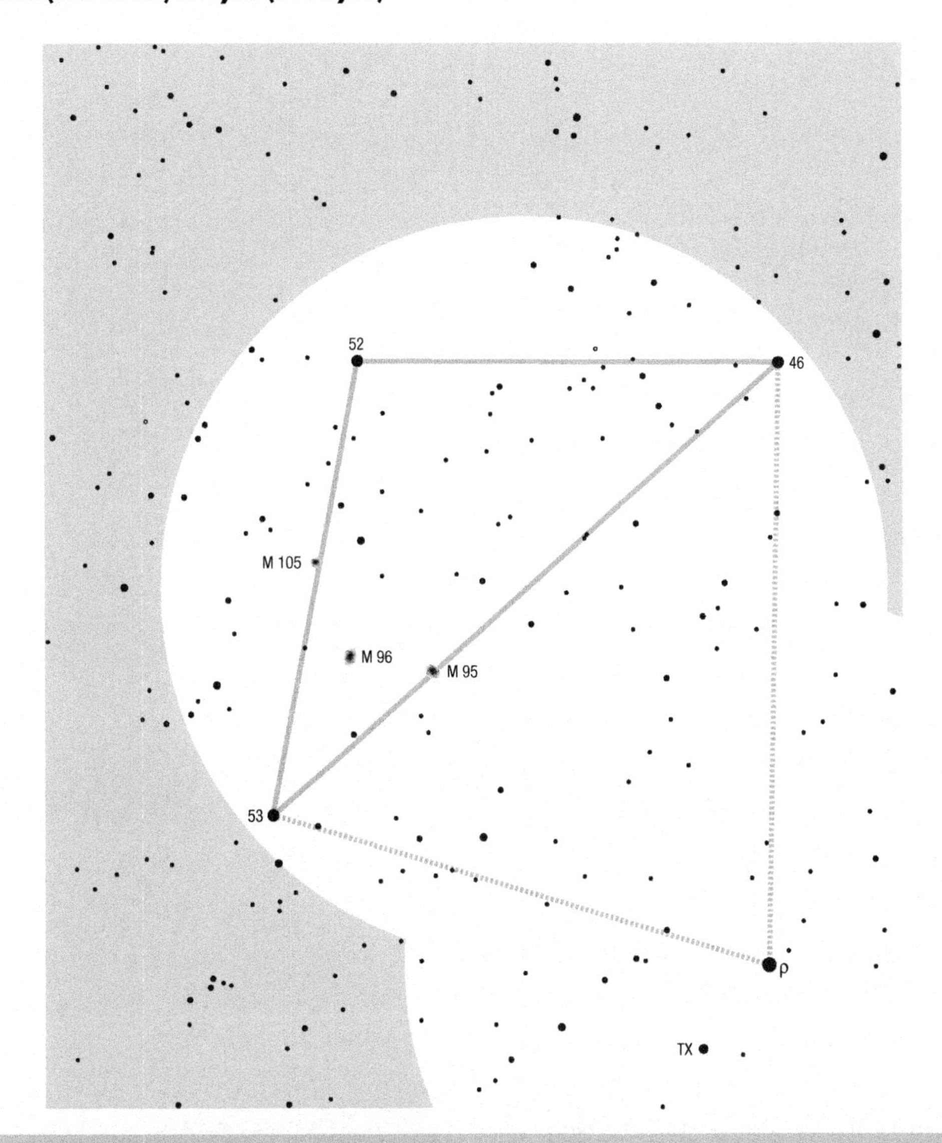

Figure 9.15 The trapezoid, formed by stars 46, 52, 53, and Rho, is clearly seen through binoculars. Between stars 52 and 53 you should notice the galaxy M 105 first, but if observing conditions allow two additional faint clouds of light can also be seen – M 95 and M 96

their exact position use the detailed chart (Figure 9.15). It is most likely that M 105 and M 96 will be the ones you glimpse first. M 95, which is a half a magnitude fainter, is seldom seen through binoculars – and only when there are perfect observing conditions. Your eyes have to be completely adapted to night vision, and most likely you will have to use averted vision. Observing these galaxies is suitable only for skilled observers.

All three are members of a smaller group, named Leo I (Leo + Roman numeral 1), which is approximately 38 million light years from us. The actual diameter of M 95 and M 96 is approximately 100,000 light years. The elliptical M 105 is smaller, with an estimated diameter of 60,000 light years.

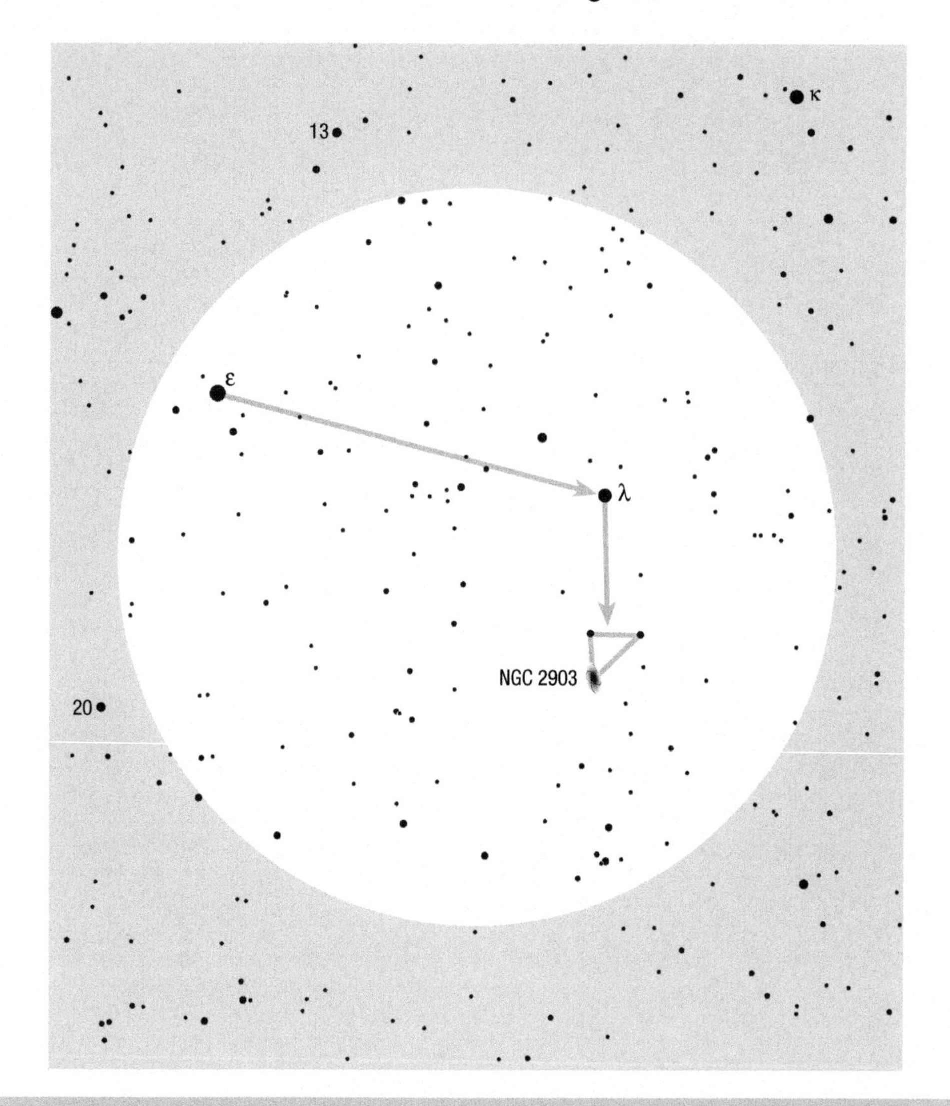

Figure 9.16 The striking spiral galaxy NGC 2903

Slightly more than 1 degree south from the bright Lambda, the spiral galaxy NGC 2903 (8m9/12′ × 6′6) awaits us. It is seen face on. Despite its promising integral brightness this is apparently a large galaxy; thus, its surface brightness is rather small. In order to observe it, we have to choose a night with excellent observing conditions. And even then it is seen merely as a faint, few arc minutes large spot of light. The exact positioning of the galaxy is rather easy, since it forms a right-angled triangle with two magnitude 7 stars, as clearly seen in the detailed chart (Figure 9.16).

NGC 2903 is approximately 20.5 million light years from us; its real diameter reaches approximately 75,000 light years (Figure 9.17).

Figure 9.17 The surroundings of the spiral galaxy NGC 2903 with the leading star Lambda

For all observers who like a real challenge, let's introduce another galaxy – the faint spiral NGC 3521 (8m9/9′5 × 5′0), which is seen edge on. Through binoculars it appears as a tenuous, few arc minutes long stripe of light, which requires excellent observing conditions to be seen. It is recommended that you search and observe it only when the galaxy is close to culmination. The leading star is Phi, which lies on the extreme southern part of the constellation. In order to locate its exact position you should recognize the trapezoid composed of stars Phi, 61, 62, and 69, clearly seen through binoculars. The galaxy lies between star 62 (which has a nearby star of magnitude 8) and a faint magnitude 8 star less than one degree east of it. As soon as you are convinced that the area with the galaxy is in the center of the field of view, use all of the tricks and techniques that you have picked up along the way. Your eyes should be completely adapted to night vision, and use averted vision. Despite the fact that the galaxy is as much as a whole magnitude brighter than M 65, it is much harder to spot, since it lies 13 degrees lower south and is always seen through a thicker layer of atmosphere.

NGC 3521 is approximately 35 million light years from us. Its diameter is almost 100,000 light years.

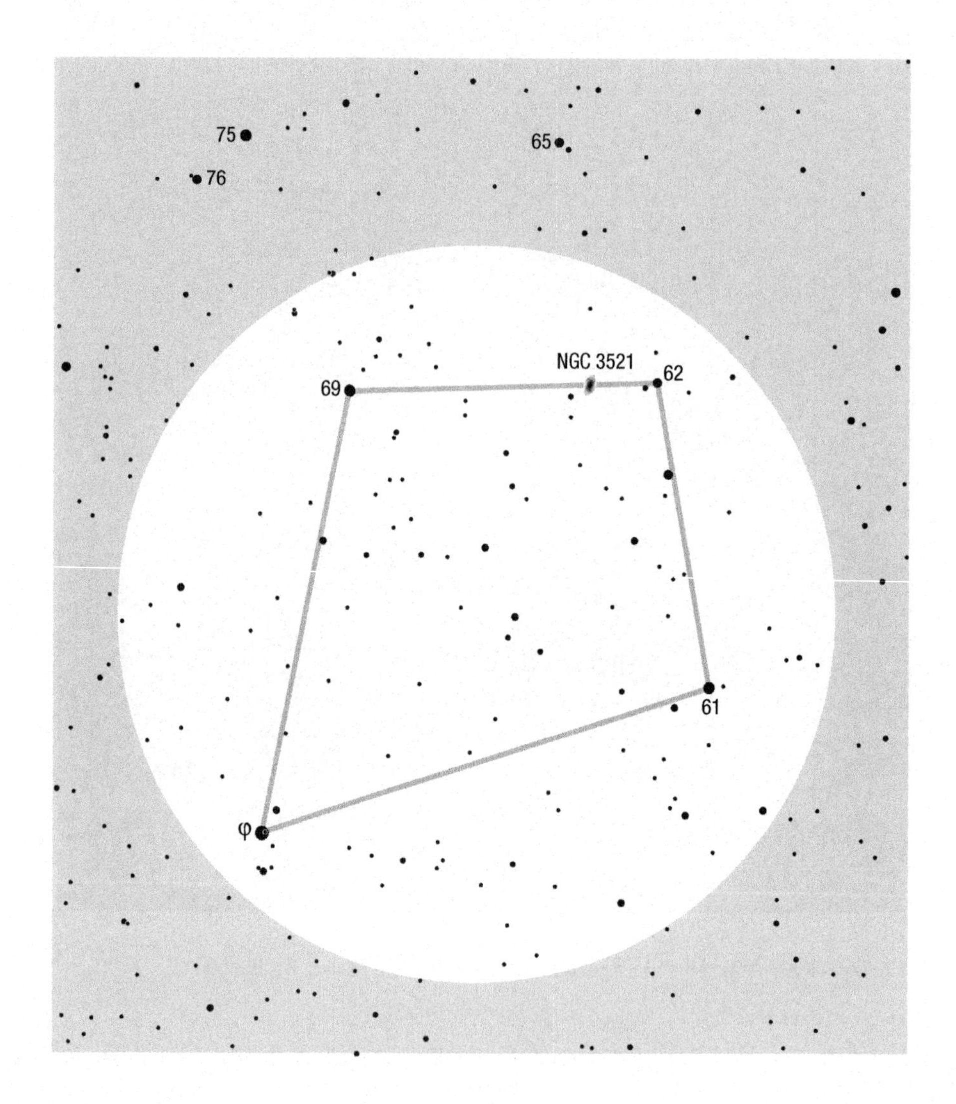

LEO MINOR (The Little Lion)

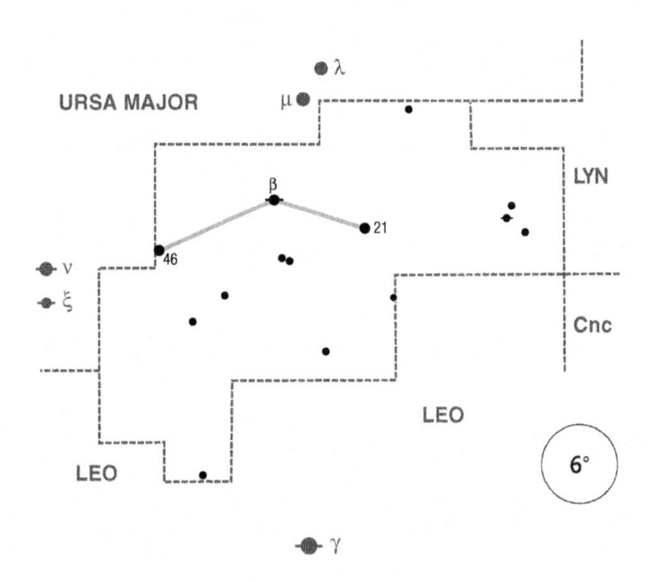

CONSTELLATION CULMINATES		
end January at 2 a.m.	end February at midnight	end March at 10 p.m.

Leo Minor lies to the southwest of Ursa Major and to the north of Leo. It is so small and so faint that some astronomers doubt it is right to call it a constellation. It was introduced by the Polish astronomer Jan Hevelius in the book *Prodromus Astronomiae*, published in 1690. The brightest stars are 46 (3m8) and the yellowish Beta (4m2), which is a close double star separated by 0.4 arcsec and with an orbital period of 38 years.

LEPUS (The Hare)

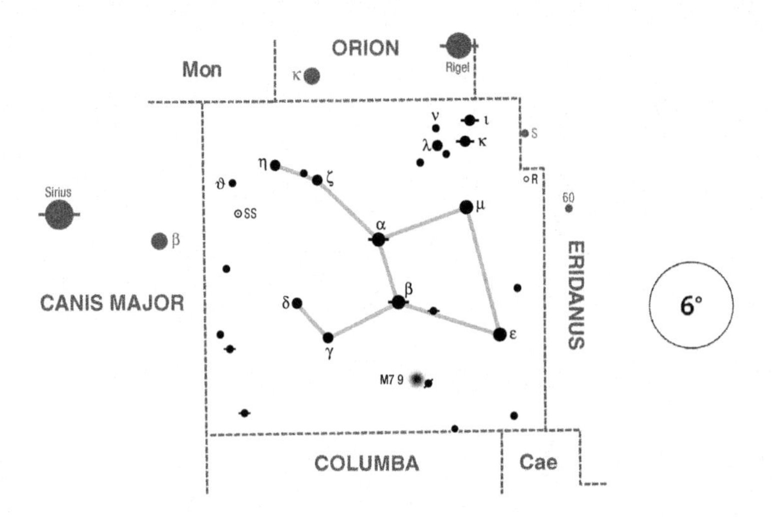

CONSTELLATION CULMINATES		
mid November at 2 a.m.	mid December at midnight	mid January at 10 p.m.

Lepus, located beneath Orion, is a small but interesting constellation, even though it is shaded by the winter sky. The brightest stars are Alpha (2m6), Beta (2m8), Epsilon (3m2), Mu (3m3), and Zeta (3m5).

Alpha is the 99th brightest star in the sky. It is a real giant, which is why it appears so bright in our sky despite its great distance of 1,300 light years. Its luminosity is as much as 11,000 times that of the Sun. The star has two faint companions. One is of magnitude 11.2 and is 35.6 arcsec away (p. a. 157°, 1999), while the other is magnitude 11.9 and is 91.2 arcsec from Alpha (p. a. 186°, 1999). Neither can be seen through binoculars.

The yellow Beta is the 133rd brightest star in the sky. It is 160 light years away from us. Its luminosity surpasses our Sun by 140 times.

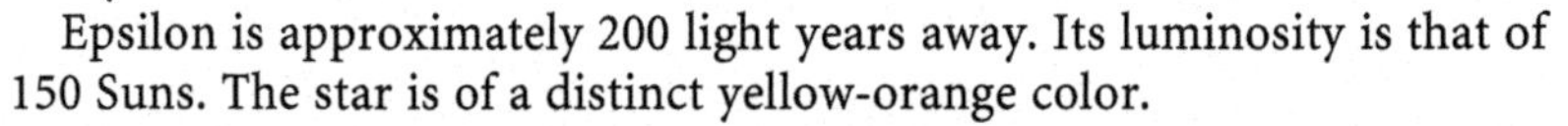

Epsilon is approximately 200 light years away. Its luminosity is that of 150 Suns. The star is of a distinct yellow-orange color.

Gamma is a double with wonderful color contrast. The stars with magnitudes 3.6 and 6.3 are separated by 97 arcsec (p. a. 350°, 2002), so they are easily split through binoculars. The colors of the stars are described differently by different observers. Most of them see the brighter (Gamma A) as white or yellowish and the fainter one (Gamma B) as orange, reddish, or even greenish. This cute couple is only 29 light years from us. The actual separation of the stars is 900 AU – enough to put 15 Solar Systems between them! The fainter star is also a double. Its companion (Gamma C) can be seen through middle-sized amateur telescopes; but, it is too faint (11m0) to be seen through binoculars. They are separated by 112.5 arcsec (p. a. 8°, 1999). While Gamma A and B are gravitationally bound and therefore represent a true binary, Gamma C lies in the same direction by mere coincidence.

R Leporis is a long-period variable star of the Mira type with a period of 427 days. At minimum, its brightness is approximately magnitude 11.7, so within every cycle it cannot be seen through binoculars for a certain period of time. At its maximum, it shines with magnitude 6.7, but occasionally it can even reach magnitude 5.9. On these occasions it can be seen in the dark sky even with the naked eye. R Leporis is a red giant, as are most variables of this type. The star is known for its strong orange-reddish color, which is clearly seen through binoculars during its maximum brightness, and this is why it was nicknamed Hind's Crimson Star. (It was named after the British astronomer J. R. Hind, who

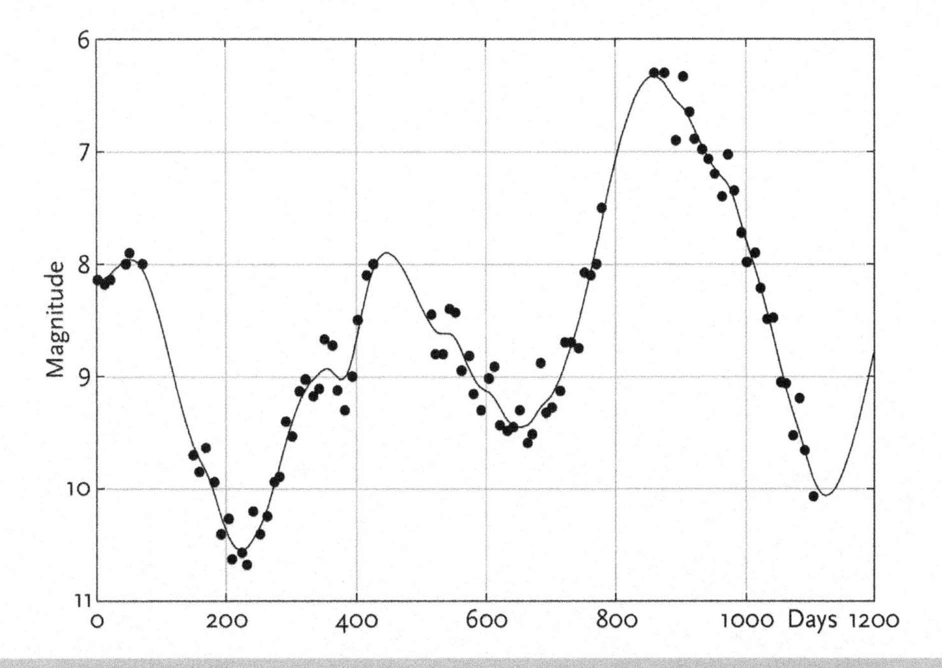

Figure 9.18 Graph of brightness changes for variable R Leporis

Figure 9.19 These images clearly show that R Leporis is of a really strong reddish color. Both images were taken on the same day. The *left* one depicts the star photographed through a red filter, and the *right* one was taken with a blue filter

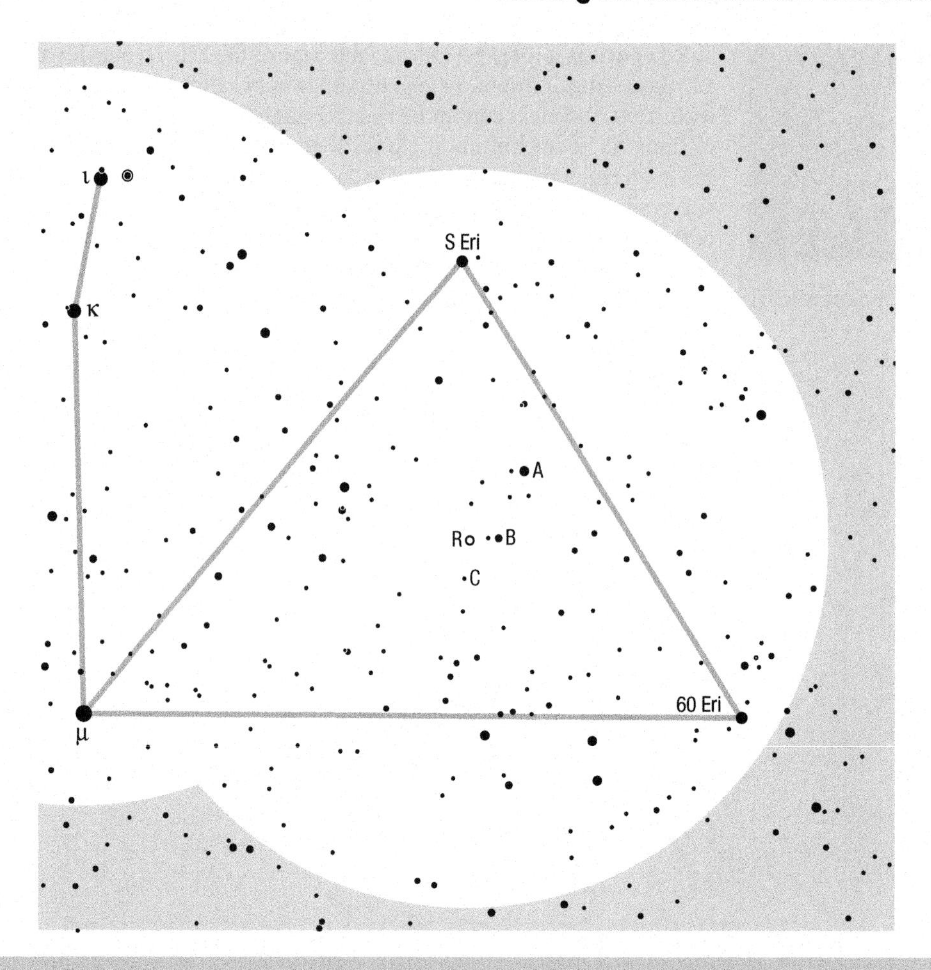

Figure 9.20 The surroundings of the variable star R Leporis with comparison stars: A (5m9), B (7m5), and C (9m1)

was the first to observe it, in 1845.) Numerous reliable observers confirm that this is the reddest star from all that can be seen through amateur telescopes. If you have the chance, look at it! Its intense reddish color is linked to its low surface temperature, which is a mere 2,500 K.

The star reached one of its maximums on December 28, 2007. From that date and the known period, one can easily calculate the dates of the next maxima. Although the color of the star is intense and is easily recognizable through binoculars, it lies in a rather empty area of the sky. The stars leading to it are the bright Mu and Kappa Leporis as well as the magnitude 5 stars labeled S and 60 Eridani.

The globular cluster M 79 (7m7/9′6) is rich with stars but rather condensed; thus, it makes a lasting impression only on observers who view it through larger amateur telescopes. It is on the border of visibility when viewed through binoculars. Seen as a few arc minutes big faint spot of light, it is easy to find, as it is in the same field of view with Beta, if the star is placed on its extreme northern edge. The bright Epsilon can also offer a helping hand in the orientation. As the cluster lies rather low above the southern horizon for mid-northern observers, you have to choose a clear, moonless night with excellent observing conditions if you want to see it.

The globular cluster M 79 is 42,100 light years from Earth. Its actual diameter is 110 light years.

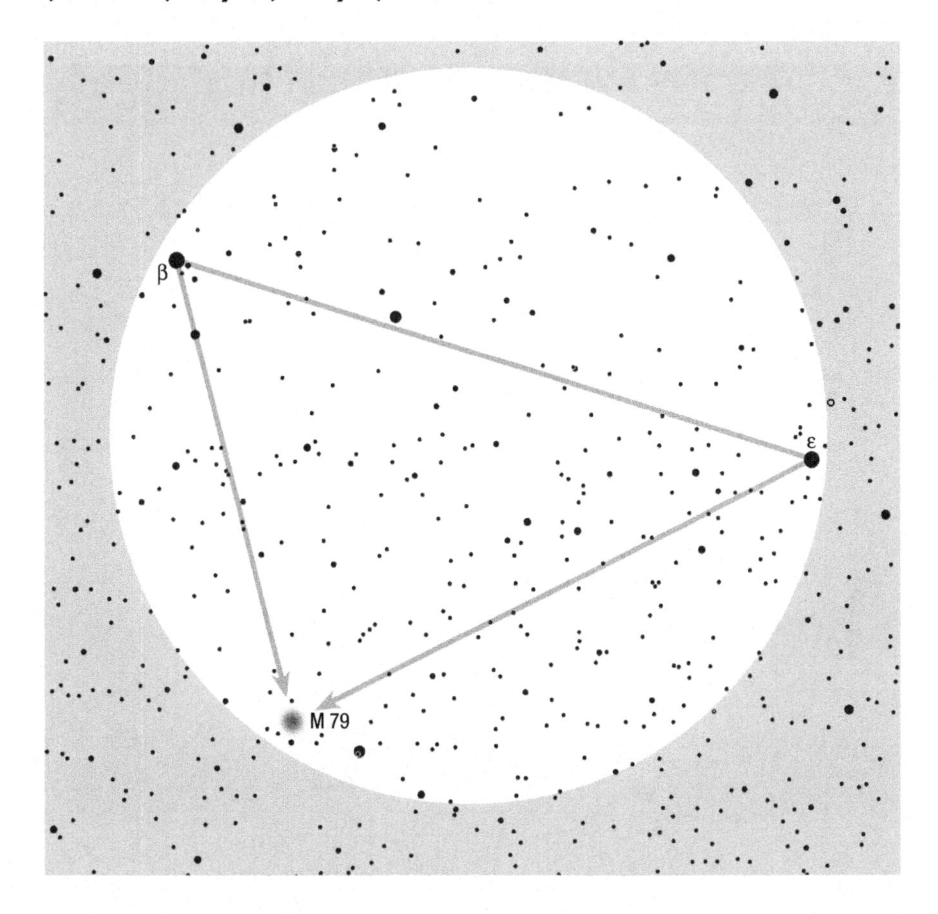
β
ε
M 79

5'

LIBRA (The Scales)

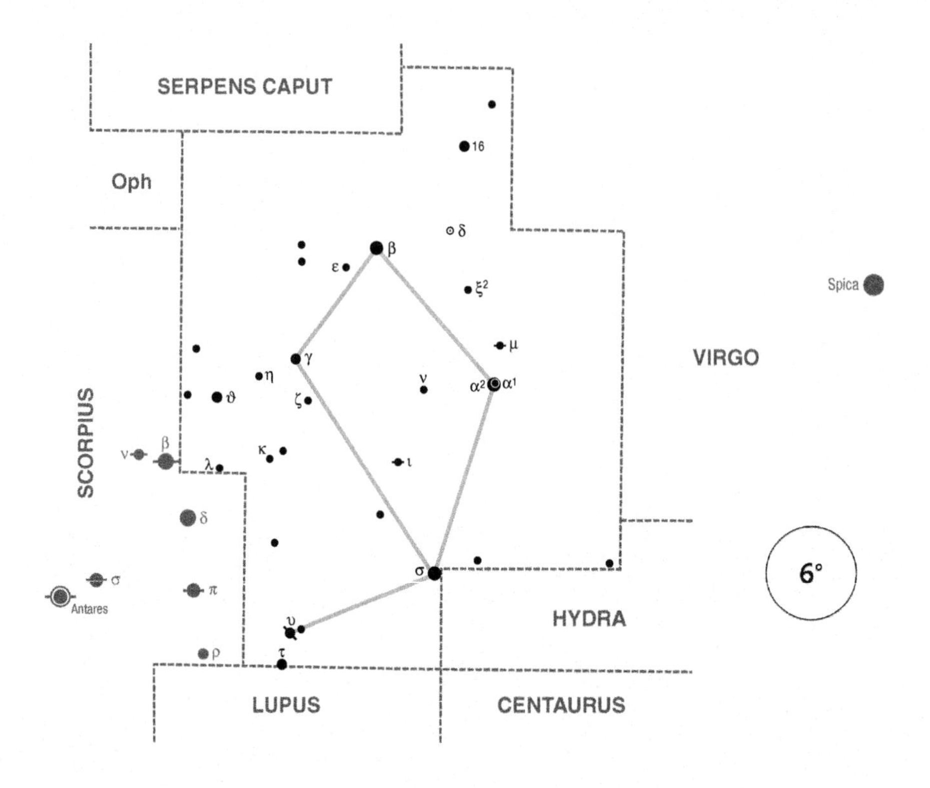

CONSTELLATION CULMINATES		
in begin. April at 2 a.m.	in begin. May at midnight	in begin. June at 10 p.m.

Libra lies below Serpens Caput and east of Virgo. It is the least distinguished constellation of the zodiac. Except for the fact that every now and then a planet strolls through it, there are no hidden treasures to be found within it – even for observers with large amateur telescopes.

The brightest stars are Beta (2m6), Alpha (2m7), and the orange Sigma (3m2). The older name for this part of the sky was the Scorpion's Claws, and it used to be an asterism of the constellation Scorpius. The brightest stars still have the Arabic names of Zubenelshemali (Northern Claws) and Zubenelgenubi (Southern Claws).

Beta is the 102nd brightest star in the sky. Its light travels 160 years to reach us. The luminosity of the star is 560 times that of the Sun. This is a star that has intrigued astronomers since ancient times. Eratosthenes described it as the brightest star in the area of Scorpius and Libra (which at the time was considered to be a single constellation). A few centuries later Ptolemy wrote that Beta Librae and Antares (Alpha Scorpii) shine with the same brightness. Today Antares (1m1) is much brighter than Beta. Because we can rest assured that the two great observers from ancient times could not make such a great mistake, there are two possible explanations: either the brightness of Antares increased or Beta's brightness decreased. Both options are rather unusual, for the change would have to appear – in astronomical terms – over a relatively short period of time. A similar problem was raised in relation to Gamma Canis Majoris. Beta is also interesting for its color. This is the only star seen with the naked eye that most observers describe as greenish in color, even though its spectrum indicates that it should be white. Check it out!

Alpha is the 122nd brightest star in the sky. Its luminosity is 34 times that of the Sun. In fact, it is a double star, 77 light years from us. The two stars with magnitudes 2.7 and 5.2 are 3.8 arcmin apart (p. a. 315°, 2002) and can be separated even with the naked eye but are more clearly seen through binoculars.

Delta is an eclipsing variable star, the same type as Algol in Perseus. Its brightness changes from magnitude 4.8 to 5.9 with a period of 2.3273543 days. The brightness falls for approximately 6 h. During this time, the larger but fainter star of the pair covers the brighter one. The stars are a mere 7 million km apart (the distance between Mercury and the Sun is 58 million km). The fainter star is not really that faint, as its luminosity is three times that of the Sun. The luminosity of the brighter star surpasses our Sun by 46 times. The stars are approximately the same size (3.4 and 3.7 times that of the Sun), while their masses are 2.7 and 1.2 times that of the Sun. This binary is approximately 200 light years from us.

The star reached one of its minimums at 19:00 UT on July 19, 2008. From these data and the known period, one can easily calculate the approximate dates of the next minima. As the Delta approaches minimum you should observe it as frequently as possible (every few hours) and compare its brightness to that of the surrounding stars. At its minimum, Delta is somewhat fainter than Mu (5m6), which lies approximately 6 degrees south-southwest. Xi–2, which is in between, is magnitude 5.5.

LUPUS (The Wolf) and NORMA (The Set Square)

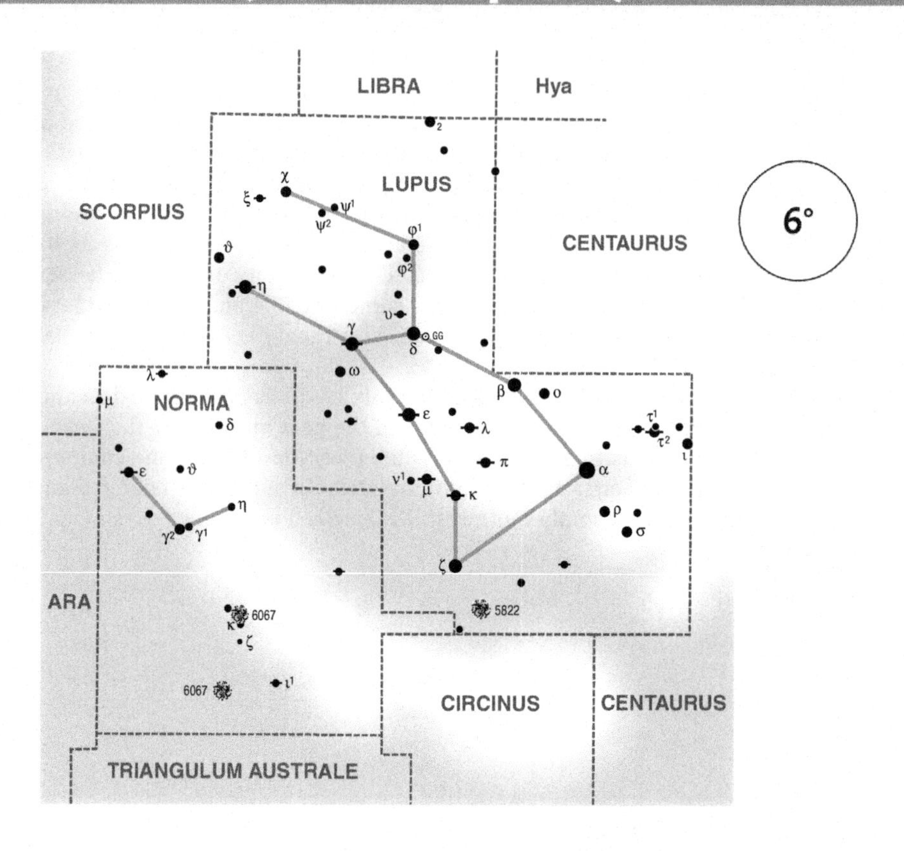

CONSTELLATIONS CULMINATE		
mid April at 2 a.m.	mid May at midnight	mid June at 10 p.m.

The constellation Lupus lies on the edge of the summer Milky Way, situated below Libra and between Centaurus and Scorpius. Observers in the mid-northern latitudes can only see its northern part, and even that is low above the southern horizon and therefore poorly seen. The brightest stars are Alpha (2m3), Beta (2m7), Gamma (2m8), and Delta (3m2).

Alpha is the 78th brightest star in the sky. It is 548 light years from us. Its luminosity is as much as 18,000 times that of the Sun, while its diameter is 10 times as large as that of the Sun, so the star is a true giant. It is also extremely massive, with between 10 and 11 solar masses. It is a Beta Cephei variable with a period of just over 6 h and 14 min. However, its brightness varies by only 0.03 magnitude – not enough to be noticed with average amateur equipment.

Beta is the 110th brightest star in the sky. Its light travels 525 years to reach us. The luminosity of the star surpasses our Sun by over 1,700 times. However, in mid-northern latitudes this star rises a degree above the horizon at culmination; thus, it is always seen through the thick layer of our atmosphere, which dims its light. This is a star for southern observers.

Gamma is the 131st brightest star in the sky. This close double is 570 light years from us. The common luminosity is 1,750 times that of the Sun.

Delta is also a giant star. Its luminosity surpasses our Sun by 2,000 times. This star is approximately 680 light years from us. For mid-northern observers, it is always low above the southern horizon. At culmination it peeks at a mere 3 degrees.

There are several interesting double stars in the constellation; however, they are more suitable for southern observers, who see Lupus higher up in the sky. We should mention Kappa Lupi, which is composed of stars with magnitudes 3.8 and 5.5, separated by 26 arcsec (p. a. 143°, 2002). When higher in the sky, this pair can be separated through binoculars.

Another object visible through binoculars is the open cluster NGC 5822 (6m5/40′), which lies inside the Milky Way. It is rich in stars, but the brightest are only of magnitude 9, so this is a cluster for southern observers.

There is not a lot we can say about Norma. It is interesting that the constellation has no Alpha or Beta. The brightest star is Gamma-2, which is only a magnitude 4.0. Together with Gamma-1 they form an easy double, separated with the naked eye. Norma is barely visible for mid-northern observers, as its northern part rises only 4 degrees above the horizon at culmination.

In more southern latitudes, observers with binoculars can see two objects of interest. They can separate Epsilon into stars of magnitudes 4.5 and 6.1, separated by 22.8 arcsec (p. a. 335°, 2002). They can also enjoy the bright but rather small open cluster NGC 6087 (5m4/12′), which is about 3,500 light years away and contains approximately 40 stars of magnitudes between 7 and 11. Its brightest member is the well-known Cepheid variable S Normae, which changes in brightness from magnitude 6.1 to 6.8 in a period of 9.75411 days. This star is located in the center of the cluster.

LYNX (The Lynx)

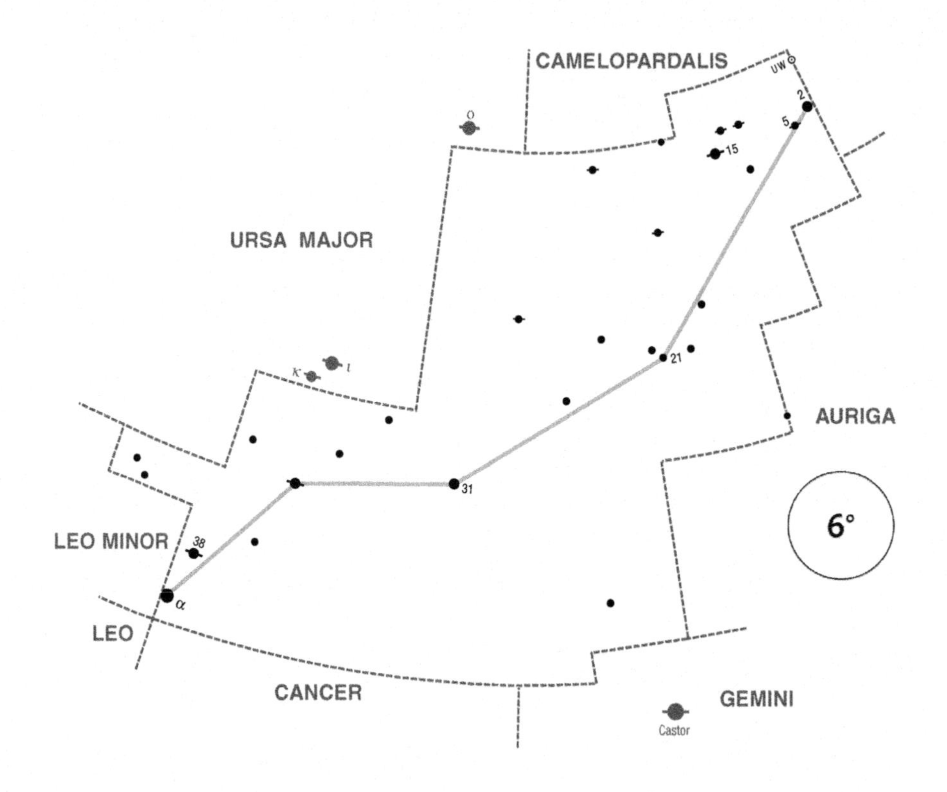

CONSTELLATION CULMINATES		
end December at 2 a.m.	end January at midnight	end February at 10 p.m.

The large and elongated, yet completely undistinguished, constellation Lynx can be found between Ursa Major, Auriga, and Gemini. The brightest stars are the orange Alpha (3m1) and 38 (3m8), which appear in the same field of view in binoculars, and 31 (4m2).

It is easy to find Alpha, since the star forms a rather large equilateral triangle together with Regulus (Alpha Leonis) and Pollux (Beta Geminorum). Alpha Lyncis is 155 light years from us. Its luminosity is 110 times that of the Sun.

Lynx is home to the famous globular cluster NGC 2419, which is nicknamed the Intergalactic Wanderer. This is one of the remotest (known today) globular clusters in the halo of our galaxy. It is 182,000 light years from us and over 210,000 light years from the galactic center. This real astronomical distance is comparable to the distances of the Magellan clouds (our satellite galaxies) and might indicate that the cluster is an independent intergalactic object. In spite of the enormous distance it is seen as a faint spot of magnitude 11, when viewed through amateur telescopes. Unfortunately, it cannot be seen through binoculars.

LYRA (The Lyre)

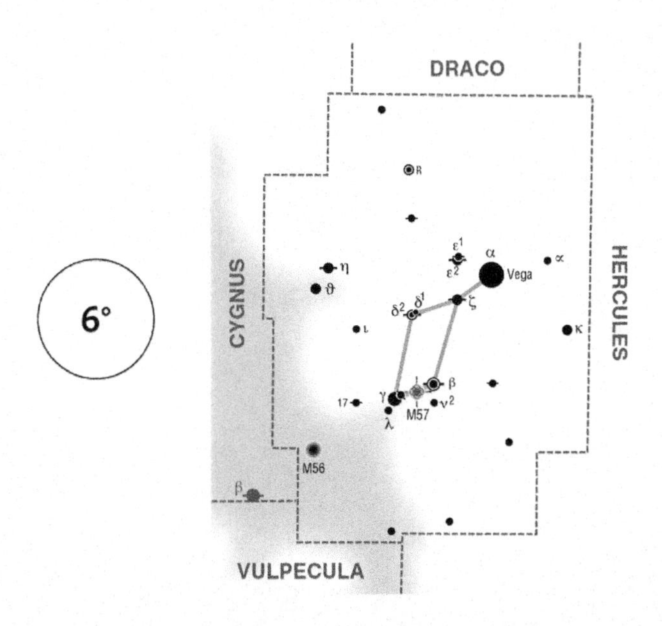

CONSTELLATION CULMINATES		
in begin. June at 2 a.m.	in begin. July at midnight	in begin. August at 10 p.m.

In summer evenings, mid-northern observers can find Lyra almost in the zenith. This small yet prominent constellation continues to fascinate us, for it contains truly interesting objects. But their true beauty is unfortunately seen only through telescopes.

The constellation is governed by Vega, a bright white-blue magnitude 0.03 star, which is the first to appear in the summer sky. Its name has nothing in common with the famous mathematician Jurij Vega but has its origins in the Arabic name Al Vaki, which means "the falling eagle." Vega is the 5th brightest star in the sky and the 3rd brightest seen from mid-northern latitudes (Figure 9.21). Only Sirius and Arcturus are brighter. It is rather close to us, at a mere 25.3 light years away, which makes it in space terms practically our next-door neighbor. Its luminosity is 45 times that of the Sun, and its surface temperature is approximately 9,200 K. Together with two other bright stars (Alpha Cygni and Alpha Aquilae), it forms the famous asterism known as the Summer Triangle, which cannot be missed in the summer night sky.

The small rectangle south of Vega consists of Gamma (3m2), Zeta, Delta (both 4m3), and the eclipsing variable Beta (3m2 to 4m4).

Astrophysicists consider Beta to be one of the most interesting celestial objects. However, an observer using the naked eye or binoculars must be satisfied with observing only its changes in brightness. The stars in this eclipsing pair are so close together that they almost touch each other. Due to their proximity, the two stars are greatly deformed. From Earth the orbit is seen at such an angle that the stars cover each other during their

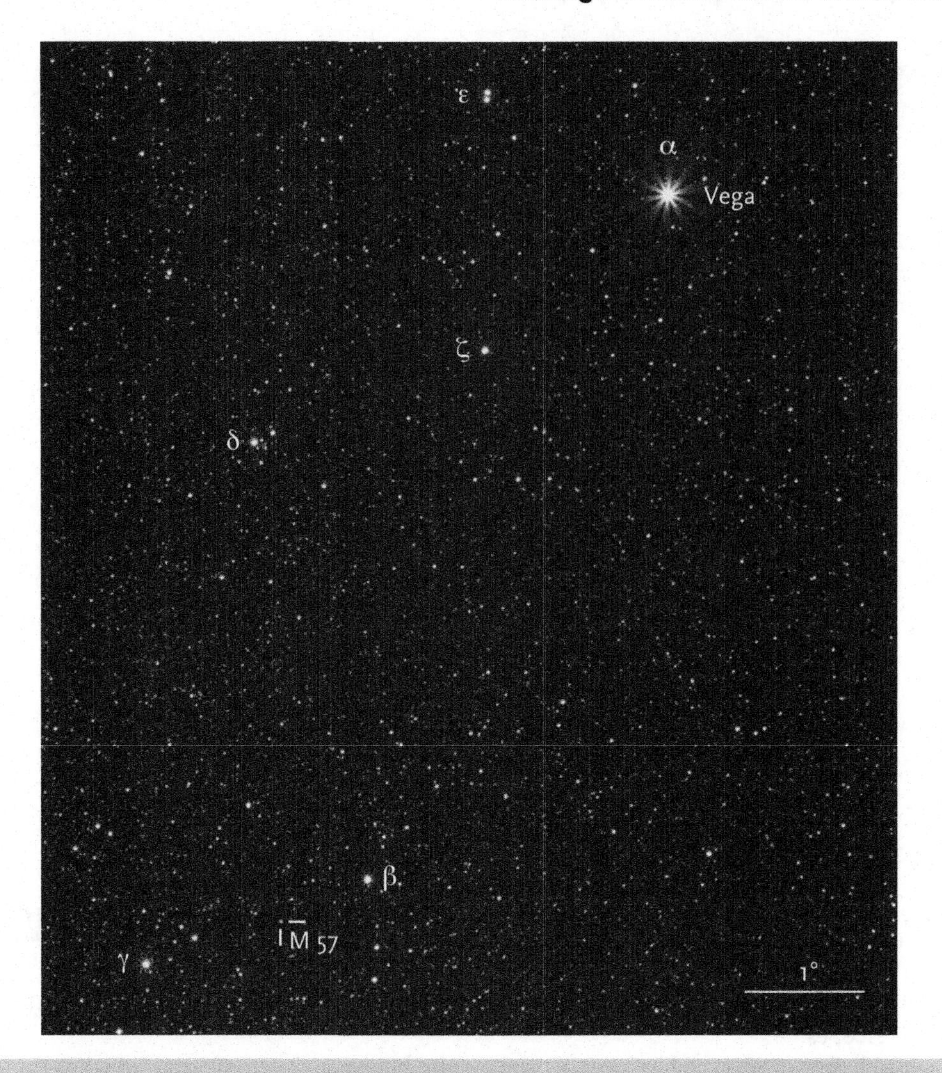

Figure 9.21 Constellation Lyra with the bright Vega

orbiting. This is the reason behind the eclipses and the changes in brightness. When Beta is at its brightest, it shines with magnitude 3.4; when it is at its faintest, it shines with magnitude 3.8 or 4.6, depending on which star is in front. The period in which these changes take place is 12.913834 days. Beta's brightness can always be compared to Gamma, as both can be seen in the same field of view of binoculars.

Delta is a wide pair of stars that can be separated with the naked eye, since they are 10.5 arcmin apart. Delta-1 is a red giant of magnitude 5.5, deep orange in color. Delta-2 is brighter, clearly bluish with a magnitude of 4.5. This scene, which is best viewed through wide-field binoculars, is complemented with the open cluster named Stephenson 1. Approximately 15 stars brighter than magnitude 10 are sprinkled around the bright Delta, but the total number of stars in the cluster reaches over 100. The two bright stars known as Delta are just the brightest members of a group approximately 800 light years from Earth (one of the closest open clusters).

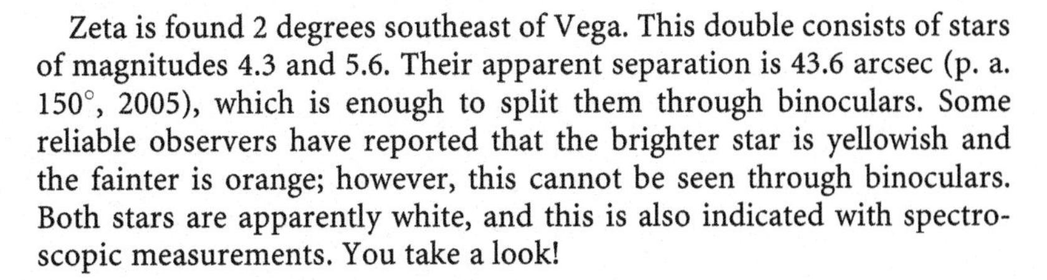

Zeta is found 2 degrees southeast of Vega. This double consists of stars of magnitudes 4.3 and 5.6. Their apparent separation is 43.6 arcsec (p. a. 150°, 2005), which is enough to split them through binoculars. Some reliable observers have reported that the brighter star is yellowish and the fainter is orange; however, this cannot be seen through binoculars. Both stars are apparently white, and this is also indicated with spectroscopic measurements. You take a look!

Epsilon is a famous double-double star (for binoculars only one double). It lies 2 degrees east-northeast of Vega and is easy to find. The main stars shine with magnitudes 5.0 and 5.2 and are 210 arcsec apart (p. a. 174°, 1998), which makes them separable through any binoculars. People with sharp eyesight are able to resolve them with the naked eye. However, if you examine Epsilon with a telescope (with at least a 10-cm objective lens), both of the two stars split into pairs, which make Epsilon a quadruple star. The brighter star has a 2.4 arcsec (p. a. 348°, 2006) distant companion of magnitude 6.1. The fainter of the main couple has a 2.4 arcsec (p. a. 80°, 2006) distant companion of magnitude 5.4. All stars are gravitationally bound, which means that they move around a common center of gravity. It is a shame that there is no color contrast between the stars. All four are white.

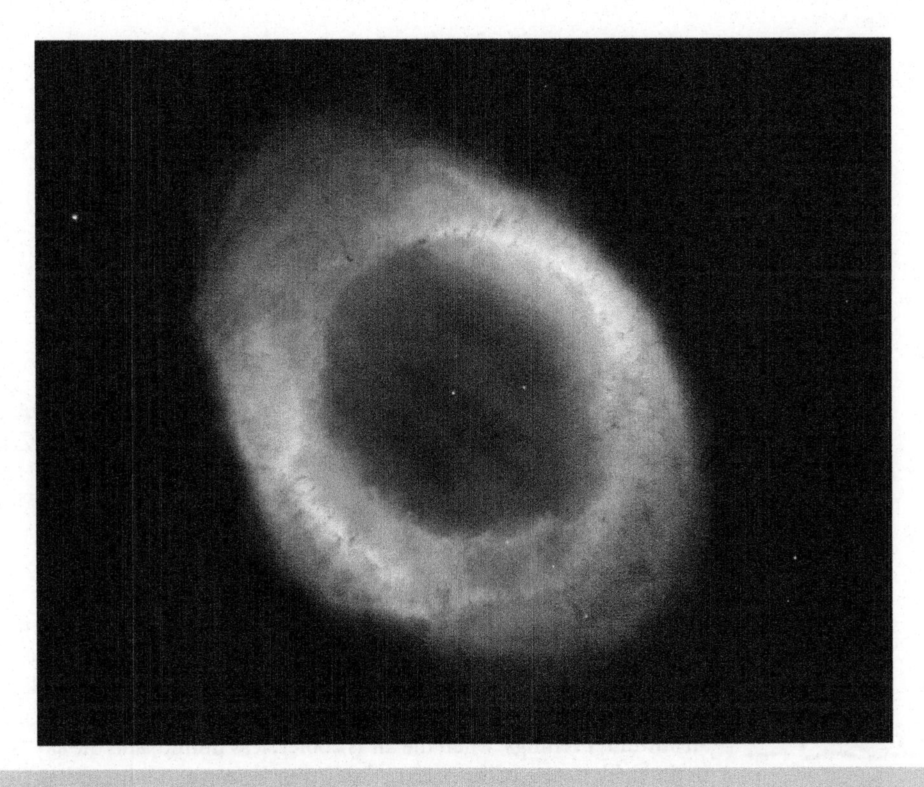

Figure 9.22 Professional image of M 57

Approximately halfway between Beta and Gamma lies the planetary nebula M 57 (8m8/86″ × 62″), also known as the Ring Nebula, which is one of the most popular objects for viewing in the summer sky. It was first noticed by the French astronomer Antoine Darquier in 1779. Unfortunately, observers using binoculars have to be satisfied merely seeing the nebula as a star-like object. In your search for it, you can get help with the image given in Figure 9.21. If you want to see the nebula as a cloud of light, you will need a larger telescope with an objective lens with a diameter of at least 10 cm. To be sure that you will see its ring structure, you will need to use at least 15-cm objective lens.

The gases in the nebula are so diluted that by Earth standards this would be considered as a vacuum. We see the nebula because the atoms within it are ionized (similar as in neon light) by the strong ultraviolet light coming from the central star, which is a very hot white dwarf. Earlier, we discussed how ionized atoms radiate during recombination with electrons. The color of the emitted light is determined by the composition of the gases in the nebula. M 57 shines in greenish (emitted by double-ionized oxygen atoms) and reddish (emitted by ionized hydrogen atoms) light. Through amateur telescopes, the planetary nebulae mainly appear as white or grayish, as we are looking at them with the color-blind rods in our eyes due to the faint light. Only the brightest nebulae can in telescopes show a greenish or bluish tint. However, wonderful vivid colors can be seen on long-exposure images, even on ones taken with modern average amateur equipment!

According to the latest measurements of the parallax of the central star, the distance to the nebula is approximately 2,300 light years. This means that its actual diameter is 0.8 light years. The star in the center blew away its outer layers in an immense outburst 6,000–8,000 years ago. The latest research indicate that the ring is most probably a ring of gases and not a spherical shell, as once thought by astronomers.

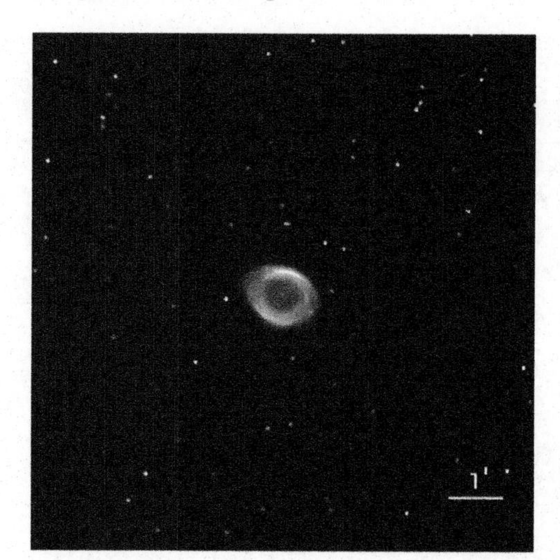

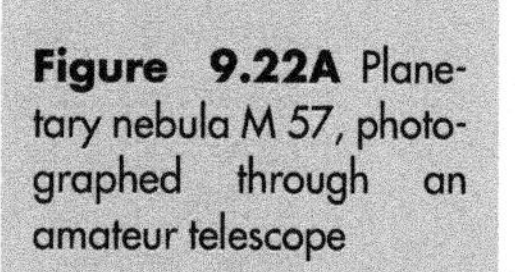

Figure 9.22A Planetary nebula M 57, photographed through an amateur telescope

The globular cluster M 56 (8m3/8′8) is rather far away from the main pattern of the constellation (Figure 9.23). It is found halfway between Gamma Lyrae and Beta Cygni. It appears on the border of visibility when viewed through binoculars; thus, you will have to choose a clear and dark summer night for observation. Or maybe you can try to find it on a spring mornings or an autumn evenings, when the air is cooler, less humid, and the sky is darker. In the field of view of binoculars the cluster is seen as a faint spot of light with a diameter of just over 1 arcmin. The leading star is the bright Gamma. If the star is moved to the northwest edge of the field of view, M 56 will appear on the southeast edge. For better orientation, one should follow the line from Gamma to the magnitude 5 star with the designation 17. Go a degree southeast to find a magnitude 6 star, labeled 19, and a degree further in the same direction to come across another magnitude 6 star that has no label but it lies merely half a degree northwest of the cluster. The other way to find M 56 is to start at Beta Cygni and continue across the magnitude 5 star labeled 2 Cygni.

M 56 is approximately 33,000 light years from us, so its real diameter is 70 light years.

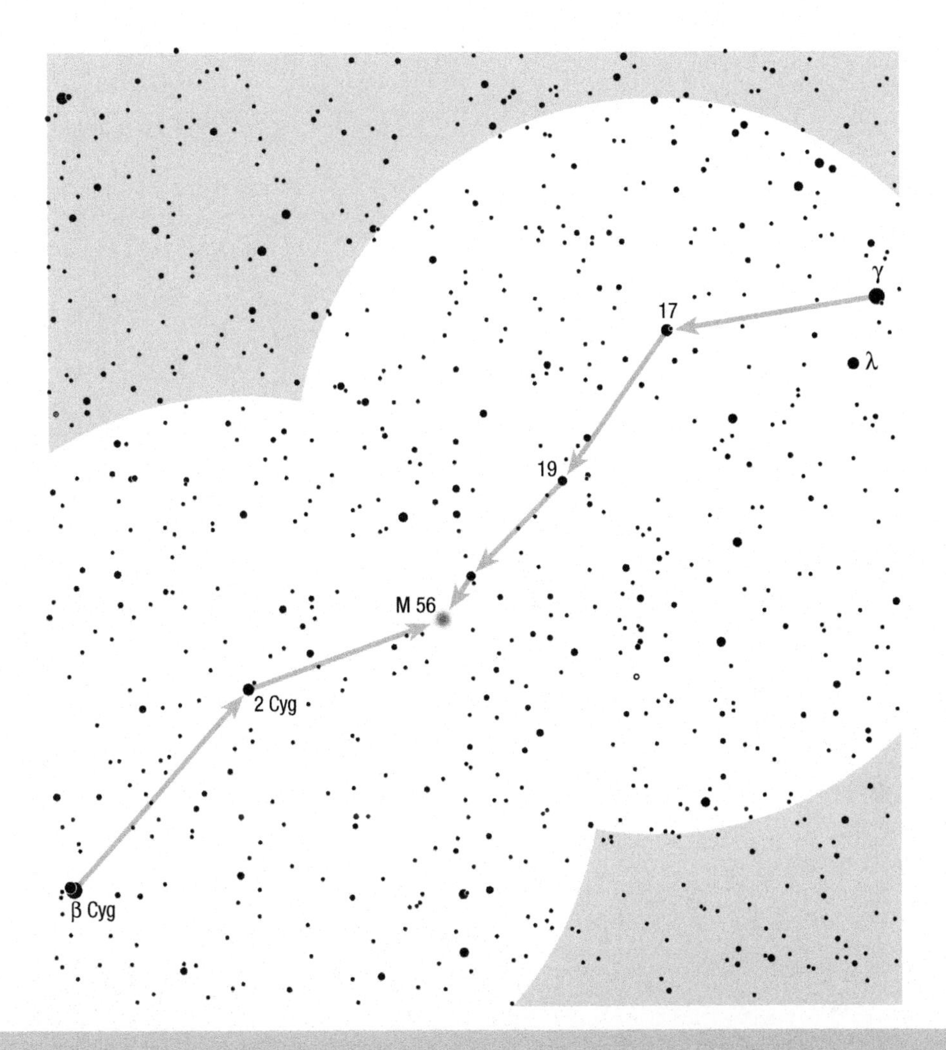

Figure 9.23 The surroundings of the cluster M 56 with the leading stars Gamma Lyrae and Beta Cygni

CHAPTER TEN

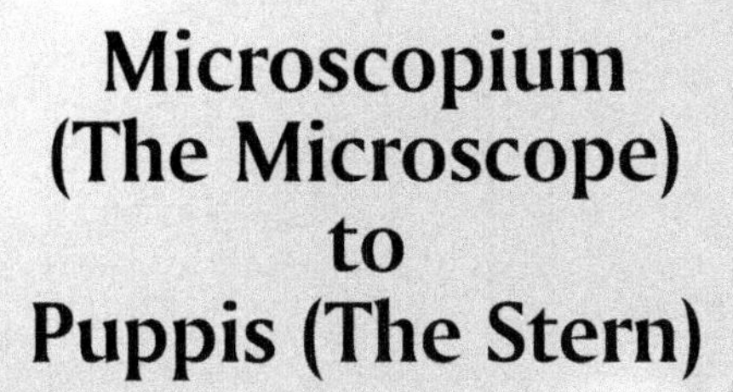

Microscopium (The Microscope) to Puppis (The Stern)

B. Kambič, *Viewing the Constellations with Binoculars*, Patrick Moore's Practical Astronomy Series,
DOI 10.1007/978-0-387-85355-0_10, © Springer Science+Business Media, LLC 2010

MICROSCOPIUM (The Microscope)

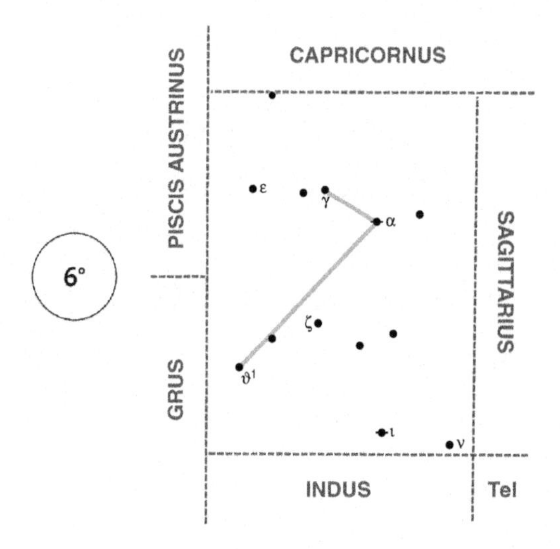

CONSTELLATION CULMINATES		
in begin. July at 2 a.m.	in begin. August at midnight	in begin. September at 10 p.m.

Microscopium is a small and completely undistinguished autumn constellation. It can be found below Capricornus. It does not include any stars that are brighter than magnitude 5; thus, it cannot be seen at all even in a minimally light-polluted sky.

The brightest star in the constellation is Gamma, which is of magnitude 4.7. In mid-northern latitudes, it rises a mere 12 degrees above the horizon at culmination.

MONOCEROS (The Unicorn)

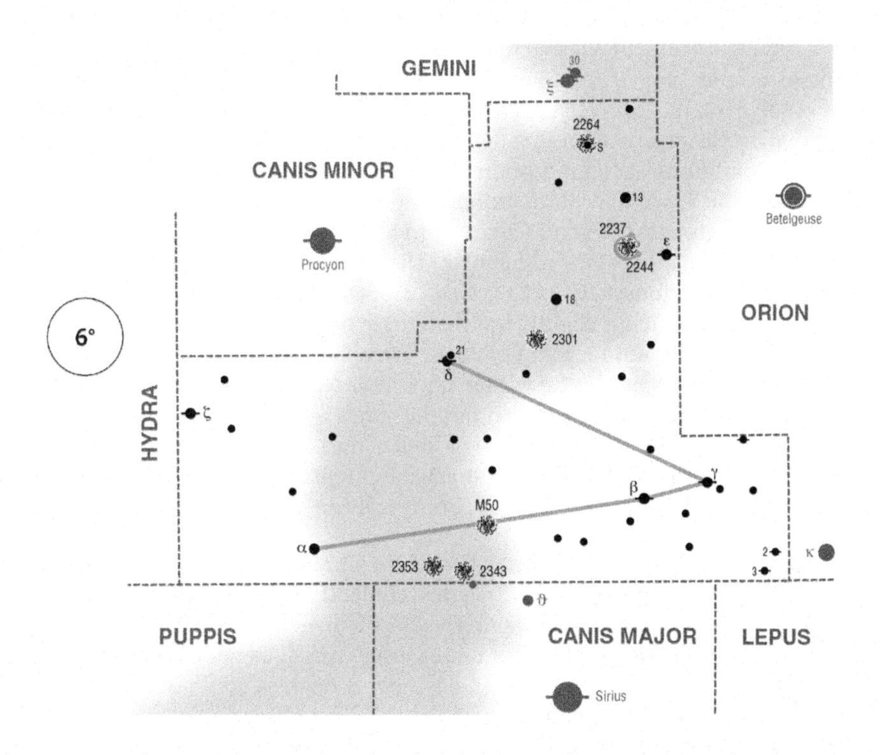

CONSTELLATION CULMINATES		
in begin. December at 2 a.m.	in begin. January at midnight	in begin. February at 10 p.m.

Monoceros is a rather large but faint winter constellation without any distinguishable shape. It fills the area within the large triangle formed by Betelgeuse, Sirius, and Procyon. The constellation lies on the celestial equator. Its brightest star is Beta, which shines with only a magnitude 3.7. The winter Milky Way lies in the background; thus, the entire area is rich with stars and open clusters and worth a panoramic survey with binoculars.

Beta is a wonderful triple star, unfortunately only seen through mid-sized telescopes. It is one of the rare ones in which all three components are of similar brightness. The stars labeled A and B are of magnitudes 4.6 and 5.0 and are 7.1 arcsec apart (p. a. 133°, 2006). The third star, labeled C, shines with magnitude 5.4 and is 3.0 arcsec away from Beta B (p. a. 108°, 2006) and 9.8 arcsec from Beta A (p. a. 125°, 2006). All three are gravitationally bound and move around a common center of gravity. This interesting triple star is 175 light years from us.

Zeta is also an interesting triple star, but only two components can be seen through binoculars, since the faintest star is only of magnitude 10.1 (separation 33 arcsec, p. a. 105°, 1998). The two stars that can be seen through binoculars are of magnitude 4.5 and 9.7, and they are separated by 64.7 arcsec (p. a. 247°, 2002). Due to the dimness of the fainter star this pair can only be viewed with good optics. The brighter star is yellowish and the fainter has an orange tint. The colors of the pair, especially of the fainter star, are better seen in telescopes with a larger objective.

Probably the most interesting object within the constellation is the open cluster NGC 2244 (4m8/24′) scattered around the yellowish star 12 Monocerotis (5m8). It is easy to find with binoculars. If you catch Epsilon (4m3), which lies slightly to the south of the connecting line between Betelgeuse and Procyon, the cluster and Epsilon will be found in the same field of view (chart on page 346) When there are excellent observing conditions, it is as a faint spot of light surrounding star 12 seen even with the naked eye. In the field of view of binoculars, the haze is broken into individual stars. There are over 1,200 stars in this region! Approximately 100 can be seen through binoculars on perfect nights and 35 when there are excellent observing conditions. At least 8 of them are rather bright, with magnitudes up to 8.

The cluster is nice, but if we had photographic eyes, we could see so much more! The gigantic Rosette Nebula extends around the stars (Figure 10.1). On long-exposure images, taken with large telescopes, this is one of the most agitated nebulae. The Rosette is so big that its brightest parts have got their own NGC designations: NGC 2237, NGC 2238, NGC 2239, and NGC 2246. Its apparent diameter is over 80 arcmin, which is slightly less than three times the diameter of a full Moon. Because of its size, the nebula is a frequent target of amateur astrophotographers. Through binoculars it can only be seen on the best nights, when the winter sky is completely dark and the atmosphere is calm. Only then can we glimpse the subtle shine around the stars.

Astronomers are convinced that the nebula is a stellar nursery, in which new stars are born from gases and dust. Its distance is estimated to be at 5,500 light years. The star 12 Monocerotis is in fact much closer to us, and therefore it is not a true, but merely an apparent, member of the cluster.

The open cluster NGC 2264 (3m9/20′) lies 6 degrees north and slightly to the east of the Rosette Nebula. If the cluster alongside Rosette is moved to the extreme southern edge of the field of view, NGC 2264 appears on the northern edge. But it is easier to use Xi Geminorum as the leading star. In this case, NGC 2264 lies only 3 degrees south-southeast of it (chart on page 346). When there are excellent observing conditions, the cluster can be completely resolved through binoculars. We can see approximately 35 stars with magnitudes between 7 and 11. Most of them are white; only three are yellowish. The brightest member is the irregular variable star S, which changes its brightness between 4.5 and 5.0.

Leland S. Copeland named this cluster the Christmas Tree. The 10 brightest stars form the shape of a fir tree. The variable S represents the trunk, while the remaining nine stars shine as lamps at the end of invisible branches. Through binoculars the tree is turned upside down. The cluster, which has an actual diameter of 15 light years, is approximately 2,400 light years from us (Figure 10.2).

The stars are wrapped in a tenuous nebula, which is connected to the Rosette. As we have already mentioned, the cluster is an easy target for binoculars. However, the unbelievably agitated nebula, even fainter than the Rosette Nebula, is seen only through large amateur telescopes, equipped with special filters, and on long-exposure images.

NGC 2301 (6m0/12′) is an open cluster that lies in the middle of the constellation. Delta is the leading star. If the star is moved to the extreme eastern edge of the field of view, the cluster appears on the extreme western edge. Even when there are excellent observing conditions this oblong group of faint stars is hard to see in binoculars. Half a dozen of the brightest stars merely reach magnitude 8. There are approximately 100 gathered within a diameter of 15 arcmin, but we can only see approximately 20 through binoculars when there are excellent conditions.

NGC 2301 is approximately 2,800 light years from us, which makes its real diameter around 12 light years.

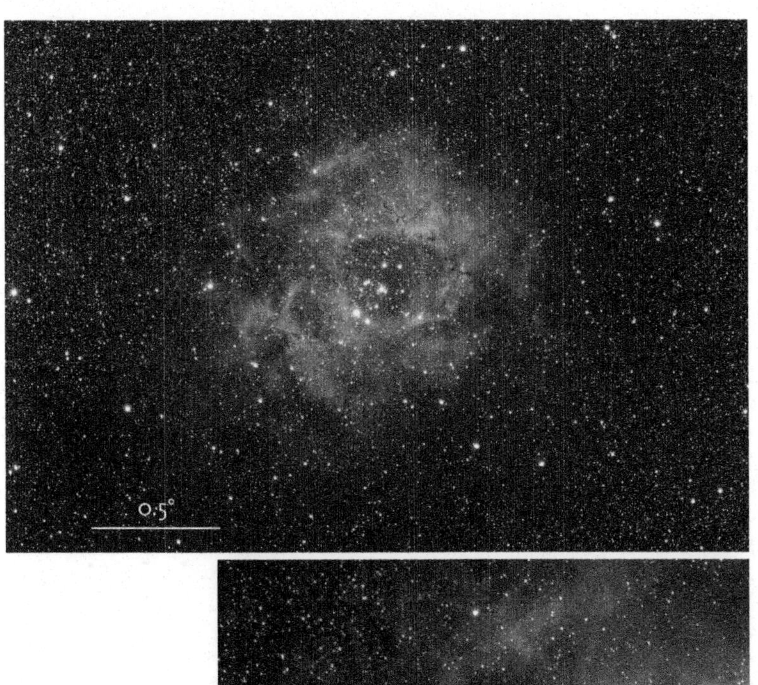

Figure 10.1 The fantastic Rosette Nebula (NGC 2237) is spread around the stars of the open cluster NGC 2244. Through binoculars the cluster with its brightest star, 12 Monocerotis (magnitude 5.8), is clearly seen, but the nebula can be glimpsed only on perfect nights, when the sky is dark and completely calm. On long-exposure images one can truly admire the wonderful turbulent nebula

In the southern part of the constellation, three open clusters await us in the same field of view. These are the wonderful and visible M 50 and the scarcely seen NGC 2343 and NGC 2353. The leading star for hunting these clusters is Theta Canis Majoris (see chart on page 351). If the star is moved to the southwestern edge of the field of

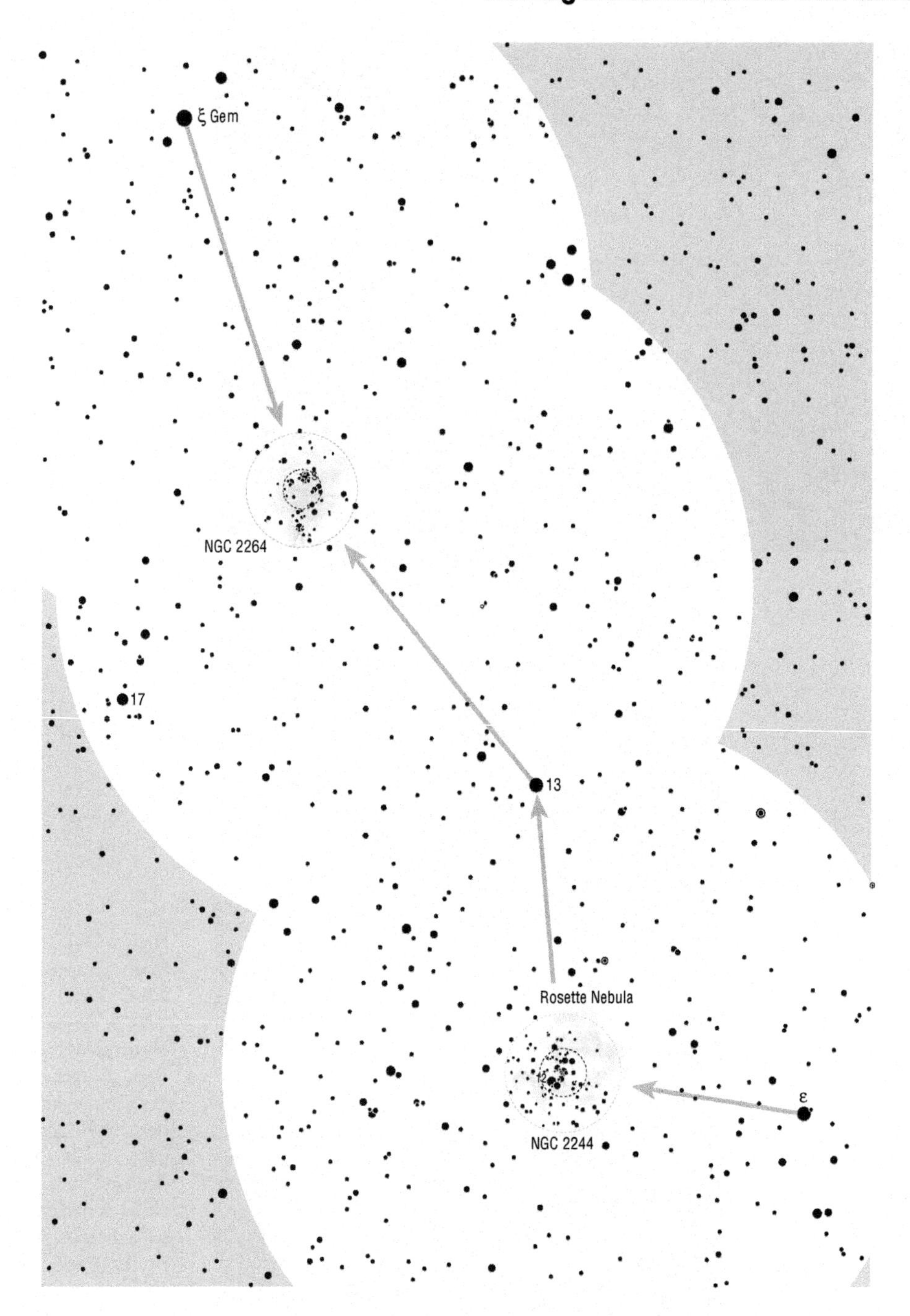

view, M 50 (5m9/16′) appears on its northeastern edge. This cluster is on border of visibility with the naked eye. In total it contains slightly less than 200 stars, out of which approximately 60 are seen in binoculars on a perfect night and approximately 30 when there are excellent observing conditions. The brightest are only of magnitude 8. The group is small and rather dense. Through binoculars the cluster can be partially separated into individual stars; however, the fainter ones are melted into a tenuous nebula (Figure 10.3).

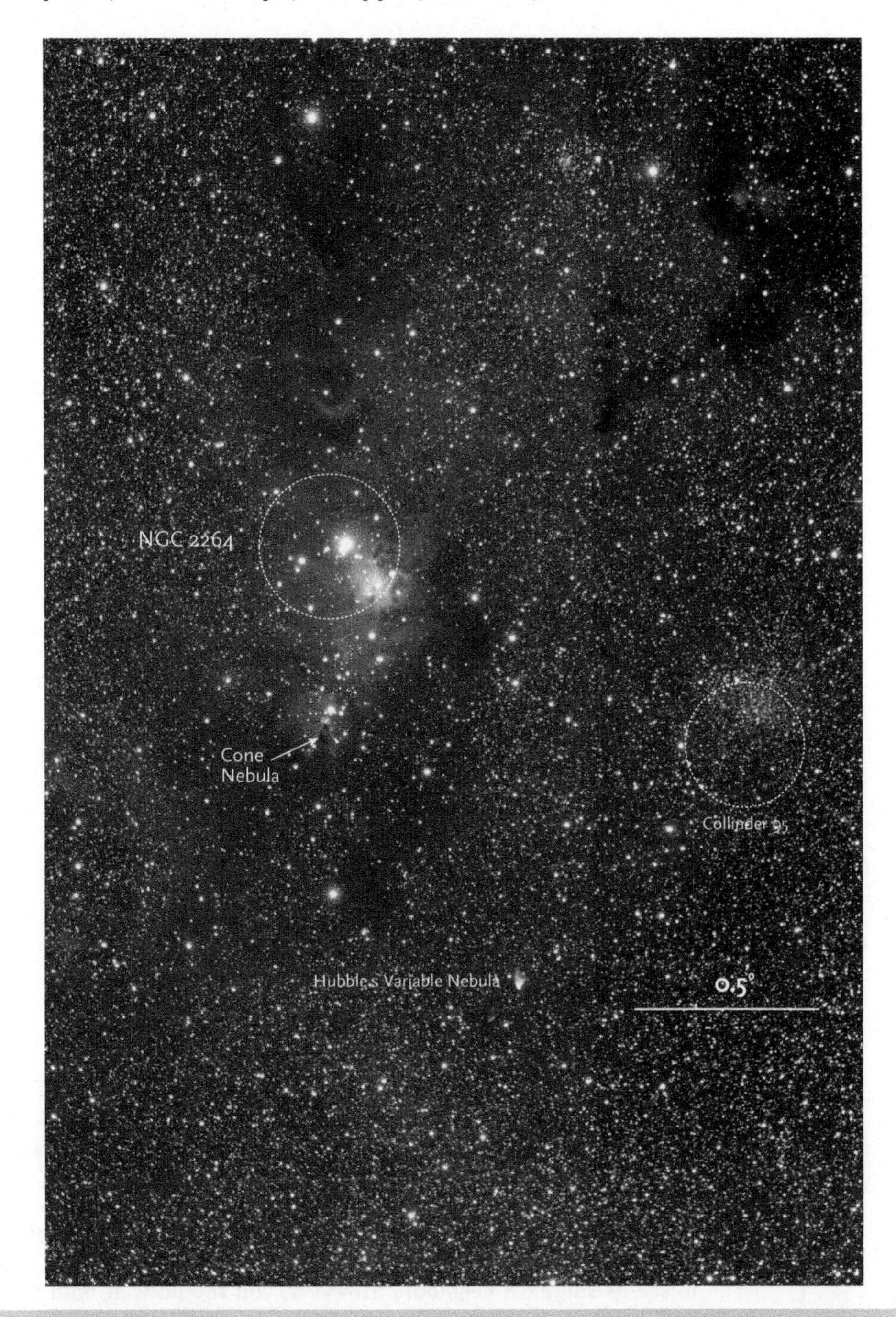

Figure 10.2 The splendid nebula around the open cluster NGC 2264 can unfortunately only be seen on long-exposure images. The dark nebula of dust and gases, which indents into the southern part of the nebula, is especially popular among astrophotographers. It is called the Cone Nebula. Right of the nebula one can see the large but faint open cluster Collinder 95

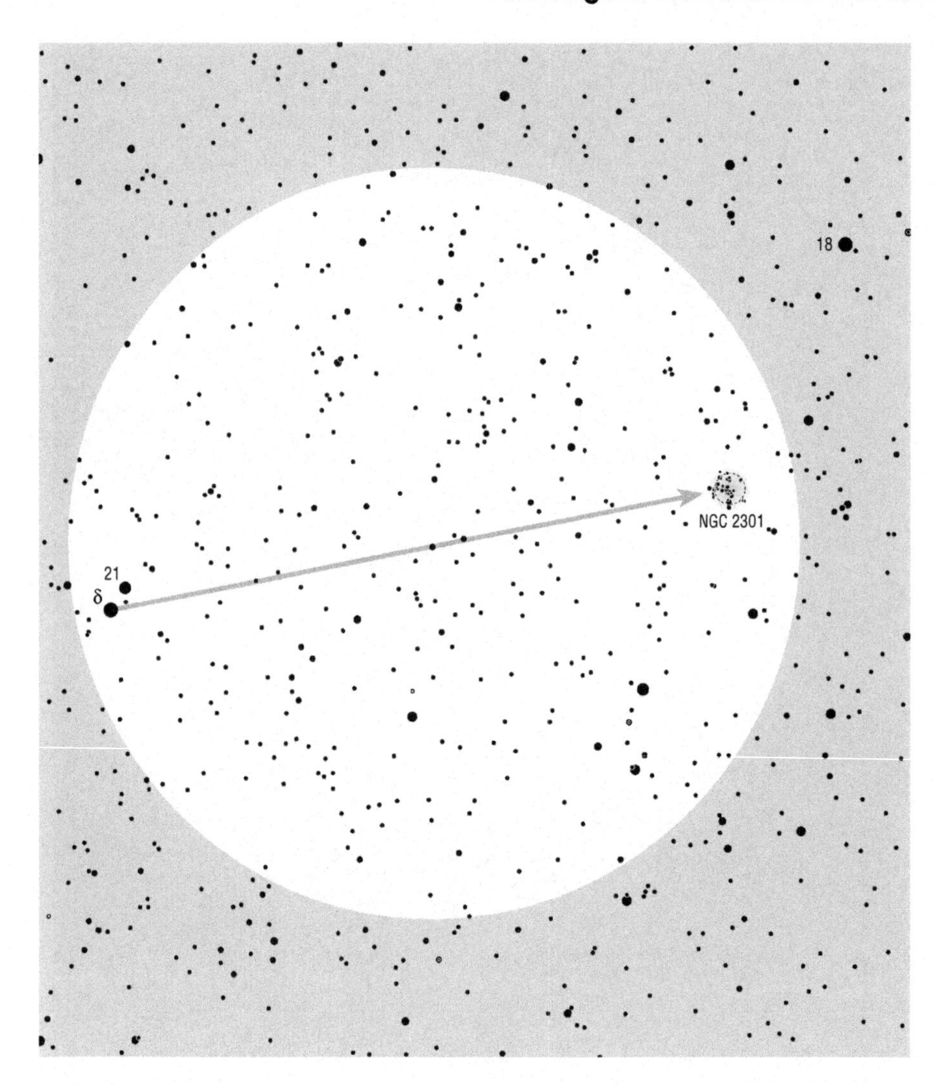

M 50 is approximately 3,200 light years from us. At this distance its real diameter reaches a mere 18 light years.

Although M 50 is easily seen (at least as a weak spot of light), you should choose a truly excellent night with a calm and dark winter sky if you want to see its southern neighbors. However, you should wait until you have had some observing experiences. Both clusters are among the smaller specimens of their kind (Figure 10.4). Approximately 60 stars are gathered in NGC 2343 (6m7/7′); most of these are very faint. Maybe 15 of them can be seen on a perfect night, and only 8 when there are excellent observing conditions. To the southwest of the cluster lie a magnitude 5 star and a few magnitude 6 stars that can help in determining the exact position.

NGC 2343 is approximately 3,400 light years from us. Its actual diameter is therefore only 7 light years.

Figure 10.3 Open cluster M 50

NGC 2353 (7ml/20′) is somewhat larger. However, it contains only about 20 stars. The brightest is of magnitude 6 and is always clearly seen, while the others are no brighter than magnitude 9. On the cluster's northern and southeastern edge we can see two magnitude 6 stars, which are good leading points when trying to determine the exact position. It is very difficult to see NGC 2353 through binoculars, so do not be overly disappointed if you fail. The cluster is approximately 3,650 light years from Earth. Its real diameter is 21 light years.

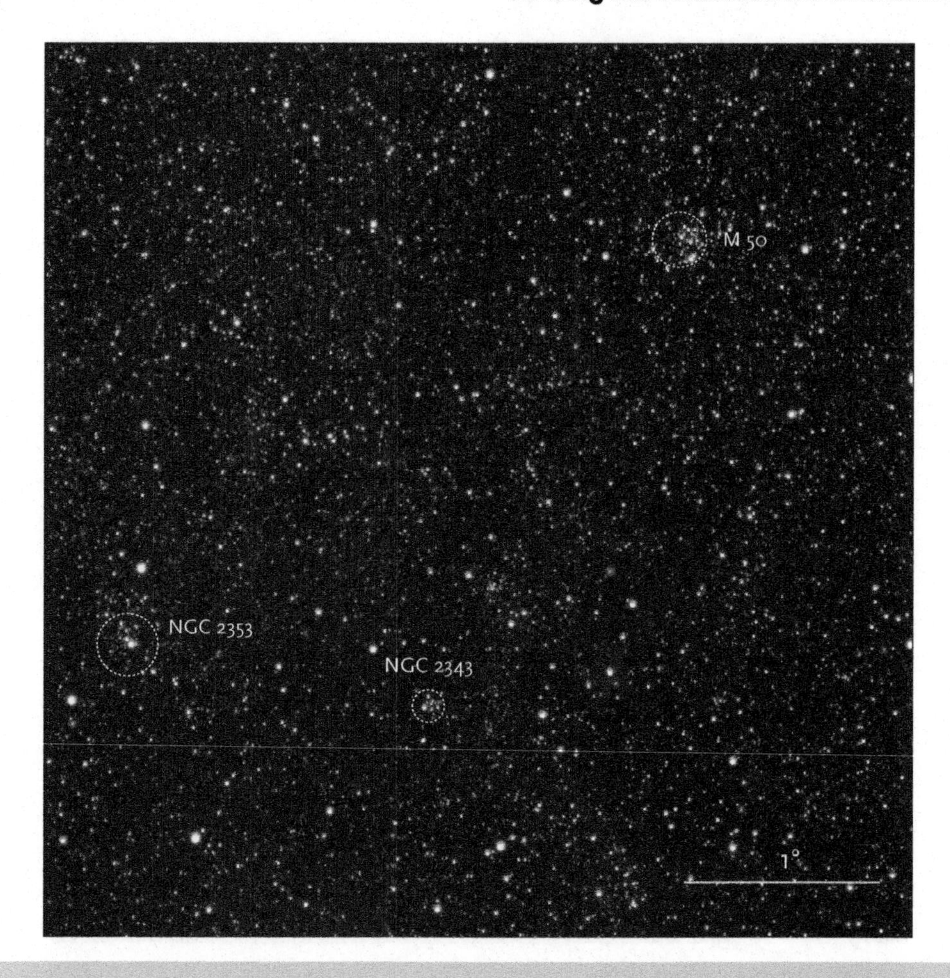

Figure 10.4 The area around the open cluster M 50, which lies in the middle of the winter Milky Way, is full of stars as well as larger and smaller groups. While M 50 is visible in any kind of observing conditions, its southern neighbors demand a night with a truly excellent sky! For searching for or recognizing the patterns of brighter stars, see the chart on next page. The image covers a field of 4.5 × 4.5 degrees, and north is at the top

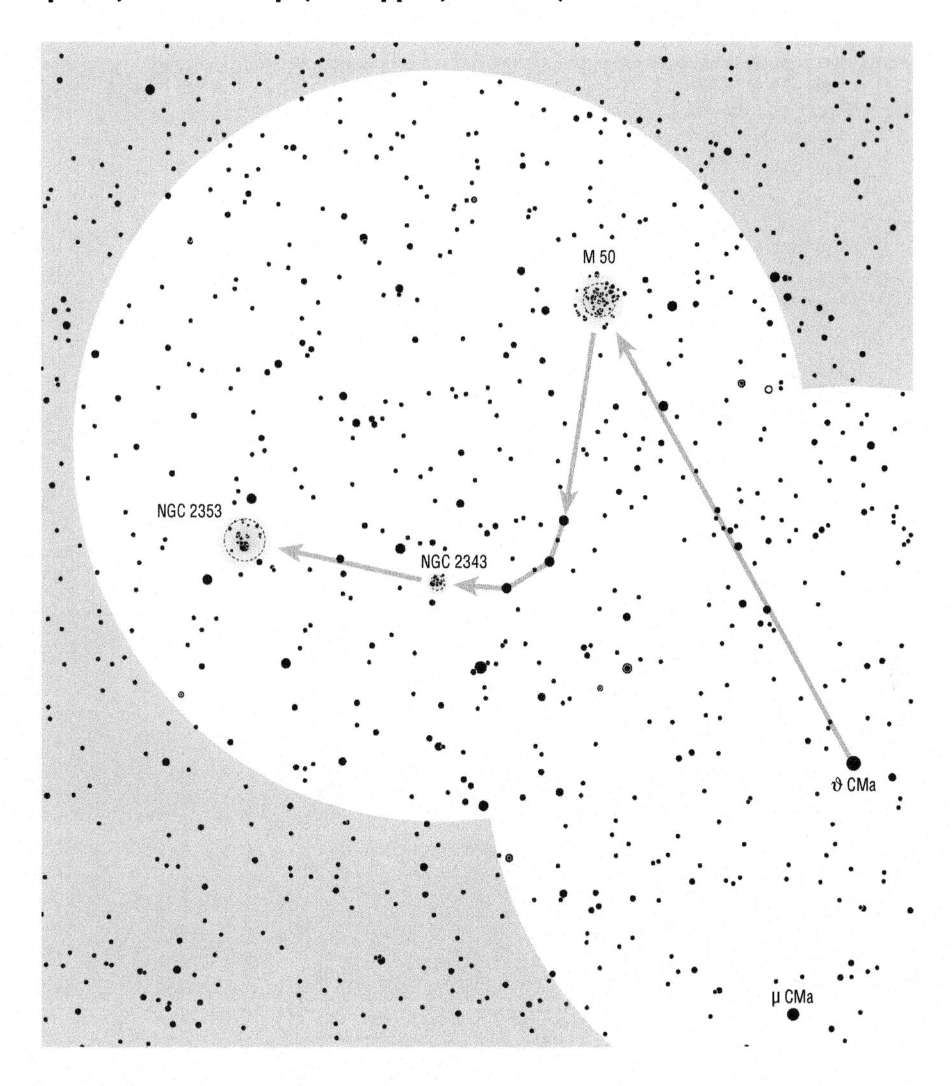
M 50
NGC 2353
NGC 2343
ϑ CMa
μ CMa

OPHIUCHUS (The Serpent Holder)

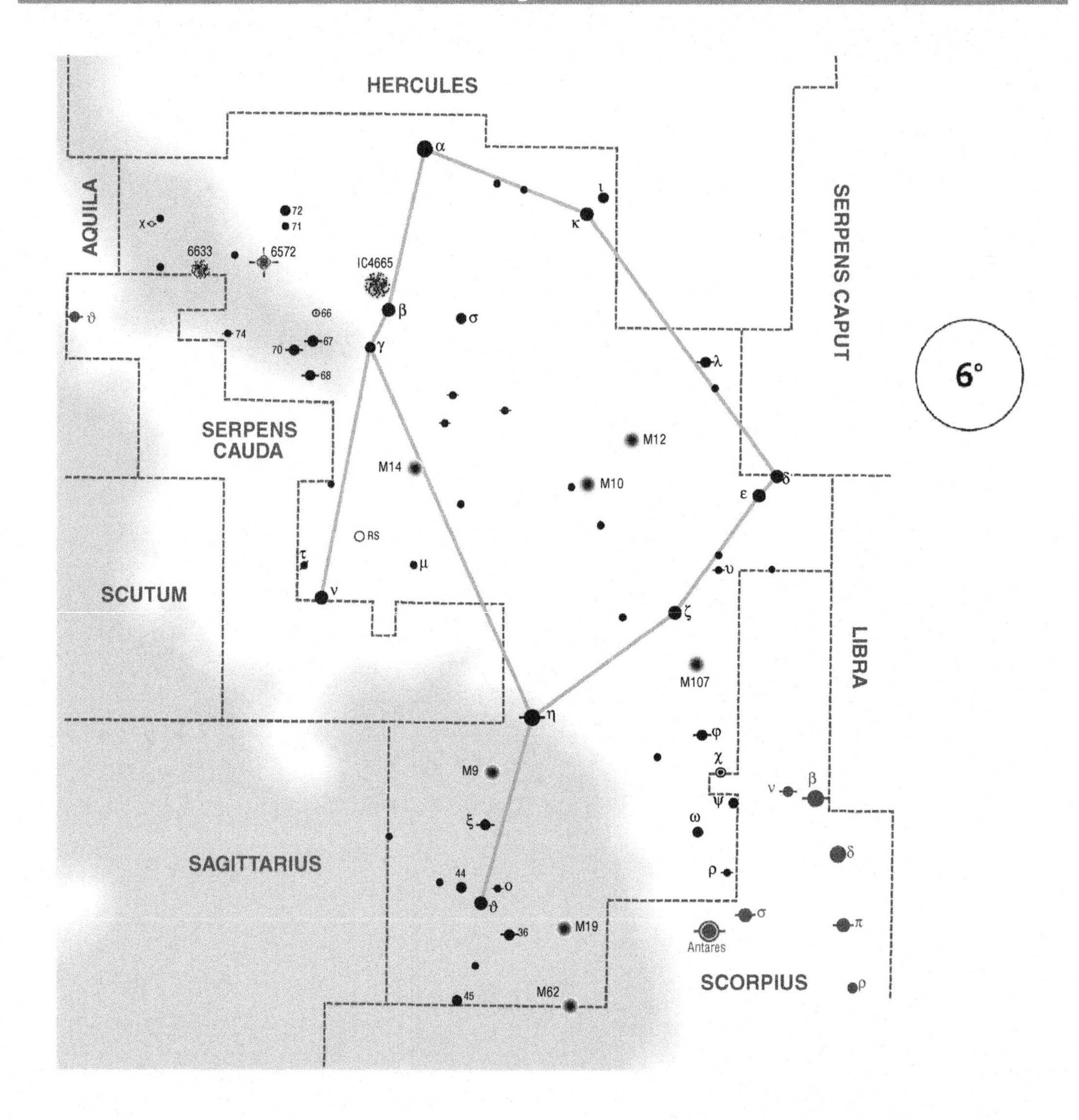

CONSTELLATION CULMINATES		
mid May at 2 a.m.	mid June at midnight	mid July at 10 p.m.

Ophiuchus, which lies below Hercules, is a large but rather undistinguished constellation. Its south part extends over the ecliptic, and each year the Sun needs approximately 20 days to travel across it. But Ophiuchus is not a zodiac constellation. On the ecliptic it divides Scorpius and Sagittarius.

None of the Ophiuchus stars are brighter than magnitude 2. The brightest are Alpha or Ras Alhague (2m1), Eta (2m4), Zeta (2m5), Delta (2m7), and Beta (2m8).

Alpha is the 59th brightest star in the sky. It is only 47 light years away from us. Its luminosity is 23 times that of the Sun. Careful measurements have shown that the star has an invisible companion, which has an orbital period of 8.5 years. Both stars' masses are estimated to be 2.4 and 0.6 solar masses. The widest separation between them is 0.4 arcsec, or a mere 6 AU (Jupiter is 5.2 and Saturn 9.5 AU from Sun).

Eta is the 86th brightest star in the sky. It is 84 light years from Earth. It is a close binary star with components of magnitude 3.0 and 3.3, separated by just 0.6 arcsec (p. a. 240°, 2008). The orbital period is 84 years. The true distance between the stars is approximately 20 AU, which corresponds to the distance between the Sun and Uranus. The common luminosity of both stars is that of about 54 Suns.

Zeta is the 93rd brightest star in the sky. It is 460 light years from us and is a true giant. Its luminosity is approximately 1,500 times that of the Sun.

Delta is the 119th brightest star in the sky, 170 light years from Earth. Its color is yellow-orange, and it has the luminosity of 170 Suns.

The orange Beta is the 125th brightest star in the sky. Its light needs 82 years to reach us. Its luminosity surpasses our star by 38 times.

X is a pulsating variable star of the Mira type, which, over an average period of 329 days, changes its brightness from magnitude 5.9 to 9.2. At the rewarding maxima it can therefore be seen with the naked eye, and it can be observed during the entire cycle through binoculars. The star is located in the northeastern part of the constellation, in an area lacking a single brighter star in the vicinity. The best star to function as the leading star is 72 Ophiuchi (magnitude 4), which can be recognized with the aid of two rather close stars. The closer one is of magnitude 6 and the other, which lies a degree to the south, is magnitude 5 and is labeled as 71 Ophiuchi. The chart on the page 357 will be of great help in your travel from this star to the variable. X is clearly recognizable for its distinct orange color, which becomes increasingly deep with its falling brightness.

This variable played an important role in the history of our knowledge of red giant stars. The story goes back to the year 1900. This star has a long and constant minimum during which it shines with a constant magnitude 9 for a period of 60 days. Astronomers suspected that X is in fact a double star and that the brightness of the variable is still falling, but the rather constant brightness is contributed by its magnitude 9 companion. In 1900, the companion was actually noticed with a 90-cm refractor at the Lick Observatory. The separation between the stars was 0.3 arcsec. During the regular observations that took place over the next decades, astronomers determined the orbit as well as the orbital period of the binary (560 years). From these data they could calculate the common mass of the stars. The result – 2 masses of our Sun – came as a shock to everybody! Until then astronomers assumed that the red giants were extremely large as well as extremely massive. With detailed studies of variable X and some other similar variable stars it was shown that these stars have a mass comparable to our Sun.

Globular cluster M 9 (7m7/12′) is found approximately 3.5 degrees southeast of Eta. Through binoculars it is seen as a small, few arc minutes large spot of light, more like a blurred star than a cluster. In order to see it, you must look when there are excellent observing conditions. M 9 is approximately 26,000 light years from us and lies in the central, denser part of our galaxy, roughly 7,500 light years from its center. The actual diameter of the cluster is approximately 60 light years.

Globular clusters M 10 (6m6/20′) and M 12 (6m7/16′) lie close together in the sky (Figure 10.5). Since they are separated by a mere 3.4 degrees, they can be found in the same field of view of binoculars. Both are clearly seen spots of light, approximately 7 arcmin in diameter. Through binoculars, M 10 is somewhat smaller. Their size and appearance depends greatly on the observing conditions. On excellent nights, you might notice that they both have a bright core. The star leading to the clusters is the bright Epsilon (the southeastern star of the pair formed together with Delta). From here you have to move your binoculars 8 degrees east across the magnitude 6 star with the designation 12 Ophiuchi and

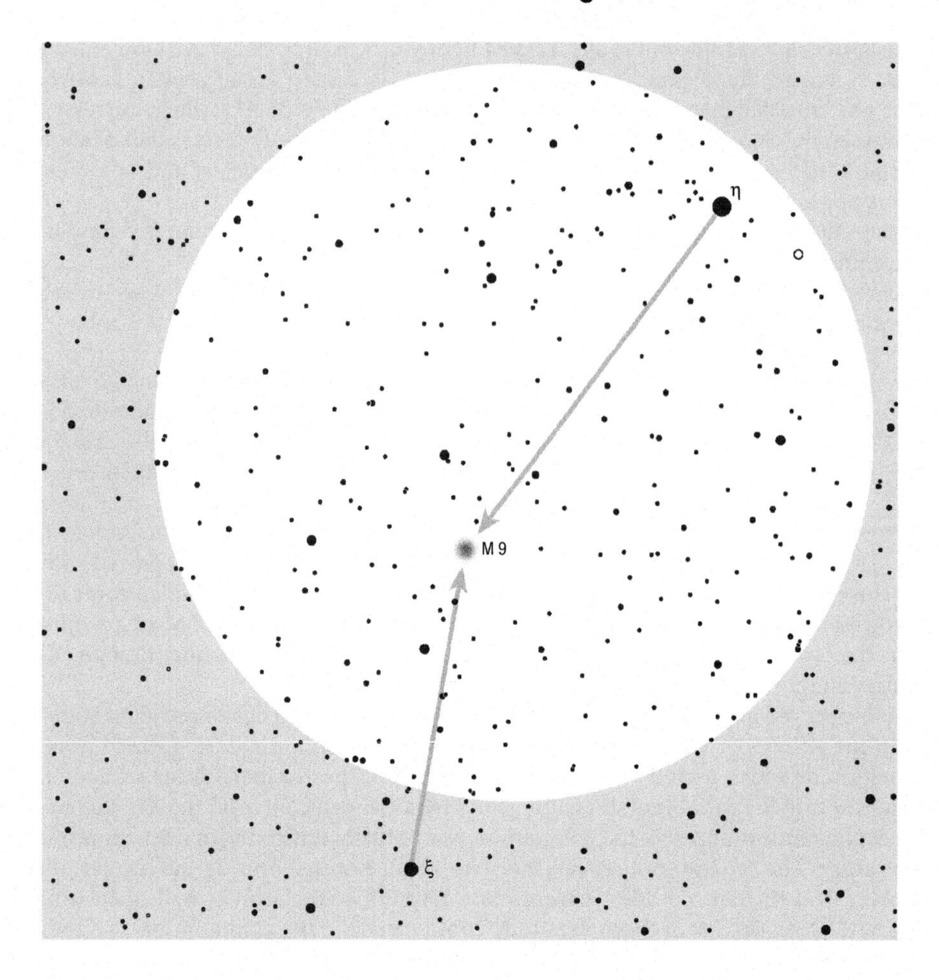

M 12 appears in the field of view. You can also find the clusters if you start at Lambda. If this star is moved to the extreme northwestern edge of the field of view, M 12 appears on the southeastern edge. Between them you should see the close magnitude 6 stars (see chart on next page).

The clusters represent an interesting pair in larger telescopes, in which we can separate individual stars on their edges. This allows us to clearly see the difference in the structures of globular clusters. M 10 is rich with stars, with a dense central part, while M 12 is slightly larger, but fainter and looser.

M 10 is approximately 14,000 light years from us, and M 12 is an additional 2,000 light years further. Their actual size is approximately 70 light years.

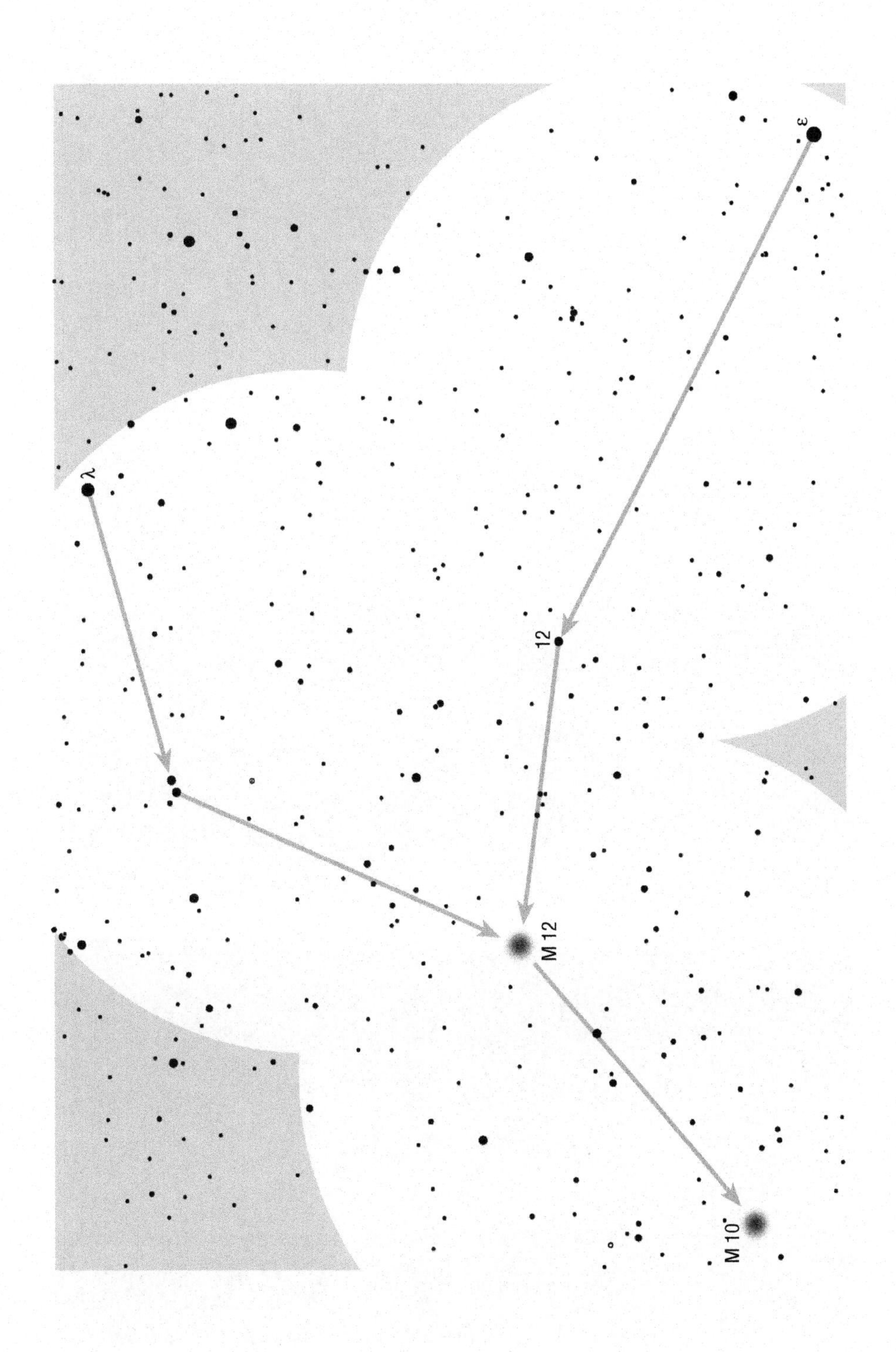

λ
ε
12
M 12
M 10

Figure 10.5 Open clusters M 10 and M 12

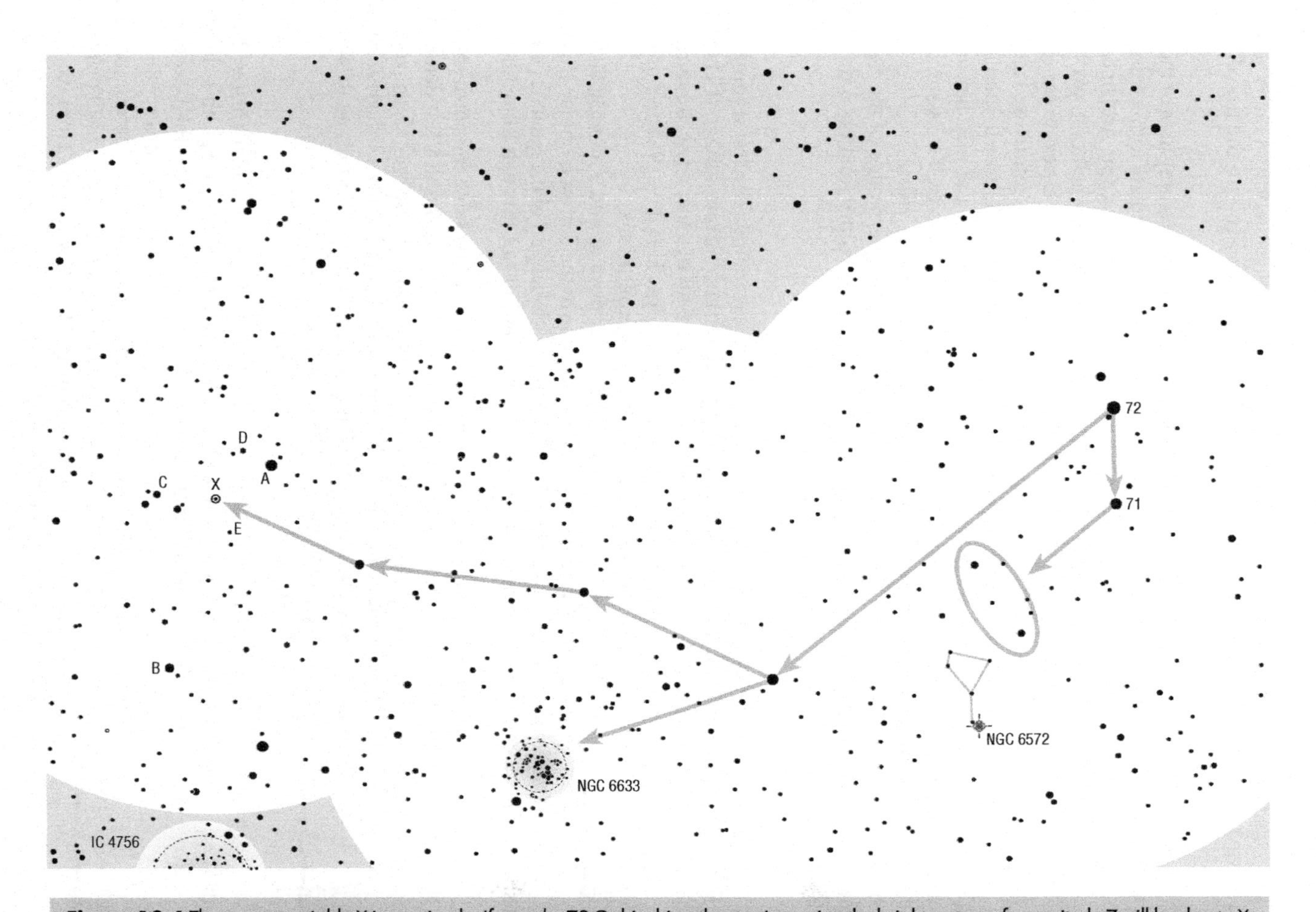

Figure 10.6 The way to variable X is not simple. If we take 72 Ophiuchi as the starting point, the brighter stars of magnitude 7 will lead us to X, which is recognizable by its distinguished orange color. Comparison stars are A (5m4), B (6m3), C (7m2), D (8m2), and E (9m5). Star 72 is leader to planetary nebula NGC 6572 too.

In the sky of a hypothetical planet residing in one of the clusters, the other cluster would be seen as a bright, approximately 2 degrees large celestial object (the diameter of the full Moon is 0.5 degrees), with an integral brightness of magnitude 2. Unfortunately, we have nothing similar in the mid-northern latitudes, but in the southern hemisphere this impression is matched somewhat by the Small Magellanic Cloud (satellite galaxy to our galaxy). The brightest globular cluster in our sky is the magnitude 4 Omega Centauri, which is seen as a star, and from mid-northern latitudes the brightest cluster is M 22 (magnitude 5) in Sagittarius, which is on the border of visibility with the naked eye.

The globular cluster M 14 (7m6/11′) lies in a rather empty area of the sky; thus it is hard to find even when viewed through binoculars with a 6 degrees field of view, let alone through an average amateur telescope. And once you finally find it, you'll probably find yourself wondering whether it was worth the effort. A small, faint spot of light is all that you can expect to see of this very remote globular cluster through binoculars. In literature, we can find numerous different estimates of its distance, but all of them are in the vicinity of 30,000 light years. This number is not really reliable, as a denser cloud of gas and dust lies in that direction, and astronomers have no idea yet as to how much light it absorbs. However, if we assume that this distance is correct, the true diameter of the cluster is approximately 120 light years.

Sigma is the leading star to the cluster. If we move it to the extreme northern edge of the field of view, two magnitude 6 stars appear on its southern edge. The cluster lies approximately 3 degrees further south Figure 10.7.

The globular cluster M 19 (6m8/17′) is easy to find. The star leading to it is the rather bright Theta and the group of stars surrounding it. If Theta is moved to the extreme eastern edge of the field of view, a few arc minutes large spot of light appears on the western edge. This is M 19. The cluster is seen clearly only with excellent observing conditions.

M 19 is approximately 28,000 light years from us and a mere 3,000 light years from the center of the galaxy. It is one of the flattest globular clusters known to us. Most probably the cause for this lies in the strong gravitational pull of the galactic center. The oblong shape of M 19 can be noticed through amateur telescopes but not through binoculars.

Slightly less than 4 degrees south of M 19 another faint and small spot of light can be found; this is the globular cluster M 62 (6m5/15′). It demands excellent observing conditions and a dark sky all the way down to the horizon. For mid-northern observers, the cluster is only 14 degrees high in the sky at culmination. M 62 is 22,500 light years from us, so its actual diameter is approximately 100 light years.

Close to Zeta, only 2.8 degrees south, lies another globular cluster. This is the small and faint NGC 6171 (7m9/13′). In modern catalogs, this cluster was added to the original Messier's list as M 107. It is on the border of visibility when viewed through binoculars. You should observe it only when there are excellent observing conditions. In order to determine the exact position of M 107 in the field of view, the detailed chart will prove of great help. The cluster is approximately 21,000 light years from us; thus, its actual diameter is 100 light years.

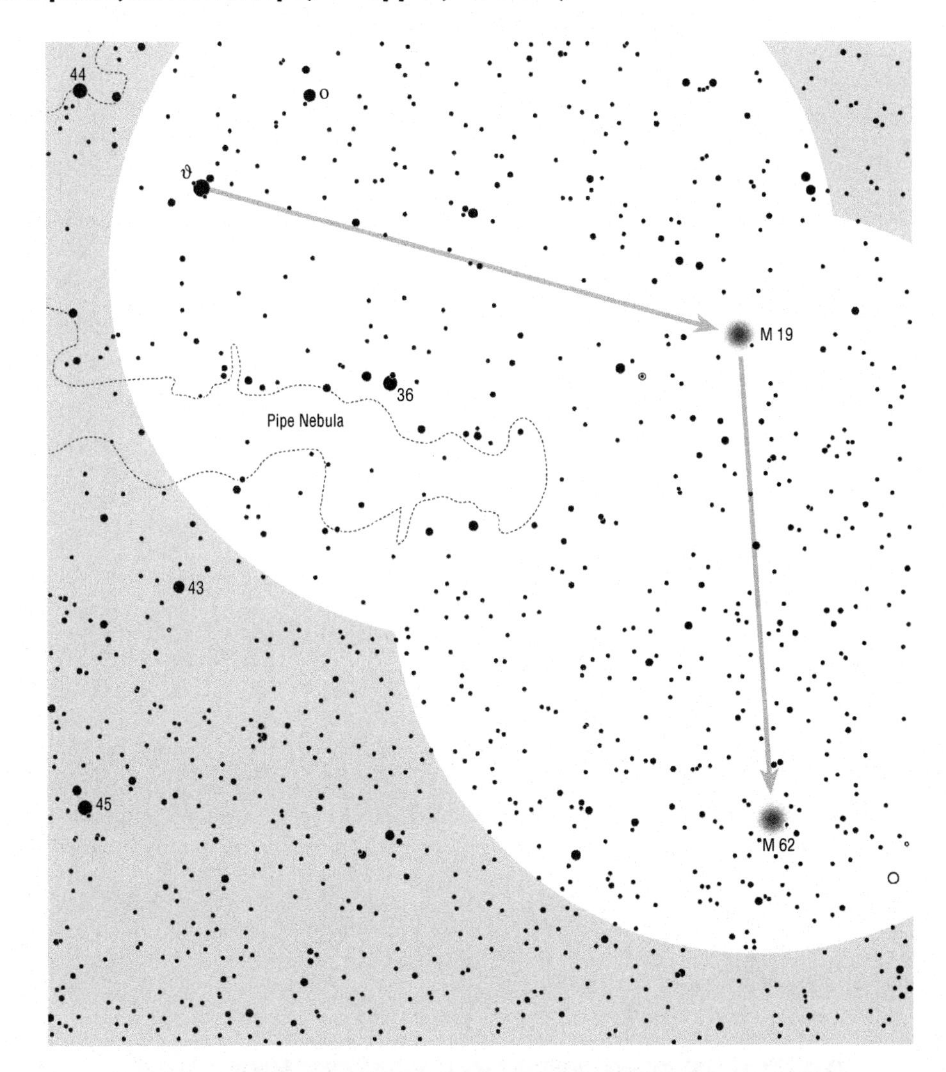

However, this is not the end of clusters in Ophiuchus, for one of the Milky Way's spiral arms stretches out into the eastern part of the constellation. Close to Beta, only 1.4 degrees northeast, lies the big and loose open cluster IC 4665 (4m2/70′). Its diameter is slightly over 1 degree; in that large area over a thousand stars are gathered, but of course most of them are faint, for the brightest are a magnitude 7. Through binoculars the cluster can easily be separated into individual stars. When there are excellent observing conditions we can see approximately 75 stars, and on a perfect night their numbers can double! This is definitely one of the most attractive objects for wide-field binoculars. A detailed chart – if it is needed at all – is found next to the globular cluster M 14 on page 363. The cluster is 1,400 light years from us.

On the border of the constellation Serpens (more precisely, with its part, called Serpens Cauda) lies another wonderful open cluster NGC 6633 (4m6/27′). The stars leading to it are three magnitude 4 stars with the designations 67, 68, and 70 Ophiuchi. If the easternmost of this triplet, 70 Ophiuchi, is moved to the extreme southwestern edge of the field of view, the first stars of the cluster appear on its northeastern edge. (By the way,

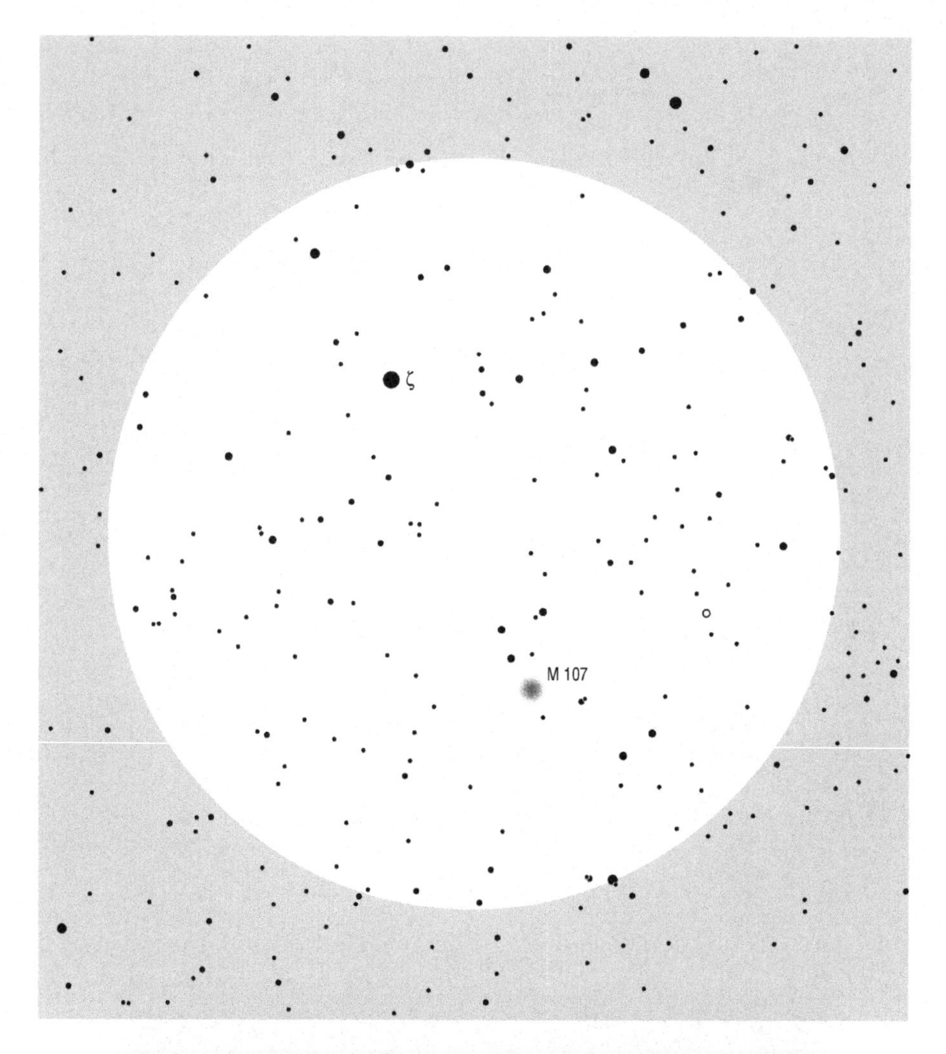

Globular cluster M 19

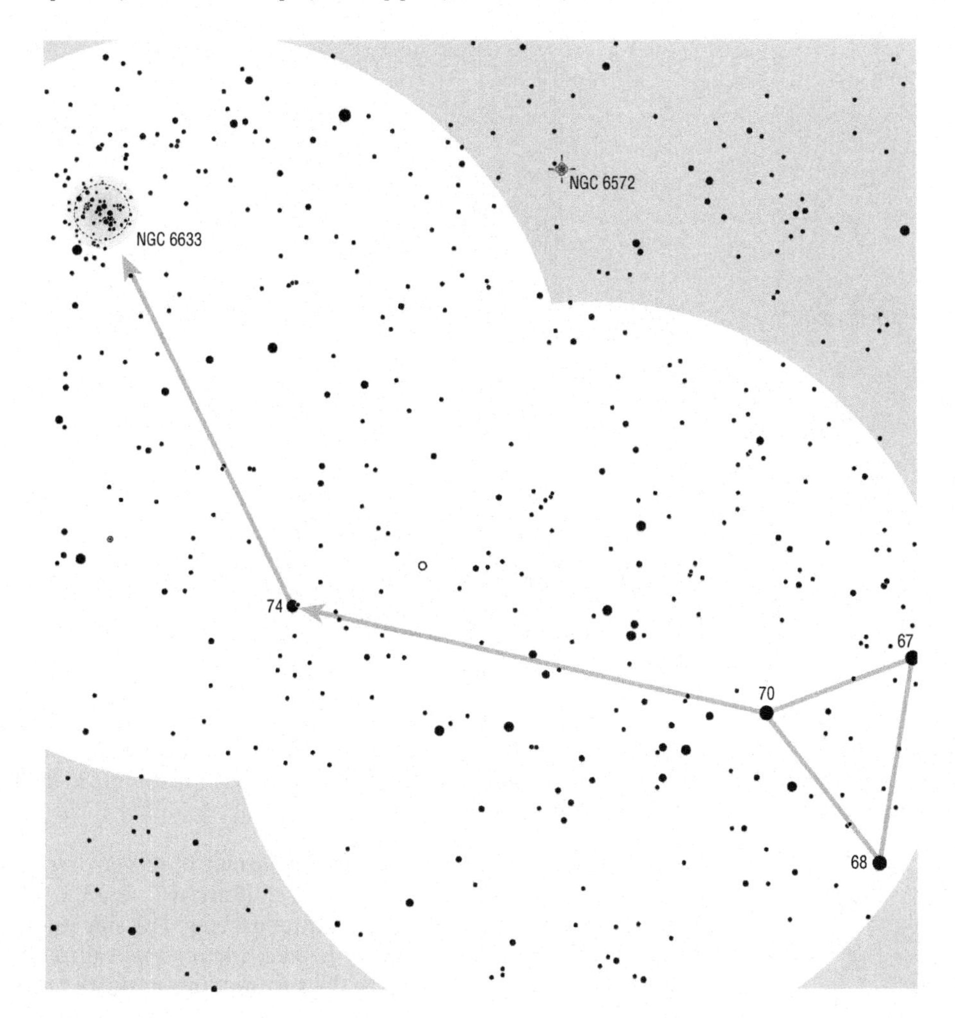

70 Ophiuchi is a splendid double star for mid-sized telescopes. Stars with magnitudes 4.2 and 6.2 are 5.3 arcsec apart (p. a. 136°, 2006); the pair is distinguished by its strong color contrast; the brighter star is yellow, while the fainter one is orange-reddish.) If you shift your binoculars just slightly in the same direction, the entire cluster will jump into your field of view. The other way is to start at Theta Serpentis and move northwest across the open cluster IC 4756 (page 348), which will bring you to NGC 6633.

The cluster, which is of a distinct oblong shape, contains approximately 300 stars. The brightest are of magnitude 7. When there are excellent observing conditions, approximately 60 stars can be seen in the area of the cluster. On a perfect night, their numbers can increase up to 110! NGC 6633 lies 1,040 light years away from us.

The dark cloud, named the Pipe Nebula, lies in the southern part of the constellation, where the density of the stars in the Milky Way is the greatest, since we are in the close vicinity of the galactic center and are looking in the direction of the galaxy's central bulge. The leading star to the nebula is Theta. The dark area, which is practically without stars, lies to the east and south of it. When there are excellent observing conditions,

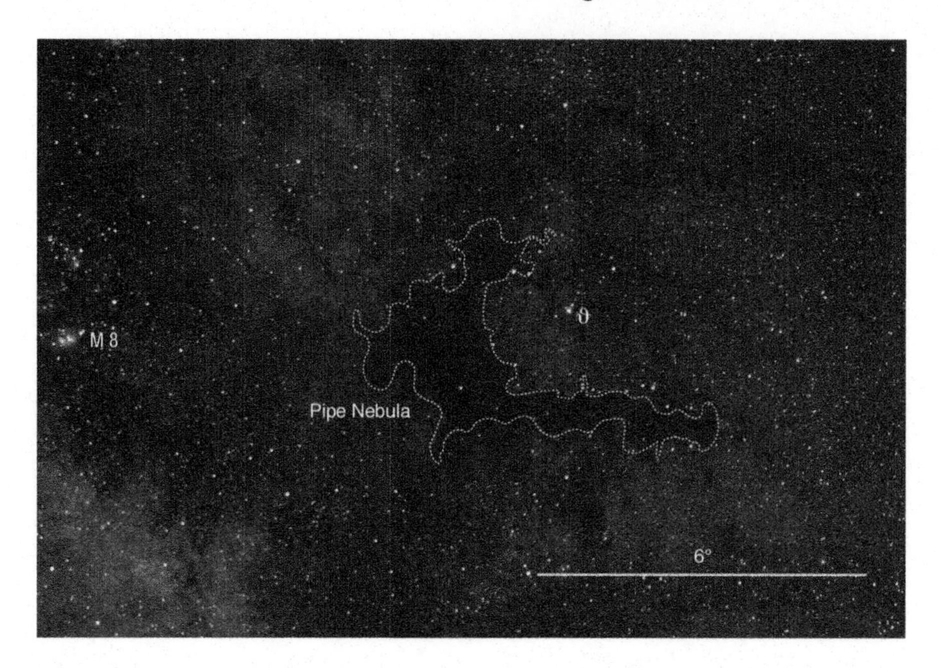

the Pipe Nebula is clearly seen even with the naked eye, and of course it can be seen even better through binoculars. Its northern edge is rimmed by bright stars that intensify the contrast with the dark bulk below them. The brightest has designation 36. Unfortunately, the Pipe Nebula in the field of view of binoculars is not as contrastive as it appears on images.

For the very end and for the most persistent observers, we can also mention the planetary nebula NGC 6572 (8m1/16″ × 13″). Through binoculars, it is seen merely as a magnitude 8 star. The star leading to it is 72 Ophiuchi. Since the planetary nebula, when viewed through binoculars, appears exactly the same as the surrounding stars, we have to use a detailed finder chart to recognize it. This chart can be found on 357. When we catch the stars 72 and 71 in our field of view, let's find two magnitude 7 stars to the southeast of them, both clearly seen through binoculars. South of this pair we can see five magnitude 9 stars, and just below them lies the planetary nebula, which is about one magnitude brighter than these stars.

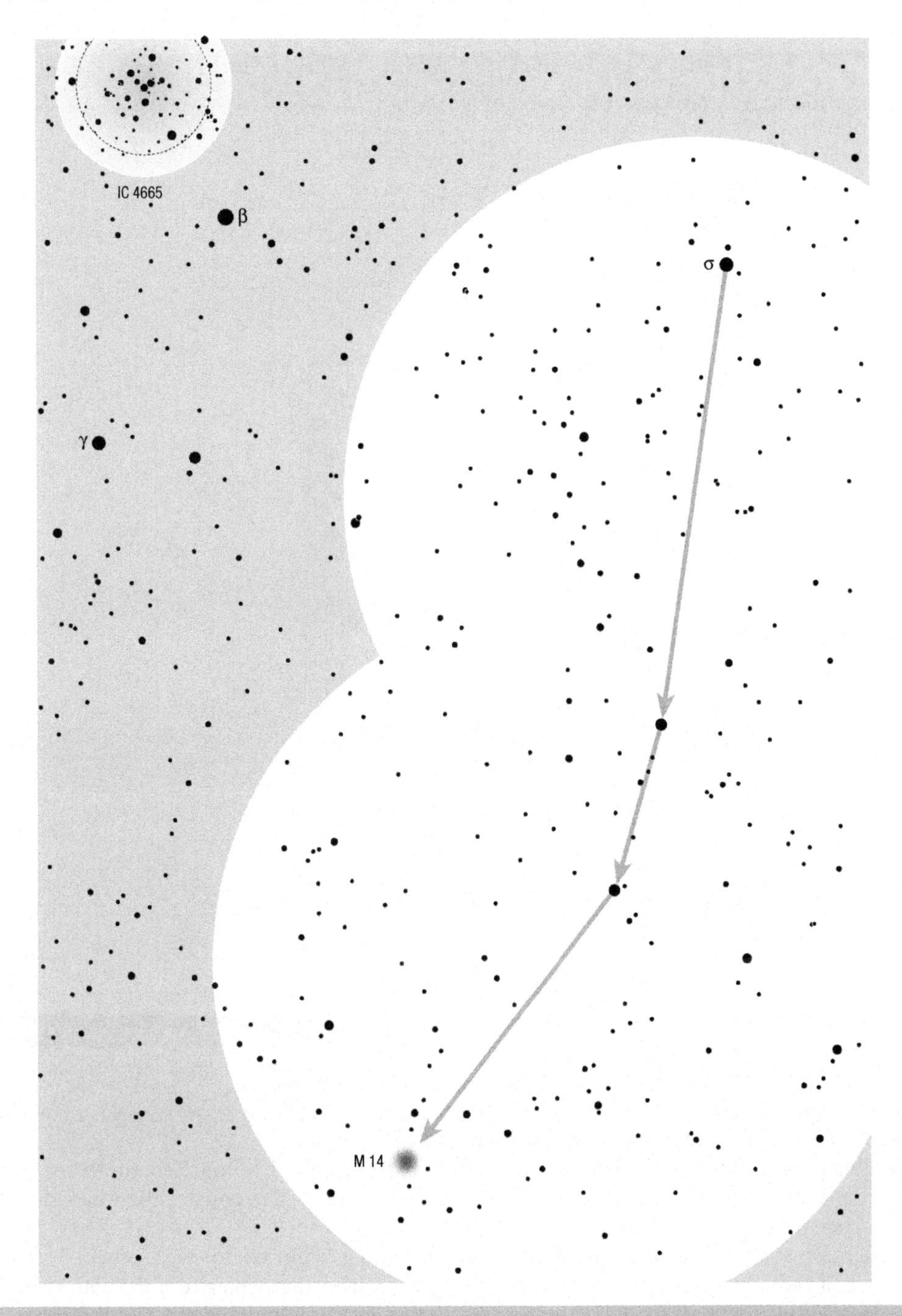

Figure 10.7 The way to the elusive globular cluster M 14 leads from Sigma across the magnitude 6 stars. At the top of the chart we can see the open cluster IC 4665, which lies in the vicinity of Beta

ORION (The Hunter)

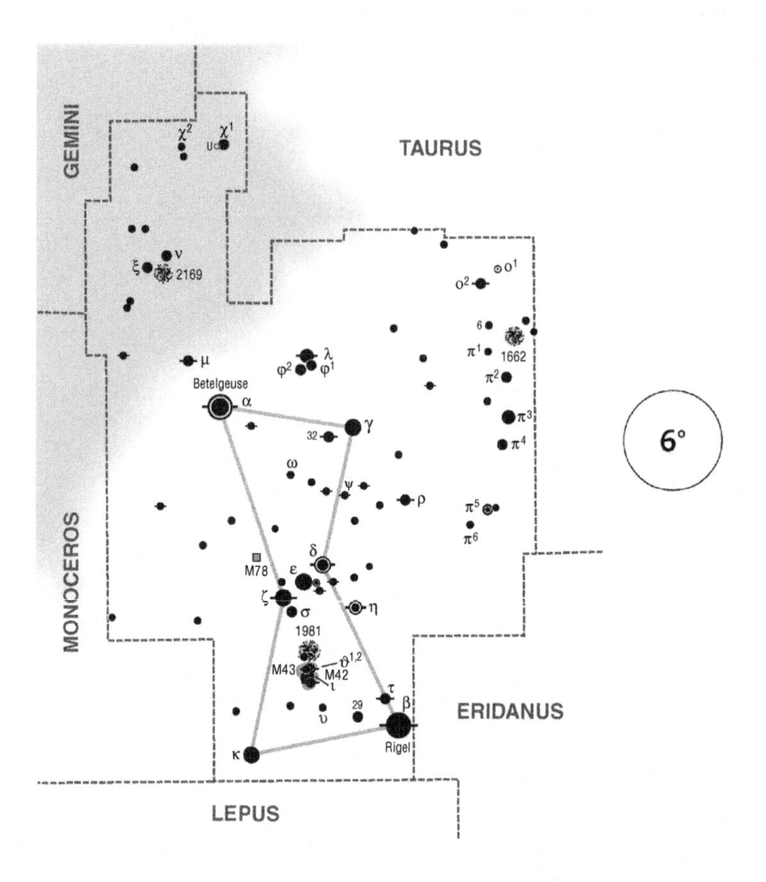

CONSTELLATION CULMINATES		
mid November at 2 a.m.	mid December at midnight	mid January at 10 p.m.

The winter Orion is undoubtedly the most splendid of all constellations, for it has eight stars that belong among the top 150 brightest in the sky. Since it is situated on the celestial equator, it can be seen from all inhabited parts of the world. The brightest stars are Beta or Rigel (0m18), Alpha or Betelgeuse (from 0m0 to 1m3), Gamma or Bellatrix (1m6), Epsilon, Zeta (both 1m7), Kappa (2m1), Delta or Mintaka (2m2, slightly variable), and Iota (2m7). The constellation has the distinct shape of a clepsydra and cannot be overlooked in the sky.

The brightest stars Betelgeuse and Rigel are in striking contrast to each other. The orange Betelgeuse is a red super giant, so large that if we placed it in the center of our Solar System, the star would span all the way to the Asteroid Belt, between Mars and Jupiter. On the other hand, the entirely white Rigel is a real cosmic searchlight, with a luminosity that surpasses our Sun by as much as 36,000 times. The difference in color is obvious even with the naked eye, while with any optical aid it only becomes more evident. All of the remaining brighter stars in the constellation are hot and white.

With a magnitude of 0.45 Betelgeuse is the 9th brightest star in the sky. It is an irregular pulsating variable star. Its brightness changes slowly, with an average period of 6.4 years. During the pulsation the size of the star also changes. When it is at its largest, its diameter is equal to the orbit of Jupiter! The star is so big that it is seen through modern interferometers as a disk with an angular diameter of between 0.034 and 0.054 arcsec. Betelgeuse is not merely one of the largest stars; its brightness is that of 8,400 Suns. The surface temperature of the star is rather low and is on average approximately 3,100 K. Only 13% of the emitted energy is in the visual part of the electromagnetic spectrum. The rest is emitted in infrared wavelengths. If we had eyes sensitive to infrared, Betelgeuse would appear as the brightest star in the entire sky. It is also at the top according to its mass. Astronomers estimate that its mass equals that of 20 solar masses. From the mass and the size we can easily calculate the average density of the star, which is – similar to Myra in Cetus – 10,000 times less than the density of the air we breathe. Of course, Betelgeuse has a dense core, but its outer layers are extremely diluted, so the star is jokingly known as the "beautiful red hot vacuum." This admirable super giant is 430 light years from us. If we could see our Sun from such a distance it would shine as a magnitude 10 star and would barely be seen through binoculars.

Rigel is the 7th brightest star in the sky. It is also a super giant, but much younger than Betelgeuse, and is therefore on the main sequence of the H-R diagram. Its surface is extremely hot (12,000 K), which makes the star entirely white. It is 770 light years from us. If it were closer, let's say at the distance of Sirius (8.6 light years), it would shine in the sky as a magnitude –10 star and would illuminate the Earth as much as one-fifth of a full Moon. Just to compare: when Venus is at its brightest, it has a magnitude of a "mere" –4.8.

Gamma is the 26th brightest star in the sky. It is 240 light years from us, and its luminosity is that of 940 Suns.

Kappa is the 54th brightest star in the sky. It is 720 light years away and is a real giant. Its luminosity surpasses our Sun by as much as 5,500 times.

Orion's Belt is composed of the stars Delta, Epsilon, and Zeta. All three are truly giants. Delta is the 73rd brightest star in the sky. Its light travels 920 years to reach us. Its luminosity is as much as 7,600 times that of the Sun. The star has a 53.3 arcsec distant companion of magnitude 6.8 (p. a. 1°, 2004). The stars can be separated through binoculars, but they are not known for any color contrast. And you might have problems seeing the fainter companion, due to the glare of the brighter one. This pair can only be separated with excellent optics!

Epsilon is the 29th brightest star in the sky, even though it is very far away – 1,350 light years from us. Its luminosity is as much as 27,000 Suns.

Zeta is the 31st brightest star in the sky. It is 820 light years from Earth. This star also has great luminosity – 9,700 times that of the Sun. In the vicinity of Zeta lies the emission nebula NGC 2024; however, it cannot be seen through binoculars. On a really clear, calm, and cold winter night, when the temperatures are deep below zero, you might be able to glimpse its brightest part with excellent optics, but only as a very faint stripe of light close to the star.

The stars below Orion's Belt represent the asterism known as the Orion's Sword. The brightest of these is Iota, which is the 124th brightest star in the sky. This is also a giant; its luminosity is as much as 10.000 times that of the Sun. And despite the fact that it is 1,330 light years away, it shines in the sky as a magnitude 3 star. Iota is the southernmost among the sword's stars. It is an interesting quadruple star with a wonderful color contrast, but through binoculars the stars cannot be seen nor separated.

Lambda (3m4) represents Orion's head and is an attractive double for small telescopes. The pair is composed of white stars of magnitude 3.5 and 5.4, separated by 4.3 arcsec (p. a. 44°, 2005). Through binoculars they are seen as a single star. Together with two nearby stars it forms a little triangle, which is just large enough to place the full Moon's disc inside it. Once you find the triangle in the sky you will be certain that we have made a mistake, but we did not. The apparent size of the Moon in the sky is – believe it or not – greatly enlarged by the processing that takes place in our brain.

Sigma is an interesting multiple star, but through binoculars only components A (3m8) and E (6m3) are clearly seen. They are 41.5 arcsec apart (p. a. 62°, 2003). Both stars are "boringly" white.

The long-period variable star U Orionis changes its brightness from magnitude 4.8 to 13 with an average period of 368.3 days (Figure 10.8). It is easy to find as it lies close to the rather bright Chi at the extreme northern part of the constellation. The variable is distinguished by its distinctive orange color, which becomes increasingly deep the lower the brightness of the star falls. U Orionis is one of the rarest variables of its kind, for it can be

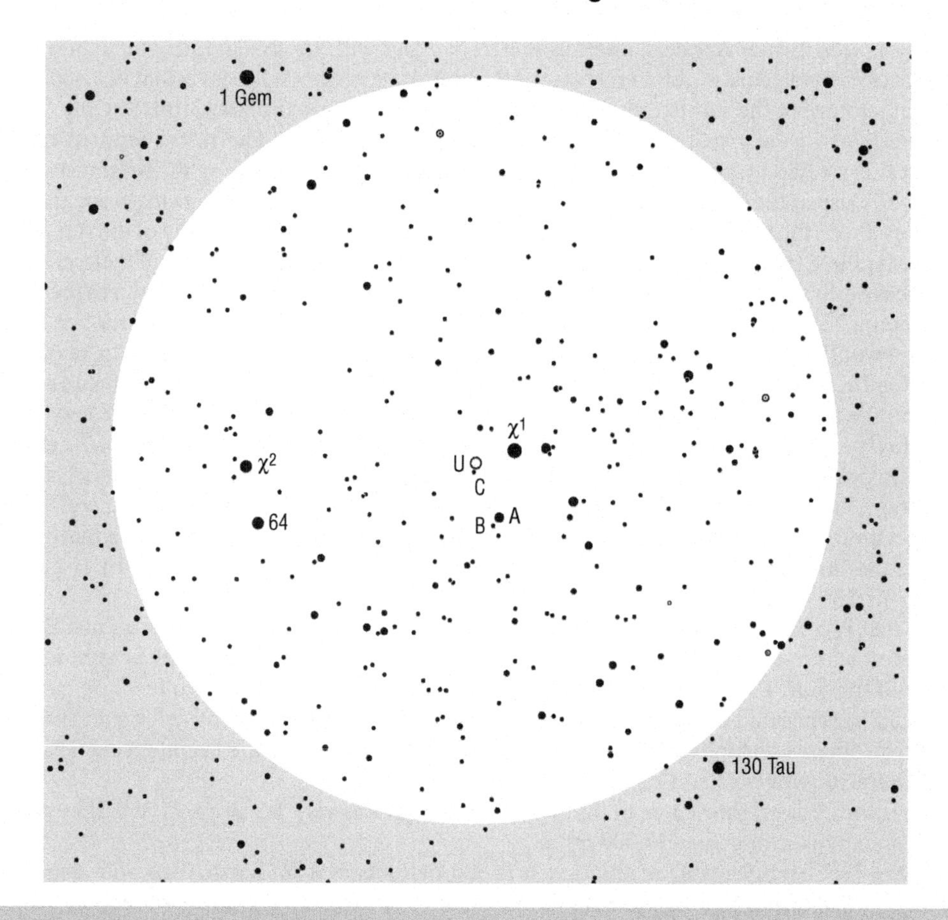

Figure 10.8 The surroundings of the long-period variable star U. The comparison stars are chosen to be as close as possible to U: χ–1 (4m4), A (5m9), B (8m3), and C (8m9)

seen with the naked eye at maximum brightness, while at minimum brightness it slips way below the reach of binoculars. The star reached one of its maximums on December 27, 2007. From this date and the known period, one can easily calculate the dates of the next maxima.

The big and bright Orion Nebula M 42 (4m0/85′ × 60′) lies in the middle of Orion's Sword. It is the brightest and the closest nebula to us, it can be seen with the naked eye, and it is truly magnificent in all instruments. These are attributes of the nebula mentioned in astronomy literature. And, in this case, the authors did not exaggerate. If you want a friend or an acquaintance to become enraptured with amateur sky gazing, you should show him the Moon, Jupiter, Saturn, and . . . the Orion Nebula through a telescope (Figure 10.9).

Approximately 4 degrees south of Zeta, right in the middle of Orion's Sword, we can see a tenuous, but clearly visible, hazy patch of light with the naked eye. Viewed through binoculars we can see its true beauty. When there are excellent observing conditions, the vast and transparent nebula can be seen. It is whitish or even slightly greenish in color, but on a perfect night we can also glimpse the reddish tint alongside the greenish hue. In this case, the nebula itself will be larger. If observing conditions are perfect we could see its fainter, marginal regions. The nebula includes a few rather bright stars, and these illuminate the clouds of dust and

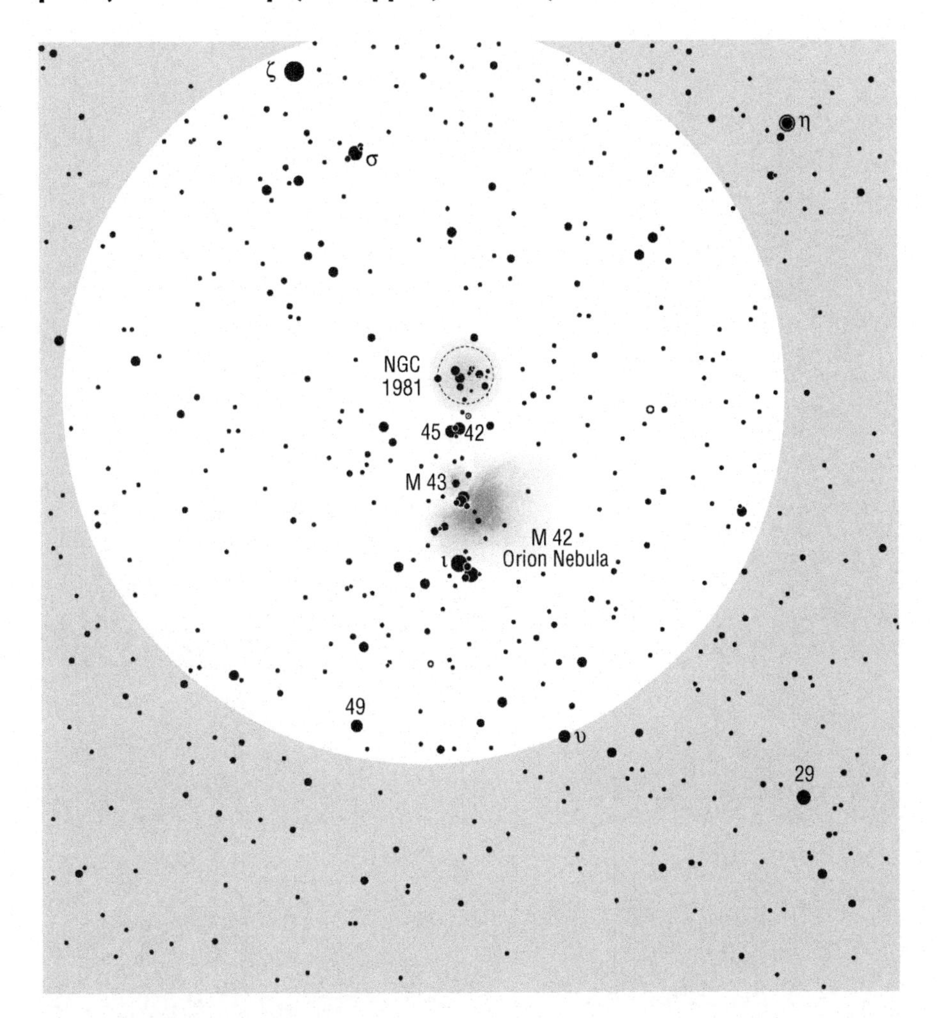

gases, which otherwise would not be visible at all. The brightest are Theta-1 and Theta-2, which are both magnitude 5 stars. These are separated by 135 arcsec (p. a. 314°), so they can easily be resolved through binoculars. Both are interesting multiple stars.

Theta-1 is definitely the most famous multiple star, popularly named the Trapezium (Figure 10.10). The four brightest stars are seen even when viewed through small amateur telescopes, but unfortunately they cannot be seen through binoculars, except maybe component D, which is 21.2 arcsec from Theta A. There are nine stars in the system of Theta-1.

Theta-2 is a triplet, composed of stars with magnitudes 5.0, 6.2, and 7.5. Theta-2A and Theta-2B are 52.8 arcsec apart (p. a. 93°, 2005). Theta-2A and Theta-2C are separated by 128.4 arcsec (p. a. 99°, 2004). We can usually separate all three stars through binoculars; however, we might have problems with Theta-C, which is rather faint and hides within the shine of the Orion Nebula.

Just what is this spectacular nebula? It is a vast, cold cloud of gases and dust that does not emit light and is composed mainly of hydrogen (91%), helium (9%), carbon (0.05%), oxygen (0.02%), and nitrogen (0.02%), as well as smaller quantities of sulphur, neon, chlorine, argon, and fluorine. The gases and dust reflect the light of the nearby stars, but in the vicinity of the hot young stars the gases in the nebula are excited by the ultraviolet light emitted by these stars and so emit their own light. Although it is seen as tenuous and transparent, it contains the matter of 10,000 Sun-like stars!

Figure 10.9 The surroundings of the Orion Nebula. Compare this image with the detailed chart on previous page

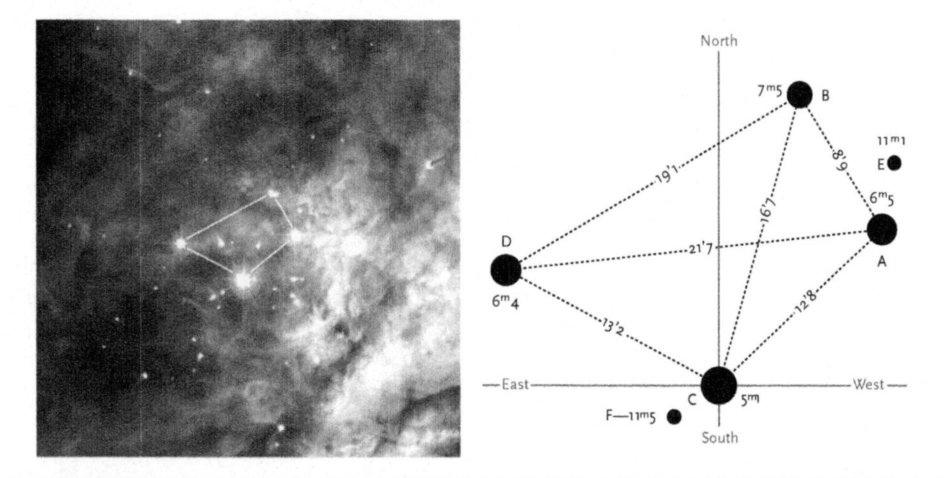

Figure 10.10 Theta–1 is probably the most famous multiple star in the sky. By observing it through binoculars we are slightly deprived, due to the poor resolution and the faintness of most of the stars. The image on the left was taken in yellow light, which minimized the disturbing influences of the nebula

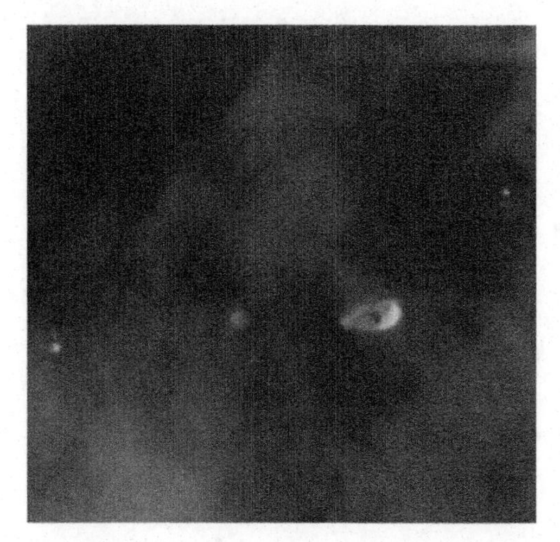

Figure 10.11 Four young stars on the edge of the Orion Nebula. Three of them are surrounded by disks of dust and gas. On the side facing the nearby, hot young stars, the gas emits light due to the atoms that are ionized by the stars (the stars are not seen on this image). The cold, reddish central protostar can be seen through the gases and dust

The Orion Nebula is not merely an ornament in the sky. It is also interesting for professional astronomers, who got the first evidence that new stars are born from the gases and dust in the nebula in 1979, through radio wave observations. This part of the electromagnetic spectrum can penetrate the thick layers of dust clouds impenetrable by visible light. In the heart of the nebula radio astronomers first of all noticed 6 and later as much as 26 denser formations. Their true nature was discovered in 1993 with the use of the Hubble Space Telescope. Images taken in near infrared light presented a true surprise for astronomers. Not only could they see over 100 denser formations on these images, for the very first time they could clearly see the embryos of stars, surrounded by protoplanetary disks. Due to atom ionization and recombination, the side that was turned toward the bright stars shone (just like the Orion Nebula itself), while the other side did not. The images depicted the star embryos in various stages of development. In some of them one could already sense the emerging star in the center.

M 42 is rather young, probably not older than 30,000 years. The nebula is 1,270 light years from us. The area that can be seen on most images measures approximately 25 light years in diameter.

The Orion Nebula is merely the brightest part of the vast interstellar cloud of gases and dust that spans the entire constellation. We can see it best on image on Figure 10.15.

Nebula M 43 (~7m/20′ × 15′) is in fact a part of the same cloud of gases and dust as the Orion Nebula. These objects are separated by a belt of opaque matter. On the images depicting the Orion Nebula, M 43 is always "there," and many observers are not even aware that it has its own designation. In binoculars, it is seen only as some kind of an outgrowth from the northeastern part of the large nebula. M 43 is also seen only because it has its own star (magnitude 8) that illuminates the clouds of cold gases and dust.

The open cluster NGC 1981 (4m6/25′) lies just 1 degree north of the Orion Nebula, and it therefore appears in the same field of view of binoculars. The cluster is very loose, composed of only 15 stars of magnitudes between 6 and 10. The detailed chart on page 367 can help you to recognize the stars within the cluster, which is approximately 1,300 light years from us.

Figure 10.12 On this detailed chart of M 78 the brightest arc of the eastern part of Barnard's Loop is also drawn. Its image can be seen on the next page. Barnard's Loop is seldom seen "live." But on some perfect winter nights, if the sky is especially dark, it is suitable for such a challenge!

Figure 10.13 On this wide-angle image of the constellation we can see the Orion Nebula (*below the center*), the nebulae surrounding stars 45 and 42 close above it, and the nebula surrounding Zeta (NGC 2024 and IC 434 with the famous dark Horsehead Nebula). We can also see the eastern arc of Barnard's Loop. Look at this image closely before you decide to search for this hard-to-see celestial veil

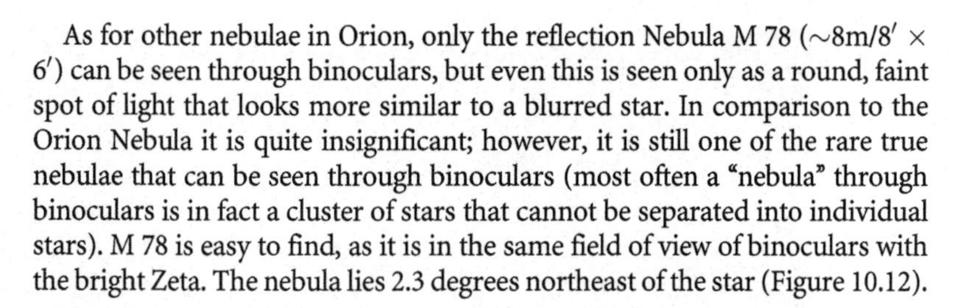

As for other nebulae in Orion, only the reflection Nebula M 78 (~8m/8′ × 6′) can be seen through binoculars, but even this is seen only as a round, faint spot of light that looks more similar to a blurred star. In comparison to the Orion Nebula it is quite insignificant; however, it is still one of the rare true nebulae that can be seen through binoculars (most often a "nebula" through binoculars is in fact a cluster of stars that cannot be separated into individual stars). M 78 is easy to find, as it is in the same field of view of binoculars with the bright Zeta. The nebula lies 2.3 degrees northeast of the star (Figure 10.12).

Barnard's Loop is a gigantic cloud of gases and dust that extends across a large area of Orion, roughly centered in the Orion Nebula (Figure 10.13). The west side contains more dust, while the east side contains more ionized hydrogen, which emits a well-defined red light. The brightest part of the loop takes the form of a large arc, but since it is really difficult to see it, it represents a great challenge even for skilled observers. A great

Figure 10.14 Probably the most famous and most photographed area in Orion (beside the Orion Nebula itself) is the Horsehead Nebula (designation B 33), which is a dark, opaque to light area in the rather bright surrounding nebula IC 434. Close to Zeta we can see the bright emission nebula NGC 2024 (seen through telescopes), which is divided in two by a dark lane of gases and dust. Unfortunately, these wonders cannot be seen through binoculars

Figure 10.15 This photograph conjures up the true beauty, motion, and extent of the interstellar clouds of gas and dust in the constellation of Orion. It was taken by an amateur astronomer named Robert Gendler, who combined the images in visible light and in the H-alpha light that is emitted by the excited hydrogen atoms

observation aid – when viewing with the naked eye or through binoculars – is a special filter that increases the contrast (UHC, or ultra high contrast) and darkens the sky. If you have the opportunity to observe it from a truly dark site, say class 1 or 2 on Bortle's Scale, you should give it a try. Otherwise the tenuous arc is best seen on images (such as the one on 371). This interstellar cloud is believed to be an ancient supernova remnant.

Approximately in the middle of the arc, composed by stars Pi–1 to Pi–4, Omicron–1, and Omicron–2, lies the open cluster NGC 1662 (6m4/20′). This is a group of rather bright stars that are seen through binoculars as a faint nebula, but with averted vision we can definitely separate it into individual stars. This is a cluster only those serious observers who want to see everything that is possible to see through binoculars would enjoy.

NGC 1662 is approximately 1,400 light years from us, which makes its real diameter a mere 5 light years.

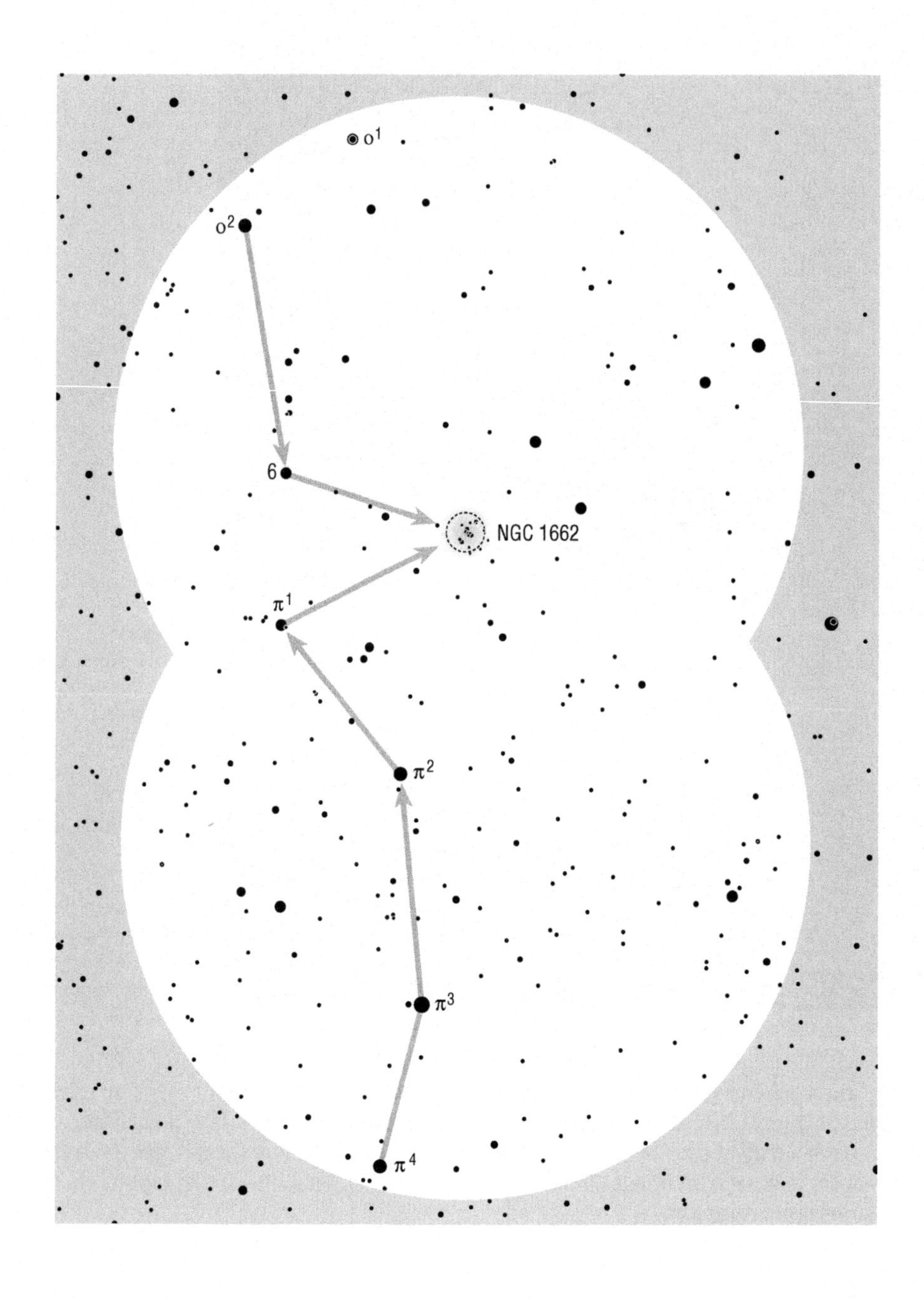

In larger telescopes with greater magnification the open cluster NGC 2169 (5m9/7′) represents a rather interesting group of 20 stars, gathered in a small area of the sky. Unfortunately, through binoculars only the four brightest stars can be seen, for they are of magnitude 7, 8, and 9. The cluster is easy to find, since it lies close to the bright Xi. But it does not include anything truly attractive; it is merely for NGC collectors.

The cluster is 3,600 light years from us, so its actual diameter is a mere 6 light years.

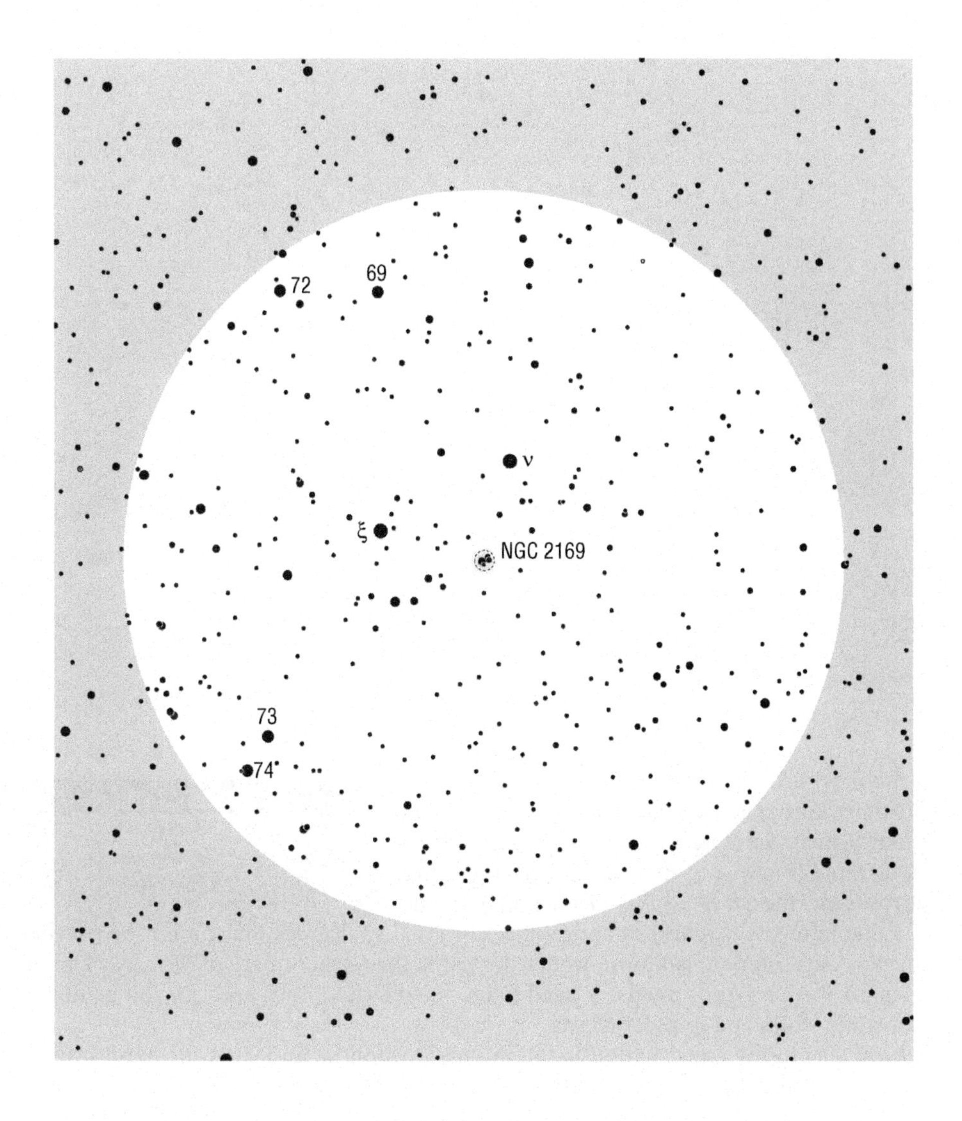

PEGASUS (The Winged Horse)

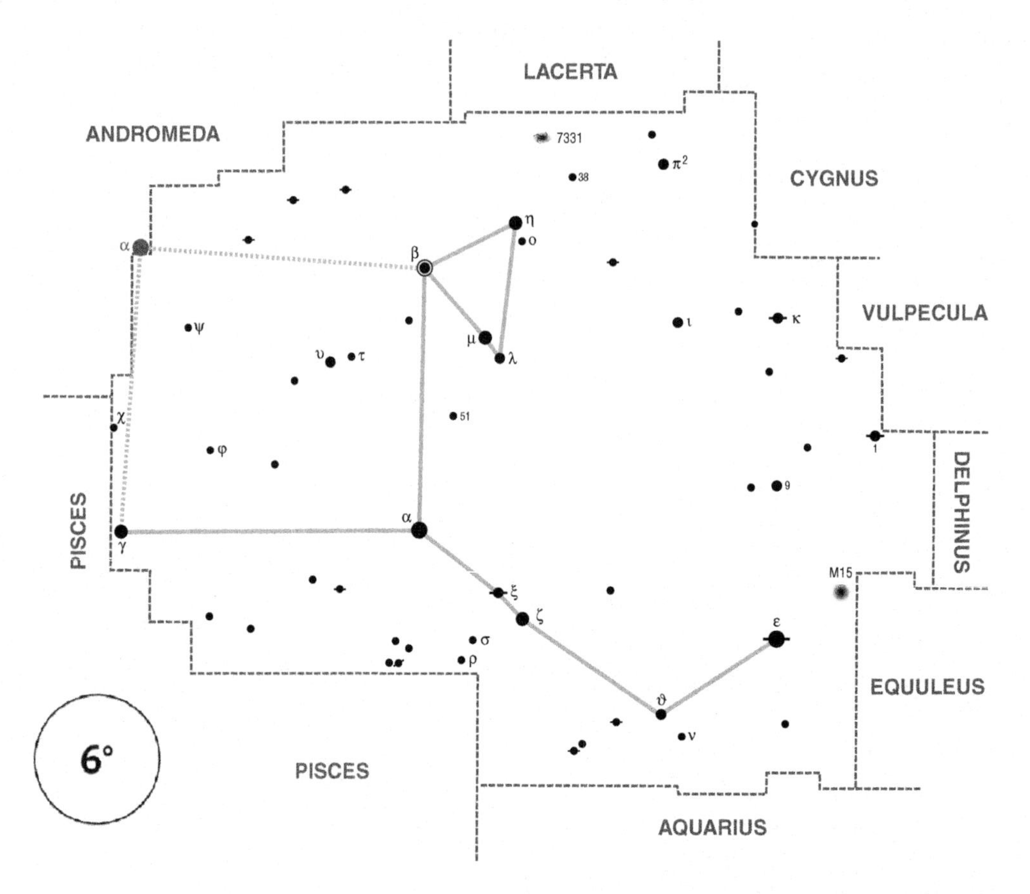

CONSTELLATION CULMINATES		
in begin. August at 2 a.m.	in begin. September at midnight	in begin. October at 10 p.m.

Pegasus is the most prominent autumn constellation seen from the northern latitudes. The brightest stars are Epsilon (2m4), Beta or Scheat (variable 2m1–3m0), Alpha or Markab (2m5), Gamma (2m8), and Eta (2m9). Together with Alpha Andromedae the stars Alpha, Beta, and Gamma form a rectangle (almost a square) of approximately 15 × 15 degrees that cannot be overlooked in the autumn sky. This asterism is known under the name the Great Square of Pegasus. The brightest star in the square – Alpha Andromedae – used to be called Delta Pegasi and can still be found under this name in some tables and in older charts.

The brightest star in the constellation is the orange Epsilon or Enif. It is the 82nd brightest star in the sky. It lies far from the Square, in the vicinity of the constellation Equuleus. Epsilon is 670 light years from us and is a red giant, more than 3,600 times as bright as the Sun. If the star was closer to us, let's say at a distance of 32.6 light years (1 parsec), it would be one of the brightest stars in the sky.

The orange Beta represents the northwestern corner of the Square. It is the 62nd brightest star in the sky (at maximum brightness), a semi-regular variable with changes in magnitude that range between 2.1 and 3.0. The period is not constant; it has an average value of 35 days. The star is similar to

Betelgeuse, except it is smaller and less luminous. It has a surface temperature of 3,100 K, a mass of 5 solar masses, and a diameter that is about 160 times bigger than that of the Sun. Its luminosity oscillates between 240 and 500 times that of the Sun. The star is 200 light years from us.

Alpha represents the southwestern corner of the Square. It is the 92nd brightest star in the sky, 140 light years from us. Its luminosity surpasses our Sun by 140 times.

Gamma is the 137th brightest star in the sky. It lies in the southeastern corner of the Square and is 334 light years from us. It is a giant, with a luminosity of 590 times that of the Sun. Gamma is a variable star with an unusually short period of 3 h 38 min. However, its brightness changes by only one-hundredth of a magnitude, which makes it impossible to detect with average amateur equipment.

Close to the western side of the Square, in about the middle of Alpha and Beta, lies a magnitude 5 star with the designation 51 Pegasi. This star is approximately 40 light years away and is similar to our Sun. In 1995, this star made history. From the periodic shifts of absorption lines in its spectrum astronomers discovered that the star is encircled by at least one planet the size of Jupiter. This first discovery of an extraterrestrial planet led to an intense search for planets around other nearby stars. So far, we know of the existence of at least 300 alien planets!

The most interesting object in Pegasus is the globular cluster M 15 (6m0/18′). The cluster is on the border of visibility with the naked eye. It lies on the extension of the line between Theta and Epsilon. If Epsilon is caught in the field of view of binoculars, and the star is moved to the southeastern edge, M 15 appears on the northwestern edge. The cluster is seen as a tiny, clearly visible nebulous spot of light with a magnitude 6 star in its immediate vicinity. Through larger amateur telescopes the edge of the cluster breaks into individual stars, while its central part cannot be separated even with the largest telescope in the world (Figure 10.16).

M 15 is one of the densest globular clusters in the sky, known for its extremely bright and dense core. Its apparent diameter is just 20 arcsec, while the entire cluster measures as much as 12 arcmin! The uniqueness of M 15 lies in its having a small planetary nebula designated as K 648, with a magnitude of 13.8 and a diameter of only 1 arcsec. The nebula lies in the northeastern part of the cluster. Measurements show that the planetary nebula is a true and not merely an apparent member of the cluster. Unfortunately, it cannot be seen through binoculars or even larger amateur telescopes.

The cluster is approximately 34,000 light years from us; its actual diameter is 175 light years, and it contains at least half a million stars. From this distance and the apparent size of the core (20 arcsec), it is not hard to calculate that the actual diameter of the core measures a mere 6 light years (the closest star to us is 4 light years away). And within this volume of space – in cosmic terms extraordinarily small – approximately 50,000 stars are arranged. It is almost impossible to imagine that in the sky of an imaginary planet circling a star in the central part of the cluster as many as 50,000 stars that surpass the brightest stars in our sky could be seen, and then there are another several hundred thousand bright stars. Just for comparison, on a clear, Moonless night on Earth, far away from light-polluted areas, we can see approximately 4,500 stars in the sky.

The spiral galaxy NGC 7331 (9m5/11′ × 4′0) represents an ultimate challenge for those observers who want to push their binoculars to the extreme limits, since it is on the border of visibility (Figures. 10.17 and 10.18). Some reliable observers claim it is visible through binoculars, if all circumstances are perfect. The night has to be especially clear and calm, completely Moonless, and the galaxy has to be close to culmination. Despite its rather low integral brightness the galaxy can boast of having a bright nucleus. The other advantage for mid-northern observers is the fact that the galaxy is only 10 degrees from the zenith at culmination. In the field of view of binoculars it will show as a tenuous, few arc minutes long stripe of transparent light. It is most likely that it will be

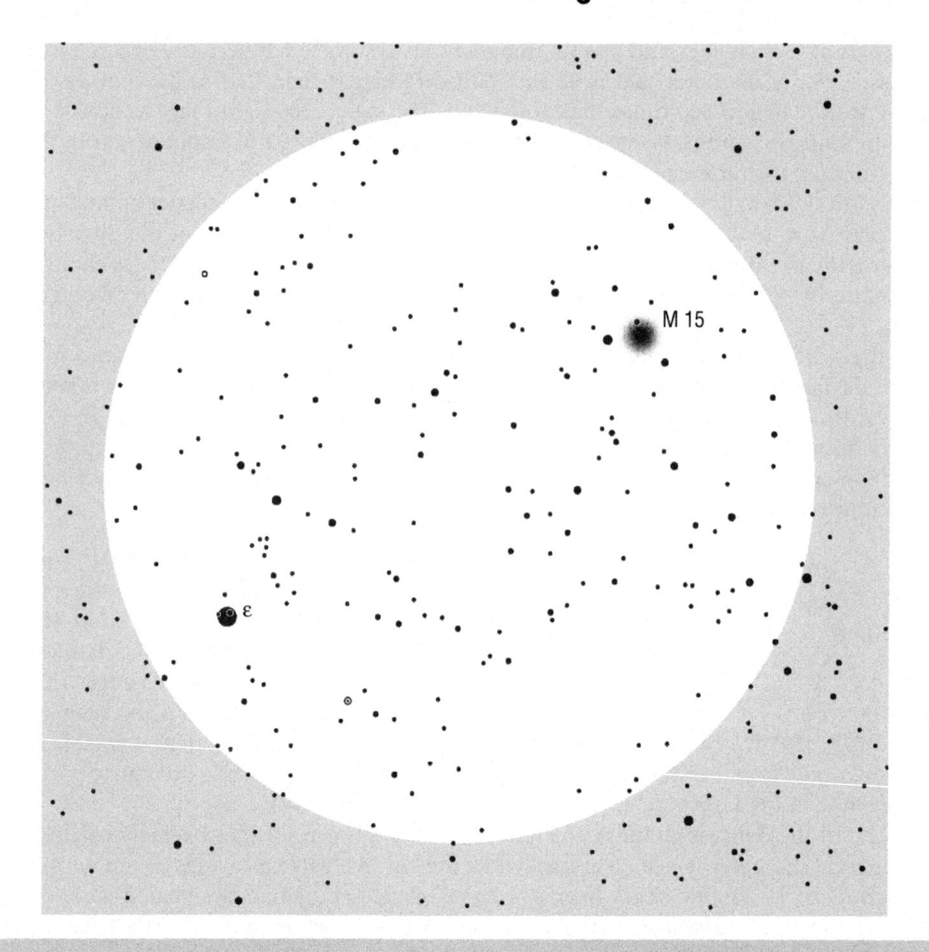

Figure 10.16 Globular cluster M 15 is easy to find, since it is in the same field of view of binoculars with the brightest star of the constellation

seen only with averted vision and by gently shaking the binoculars, as human eyes are more sensitive to moving objects. If you do not enjoy in these types of observations, there is a very simple solution: use 20 × 50 or even 20 × 60 binoculars or a telescope.

The galaxy is seen almost edge on. It is inclined by merely 15 degrees from its equatorial plane toward Earth. It is estimated to be 46 million light years from us, and its actual diameter is estimated at 65,000 light years.

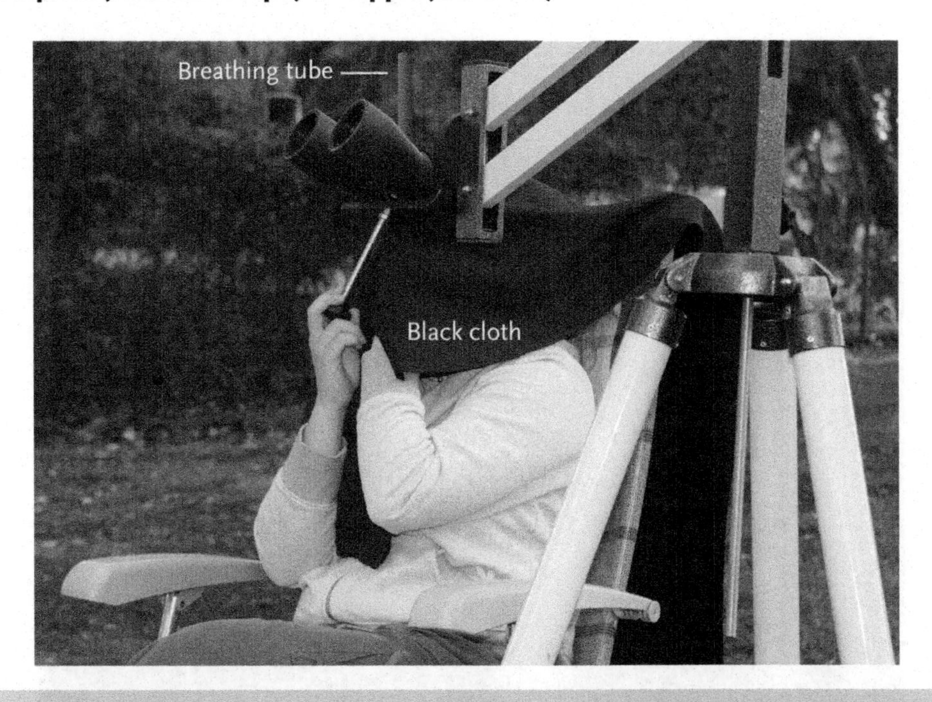

Figure 10.17 Here is the best way to observe such elusive objects as the galaxy NGC 7331. A black cloth over your head and binocular eyepieces prevent peripheral light from disturbing your observations. Only in complete darkness can your eyes fully adapt to night vision. A breathing tube is indispensable for preventing dew forming on the eyepieces, which would be caused by your exhaling. In fact, all celestial objects should be observed like this. But if you do this make sure you tell your neighbors what you are going to do first; otherwise, they just might think you have lost it

Figure 10.18 The faint spiral galaxy NGC 7331 represents a true challenge for all serious observers using only binoculars. In the lower right corner of the image we can see the group of five galaxies known as the Stephan's Quintet, which represents a real challenge for larger amateur telescopes

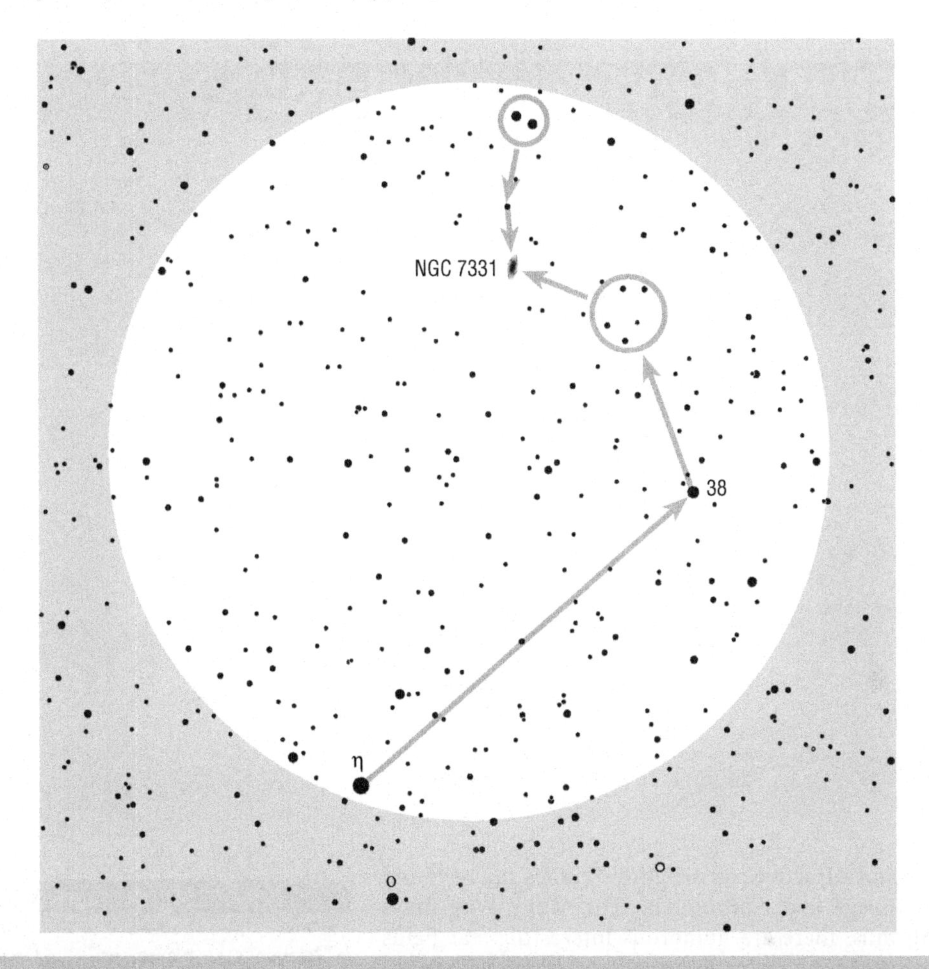

Figure 10.19 The star leading to the galaxy is the bright Eta. If it is moved to the southern edge of the field of view, two close magnitude 6 stars appear on its northern edge. For basic orientation you can also see the magnitude 5 star with the designation 38. The exact position of the galaxy can be determined with the help of the magnitude 8 star south of the bright couple or with a trapezoid of magnitude 8 stars northeast of star 38. If the magnitude 8 stars are not seen clearly, you have to try finding the galaxy at some other time. You can try later if observing conditions improve or on another night

PERSEUS (The Victorious Hero)

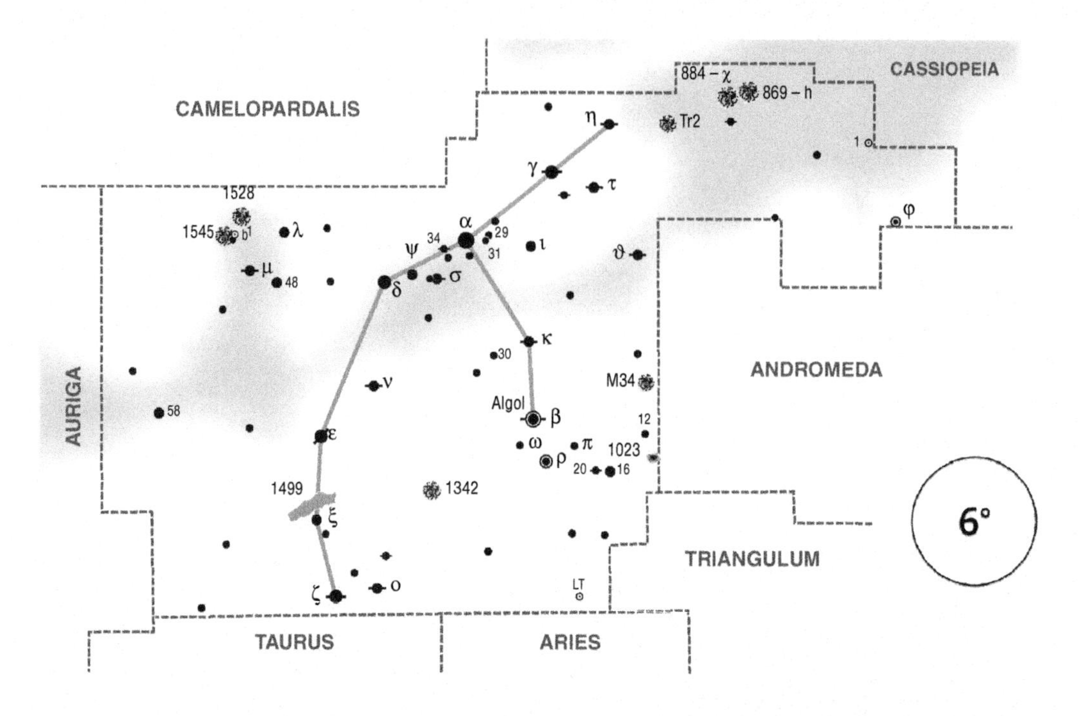

CONSTELLATION CULMINATES		
mid October at 2 a.m.	mid November at midnight	mid December at 10 p.m.

The large and attractive constellation Perseus can be found between Auriga and Andromeda. The Milky Way flows across it; thus, there are numerous interesting star fields in this constellation for panoramic observation with binoculars. The brightest stars are Alpha or Mirfak (1m8), Beta or Algol (variable, 2m1 at maximum), Zeta (2m8), Epsilon, and the yellowish Gamma (both 2m9), Delta (3m0), and Rho (variable, 3m2 at maximum).

Zeta is the 142nd brightest star in the sky. At 980 light years, it is very far from us. The star is a true giant, with a luminosity as much as 5,000 times that of the Sun. Zeta is the brightest member of the very large and loose open cluster named II Perseus (Roman numeral 2 + Perseus), which contains young, bright, hot stars. Beside Zeta the brightest members are Omicron, Xi, 40, and 42. The cluster is surrounded by a nebula of gases and dust, which is only revealed on long-exposure images. The brightest and elongated part lies only a degree north of Xi and, due to its shape, was named the California Nebula (NGC 1499; ~5m/160′ × 40′). The California Nebula is a rather large and bright celestial object, so it is a good target for amateur astrophotography (Figure 10.20). On color images, California is distinctly reddish. This light is emitted by ionized hydrogen atoms; hydrogen is the prevailing element within the nebula.

Alpha or Mirfak is the 35th brightest star in the sky. It is a giant, 590 light years away, with a luminosity of over 4,800 Suns. In its spectrum, astronomers see periodic shifts of absorption lines, but so far it is not clear whether the star is a close binary or the shifts occur due to the pulsating of the outer layers of the star.

Mirfak is the central star of a large and very loose open cluster, which is best seen through wide field binoculars. In the star charts it has the name Melotte 20 (1m2/2°8). The brightest members beside Mirfak are Psi, 30, 34, 29, and 31; at least 30 stars reach a magnitude of 9 and are therefore always clearly seen through binoculars. In excellent observing conditions their number increases. The cluster is as distant as Mirfak. Measurements show that the stars are moving away from us in the direction of Beta Tauri with the speed of 16 km/s. The bright Delta and Epsilon Persei have similar proper motion through space, so maybe these two stars are also distant members of the cluster.

Figure 10.20 The California Nebula with bright star Xi

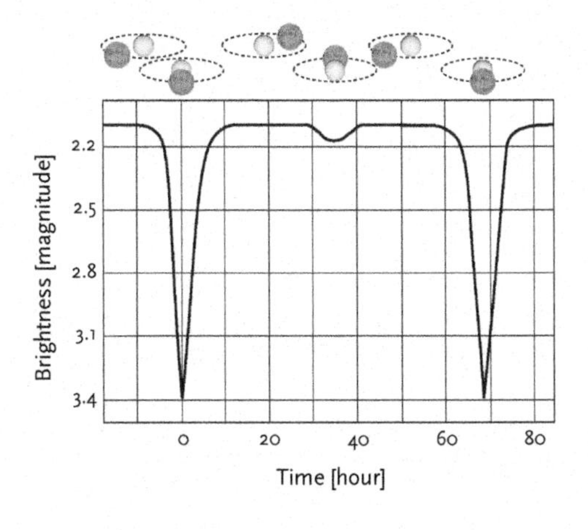

Beta or Algol (which means the "Demon Star") is the prototype of eclipsing variable stars. It has been known since time immemorial that this star changes its brightness. Today we know that Algol is a close binary star, so close that the two stars cannot be separated even through the largest telescopes in the world. The orbit is seen edge on, so in a period of 2.867 days we can see alternate eclipses of both stars. One is small, luminous, and blue; the other is larger, fainter, and orange. The brightness is the greatest (2m1) when both stars can be seen. The greater minimum (3m4) appears when the fainter star eclipses the brighter one, and the minor minimum (2m2) appears when the brighter star eclipses the fainter. The star reached one of its minimums on September 21, 2008, at 10:34 UT. From this date and with the known period, one can easily calculate the dates of the next minima. The fall in the brightness lasts for approximately 10 h. It's totally worth seeing!

The system of Algol also includes a third star, Algol C, which has been discovered by spectroscopic research. Its orbital period around the A–B pair is 1.862 years at an average distance of 80 million km (Mercury is 58 and Venus 108 million km from the Sun). This interesting triple star is 93 light years

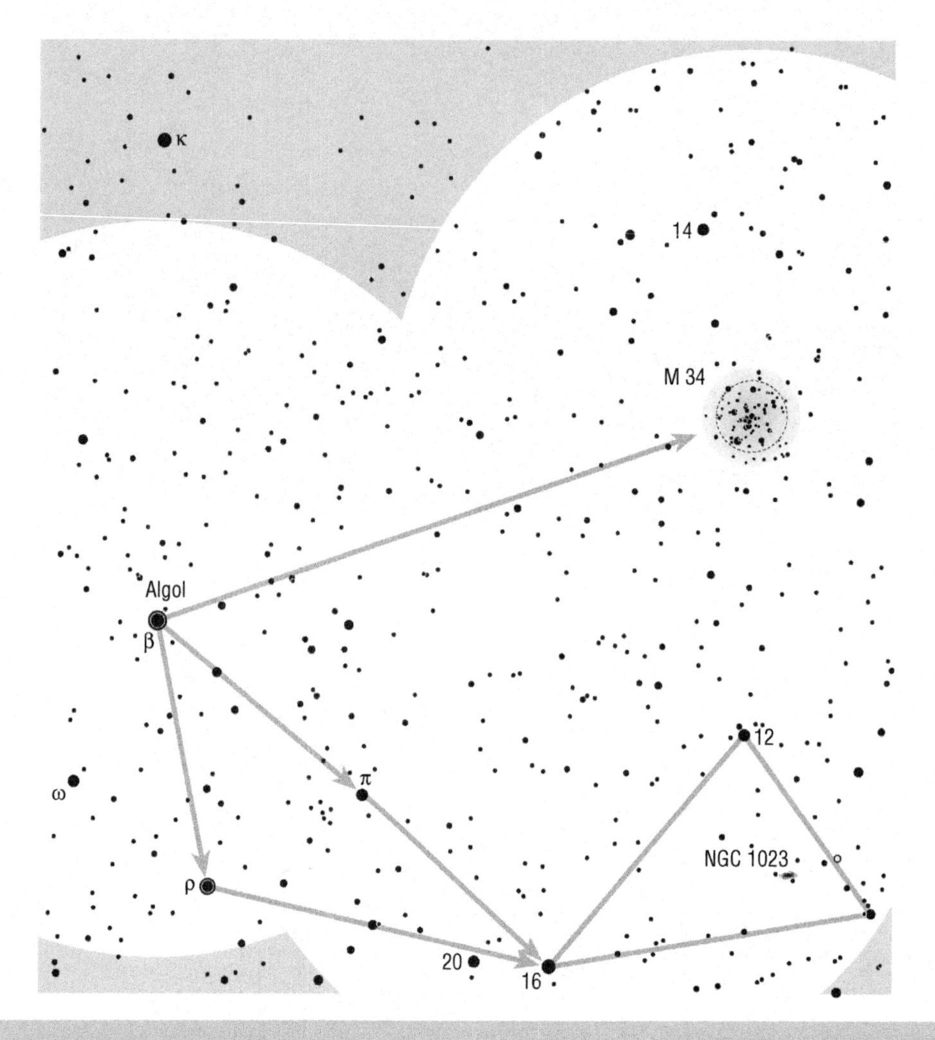

Figure 10.21 Algol is the leading star to the open cluster M 34 as well as to the barred spiral galaxy NGC 1023

from us. At its brightest, Beta is the 60th brightest star in the sky. The common luminosity of all three stars is 90 times that of the Sun.

Approximately halfway between Algol and Gamma Andromedae lies the open cluster M 34 (5m5/35′). If the binoculars are pointed toward Algol and the star is moved to the extreme eastern edge of the field of view, M 34 appears on the extreme western edge.

In total, the cluster has a few hundred members, of which the brightest shine with magnitude 8. Through binoculars approximately 20 stars can be seen when there are good observing conditions, approximately 40 when there are excellent conditions, and as much as 100 during perfect conditions! Of course, it always helps if you use averted vision.

M 34 is approximately 1,450 light years from us. Its densest part has a diameter of 9 arcmin (which corresponds to 15 light years), but the most distant members can be found as far as 25 arcmin (10 light years) from the center of the cluster. M 34 is estimated to be only 100 million years old.

The barred spiral galaxy NGC 1023 (9m4/8′7 × 3′3) is on the border of visibility when viewed through binoculars, so in order to glimpse it the observing conditions should be excellent if not perfect. The leading star is Algol (Figure 10.21). From it we move across Rho or Pi to the stars with the labels 20 and 16, which form the eastern corner of the triangle. The star in the northern corner is labeled 12. The galaxy lies within the triangle. In order to locate its exact position you can use the two magnitude 9 stars close to it. Through binoculars it is seen as a faint, slightly oval spot of light.

Figure 10.22 The eye-catching Double Cluster

NGC 1023 is the brightest member of the smaller nearby cluster of galaxies, which is 20.6 million light years from us.

In Perseus, we can find two magnificent open clusters: NGC 869 (4m3/30′) and NGC 884 (4m4/30′). These clusters are better known as the Double Cluster, or sometimes as Chi-h Persei. They are seen together in the same field of view of binoculars (Fig. 10.22). The clusters are easy to find, since they lie between Mirfak and the constellation Cassiopeia and on a clear, Moonless night are quite visible even with the naked eye. The leading star is the orange Eta (3m8).

These clusters have been known since antiquity. They were mentioned by the Greek astronomers Hipparchus and Claudius Ptolemaeus. Of course, in those times astronomers did not know of clusters, so they named them nebulae, or clouds. The true nature of these "clouds" was revealed in the seventeenth century, when astronomers examined them with the first telescopes. The clusters were first photographed as far back as 1890, and ever since they have remained popular targets of professional and amateur astronomers.

If you have the time and enjoy counting stars, you will be able to count as many as 120 stars up to magnitude 11 in NGC 869, and up to 140 stars in NGC 884. Together the two clusters include at least 3,000 members! The two brightest stars are magnitude 6, five are magnitude 7 and three are magnitude 8. All of these stars are super giants. The most luminous of them surpass our Sun by as much as 60,000 times! It is interesting that there are many red giants visible through binoculars and on color images. Today we know that massive stars live a fast and lavish life and that they grow old very quickly. The clusters are – despite the presence of the red giants – among the youngest of their type. The estimated age of NGC 869 is only 5.6 million years, while NGC 884 is considered to be a mere 3.2 million years old. It is also interesting to note that despite the young age, there are no traces of nebulosity (remains of star-building material) between the stars. Most probably, the giant stars with their powerful stellar winds have blown the rest of the material far off into interstellar space.

The clusters are 7100 (NGC 869) and 7400 (NGC 884) light years from us. They are therefore a close space neighbors. Their actual diameter is approximately 75 light years. If we were to live on a planet in one of these clusters, the other would be seen in the night sky as a 14 degrees big celestial object! This would definitely be a breath-taking sight!

The open cluster Trumpler 2 (5m9/20′) lies 2 degrees west from the orange Eta, which is also the leading star to the Double Cluster (Figure 10.24). When there are good observing conditions it is barely visible, because only half a dozen faint stars can be seen. On an excellent night, the number of stars could increase up to 20, but the impression of the cluster is not much stronger. Trumpler 2 is 2,100 light years from us.

Open cluster NGC 1342 (6m7/14′) lies in a rather empty area of the sky. The best way of finding it is by using Beta as the leading star. From Beta you have to move across Omega to Zeta. Halfway through, 6.5 degrees southeast of Beta, you will stumble across the cluster. When there are good observing conditions you can see only a few faint stars in the field of view of binoculars, around which you can glimpse a tenuous nebula of the other members. With perfect observing conditions, the number of stars increases to 20. In total, the cluster contains approximately 50 stars. NGC 1342 is 2,200 light years from us.

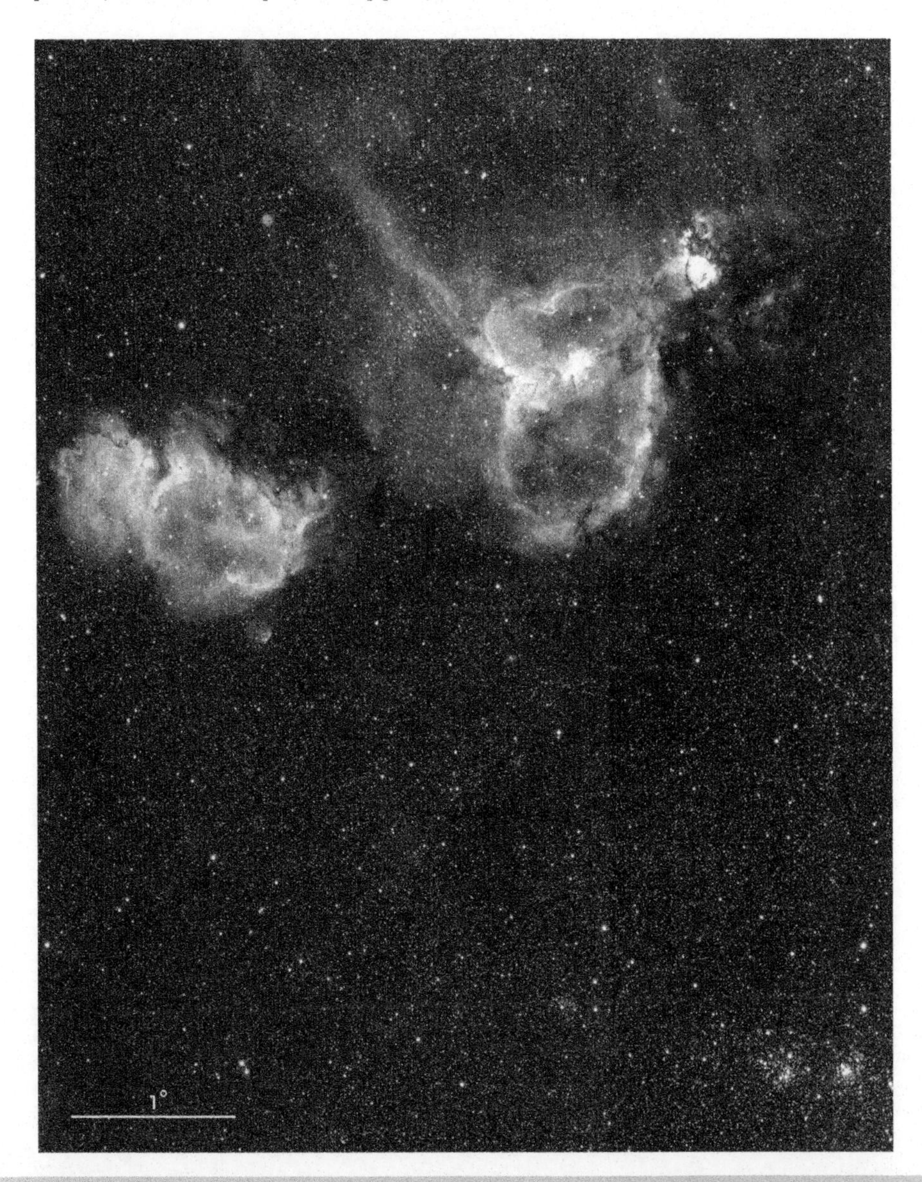

Figure 10.23 The area surrounding the Double Cluster is an excellent hunting ground for skilled astrophotographers. This image shows the Double Cluster (in the *right lower corner*) as well as the nebula IC 1805 surrounding the open cluster Melotte 15, the open cluster NGC 1027, and nebula IC 1848. These four objects all lie in the constellation Cassiopeia (all can be found in the chart on page 242). Hidden behind the right edge of the image is the large open cluster Stock 2 (page), also found in Cassiopeia

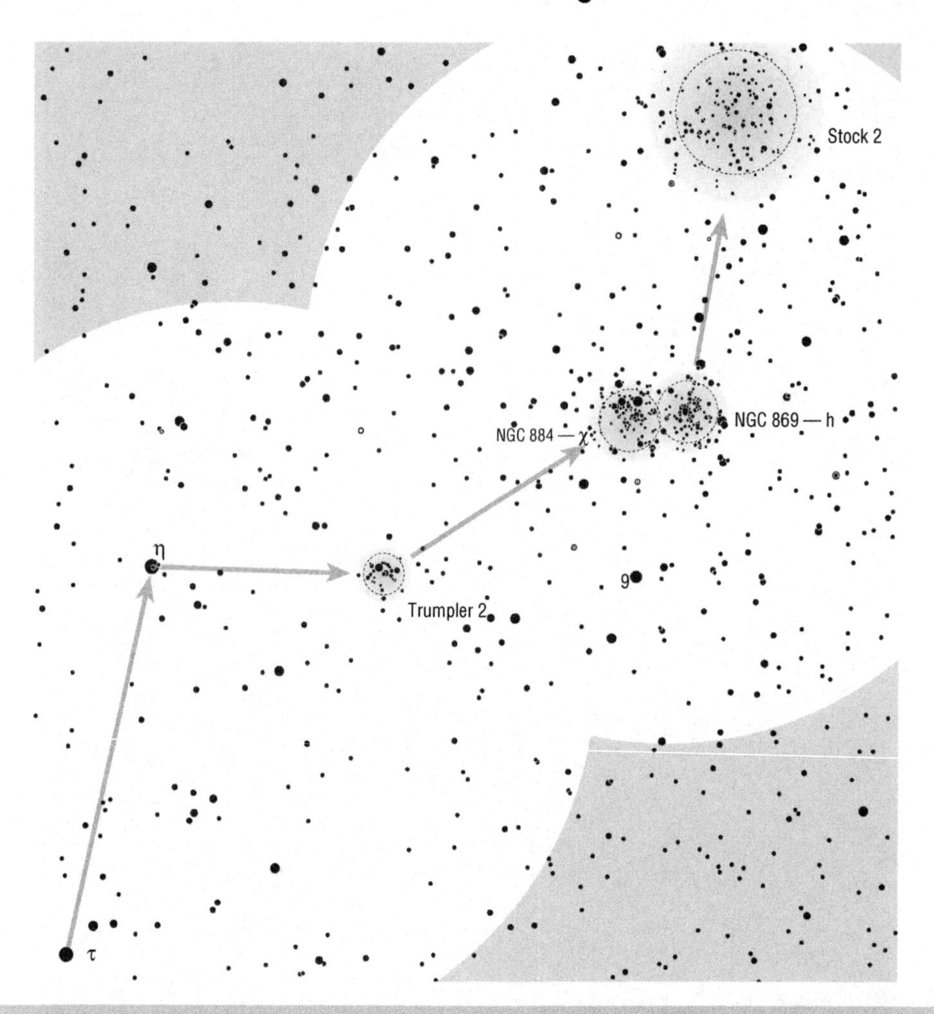

Figure 10.24 The position of the large and bright Double Cluster in Perseus. The leading star is the orange Eta. Halfway between the two we can find another open cluster – Trumpler 2. The open cluster Stock 2 at the top of the chart lies in constellation Cassiopeia (page 242)

In the northeastern part of the constellation another two open clusters await us: NGC 1528 (6m4/24′) and NGC 1545 (6m2/18′). These are seen together in the same field of view of binoculars, but they are nowhere near as attractive as the Double Cluster. The star leading to them is Lambda. NGC 1528 consists of approximately 60 stars, of which the brightest shine with magnitude 8. Through binoculars we can see approximately 30 faint stars that tend to melt into a nebula, but with the aid of averted vision we can definitely separate them. The other cluster is smaller and not as rich in stars. From the total of 40 members we can see 20 of them through binoculars. The three brightest stars are orange and yellow.

NGC 1528 is approximately 2,500 light years from us and NGC 1545 is 2,300 light years away.

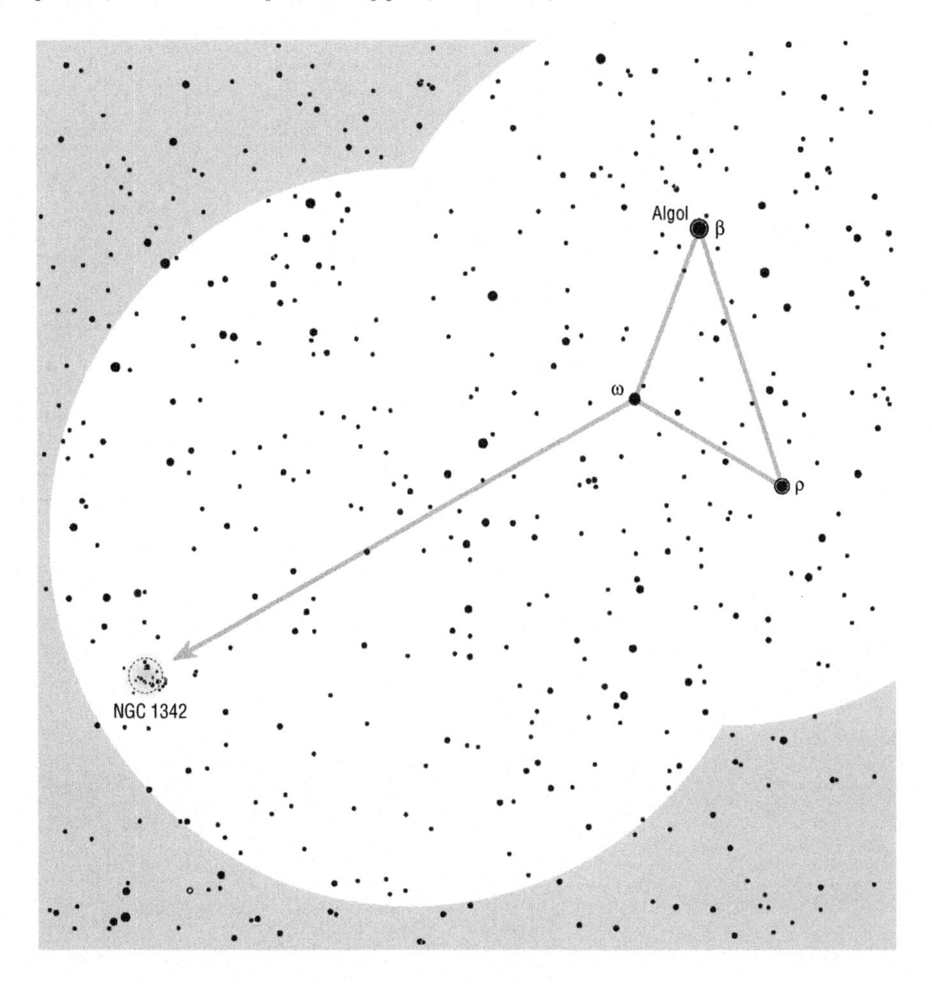
Algol
β
ω
ρ
NGC 1342

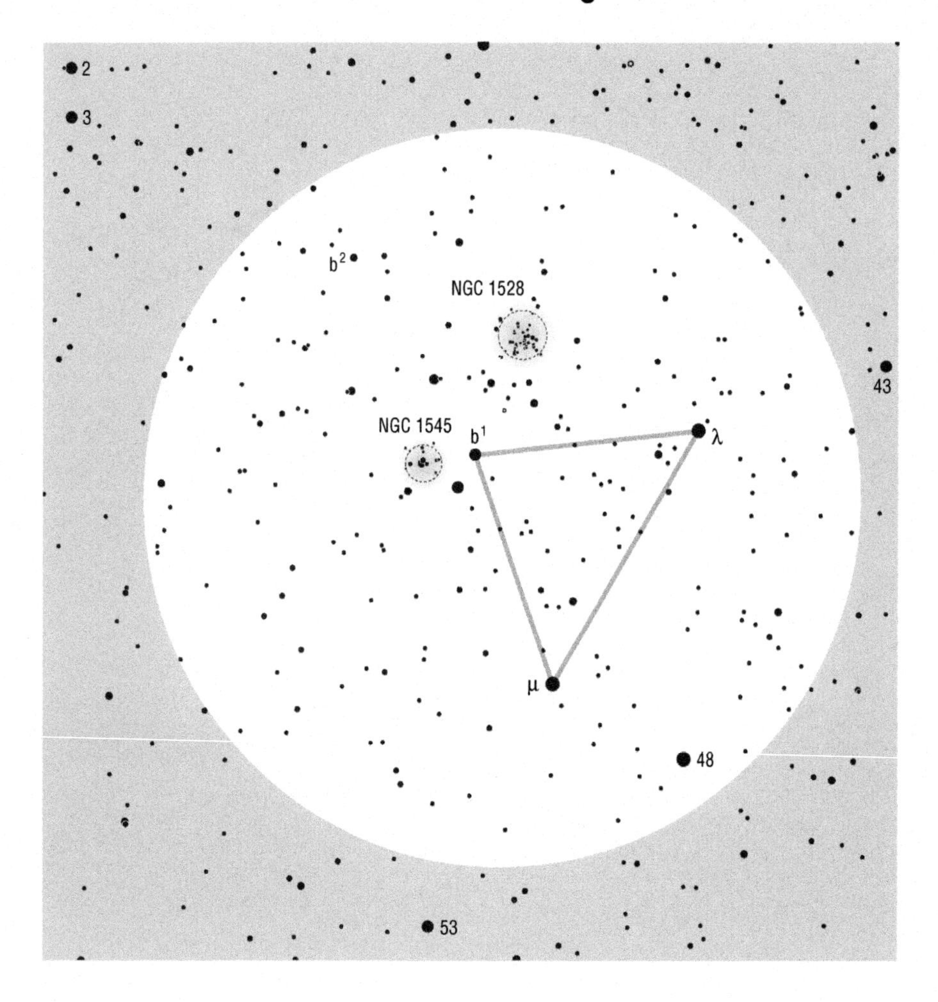
2
3
b²
NGC 1528
43
NGC 1545
b¹
λ
μ
48
53

PISCES (The Fishes)

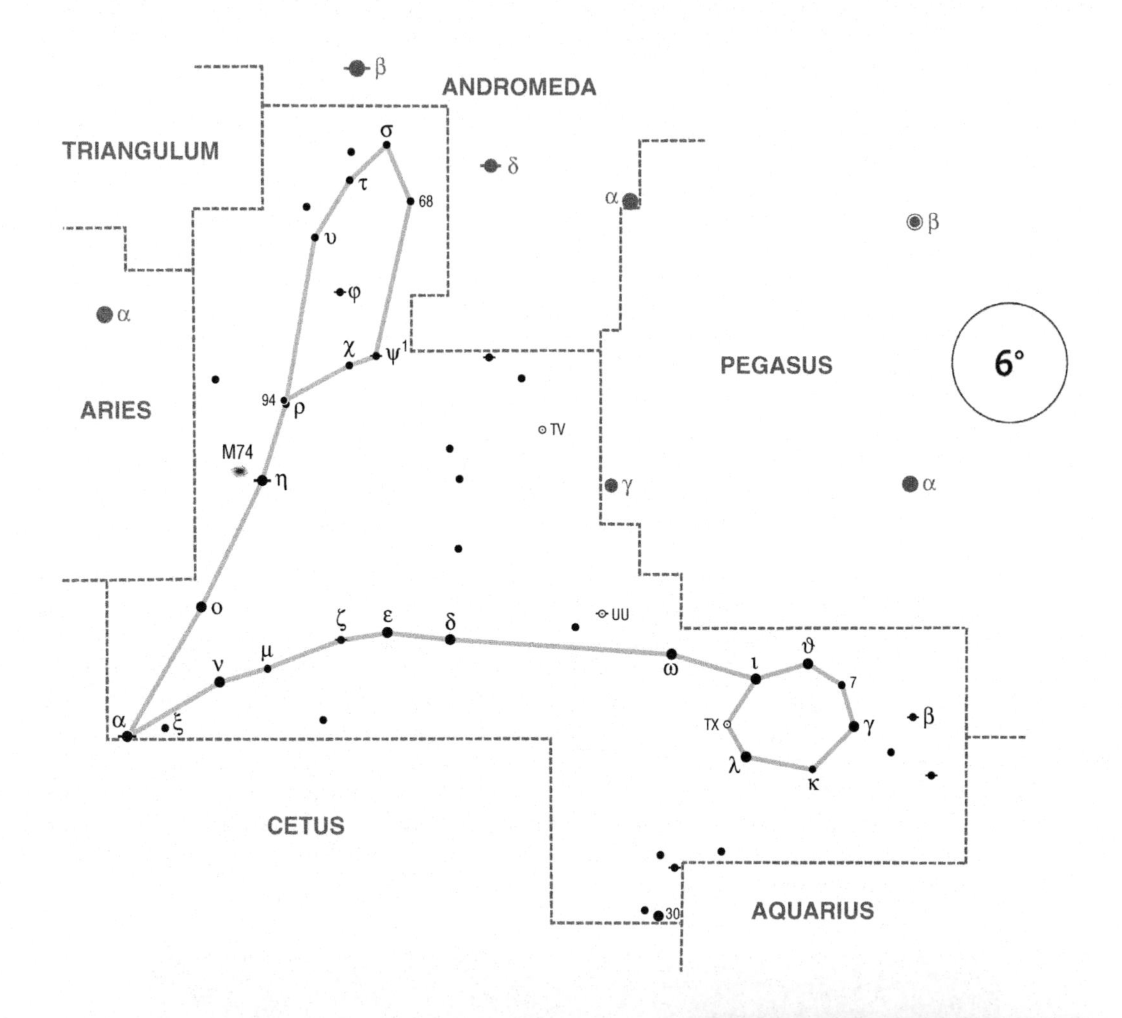

CONSTELLATION CULMINATES		
in begin. September at 2 a.m.	in begin. October at midnight	in begin. November at 10 p.m.

The large constellation of the zodiacal Pisces is located under Andromeda and east of Pegasus. It consists of rather faint stars. The constellation represents two fish. One extends from the head of Cetus to Aquarius, while the other spreads out from the head of Cetus in the direction of Andromeda. They are connected by their tails.

The brightest stars in this group are only magnitude 4: Eta (3m6), Gamma (3m7), Alpha (3m8), and Omega (4m0). In order to see this constellation you have to choose a clear, Moonless night, far from light-polluted areas; otherwise, you will not see it at all. The best places for observation are the countryside or high in the hills.

Alpha is the connecting point between the two fish; one of the fish has its head by Andromeda, the other under Pegasus. Alpha is a close binary star, suitable for testing medium-sized amateur telescopes. The last data on the separation between the stars was 1.8 arcsec (p. a. 269°, 2006), which means they cannot be split through binoculars. Alpha A is of magnitude 4.1, and Alpha B is of magnitude 5.2. The orbital period is estimated to be 930 years. The stars will be at their periastron (closest apparent proximity) in 2060, when the separation will be a mere 1.1 arcsec. Alpha is 140 light years from us.

Psi-1 is a pair of stars with similar brightness (5m3/5m4) that are 30.3 arcsec apart (p. a. 159°, 2006). The stars can be separated through binoculars, but the pair has no observable color contrast.

This constellation also includes the splendid spiral galaxy M 74 (9m4/10′2 × 9′5), which is a bit too faint to be glimpsed through binoculars (Figure 10.25). The galaxy is seen face on. On long-exposure images, it shows up as a symmetrical spiral galaxy with a small nucleus and very loose arms. For all observers who want to try to find it through binoculars, there is a detailed chart at the end of the constellation description (Figure 10.27).

Zeta is a double star with components of magnitudes 5.2 and 6.1, separated by 23.3 arcsec (p. a. 63°, 2006), which makes the pair appear through binoculars as very close. Some reliable observers report an interesting color contrast; however, the star colors are described quite differently. Some see one star as yellowish and the other as pale violet, while others describe them as white and gray, or yellowish and pink. See for yourself! The colors appear stronger when the stars are slightly blurred!

This double star is approximately 140 light years away, and the actual distance between the stars is 1,000 AU, which means that we could put 17 Solar Systems between them! This shows us just how difficult it is to see planets surrounding other stars. And by galactic standards the star is our close neighbor! Zeta lies almost on the ecliptic; thus the Moon and the planets frequently travel passed it.

The orange-reddish variable star TX, or 19 Piscium, changes its brightness from magnitude 4.8 to 5.2 irregularly. In order to find it you should first look for a circle of stars that represent the head of the western fish. These are Gamma, Kappa, Lambda, Iota, and Theta (all with magnitudes close to 4). The variable TX lies between Iota and Lambda. The star is a red giant with extremely low surface temperatures and in the dark sky resembles a glowing piece of charcoal. You cannot miss it; however, noticing its brightness changes is another thing. The amplitude is rather low, only 0.4 magnitudes, and the comparative stars are rather far away. The strong reddish tint is a result of the low surface temperature of the star. Astronomers have discovered large quantities of carbon in its atmosphere, so we could legitimately call it the Carbon Star. The variable TX is approximately 400 light years from us.

Figure 10.25 Spiral galaxy M 74

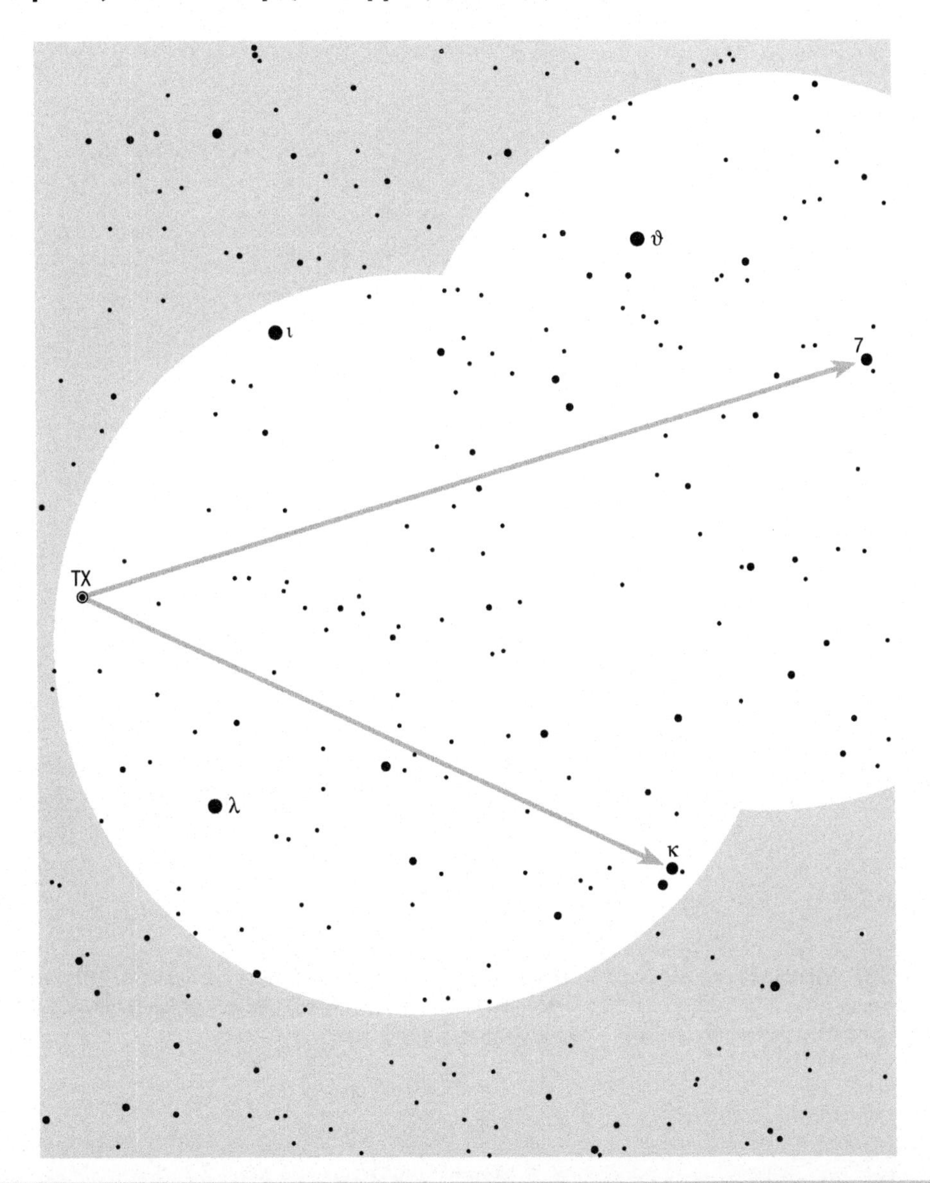

Figure 10.26 The star field surrounding variable TX, with almost all stars (except Gamma) that represent the head of the western fish. The suitable comparison stars are Kappa (4m9) for maximum brightness and 7 Piscium (5m2) for the minimum. In the vicinity there is no other star with similar brightness

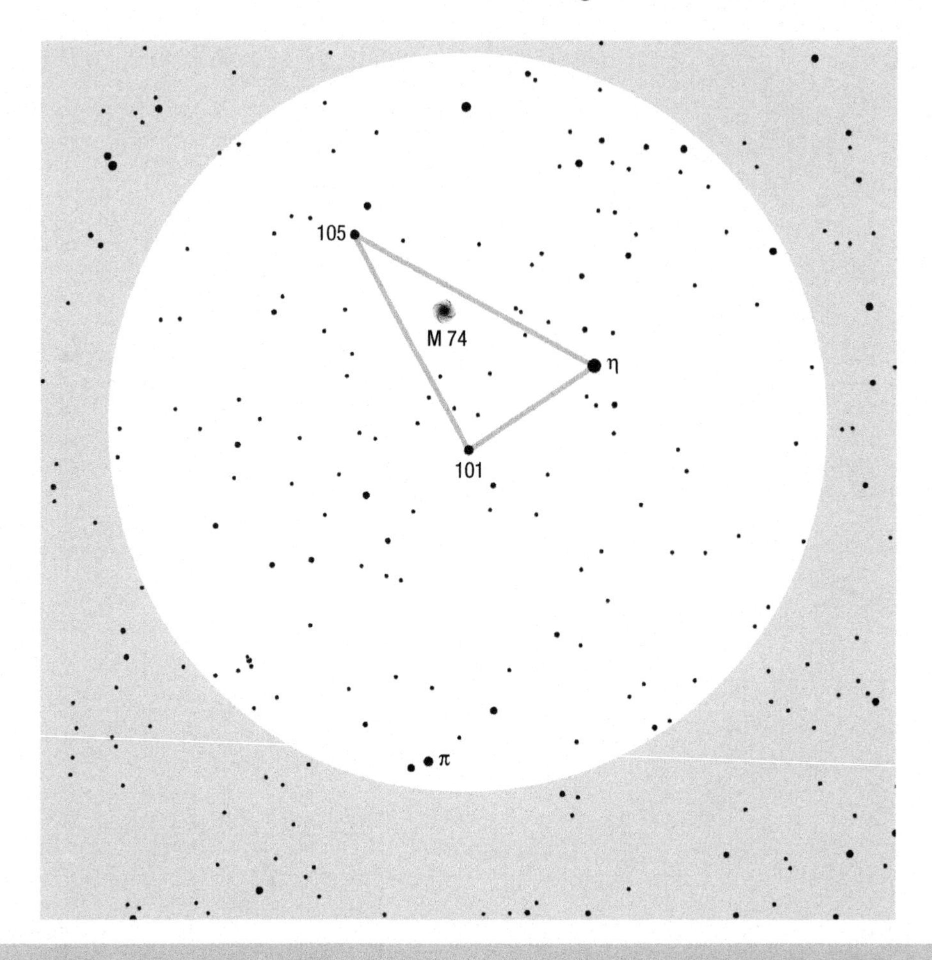

Figure 10.27 A detailed star chart to aid the search for the elusive spiral galaxy M 74. Stars Eta, 101, and 105 are always visible, as is the pattern of fainter stars surrounding them. The galaxy will appear in high-quality binoculars and when there are perfect observing conditions. Anyone up for a challenge should give it a try!

Piscis Austrinus (The Southern Fish) and GRUS (The Crane)

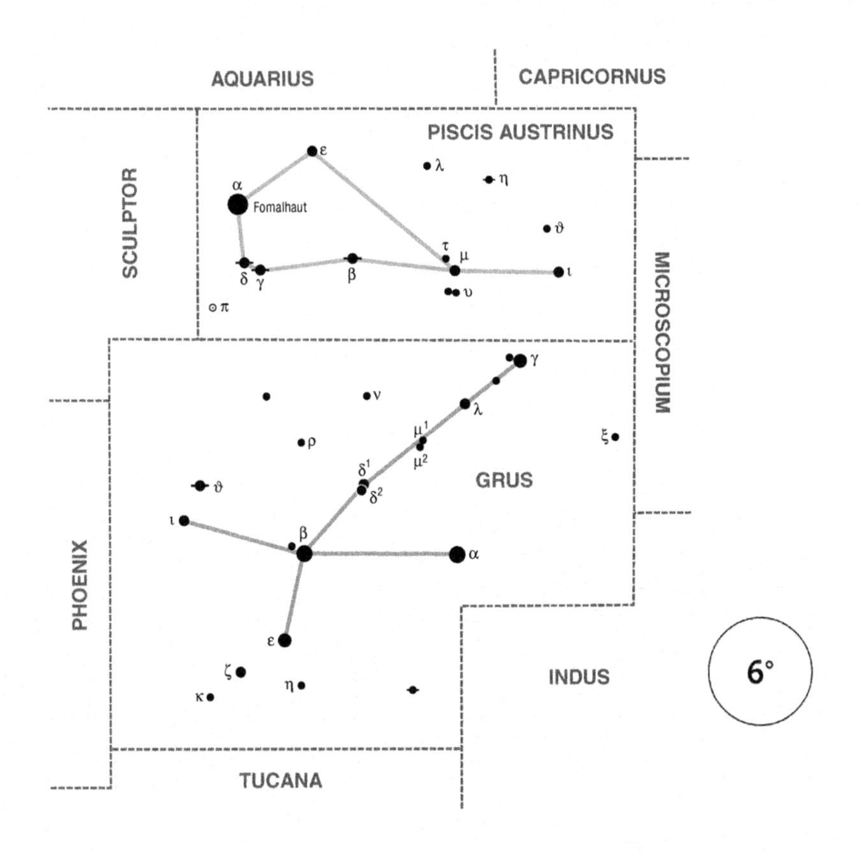

CONSTELLATIONS CULMINATE		
in begin. August at 2 a.m.	in begin. September at midnight	in begin. October at 10 p.m.

The autumn constellation Piscis Austrinus lies south of Aquarius, low above the southern tract of the horizon. It contains only one truly bright star. This is Alpha or Fomalhaut (1m2); all others are fainter than magnitude 4.

Fomalhaut is a very bright and distinctively white. It is the 18th brightest star in the sky. Its distance from us is a mere 25.1 light years. The star is twice as large as our Sun, it has 15 times bigger luminosity and its surface temperature reaches 9,000 K. Because of the position of the constellation, Fomalhaut is always low above the horizon when viewed from mid-northern latitudes. Thus its brightness is suppressed, except on those autumn nights when the sky is clear and dark all the way down to the horizon. At times like these it can take us by surprise that such a bright star is usually not seen. Its light must travel through a thick layer of atmosphere to reach us, and that dims it for at least half a magnitude, and on top of this you have to also take into account the mistiness of the horizon. At sites close to the equator, where Fomalhaut is high in the sky at culmination, the star is as bright as Deneb in Cygnus is to us!

According to astronomic criteria, Fomalhaut is a very young star. It is estimated to be only 200 million years old. In 2002, Scottish astronomers discovered a disk of dust surrounding it. The diameter of the disk is equal to that of ours Kuiper belt. The disk is composed of millions of comet nuclei that are clumped together and starting to form the first proto planets.

When viewed through binoculars Beta is a close double star. Its components are magnitudes 4.3 and 7.1 and are 30.6 arcsec (p. a. 173°, 2006) apart. This pair, which is hard to separate due to its low position in the sky and the faintness of the companion, is not known for any particular color contrast. Both stars are plain white.

Even lower, below Piscis Austrinus, lie the most northern stars of Grus. For mid-northern observers most of the constellation remains below the horizon together with the brightest stars Alpha (1m7 – the 30th brightest star in the sky) and the orange Beta (slightly variable: 2m0 to 2m3 – at magnitude 2.1 the 56th brightest star in the sky). Observers from southern Europe or the southern United States can observe the double stars Delta and Mu, which can all be separated with the naked eye. The first pair is composed of a yellow and an orange star, both of magnitude 4, separated by 16 arcmin. Mu is somewhat fainter. The yellow stars with magnitudes 5 are separated by almost 20 arcmin. Otherwise the constellation does not contain any objects that might be of interest to observers with binoculars.

PUPPIS (The Stern)

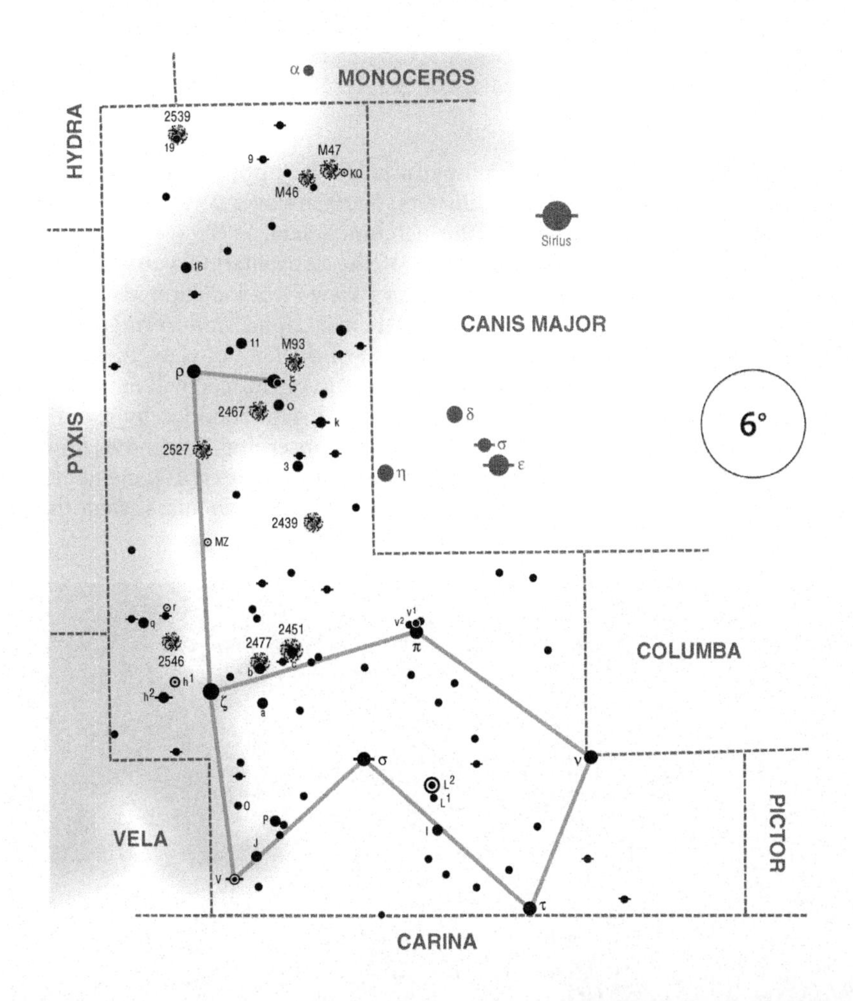

CONSTELLATION CULMINATES		
mid December at 2 a.m.	mid January at midnight	mid February at 10 p.m.

The winter constellation Puppis lies east and south of Canis Major. It is a part of the formerly largest constellation Argo Navis, which has been divided into Carina, Puppis, and Vela.

The brightest stars in Puppis are Zeta (2m2), Pi (2m7), Rho (2m8), Tau (2m9), Nu, Sigma (both 3m2), and Xi (3m3). For observers in mid-northern latitudes, the extreme southern part of the constellation is not visible, but if you are observing from 40 degrees north or below, the entire constellation is above the horizon.

Zeta is the 65th brightest star in the sky. It is a super giant with a luminosity that is approximately 18,000 times that of the Sun! The star is very far away, at 1,400 light years from us, but it still shines bright in our sky. When viewed from mid-northern latitudes it is always low in the south (it rises only 4 degrees at culmination), so its light is weakened by the hazy horizon, which is why it is not as prominent for us as it is for southern observers.

The small group around Pi is known for its wonderful color contrast. Stars v-1 (4m6) and v-2 (5m1) are blue, while Pi is – as are all red giants – orange. When viewed from the mid-northern latitudes this lovely trio is not as colorful, because the stars lie low in the south. However, if the observing conditions are perfect or if our observing site is further south, it is worthwhile to take some time to view this group. All three stars can easily be separated through binoculars. Pi is the 115th brightest star in the sky, a giant 1,100 light years from us. Its luminosity is as much as 7,100 times that of the Sun.

Rho is the 138th brightest star in the sky. It is only 63 light years from us, and its luminosity is that of approximately 21 Suns.

Xi is a yellow super giant, 1,200 light years from us. Its luminosity is also great – approximately 6,000 times that of the Sun.

When we enter the constellation Puppis it is as if we have just entered a kingdom of open clusters. There are over 25 very bright clusters, some of them visible even through binoculars.

In the northern part of the constellation you will come across two clusters in the same field of view of binoculars: M 46 and M 47 (Figure 10.28). The star leading to them is Alpha Monocerotis. If the star is moved to the extreme northern edge of the field of view, the clusters will appear on the southern edge. The first to be noticed will most definitely be M 46 (6m1/27′). Through binoculars it is clearly seen as a rather bright round nebula, out of which some individual magnitude 9 stars shine when there are excellent observing conditions; however, these are not true members of the cluster. Astronomers recognized approximately 100 members in M 46, 70 of which are brighter than magnitude 11 and can be seen through binoculars when there are perfect observing conditions.

Figure 10.28 M 46 and M 47 are wonderful examples of their kind. Even though both are open clusters, they are quite dissimilar. M 46 is densely populated with predominantly fainter stars, which melt into a faint haze when viewed through binoculars. M 47 is somewhat larger and contains half a dozen bright stars (three of which are magnitude 6 and are visible with the naked eye) as well as numerous fainter ones. On the image above M 47 we can see the open cluster NGC 2423 (6m7/19′), which is not visible through binoculars. The size of the field on the image is approximately 4.5 × 2 degrees

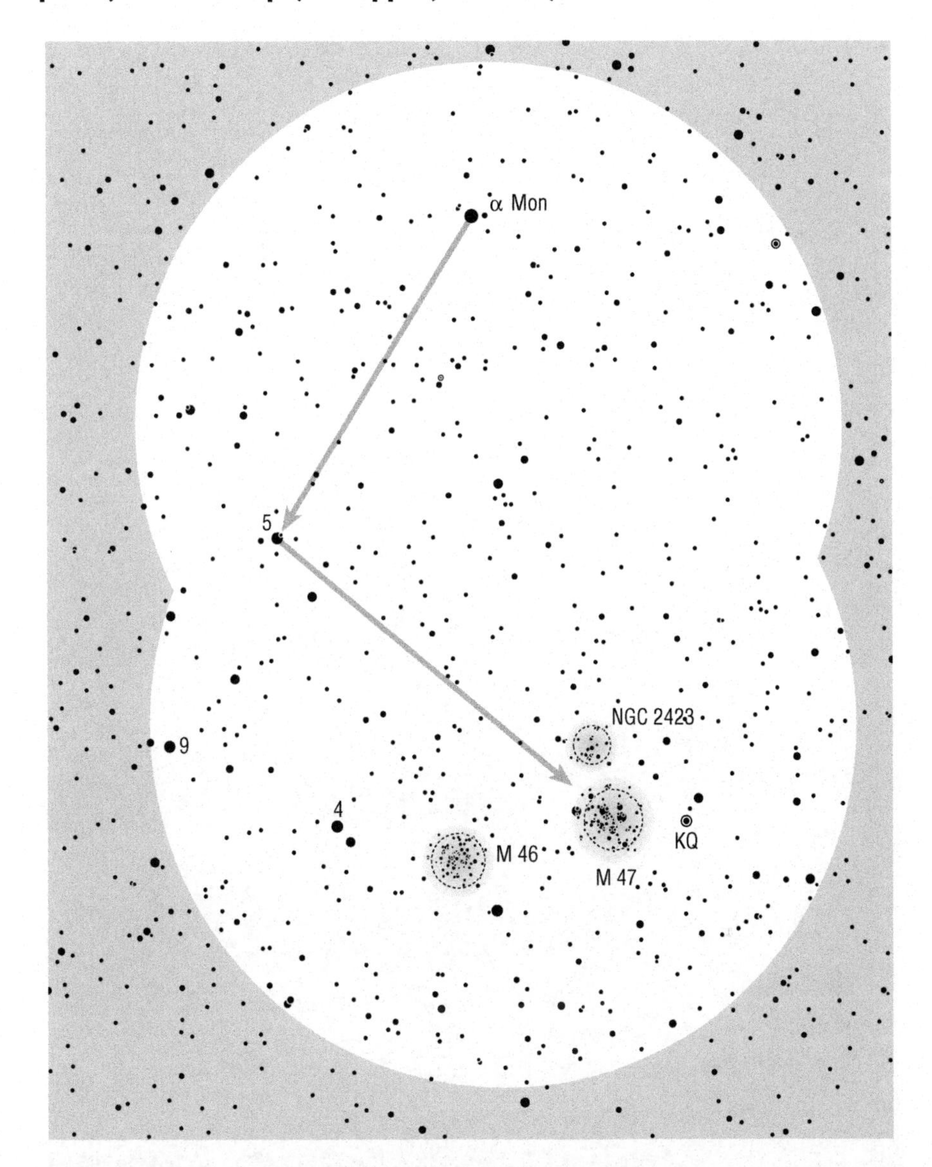

A special point of interest is the tiny planetary nebula that cannot be seen through binoculars. However, precise measurements show that the planetary nebula is not a true member of the cluster. It is just by coincidence that it lies in the same direction when viewed from Earth. The open cluster is 5,400 light years from us; at 3,300 light years, the nebula is much closer. The true diameter of M 46 is approximately 30 light years.

Slightly over a degree west from M 46 we can find the open cluster M 47 (5m2/30′), which, when there are good observing conditions, is seen with the naked eye as a faint spot of light against the Milky Way background (Figure 10.29). M 47 is different from its neighbor. Though it is bigger and its stars are brighter (three of them, which are magnitude 6, are seen with the naked eye), it is not as dense. So far measurements have confirmed that there are 500 members, 120 of which are brighter than magnitude 11.

Figure 10.29 Open clusters M 46, M 47, and NGC 2423

M 47 is approximately 1,600 light years from Earth. Its true diameter is thus approximately 20 light years.

Open cluster NGC 2539 (6m5/22′) is found close to a magnitude 5 star with the designation 19 (in fact, this is a quadruple star, composed of one bright and three faint members that cannot be seen through binoculars). It is even better if you start your search at the open cluster M 46, cross the magnitude 5 star with the designation 9, and thus reach NGC 2539 (see chart below). The cluster lies outside the denser part of the Milky Way. Through binoculars it is seen as a faint spot of light. NGC 2539 is a rich and dense cluster. So far astronomers have confirmed approximately 100 members, of which the brightest are a mere magnitude 11. The cluster is approximately 4,450 light years from us and has an actual diameter of 28 light years.

Open cluster M 93 (6m2/12′) is found only a degree and a half northwest of the bright Xi (Figure 10.30). This is a small but rather bright cluster, with slightly over 25 members with magnitudes of between 8 and 11. It is approximately 3,400 light years from us and has a true diameter of 12 light years. Through binoculars it is seen as a bright patch of light with a few individual brighter stars shining from within. Some reliable observers report that under perfect observing conditions the cluster is seen with the naked eye. It could be possible!

To the southeast of M 93, two more open clusters await us; however, they are not as bright, and because of their lower position in the sky they are further weakened. They can be observed only on nights with excellent viewing conditions.

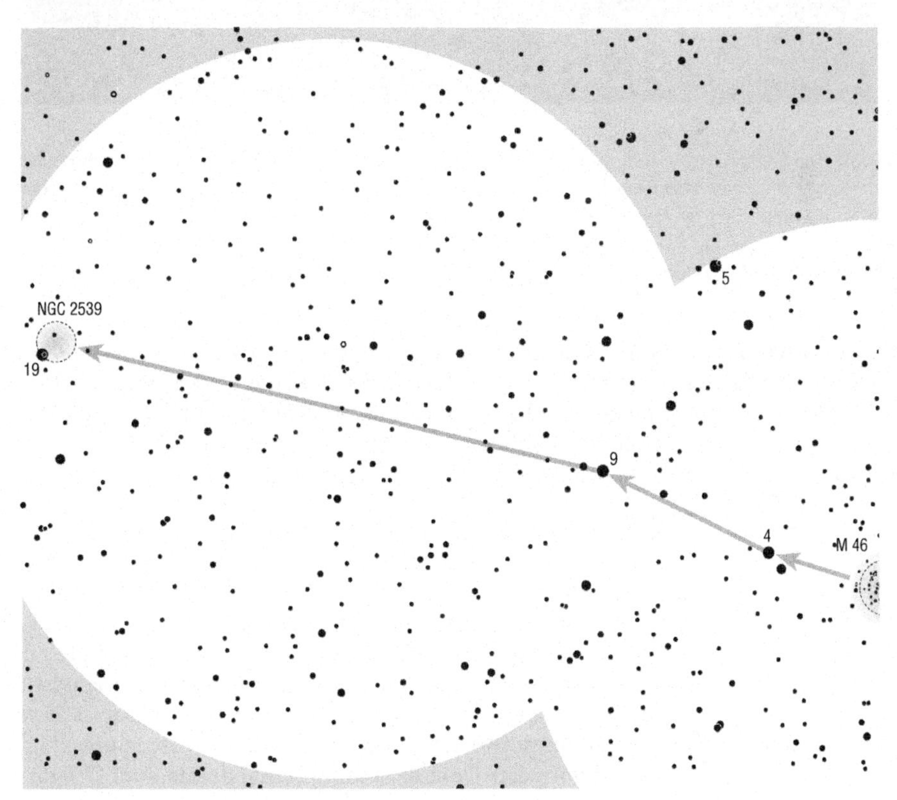

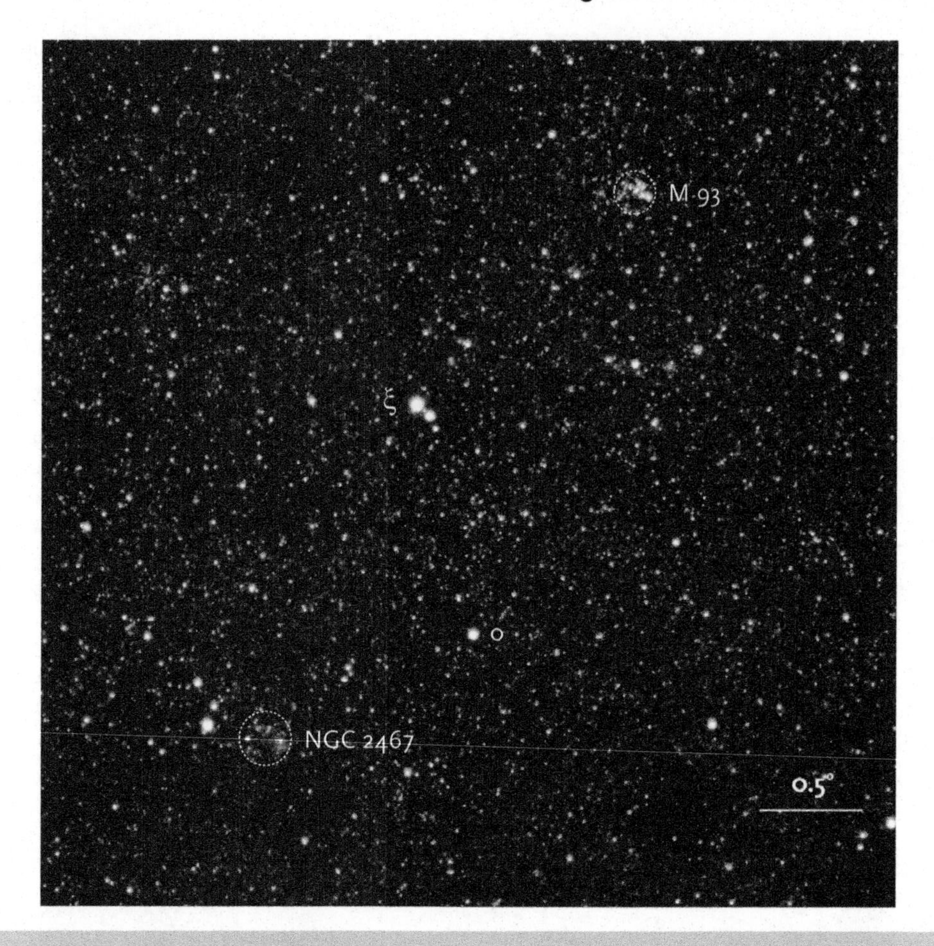

Figure 10.30 The bright Xi is the leading star to the three open clusters. M 93, which is clearly seen through binoculars, lies approximately 1.5 degrees northwest of the star. About 2 degrees to the south-southeast, we can find the faint NGC 2467. Its stars are wrapped in a tenuous nebulosity, which can be seen on the image but not through binoculars. NGC 2527 is on the border of visibility. It lies 3 degrees southeast of NGC 2467 and does not show in this image. Its precise position is drawn in the detailed chart on page 403. The size of the field on the image is approximately 4 × 3.5 degrees

Approximately 2 degrees south-southeast from Xi lies the open cluster NGC 2467 (~7m/16′ – the cluster can be seen in the image on page 401). Through binoculars it is seen as a small, elongated spot of light with its brightest stars of magnitude 8 and 9 shining out from the haze. The stars in the cluster are wrapped in a nebulosity that appears only on long-exposure images. NGC 2467 is approximately 4,400 light years from us, and it has a real diameter of 20 light years.

Approximately 3 degrees southeast of NGC 2467 or 5 degrees southeast from Xi (they can be seen together in the same field of view) we come across the open cluster NGC 2527 (6m5/22′). This cluster is on the border of visibility when viewed through binoculars. In the field of view, it is seen as a faint cloud of light. With averted vision and when there are excellent observing conditions you can glimpse some of the brightest individual stars, but not all of them are true members of the cluster. In total, the cluster is comprised of approximately 30 stars, but only the brightest two are of magnitude 9; all rest are fainter (Figure 10.31).

NGC 2527 is approximately 2,000 light years from us, and its actual diameter is 13 light years.

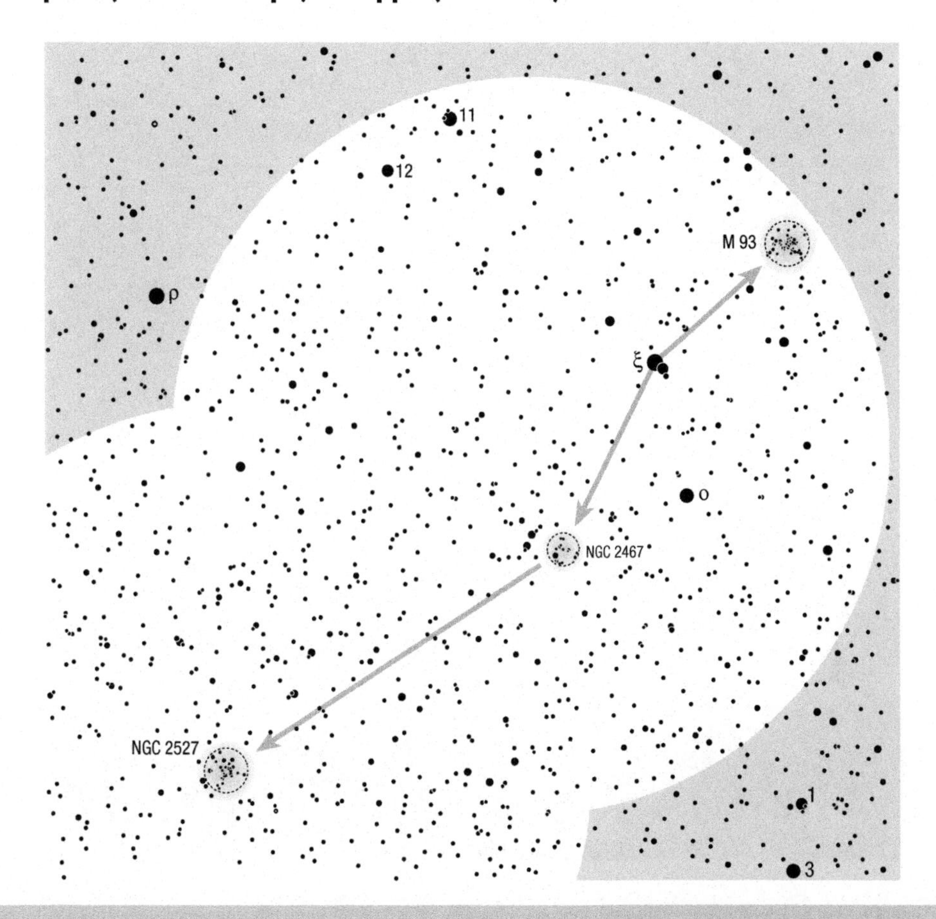

Figure 10.31 If you choose Xi as the leading star, you will be able to tick off three objects from your list. If the star is in the center of the field of view, you can see M 93 and NGC 2467. NGC 2527 lies outside the field of view, southeast of Xi, but it is not hard to find when observing conditions permit

NGC 2477 (5m8/27′) lies in the same field of view as the bright Zeta. It is found approximately 3 degrees northwest of the star. This bright object is, in the opinion of numerous observers, the most beautiful open cluster in the constellation, but for observers from mid-northern latitudes it is too far south to reveal all of its charm. We can see it only on those nights when the sky is clear and dark all the way down to the horizon. At culmination, the cluster is a mere 5.5 degrees above the horizon (Figure 10.32). However, for southern observers it is a true celestial gem!

The central part of the cluster is slightly smaller than at M 46 but richer in stars and denser. There are approximately 120 members within the cluster, the brightest of which are magnitude 10. Just below the cluster is a magnitude 4 star with the designation b Puppis, which is a good pointer that shows us where to look for the faint nebulosity. NGC 2477 is 4,200 light years from us; its densest part measures 33 light years in diameter.

Figure 10.32 This wonderful group of stars is the open cluster NGC 2477. In mid-northern latitudes, the cluster rises a mere 5.5 degrees above the horizon at culmination; therefore, it loses much of its shine

Approximately 2 degrees northwest of open cluster NGC 2477, we can find a small heap of bright stars (the brightest is magnitude 4 and has the designation c) that form the very loose open cluster NGC 2451 (2m8/45′). During the best nights the stars can be seen with the naked eye even from mid-northern latitudes, but with lesser observing conditions they appear only through binoculars (if at all). The cluster is clearly visible to southern observers.

NGC 2451 is approximately 1,000 light years from us and has a true diameter of 13 light years (Figure 10.33).

About 2.5 degrees northeast of Zeta we come across the open cluster NGC 2546 (6m3/40′). In excellent observing conditions it is seen as a patch of light with two brighter magnitude 6 stars to the north and south of it, but they are not its true members. The cluster contains approximately 100 stars, the brightest of which are of magnitude 8.

NGC 2546 is roughly 3,000 light years from us, and its actual diameter is over 35 light years.

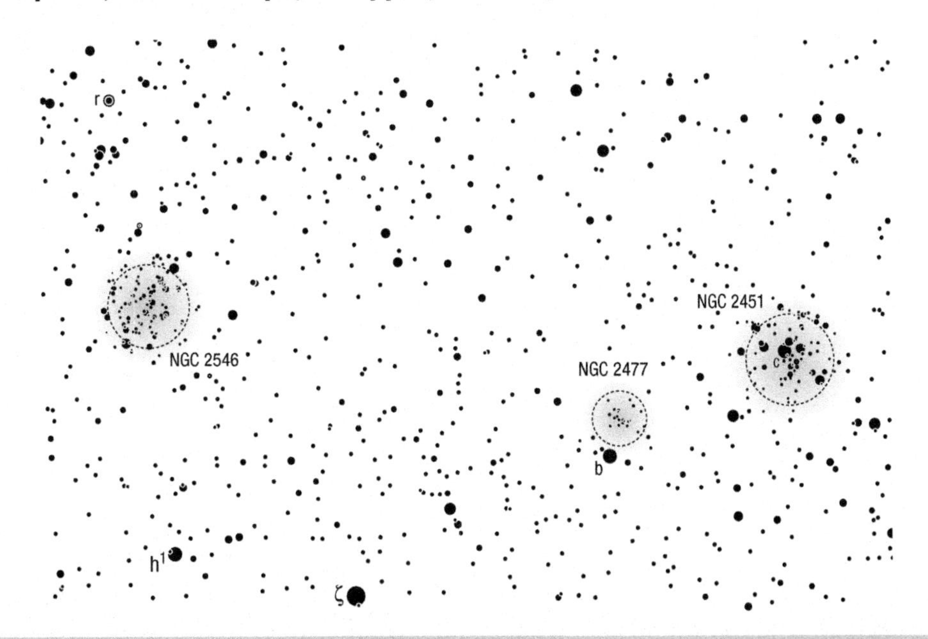

Figure 10.33 Only 7 degrees above the horizon (for mid-northern observers) we can see three open clusters, which are clearly visible from the southern latitudes, but for mid-northern observers they are reserved only for nights with dark skies all the way down to the horizon

The last cluster in Puppis can be found with the aid of Eta Canis Majoris. This is NGC 2439 (6m9/10′), which lies outside of the main stream of the Milky Way in a rather empty area of the sky, some 4.5 degrees southeast of Eta. The star and the cluster therefore appear together in the same field of view of binoculars. In good observing conditions, we will probably see only the brightest member named R Puppis, but when there are excellent observing conditions and the sky is dark all the way down to the horizon, a small, faint nebulosity surrounding R will appear. The cluster contains approximately 50 stars, the brightest of which are magnitude 9. It is about 12,500 light years from Earth and measures 33 light years.

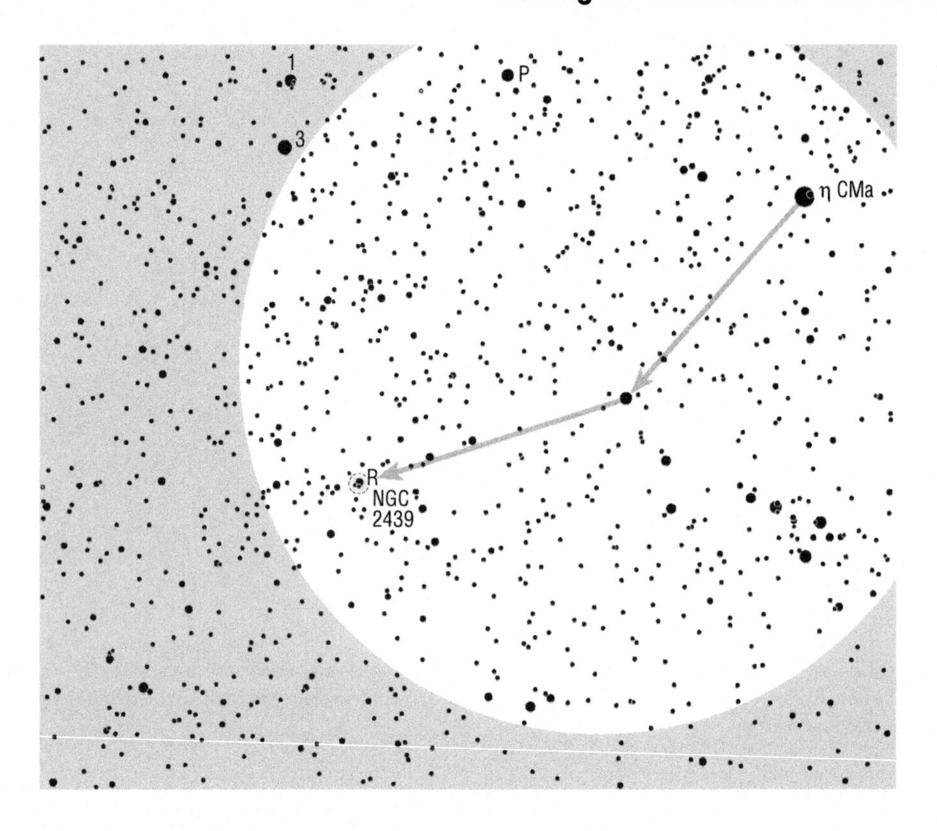
1
P
3
η CMa
R
NGC
2439

CHAPTER ELEVEN

Sagitta (The Arrow) to Vulpecula (The Fox)

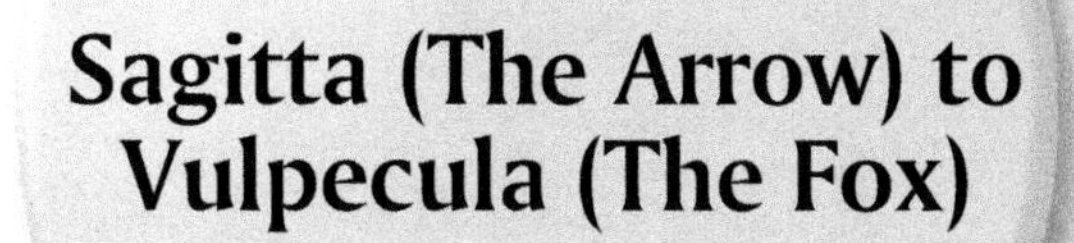

B. Kambič, *Viewing the Constellations with Binoculars*, Patrick Moore's Practical Astronomy Series,
DOI 10.1007/978-0-387-85355-0_11,

SAGITTA (The Arrow)

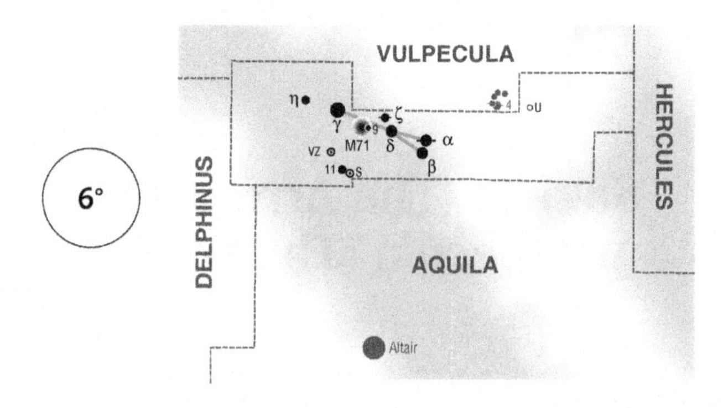

Sagitta is a summer constellation situated south of Vulpecula and north of Aquila. It is small, but it has a distinctive shape, so it is easy to recognize. Its brightest stars are Gamma (3m5), which has a beautiful orange color, Delta (3m8), Alpha, and Beta (both 4m4).

CONSTELLATION CULMINATES		
mid June at 2 a.m.	mid July at midnight	mid August at 10 p.m.

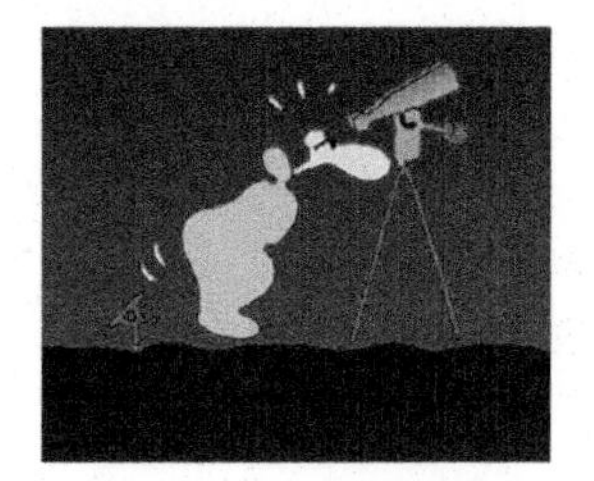

This constellation is home to one of the most interesting eclipsing variable stars known to us – U Sagittae. In this pair of stars, the brighter one is completely covered by its larger and fainter companion. The changes in brightness can also be seen through binoculars. Most of the time the star shines with a magnitude of 6.4. Every 3.38061933 days (3 days, 9 h, 8 min and 5 s) we can see the eclipse. For approximately 1 h and 40 min the brightness of the star falls to magnitude 9.3, thus appearing almost on the border of visibility when viewed through binoculars. Because the eclipses are total at this binary, the fall in brightness is much faster and more complete than in the more famous Algol (Beta Persei), where the eclipses are merely partial. Halfway through the period you may notice another small drop in the brightness, which occurs when the brighter star covers the fainter one. However, this cannot be seen through binoculars, as the fall in the brightness is merely one-tenth of a magnitude.

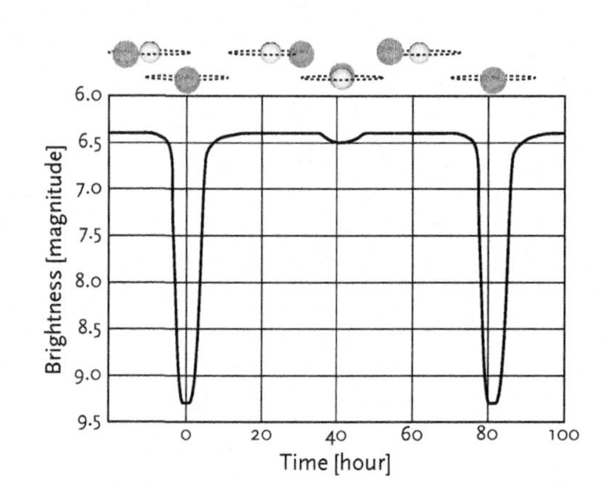

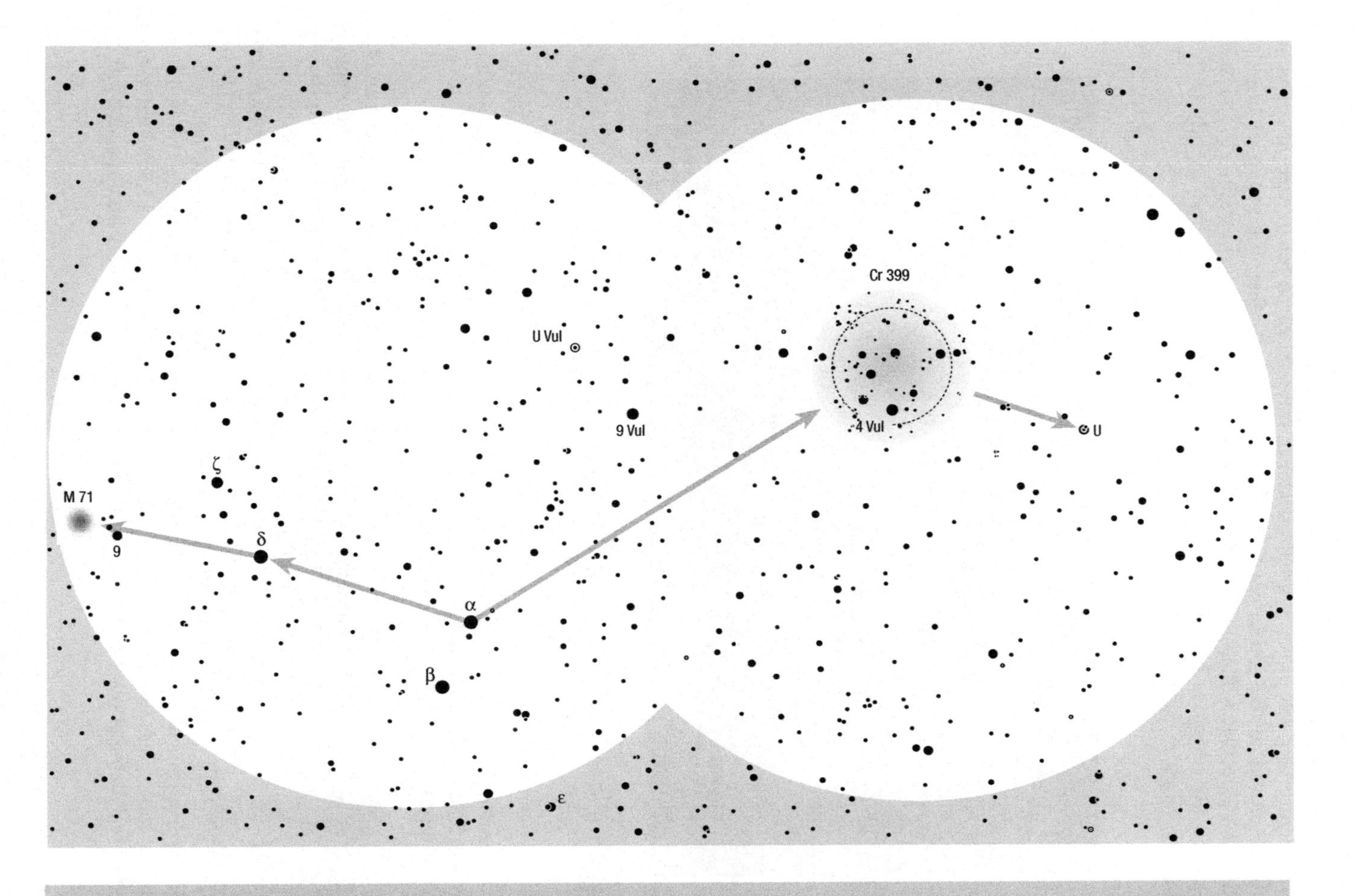

Figure 11.1 The positions of the globular cluster M 71 and the eclipsing variable star U Sagittae

The brighter member of the pair is a normal main-sequence star. Its luminosity is approximately 120 times that of the Sun. The fainter companion is a larger and colder subgiant. Its luminosity surpasses our star by merely 10 times. This interesting binary is approximately 750 light years from us.

The star leading to the variable U Sagittae is, in fact, a small group surrounding the star 4 Vulpeculae. Figure 11.1 will prove of great assistance in your search for it. Start at Alpha Sagittae. If this star is moved to the southeastern edge of the field of view, the group of brighter stars surrounding 4 Vulpeculae will appear on its northwestern edge. In the chart, the group is labeled as open cluster Cr 399 (image is on page 498). Slightly over 1 degree west and slightly to the south lies U Sagittae (Figure 11.2).

Figure 11.2 U Sagittae at maximum and minimum brightness

Until you notice the fall in brightness you should observe the star frequently. It reached one of its minimums on August 30, 2008, at 20.4 UT. From these data and from the known period we can easily calculate the approximate dates of the next minima.

Globular cluster M 71 (8m2/7′2) lies halfway between Delta and Gamma and in the same field of view with Alpha (chart on previous page) and is therefore easy to find. It is a rich and very condensed cluster, with a magnitude 6 star on its western edge. Through binoculars it is shown as a few arc minutes large, tenuous spot of light. When there are excellent observing conditions you will see it clearly, but on sultry summer nights it will appear fainter, smaller, or it will not be visible at all. It is best to observe it on spring mornings or autumn evenings, when the constellation is higher up in the sky and the nights are cooler, less humid, and darker. The cluster is approximately 13,000 light years from Earth, and its real diameter reaches only 27 light years. This is truly not very big for a globular cluster.

SAGITTARIUS (The Archer)

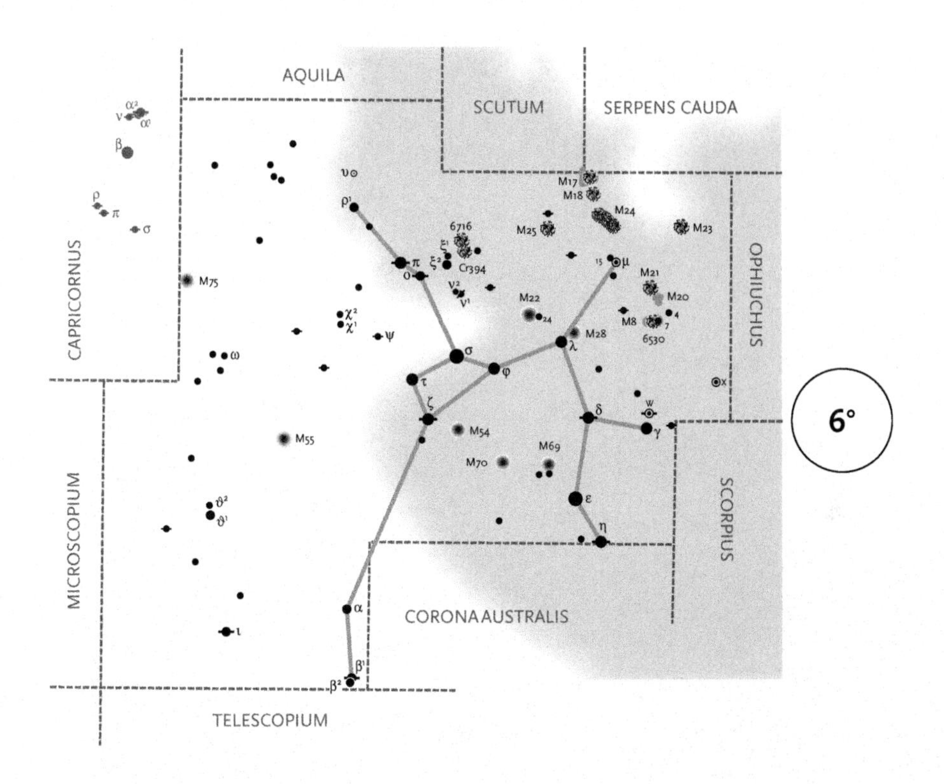

South of Aquila and Scutum, between Scorpius and Capricornus, lies the southernmost zodiac constellation of Sagittarius. Its brightest stars are Epsilon (1m8), Sigma (2m0), Zeta (2m6), Delta (2m7), Lambda (2m8), Pi, and Gamma (both 3m0). It is interesting that Alpha and Beta are faint (both 4m0) and slightly removed from the rest of the constellation. From the mid-northern latitudes Beta never reaches above the horizon, while Alpha is always low above the southern horizon – at culmination, it peaks at a mere 3.5 degrees above the horizon in our sky.

CONSTELLATION CULMINATES		
in begin. June at 2 a.m.	in begin. July at midnight	in begin. August at 10 p.m.

Epsilon is the 34th brightest star in the sky. It is 145 light years from us. Its luminosity is 290 times that of the Sun.

Sigma is the 52nd brightest star in the sky, 225 light years from Earth. It is a giant star with luminosity that surpasses our Sun by 550 times.

Zeta is the 101st brightest star in the sky. It is only 89 light years from us, and its luminosity is that of 52 Suns.

The yellow Delta is the 117th brightest star in the sky. Its light travels 306 years in order to reach us. It is a giant with a luminosity of 550 times that of the Sun. Its color is vivid and is even stronger when viewed through binoculars.

The yellow Lambda is the 136th brightest star in the sky. It is 77 light years from us, and its luminosity surpasses our Sun by 32 times.

Figure 11.3 The Lagoon Nebula is a wonder to see even on amateur images

Stars Xi-2 and Xi-1 (5m1) form a pair that can easily be separated with the naked eye. Xi-2 is brighter (3m5) and yellow, while the half a degree north Xi-1 is bluish. Their colors become even more evident when viewed through binoculars.

W is a Cepheid variable star with changes in brightness of between 4m3 and 5m1, with a period of 7.59503 days. The star can be observed with the naked eye during its entire cycle.

Sagittarius is home to more Messier objects than any other constellation. It is swarming with nebulae, as well as open and globular clusters. Many of them can be seen with the naked eye, and almost all are visible through binoculars.

The splendid nebula M 8 (4m6/14′), called the Lagoon Nebula, is as faint spot of light visible even with the naked eye. It is found some 5.5 degrees west and slightly north of the bright Lambda (see chart on page 415, image on page 412), so they appear together in the same field of view. Through binoculars this is quite a sight, one you will not forget for a long time! Stars in the open cluster NGC 6530 are sunken in a tenuous, transparent nebula. Yes, this is a true nebula, an interstellar cloud of gases and dust and not a nebula composed of stars that our binoculars cannot resolve. The diameter of the nebula is at least 10 arcmin, but – by now you are undoubtedly already well aware of this – its size depends very much on observing conditions. The entire beauty of the Lagoon cannot be appreciated with the naked eye or even through a very large telescope. The magic mixture of the bright and dark clouds of interstellar matter appears only on long-exposure images.

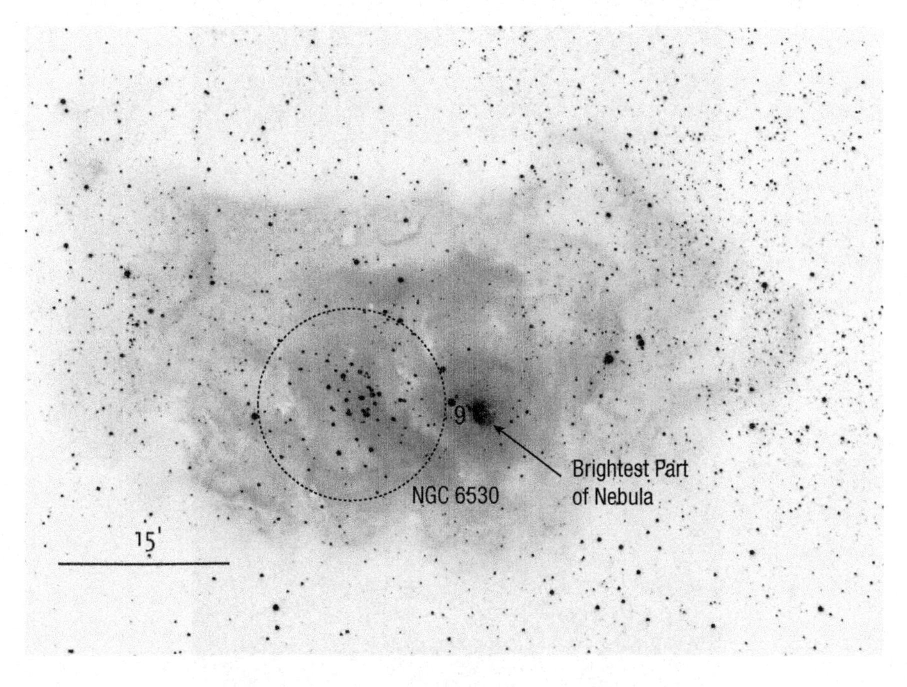

On the west side of the nebula we can see two brighter stars. The southern, with designation 9, is of magnitude 6, and this is most probably the star that illuminates the gases and dust in the nebula so that we can see it. Besides this star there are probably other young and hot stars hidden in the dense clouds. Only 3 arcmin west-southwest of star 9 lies the brightest part of the nebula. This is also the source of radio waves that astronomers can "see" with radio telescopes.

The densest part of the open cluster lies in the eastern part of the nebula. There are as many as several thousand stars in the broader area surrounding it (the cluster lies in the Milky Way), but these are mostly very faint and are not seen even through larger amateur telescopes. When there are excellent observing conditions, we can see approximately 30 stars through binoculars, and on the best nights close to 50 appear in sight. The brightest are of magnitude 7.

The exact distance to the Lagoon Nebula has not yet been established, since astronomers do not know how much the dark clouds of matter, lying between the nebula and us, dim the light of the stars. The best estimates predict that the nebula is approximately 5,000 light years from us. If this is true, then the actual size of the central part of the nebula is 60 × 44 light years, but its faintest regions are as much as twice the size. Most probably the Lagoon Nebula is connected to the northern Trifid Nebula (M 20), since both are approximately the same distance from us.

The emission nebula M 17 (6m0/25′), which extends around the open cluster of stars, lies in the northeastern part of the constellation, some 3 degrees north of Star Cloud M 24 and 1 degree north of open cluster M 18 (see the chart on page 419, image on page 418). Through binoculars the brightest, elongated part of the nebula can be glimpsed when there are excellent observing conditions. Because of the appearance of its stirred clouds of dust and gases, the nebula has a number of names: Omega, The

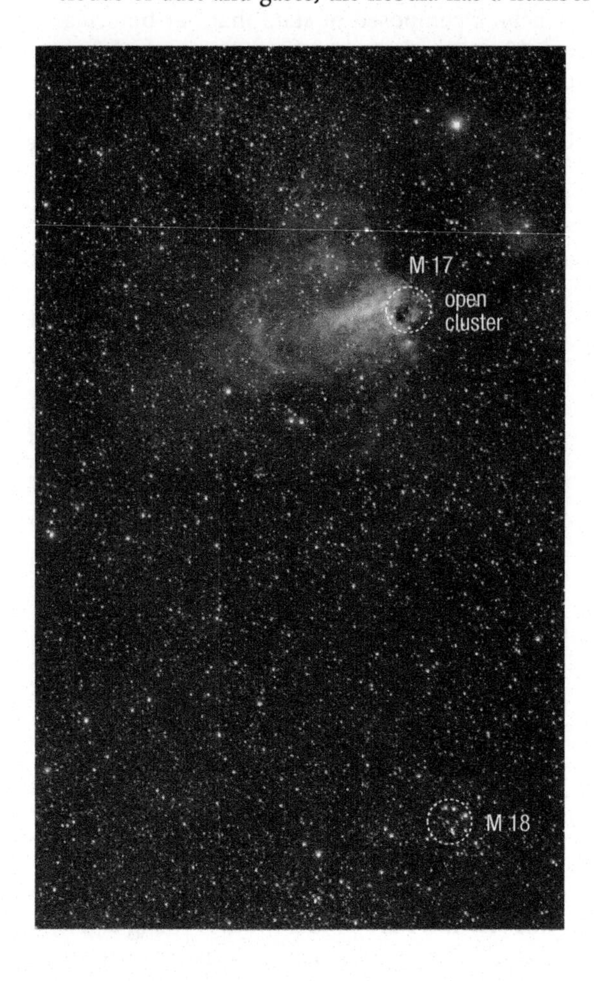

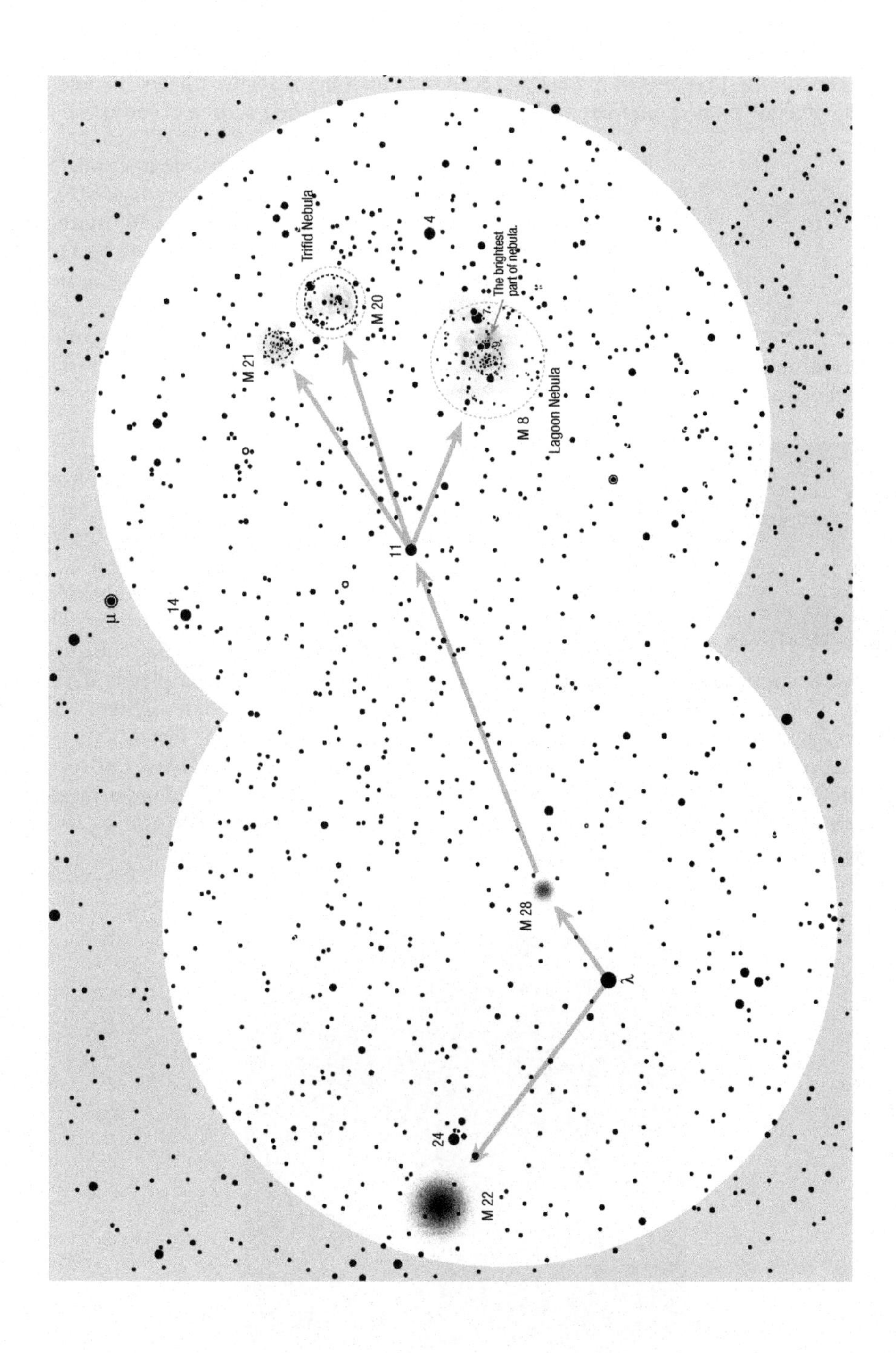

Trifid Nebula
M 21
M 20
4
The brightest part of nebula.
7
M 8
Lagoon Nebula
11
14
μ
M 28
λ
24
M 22

Swan, The Horseshoe, and The Lobster Nebula. The clouds of matter contain a mass of 800 Sun-like stars. In contrast to M 8 there are no brighter stars in M 17. The open cluster is composed of approximately 350 stars, but only 15 are up to magnitude 11 in brightness. The brightest one is merely magnitude 9.

M 17 is approximately 5,000 light years from us. The brightest part of it measures approximately 12 light years. The dimmest regions of the nebula, seen only on long-exposure images, extend across an area covering 40 light years in diameter. The nebula is a rather strong source of radio waves.

Open cluster M 18 (7m5/9′) lies approximately 2 degrees north from the Star Cloud M 24 (see the chart on page 419, image on page 418). It is seen as small and faint spot of light, composed of a mere 100 stars. Through binoculars, approximately 20 of these can be seen on the best nights, but usually we see a mere dozen stars, and even to see these we need to use averted vision. The brightest star shines with magnitude 9. In order to observe M 18, we have to choose a clear and dark night with excellent observing conditions. The distance to the cluster is not entirely known; however, most estimates predict a distance of approximately 5,000 light years.

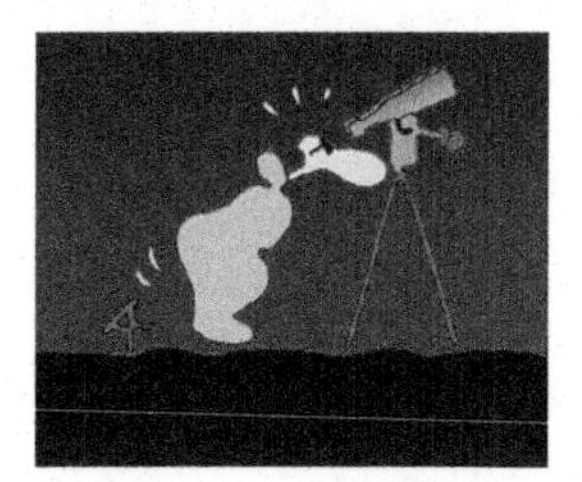

The Trifid Nebula, or M 20 (6m3/28′ × 20′), is one of the most beautiful celestial objects. It is found approximately 1.5 degrees north-northwest of the Lagoon Nebula (M 8), and therefore they appear in the same field of view of binoculars. Probably the Trifid is just a part of the vast cloud of interstellar matter in that area (see the chart on page 415, image on page 418). Unfortunately, observers using binoculars are not able to see its real beauty. When there are excellent observing conditions, the field of view reveals a rather large but faint transparent nebula. The quality of the sky is of special importance when observing this nebula. If the sky is not completely dark, it is not visible at all. Now you know why astronomers are among the most passionate fighters against light pollution. And if you become a serious observer, you will certainly join their forces!

On long-exposure images the nebula shows its true beauty. The bright and dense bulk of gases and dust, illuminated by the central giant sextuple star, is covered by dark, dense, and impermeable clouds and stripes of interstellar matter. Four almost straight lines are especially dark, and these divide the bright nebula into four clearly separate parts.

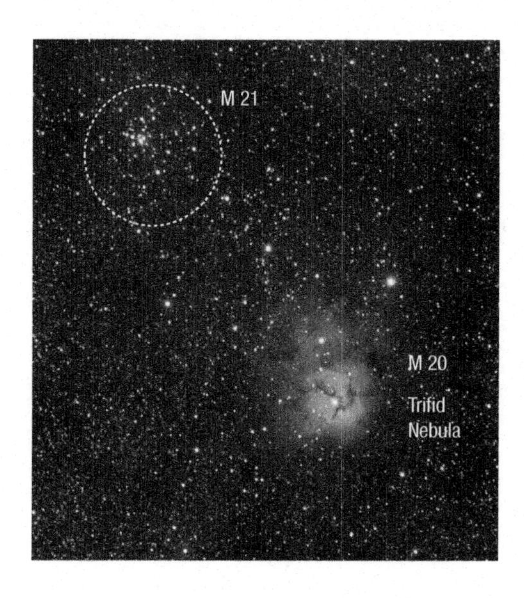

The open cluster within the nebula contains a few hundred young stellar objects. When there are excellent observing conditions, approximately 20 stars can be seen. On a perfect night, when the nebula appears larger and brighter, the number of stars can increase to approximately 30. What a sight this is!

The exact distance to the nebula is not yet known. The best estimates place it at approximately 5,200 light years from us. The Trifid Nebula is also a strong source of radio waves.

The open cluster M 21 (6m5/13′) is in the same field of view with the Trifid Nebula. It lies a mere 0.7 degrees northeast (see the chart on page 415, image on page 418). It can be seen through binoculars. However, it includes only one magnitude 7 star and one magnitude 8. In total, the cluster consists of approximately 400 members. On the best nights, some 40 of them with magnitudes up to 11 can be seen. The cluster is somewhat closer to us than the Trifid, for it is approximately 3,900 light years from us. The stars were most probably born in the area of dust and gas within the vast and complex cloud that also contains M 8 and M 20. The true diameter of the cluster is 16 light years.

The globular cluster M 22 (5m1/32′) is easy to find. It lies approximately 2 degrees northeast of the bright Lambda, which means that they both appear in the same field of view of binoculars (see the chart on page 415, image on page 418). This cluster is one of the most beautiful and brightest of its kind. If it were higher in the sky, it would completely overshadow the more famous M 13 in Hercules. M 22 is third brightest cluster - only behind Omega Centauri and 47 Tucanae, which are not seen from the mid-northern latitudes. Fourth in line is M 13.

Through binoculars M 22 is seen as a bright, approximately 10 arcmin big spot of light with a clearly seen brighter central part. The size of the spot depends greatly on the observing conditions, which can be poor on sultry summer nights. It is much better to observe the cluster on April or May mornings.

Figure 11.3A Magnificent globular cluster M 22

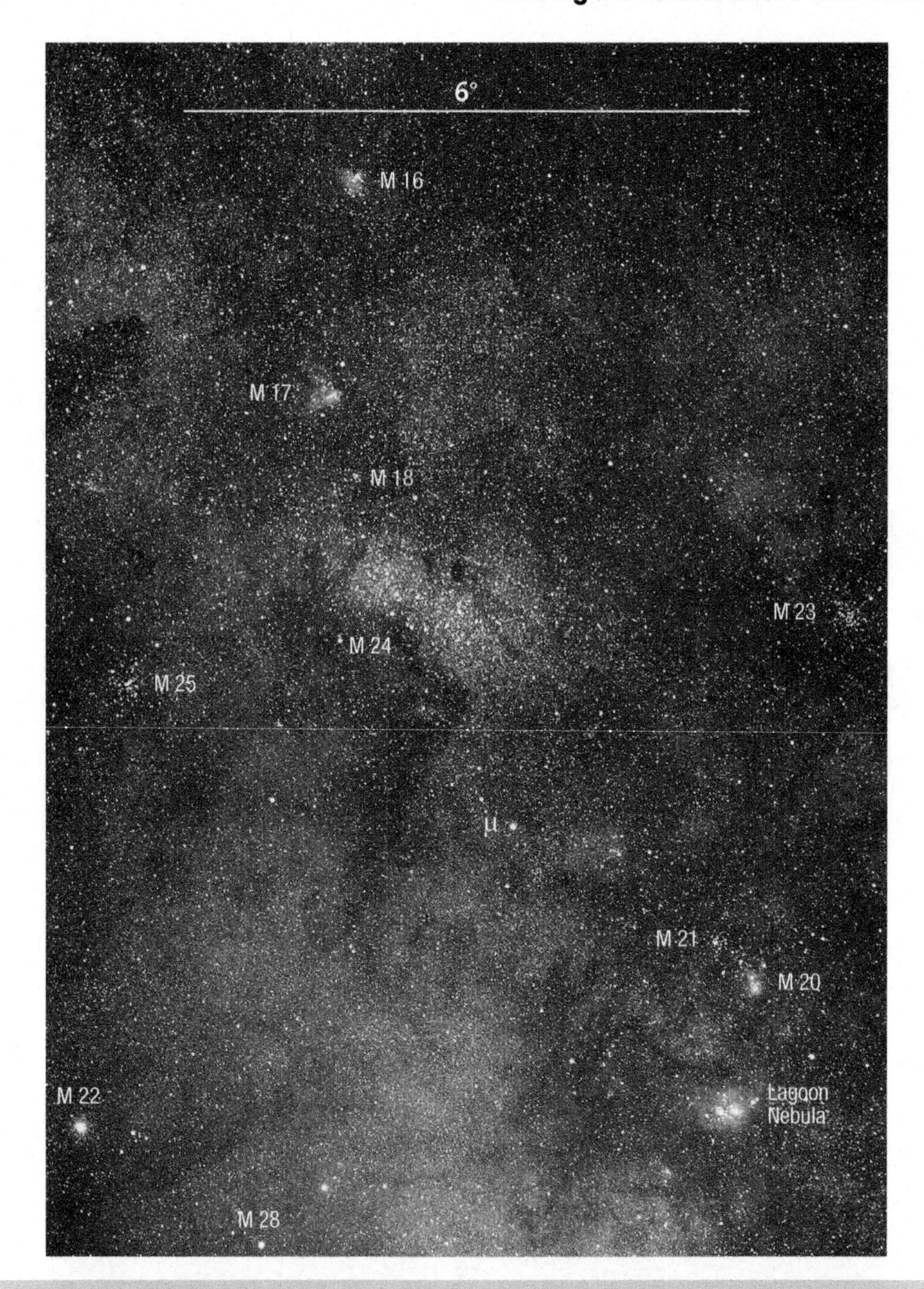

Figure 11.4 The wealth of Sagittarius in a single shot

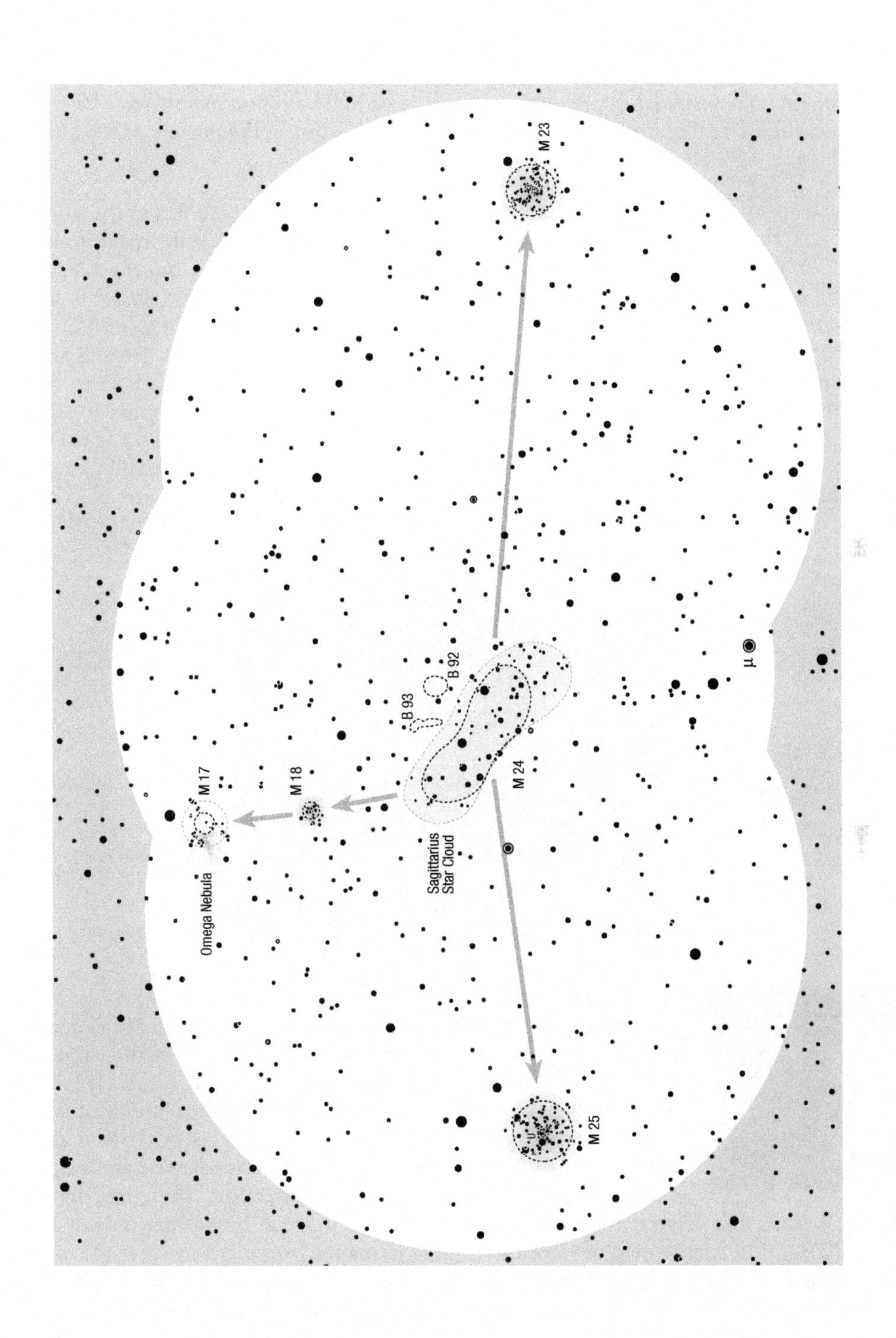
M 23
B 92
B 93
μ
M 17
M 18
M 24
Sagittarius Star Cloud
Omega Nebula
M 25

M 22 is approximately 10,400 light years from us and is one of the nearest globular clusters. It is hard to determine its actual diameter, as the stars become rarer and rarer toward the edge. Measurements are also made more difficult by the immense number of stars from the Milky Way that by coincidence are in the same direction and are of course not members of the cluster. The central, densest part measures approximately 50 light years in diameter. On photographic plates, astronomers are able to count 75,000 stars, but in fact there are at least half a million members.

M 22 is less than 1 degree from the ecliptic, and sometimes one of the planets or the Moon travels close by. This is a sight worth seeing!

The remarkable open cluster M 23 (5m5/27′) lies in the northwestern part of the constellation. The star leading to it is the magnitude 4 Mu, or even better the Star Cloud M 24, which lies approximately 4.5 degrees east (see the chart on page 419, image on page 418). The cluster is clearly seen through binoculars. When there are good observing conditions it is an approximately 15 arcmin big uniform spot of light. During an excellent night, the spot increases and "crumbles" into individual stars. The separation is even better with averted vision. The cluster contains approximately 100 stars. In a perfect night, you can see some 40; however, the brightest are only magnitude 9. The brighter star on the northwestern edge of the cluster and some magnitude 8 stars are not true members and by sheer coincidence lie in the same direction of the sky. M 23 is approximately 2,150 light years from Earth, so its actual diameter is 15 light years.

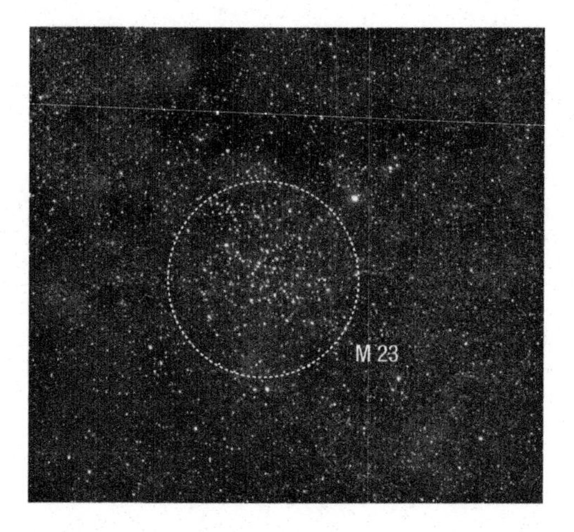

The large cloud of stars in the middle of the Milky Way, denser and brighter than its surroundings, has its own number in Messier's catalog; it is known as M 24 (4m6/2 × 0°9). It is also known as the Sagittarius Star Cloud, which is sometimes shortened to the Star Cloud. It lies approximately 3 degrees north of the magnitude 4 star Mu (see the chart on page 419, image on page 418). But in fact we do not need Mu to lead us, since the cloud is clearly visible with the naked eye. The densest and brightest part extends in the northeast–southwest direction. The cloud includes two magnitude 6 stars. But if you manage to catch the cloud in your binoculars, it will be a truly breathtaking sight! In a large field of view, the cloud charms the beginner as well as the experienced observer, regardless of the size and quality of the instrument. The stars in the Milky Way are especially dense in this area, and they melt into a nebula from which countless stars shine. The aforementioned magnitude 6 stars are accompanied by four magnitude 7 stars, five magnitude 8 stars, a dozen magnitude 9 stars, as well as countless fainter stars all down to the limit of the binoculars and the observing conditions.

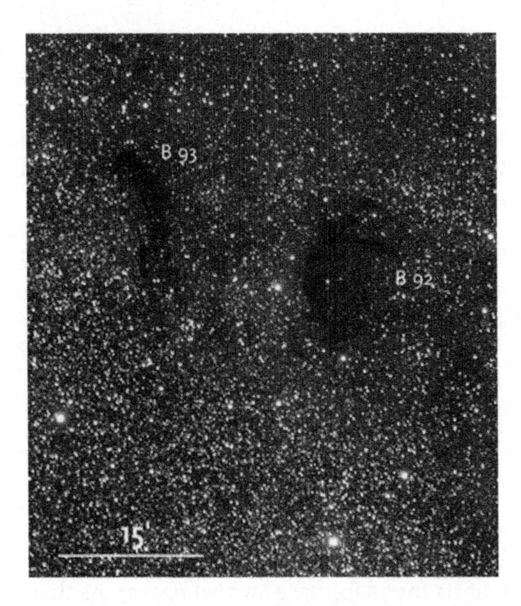

Approximately half a degree above the center of the cloud, there are two wonderfully contrasting dark nebulae with the designations B 92 and B 93. Unfortunately, they are not seen nearly as clearly through binoculars as they appear on images.

The Star Cloud is approximately 10,000 light years from us; its actual size is almost 350 light years!

Open cluster M 25 (4m6/ 32′) lies approximately 3.5 degrees east of the Star Cloud M 24 (see the chart on page 419, image on page 418). It is bright, but not very rich in stars. There are some 50 members that are brighter than magnitude 11 and some tens of fainter ones. The brightest star in the cluster shines with magnitude 6. This is the Cepheid variable star U, which changes its brightness from magnitude 6.3 to 7.1 over a period of 6.744925 days. Through binoculars we can see three brighter stars and a stack of fainter ones, gathered in an area of 20 arcmin in diameter.

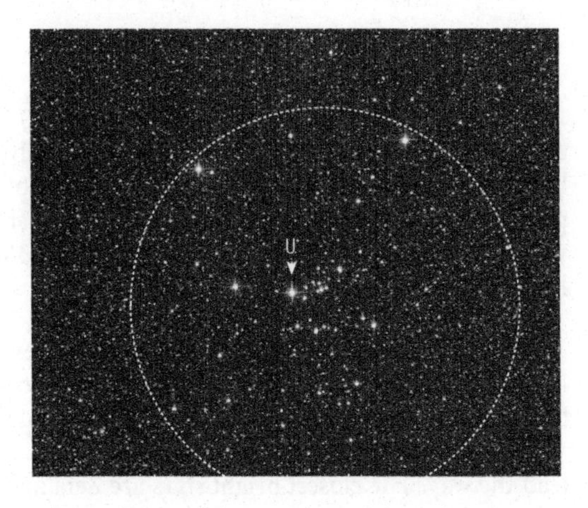

The cluster is approximately 2,000 light years from us, and its true diameter is 20 light years.

Globular cluster M 28 (6m8/11′) lies on the opposite side of Lambda to its brighter neighbor M 22 (see the chart on page 415, image on page 418). Through binoculars it is seen as a small, uniform spot of light, only a few arc minutes in diameter. When there is excellent night it is clearly seen, while in poorer observing conditions it can melt completely with the sky and thus becomes invisible.

What makes this cluster truly interesting (but not for amateur observers) is the presence of a millisecond pulsar, which was discovered in 1987. This pulsar rotates around its axis once every 11 ms, or to put it another way, it spins at approximately 90 rotations per second! This was the second millisecond pulsar to be found in any globular cluster (the first one was discovered the same year in the globular cluster M 4 in Scorpius). Since then astronomers have discovered an additional 7 millisecond pulsars in M 28.

Globular cluster M 28 is 18,300 light years from us; thus, its actual diameter is approximately 60 light years.

In the south part of the constellation, another four globular clusters await the serious observer. These clusters can only be seen when there are excellent observing conditions. They are M 54, M 55, M 69, and M 70 (charts for all four are on next pages). As they are a mere 12 degrees above the southern horizon at culmination for mid-northern observers, you should attempt to see them only on those summer nights in which the sky is dark all the way down to the horizon. However, you have greater chances of succeeding on early mornings in April and May.

M 54 (7m6/12′) lies 1.5 degrees west-southwest of Zeta. It does not appear terribly attractive through binoculars, since it is seen as a small, uniform spot of light. But beware! The newest research shows that M 54 is most probably a member of the dwarf elliptical galaxy with the designation SagDEG, which is our satellite galaxy. In this case, the cluster is as much as 87,400 light years from us, which means that its actual diameter is over 230 light years – a truly admirable object and the only extragalactic globular cluster seen through binoculars! Viewing is highly recommended.

Globular cluster M 69 (7m6/9′8) is very similar to M 54 in appearance. It lies approximately 3 degrees northeast of Epsilon, so they appear in the same field of view. The cluster is 29,300 light years from us, and its real diameter is approximately 60 light years.

Even a bit more fainter globular cluster M 70 (7m9/8′) lies between M 54 and M 69. Therefore it is on the border of visibility when viewed through binoculars – but it can be seen! This is one of the smallest and faintest globular clusters in the Messier catalog, so we should not expect too much when viewed through binoculars. M 70 is 29,300 light years from us and has a real diameter of 68 light years.

Globular cluster M 55 (6m3/19′) is the brightest of the four mentioned and is the most clearly seen though binoculars. Its only "imperfection" is that it lies in a rather desolate area of the sky, outside of the denser part of the Milky Way, which means that there are no brighter stars in the vicinity that would lead the way. The closest bright stars are Zeta and Tau, but they are as far as 8 degrees west (see chart on page 424). In excellent observing conditions, the cluster is seen as a small and faint spot of light through binoculars. It is 17,300 light years from Earth, and its real diameter is approximately 100 light years.

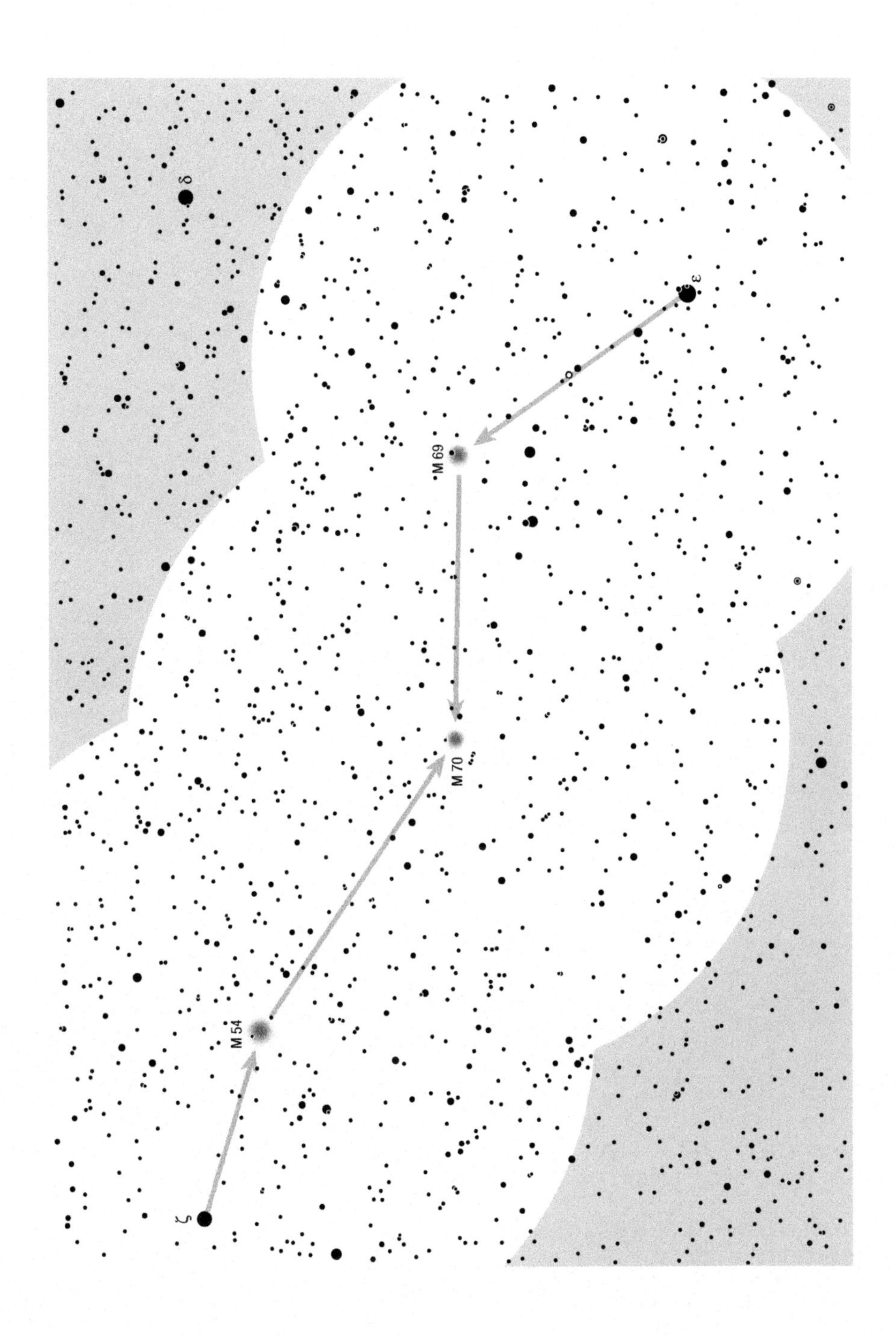
δ
ε
M 69
M 70
M 54
ζ

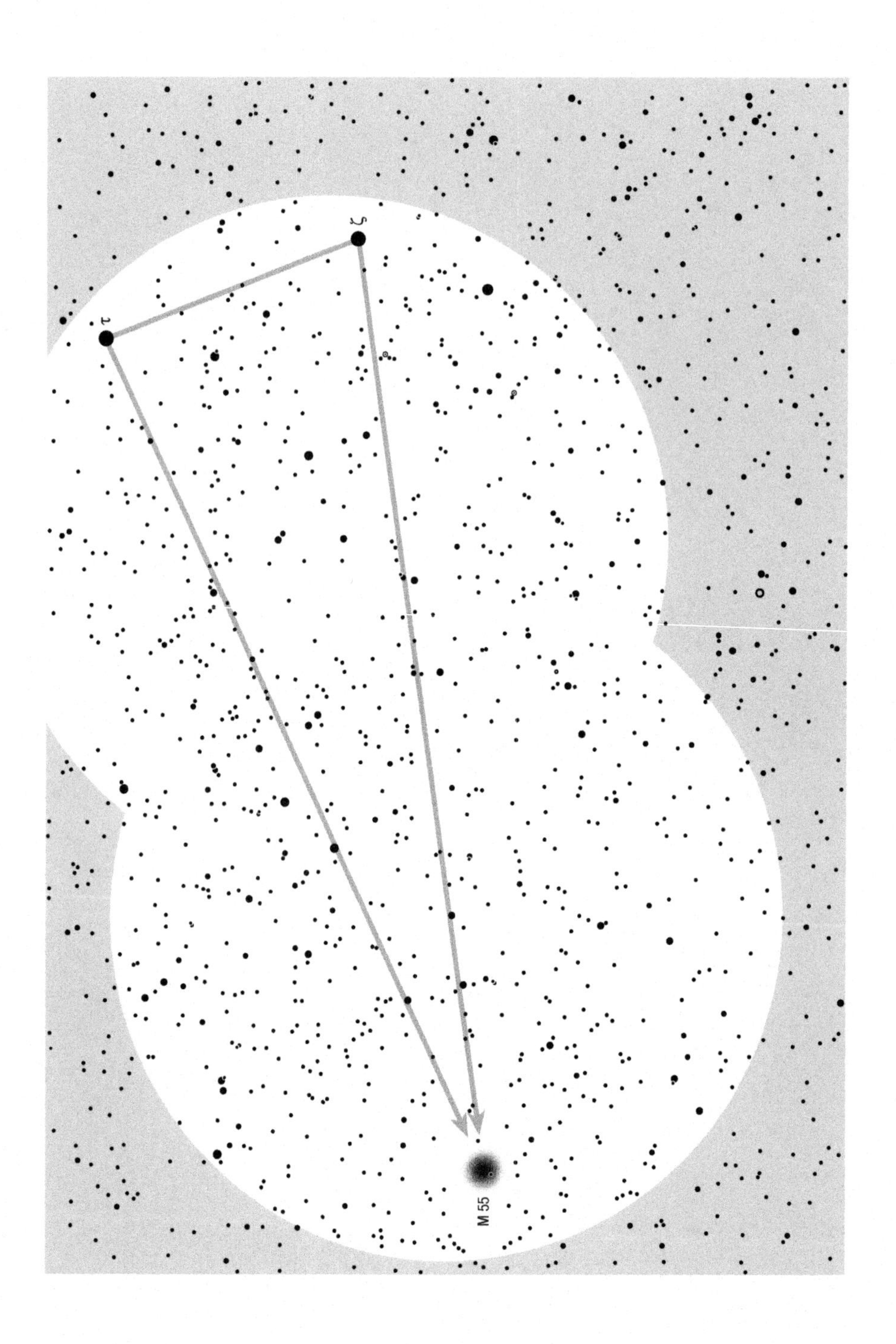
ζ
τ
M 55

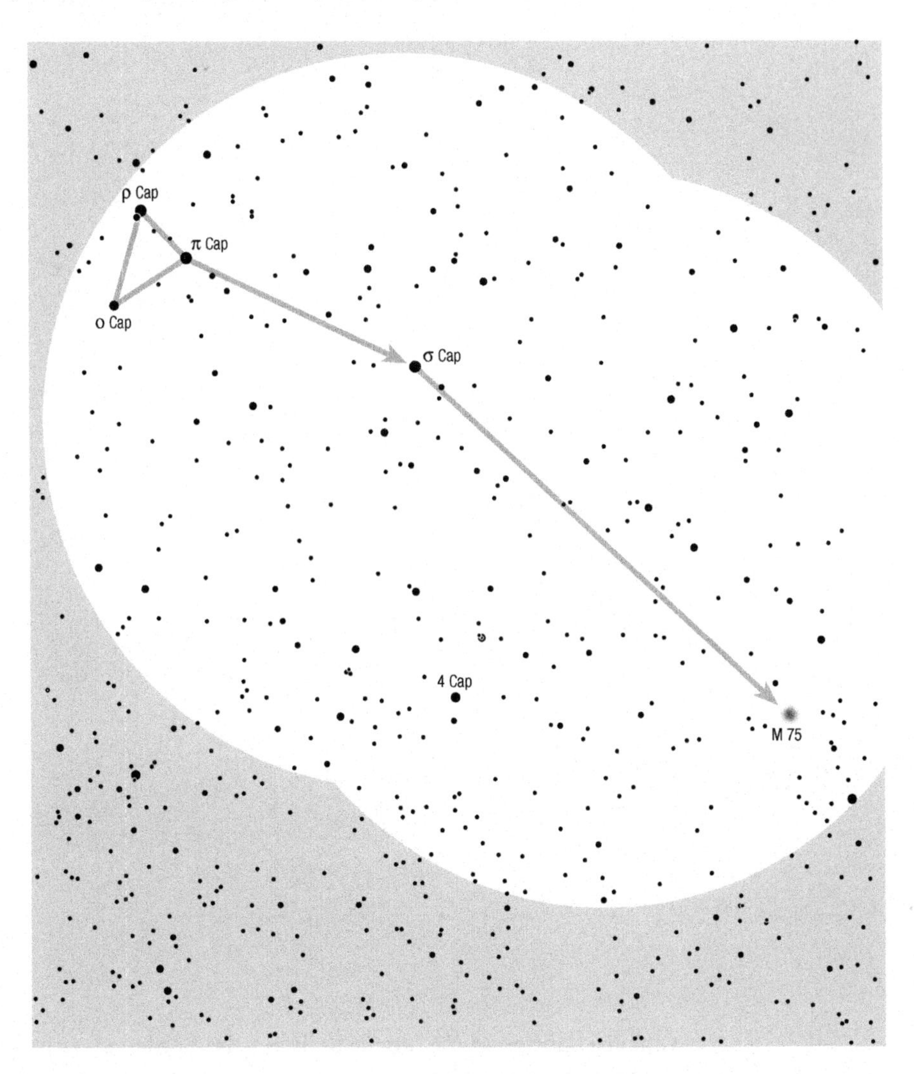
ρ Cap
π Cap
ο Cap
σ Cap
4 Cap
M 75

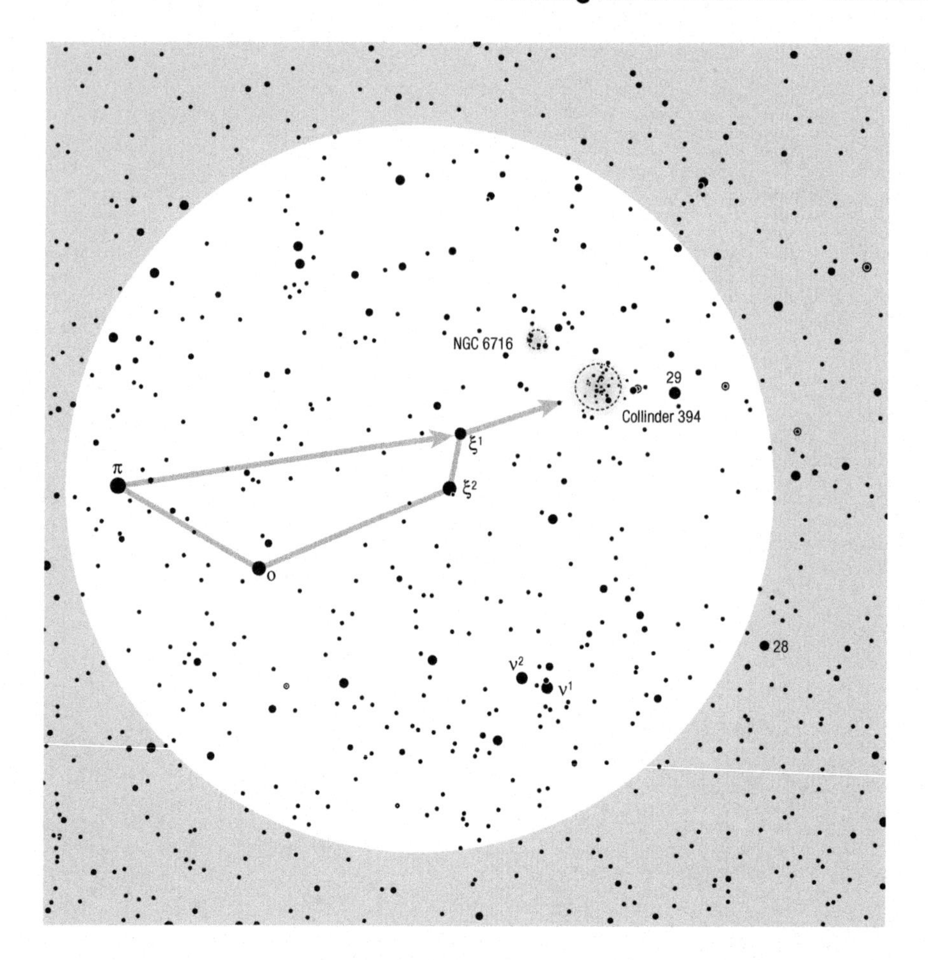

Globular cluster M 75 (8m5/6′8) is on the border of visibility when viewed through binoculars. It is found in the extreme eastern part of the constellation, on the border of Capricornus. The stars leading to the cluster are Beta, Rho, Pi, and Sigma Capricorni; for more precise positioning look at the detailed chart. M 75 is one of the densest and most remote globular clusters known to us. Its light travels up to 67,500 years in order to reach Earth. Its true diameter is therefore approximately 130 light years.

Some 4 degrees west-northwest of Pi, among a crowd of stars in the Milky Way, we come across two open clusters – NGC 6716 (7m5/10′) and Cr 394 (6m3/22′). The first lies less than a degree northeast of the second, so they appear together in the same field of view of binoculars. In such a rich constellation as Sagittarius, these two clusters are mentioned last, even though they are quite nice, especially Cr 394. When there are perfect observing conditions, we can see approximately 50 stars within this cluster. It is approximately 2,200 light years from us; thus, its real diameter is 15 light years. Its neighbor is a slightly further away (2,600 light years) and half the size (7.5 light years).

For mid-northern observers the most attractive regions of the summer Milky Way can be found in Cygnus, Scutum, Aquila, and Ophiuchus. Those in Sagittarius are a bit too far south. But on those nights when the sky is dark and clear all the way down to the horizon, the Milky Way in Sagittarius is truly magnificent. The brightest and densest part lies 26,000 light years from us, in the vicinity of Gamma, where one can find the center of the galaxy. The interstellar dust and gases are so thick here that they completely cover the sight right to the very center. Despite this, the field of view is full of brighter and fainter stars, larger or smaller condensations, covered with darker regions. As we stand under the sky on a clear summer night and the Milky Way spans above us from Perseus and Cassiopeia over Cygnus and Scutum all the way down to Sagittarius, we can easily imagine its true structure. In our minds, we can travel thousands of light years away and look back at the galaxy. Wow!

Anyone can notice that the Milky Way seems to be divided by a wide, dark zone between Cygnus and Sagittarius. This zone, which is just a denser region of opaque clouds of interstellar matter, is called the Great Rift. Similar equatorial lanes can also be seen in other galaxies, galaxies that we can see from "outside." The best examples of these are the Sombrero in Virgo, NGC 891 in Andromeda, and NGC 4565 in Coma Berenices.

The very center of our galaxy lies in the constellation Sagittarius, close to Gamma. This mysterious region, which is hidden from our eyes by dense clouds of dust and gases, can be "seen" in all parts of the electromagnetic spectrum – except visible light! When we look at the center with our binoculars, only imagination and the introductory chapter of this book (in which we described the exciting events that take place in this small part of the sky, no larger than our Solar System) can help us see the real picture. It is here, behind this thick curtain that will never rise for visual observers, that a show takes place that is incredibly hard to imagine with our Earthly experience. In the center of the galaxy lies a gigantic black hole – an infinitely small singularity with such gravitational pull that it gobbles up everything that comes its way. Giant rivers of gases descend into the bottomless pit, and as they spin into a vortex for the very last time before they disappear from the visible horizon, they emit their last, silent X-ray screams.

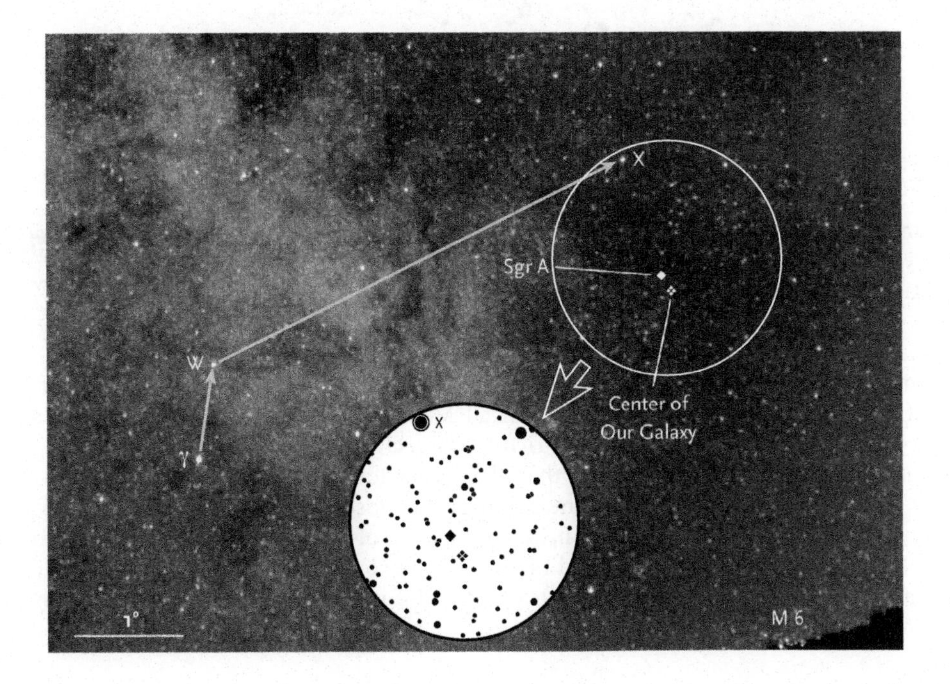

SCORPIUS (The Scorpion)

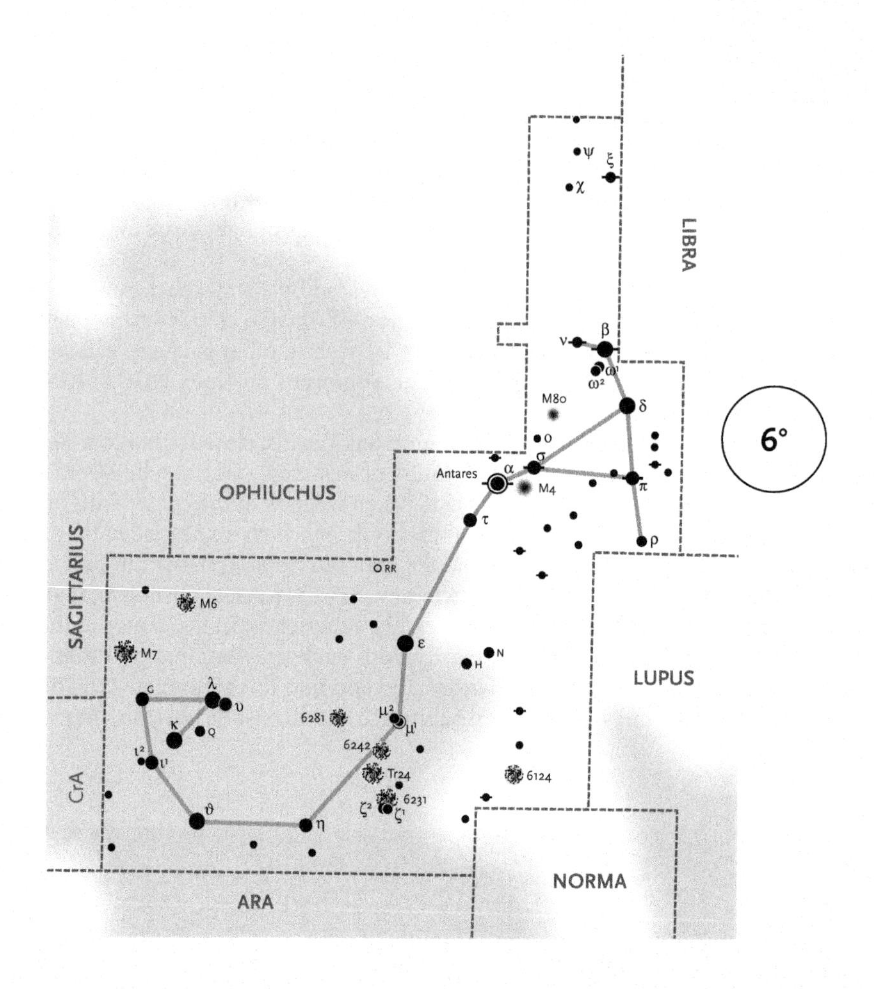

The summer constellation Scorpius is south of Ophiuchus, low above the southern horizon. It is one of the biggest constellations in the sky and certainly one of the most beautiful. Only Orion can compare with it in beauty. Scorpius is one of those rare constellations for which one does not need much imagination to see its shape in the star pattern. It is so distinct that it is one of the most important constellations in the mythologies of all Mediterranean nations. The Milky Way winds across the eastern part, so the entire area is rich in stars, clusters, and all types of nebulae.

CONSTELLATION CULMINATES		
in begin. May at 2 a.m.	in begin. June at midnight	in begin. July at 10 p.m.

Scorpius is one of those constellations that always lies low above the horizon for mid-northern observers. If a large town is located to the south of our observing site, light pollution is likely to completely override the entire constellation, except for a few of the brightest stars. If you really want to enjoy the hidden treasures that are offered here, you have to find an observing site with a clear and dark sky all the way down to the horizon. Or you could travel down south, from where you can see Scorpius high in the sky!

The brightest star in Scorpius is Antares (1m1), a red super giant with a luminosity that is over 10,000 times greater than that of our Sun. There is no chance that we could miss it in the sky, as it shines in a distinct orange-reddish color, true to its name, which means "(holds) against Ares (Mars)." In the figure of the scorpion, it represents its heart. Antares is the 16th brightest star in the sky, 600 light years away. It is one of the largest known stars. It has a diameter believed to be approximately 980 million km, and if we placed it in the center of our Solar System, the star would extend all the way to the Asteroid Belt, which means that the orbits of all the inner planets, including Mars, would be within the star!

Antares is a double star. Its companion is, as is the case with Alpha Herculis, pale green. However, this is most probably due to the strong orange color of Antares and the processing that takes place in our brains. The stars are separated by 2.5 arcsec (p. a. 274°, 1997). The pair is hard to observe even though the companion is of magnitude 5.4. When there are truly excellent observing conditions, and when the air is completely calm, this greenish star can be seen with a 15-cm objective lens and high magnification on the edge of the light emitted by the orange Antares.

Like most red giants, Antares is a pulsating variable star, but its brightness changes only by a few tenths of a magnitude. Not enough to notice visually.

The head of the Scorpius is represented by the stars Beta or Akrab (2m6), Nu (2m7), and Omega (3m9), while its tail consists of Lambda or Shaula (1m6), Kappa (2m4), Upsilon (2m7), G (3m2), and Q (4m3). However, this constellation has a number of other bright stars, such as Theta (1m9), Delta (variable), Epsilon (2m3), and Tau (2m8).

Following Antares, Lambda is the second brightest star in the constellation and the 25th brightest star in the sky, immediately behind Castor and Gamma Crucis. It is a 700 light years distant giant with a luminosity that surpasses our Sun by 8,000 times. In the shape of the scorpion it represents the stinger, and this is also the translation of its Arabic name Shaula (Al Shaulah).

Theta is the 39th brightest star in the sky. It is 270 light years from Earth. This giant star has a luminosity 960 times greater than that of our Sun.

The yellow-orange Epsilon is the 76th brightest star in the sky. Its light travels a mere 66 years in order to reach us. The luminosity of the star is that of 37 Suns.

Kappa is the 83rd brightest star in the sky. It is 465 light years from us, and its luminosity surpasses our Sun by 1,700 times.

Beta is the 96th brightest star in the sky. It is an especially interesting multiple star. Two of its components are resolvable through any telescope but unfortunately not through binoculars (due to their low magnification). The stars with designations A and C shine with magnitudes 2.6 and 4.5 and are 13.1 arcsec apart (p. a. 24°, 2005). This pair is accompanied by star B, which cannot be seen in amateur telescopes, since it is a mere 0.3 arcsec from the brighter star in the pair. However, this is not all! The brightest star in the system is a spectroscopic binary, with an orbital period of 6.8 days, but most probably Beta C is also a close binary with the separation of a mere 0.1 arcsec. This interesting quintuple star is approximately 530 light years from us. The luminosity of its brightest member is as much as 1,900 that of our Sun.

Nu is the 114th brightest star in the sky, some 520 light years from us. It lies approximately 1.5 degrees east and slightly to the north of Beta. Nu is also a quintuple star, one of the most beautiful in the sky. Only the pair A–C can be separated through binoculars. The two stars with magnitudes 4.2 and 6.6 are 41.3 arcsec apart (p. a. 338°, 2005). Through a telescope with higher magnification it would be possible to split star C in the pair with the designation C–D. The stars are 2.4 arcsec apart (p. a. 54°, 2003), and their magnitudes are 6.6 and 7.2. Pair A–B is a close binary that can be separated only in large amateur telescopes. Stars with magnitudes 4.3 and 5.3 are 1.3 arcsec apart (p. a. 2°, 2003). The fifth star is a member of the spectroscopic binary Nu A. All five stars are white.

Tau is the 134th brightest star in the sky. It is 430 light years from us. This star is also a giant with a luminosity of 1,000 Suns.

Iota is another super giant with a luminosity that surpasses our Sun by over 60,000 times! It shines as a magnitude 3 star in our sky, in spite of that is over 3,400 light years from us.

Delta is a very interesting star. It is a giant approximately 400 light years from us. Its luminosity surpasses our Sun by approximately 1,400 times. Throughout the entire history of astronomical observations it has shone at a constant magnitude of 2.3, and as such it is listed as the 75th brightest star in the sky. Suddenly, in June 2000, a slight change in its brightness was noticed for the very first time. The star was only 0.1 magnitude brighter than usual; however, its brightness has varied ever since, and it has occasionally gone as high as a magnitude of 1.6. The unusual behavior attracted the attention of astronomers, who discovered that the star is shedding luminous gases in violent outbursts. It has varied in brightness ever since it began this shedding, but the star constantly remains well above its previous magnitude.

Zeta is a double star that can be separated even with the naked eye. The distance between the stars is 6.8 arcmin. The eastern one, known as Zeta-2, is brighter and is of magnitude 3.6. It is an orange subgiant, 155 light years from us. The western one, known as Zeta-1, is much further away (approximately 5,700 light years) and is as white as Rigel in Orion. If the estimate of its distance is correct, then this is a true super giant star, with a luminosity that surpasses our Sun by over 100,000 times. The color contrast, which is even more emphasized through binoculars, is apparent for mid-northern observers only during those nights in which the sky is dark all the way down to the horizon. For those observers, Zeta is only 1.5 degrees above the southern horizon at culmination. It cannot be seen from England at all, but it is clearly visible from Florida!

Mu form a wide pair, clearly resolvable with the naked eye. Mu-1 is magnitude 3.1 while Mu-2 is slightly fainter - magnitude 3.6. The two stars are 346 arcsec (5.8 arcmin) apart. Most probably they are gravitationally bound, although the actual distance between them is as much as 55,000 AU, or 0.5 light years. The pair is 520 light years from us. In comparison to our Sun both stars are true giants. The luminosity of Mu-1 is as much as 10,000 times, and the luminosity of Mu-2 is as much as 700 times that of the Sun. Besides, Mu-1 is also a spectroscopic binary. Do we, here in our Solar System, truly live in the most boring part of the universe?

Sigma is an easy and splendid double star for small telescopes, but a little bit harder to see for observers with binoculars. The stars with magnitudes 2.9 and 8.4 are 20 arcsec apart (p. a. 273°; 1999), so they are seen as a close pair in the field of view. They have a nice color contrast; the brighter is of a distinct yellow color, and the fainter is white. However, this pair demands the best possible optics! It is mentioned at this point merely as a challenge.

Globular cluster M 4 (5m6/36′) is easy to find, since it lies slightly over 1 degree west of Antares, which means that they both appear in the same field of view. It is clearly seen through binoculars, but only as a 15 arcmin spot of light with a slightly brighter core. Its appearance depends greatly on observing conditions, as the cluster is always low above the horizon (18 degrees at culmination) for mid-northern observers. Reliable observers have reported that it can be seen with the naked eye from southern observing sites, where it is high in the sky.

M 4, one of the closest globular clusters, is 7,200 light years from us. If this is correct, then its actual diameter measures "a mere" 54 light years (Figure 11.6).

Globular cluster M 80 (7m3/10′) is found halfway between Antares and Beta. It is bright but small, so it resembles a blurred star rather than a cluster of stars in the field of view of binoculars. As it can be found in the same field of view of binoculars with M 4, it offers a rather interesting sight. While M 4 is striking to the eye, excellent observing conditions, a detailed chart (Figure 11.5), and experience are necessary in order to recognize and differentiate M 80 from the surrounding stars. The cluster is over 32,600 light years from us. Its real diameter is estimated to be 85 light years.

When there are good observing conditions, open cluster M 6 (4m2/25′) is seen even with the naked eye, which is why it appeared on star charts even before the telescope was invented. Through binoculars it can be broken into individual stars. The brightest are of magnitude 6, while the fainter ones melt together into a soft nebula. The brightest dozen or so stars are arranged in the shape of a butterfly with open wings, which is why the cluster is popularly known as the Butterfly Cluster. When there are excellent observing conditions one can see at least 50 stars, while on a perfect night their number can increase to over 70. In total, there are 100 stars in the cluster. The brightest star is orange, while all the others that can be seen through binoculars are white.

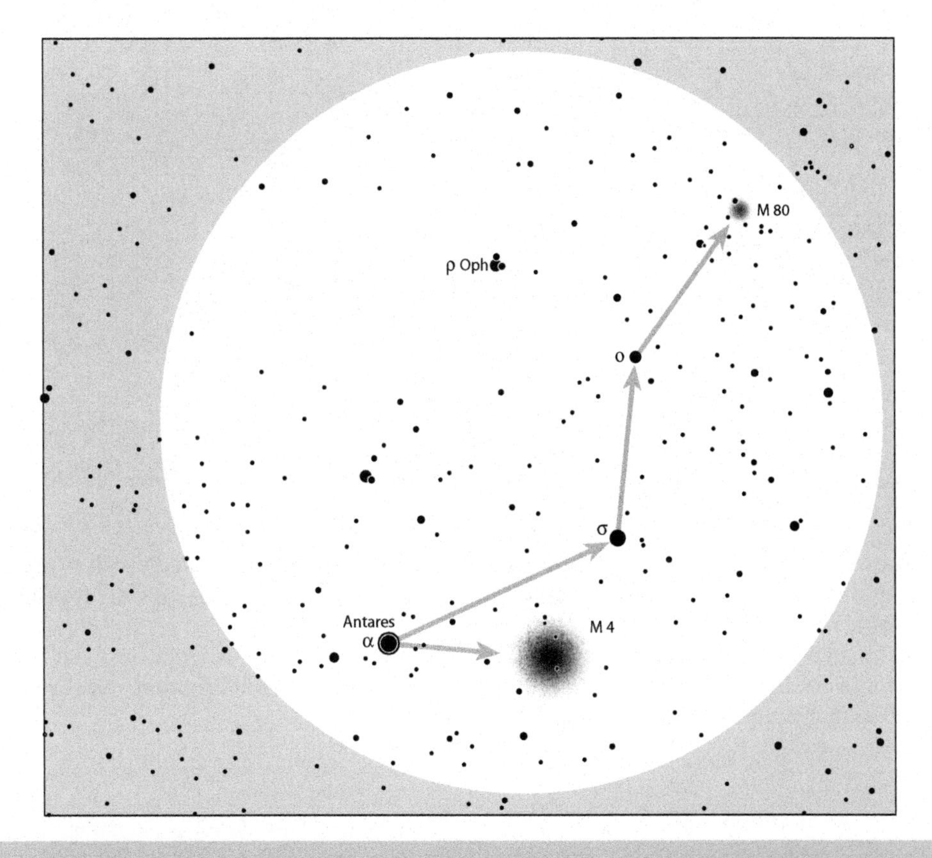

Figure 11.5 Near Antares there are two completely different globular clusters: M 4 is large and clearly seen, while M 80 is bright but small, so a superficial observer can easily mistake it for a star

Even if observing conditions do not allow us to see the cluster with the naked eye, it is not hard to find, as it lies only 5 degrees north-northeast of the bright Lambda, and so they appear together in the same field of view of binoculars.

The exact distance to M 6 has not yet been fully determined. It is approximately 1,600 light years from us. At this distance its real diameter reaches 15 light years, and the brightest stars are gathered within an area that covers only 9 light years in diameter.

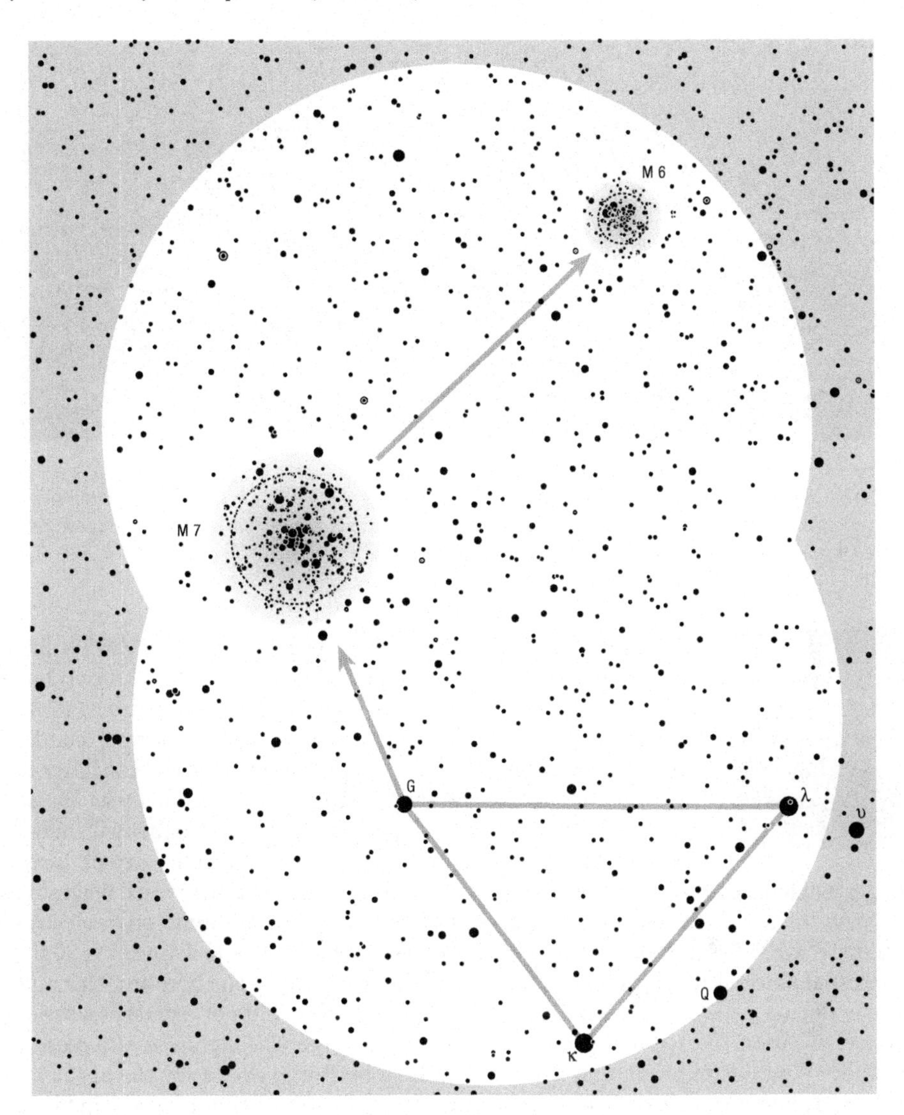
M 6
M 7
G
λ
υ
Q
κ

Figure 11.6 M 4 (*left*) and M 80 are two completely different globular clusters. M 4 is one of the least dense known to us. Its central parts can be totally resolved through large professional telescopes. The stars in M 80 are so close together toward the center that they could not be separated even if the cluster was closer to us

If M 6 is fantastic, then its close neighbor M 7 (3m3/80′) is breathtaking when viewed through binoculars. This cluster can also be found with the aid of Lambda. Their separation is slightly less than 5 degrees, so they appear together in the same field of view. With good observing conditions, it is easily seen with the naked eye. When viewed through binoculars it crumbles into a number of bright and less bright stars. As many as 130 stars are of magnitude up to 10, but on a perfect night their number will increase to over 200! In total, there are approximately 400 stars in the area of the cluster. M 7 is much larger and looser than its neighbor; its stars are slightly brighter, as they reach magnitude 5.5. It is at its most beautiful in wide-field binoculars. The scene is supplemented by numerous faint stars from the summer Milky Way. M 7 is approximately 800 light years away, and its real diameter is 20 light years.

Which celestial object is most astonishing? Most probably, it is M 7. The first time this author saw it was years ago, during a late spring night. At the time light pollution did not represent a problem, and I was far from civilization, in a region of Slovenia called Bela Krajina. The night was Moonless and totally dark. The Milky Way was so bright that it cast shadows! John Bortle would say class 1–2. The crickets were singing and glowworms tried to chase the night away. Something primordial was in the air. I found Lambda through my binoculars and then shifted them slightly to the northeast, and the cluster appeared in the field . . . WOW!!! . . . I was enchanted, my skin tingling. I could not move from the eyepieces for hours! I just stood there and stared. This was one of the most beautiful sights I had ever seen.

For those of you who like a challenge every now and then, try to find the open clusters in the extreme southern part of the constellation. These clusters are clearly seen from the southern latitudes, but for mid-northern observers they are too low and can be seen only on those rare nights when the sky is dark all the way down to the horizon. The most promising are NGC 6281 (5m4/8′ – approximately 40 stars up to magnitude 11), NGC 6242 (6m4/9′ – approximately 30 stars up to magnitude 11), Trumpler 24 (~5m/60′ – approximately 190 stars up to magnitude 11), NGC 6231, and NGC 6124 (5m8/29′ – approximately 60 stars up to magnitude 11) (Figure 11.7).

Figure 11.7 The rich area of open clusters, which is more suitable for southern observers, extends to the north of Zeta (the image is slightly turned, so the north lies diagonally – compare it with the chart on page 435). *Below left* is NGC 6231, the large and very loose Trumpler 24, which is wrapped into a soft nebula, is in the center, and NGC 6242 lies *top right*

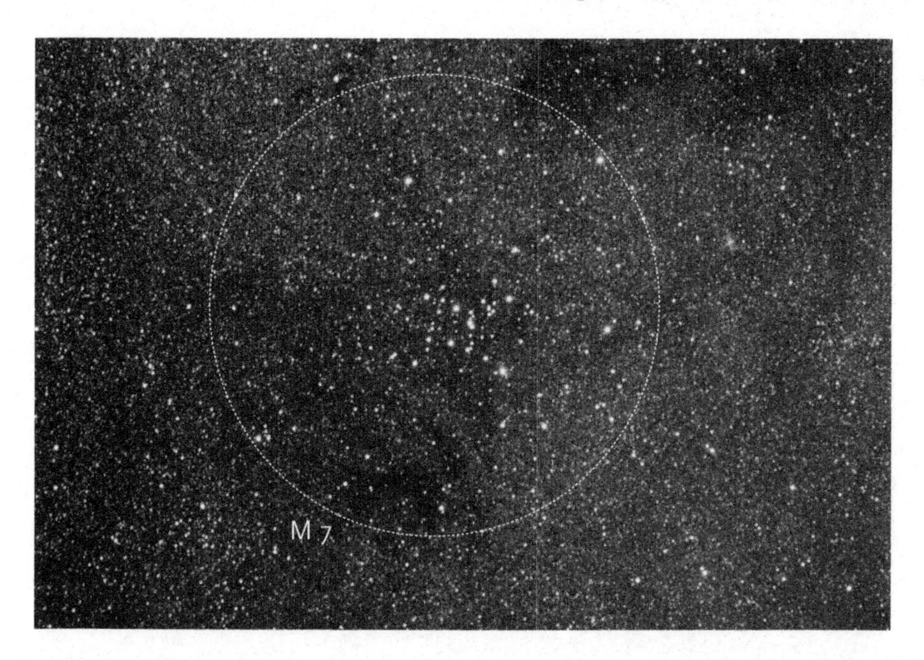

Even though it is the most southern one, NGC 6231 (2m6/15′) is the most clearly seen among these clusters. It is composed of a magnitude 5 star, six magnitude 6 stars, eight magnitude 7 stars, and at least 100 fainter ones that melt into a tenuous nebula. The cluster is only 2 degrees above the horizon at culmination (for mid-northern observers, of course), so you need an almost mathematical horizon for viewing (by the sea, for example). Anyone who wishes to try spotting such an elusive object, should try in an early spring morning (March or April), when the nights are still cold and not as humid. Why don't you try it? After all, it is free entertainment!

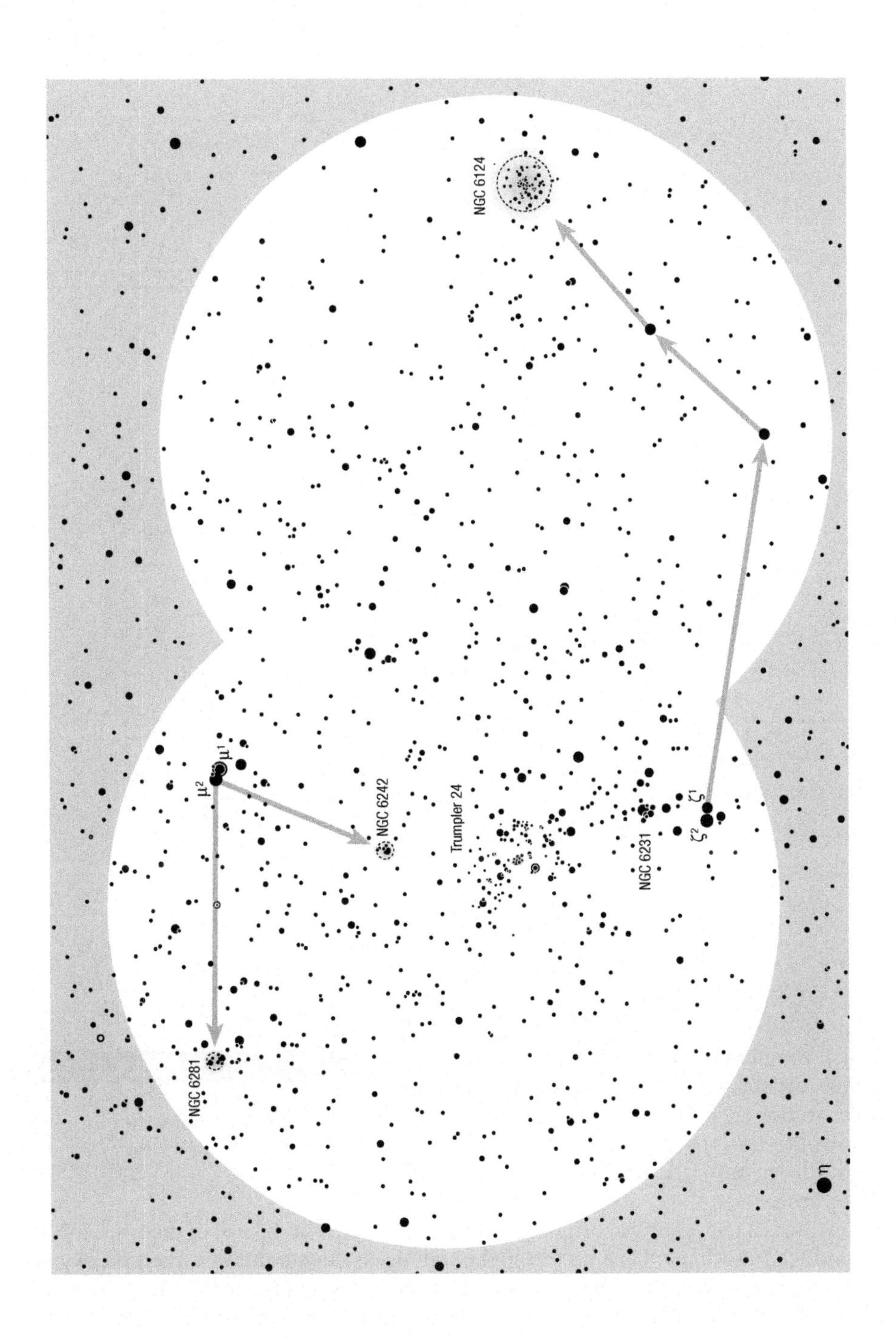
NGC 6124
μ1
μ2
NGC 6242
Trumpler 24
NGC 6231
ζ1
ζ2
NGC 6281
η

SCULPTOR (The Sculptor) and PHOENIX (The Phoenix)

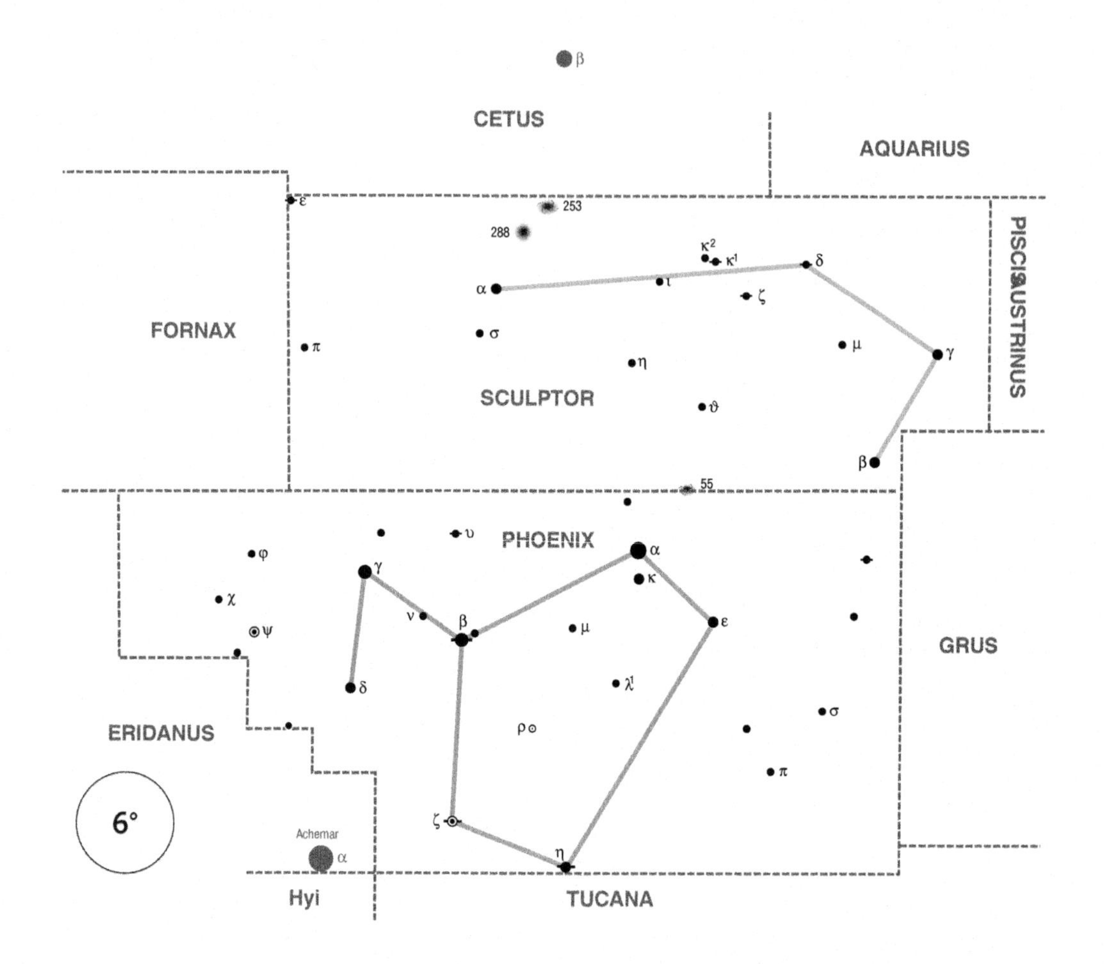

CONSTELLATIONS CULMINATE		
end August at 2 a.m.	end September at midnight	end October at 10 p.m.

Below Cetus, low above the southern horizon, lies the barely visible constellation Sculptor. Its brightest star Alpha shines with a magnitude of only 4.3 and is only 15 degrees above the horizon at culmination when viewed from mid-northern latitudes. In light-polluted areas it can hardly be seen.

Sculptor includes a few large and bright galaxies, but except NGC 253 all of them are too low south to be seen through binoculars. On a perfect night high up in the mountains, when the sky is dark all the way down to the horizon, you can try to catch a glimpse of NGC 55 (7m4/ 32′4 × 5′6), a large spiral galaxy that is seen edge on. The star leading to the galaxy is Alpha Phoenicis. The star and the galaxy appear together in the same field of view of binoculars. From mid-northern latitudes the galaxy is only 5 degrees above the horizon at culmination, which makes this object more suitable for southern observers.

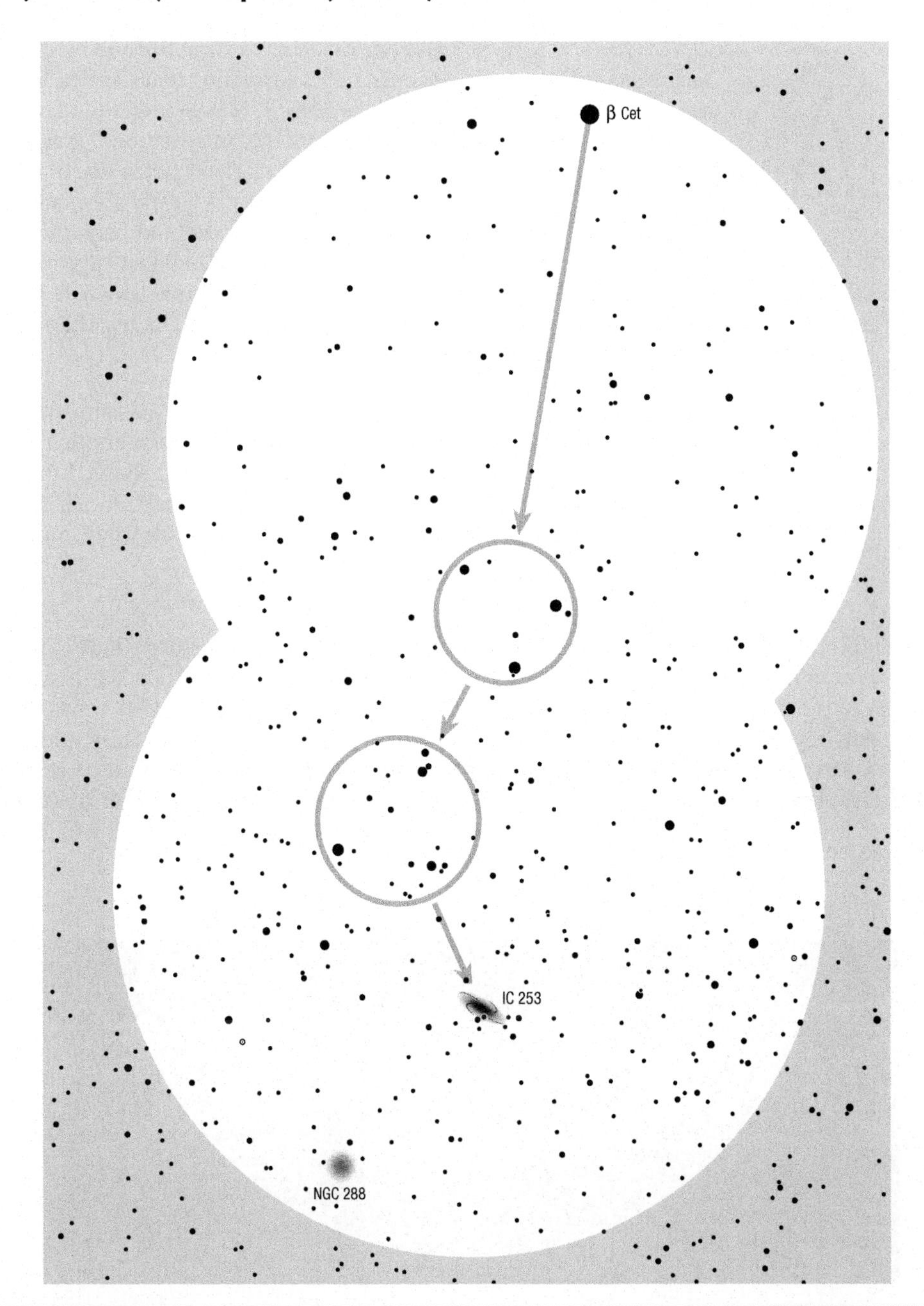

Figure 11.8 The stars leading to the galaxy NGC 253 and the globular cluster NGC 288 are the bright Beta Ceti and the line of clearly visible brighter stars below it

NGC 253 (7m1/25′ × 7′4) is one of the most attractive and brightest spiral galaxies visible through binoculars. It is surprising that Charles Messier did not include it in his list of nebulous objects. Maybe it was too far south for Parisians? The galaxy lies on the border of the constellation Cetus and is easy to find. The star leading to it is the bright Beta Ceti and the line of magnitude 5 and 6 stars below it (Figure 11.8). The galaxy is clearly seen, but only as a faint but distinct elongated stripe of light. Its size and brightness depend mainly on the observing conditions. The galaxy is elongated because we see it almost edge on.

NGC 253 is the brightest member of the smaller galaxy cluster in Sculptor, which is one of the nearest to our Local Group. It is estimated to be 10 million light years from us. The real diameter of the galaxy is approximately 70,000 light years (Figure 11.9).

When there are excellent observing conditions and a calm atmosphere, you will notice between the galaxy NGC 253 and Alpha another faint spot of light - globular cluster NGC 288 (8m1/14′). Through binoculars it appears more like a blurred star than a globular cluster. This is not an interesting or attractive cluster, merely one for the collectors of NGC objects.

Below Sculptor, just above the horizon when viewed from the mid-northern latitudes, lies the northern part of the constellation Phoenix with the bright Alpha. With magnitude 2.4 this is the 84th brightest star in the sky. The star is 78 light years from us and has a luminosity of 47 times that of the Sun. It is a relatively prominent star when viewed from the southern latitudes, but from mid-northern latitudes it is only a few degrees above the horizon at culmination and thus loses all its shine due to the thick atmosphere. The constellation does not contain any interesting objects for observers with binoculars.

Figure 11.9 Because the spiral galaxy NGC 253 is large and bright in the sky, it shows great detail even on amateur images

SCUTUM (The Shield)

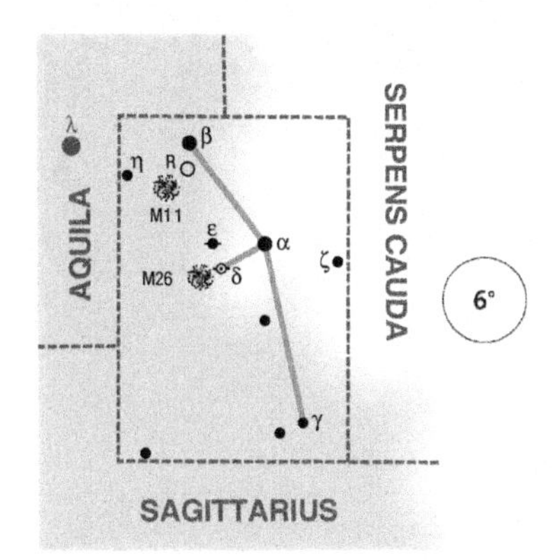

CONSTELLATIONS CULMINATE		
in begin. June at 2 a.m.	in begin. July at midnight	in begin. August at 10 p.m.

The summer constellation Scutum lies north of Sagittarius and southwest of Aquila, close to its bright star Lambda.

Alpha (3m8) is the only star brighter than magnitude 4. The constellation is small and only covers two fields of view of binoculars, but in the summer Milky Way it is well worth a panoramic survey, the results of which cannot be surpassed by the view even through the largest telescopes! Scutum is namely the home to one of the densest, brightest, and most beautiful star clouds, full of bright stars and covered with dark spots and stripes – opaque clouds of cold gases and dust. With excellent observing conditions, this condensation can even be seen with the naked eye!

Variable R is found just 1 degree south of Beta (see the chart on page 442). R is quite an unusual star. Its basic period of variability is 146 days. However, the curve is not regular (sinusoid), and maximum and minimum are also not always the same (Figure 11.10). At its maximum brightness, the star reaches magnitude 4.2 and is therefore visible with the naked eye. Every 146 days it falls into its deepest minimum of magnitude 8.6. In between it oscillates with a period of 33 days, with changes in brightness between magnitude 4.2 and 6.0. It seems that the star is a Cepheid as well as a red giant pulsating variable of the Mira type at the same time; thus, its brightness oscillates with a combined period. This is probably the main reason that the deepest minimum sometimes fails to come. Despite the fact that the star is a red giant, it does not have a truly distinctive color. It has an orange hue, but this hue is no stronger than the orange of Alpha or of the yellow-orange Beta, which are also both giants.

The star can be seen through binoculars during the entire period. One of its deepest minimums was reached on March 7, 2008. From this date and the known period, one can easily calculate the approximate dates of the next minima.

The variable is approximately 2,750 light years from us. At its brightest, its luminosity surpasses our Sun by as much as 8,000 times. Yes, it is a true giant!

When viewed through medium-sized amateur telescopes, M 11 (6m3/14′) is one of the most beautiful open clusters in the sky (Figure 11.11). Unfortunately, observers using binoculars will have to be satisfied with rather the bright, approximately 15 arcmin large and clearly seen spot of light. With average observing conditions you can notice two magnitude 8 stars on the southeastern edge of the cluster; however, these are not its true members.

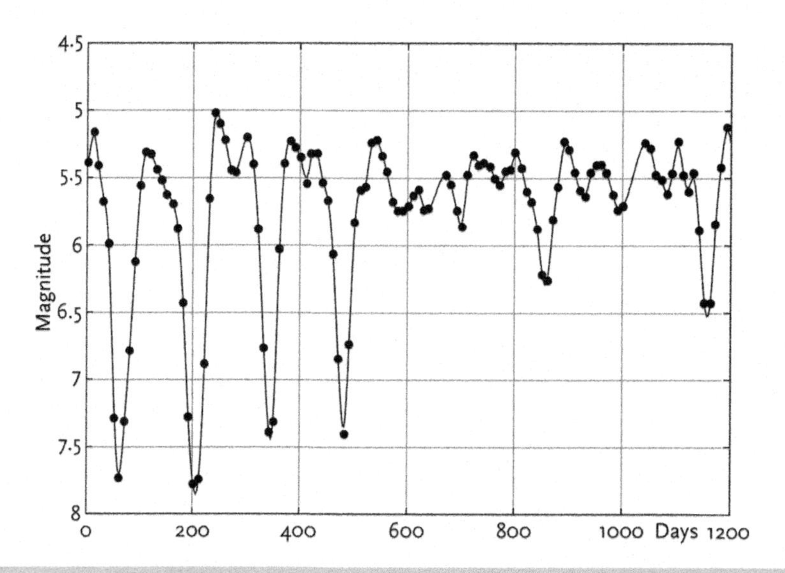

Figure 11.10 Light curve for variable R

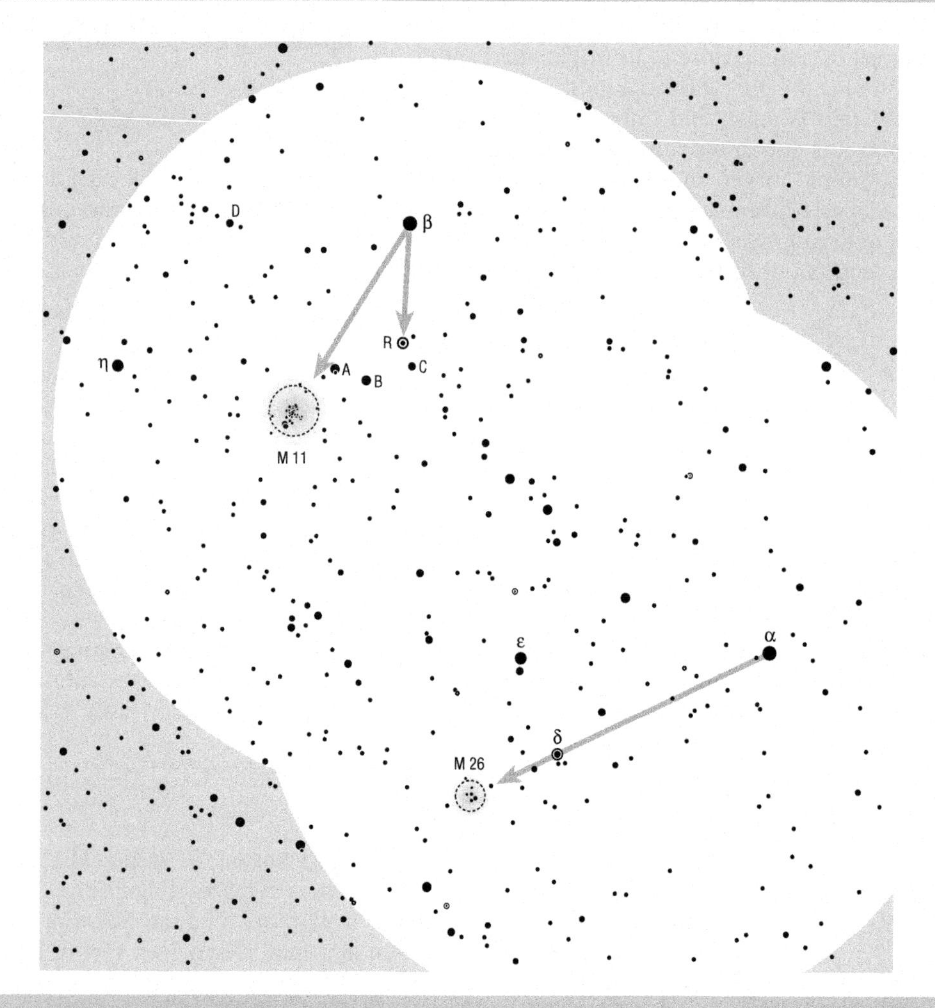

Figure 11.11 The surroundings of variable R with comparison stars Beta (4m2), Eta (5m0), A (6m1), B (6m7), C (7m1), and D (7m5). On the image of the open cluster M 11 (found on next page), one can see the variable R and comparable stars

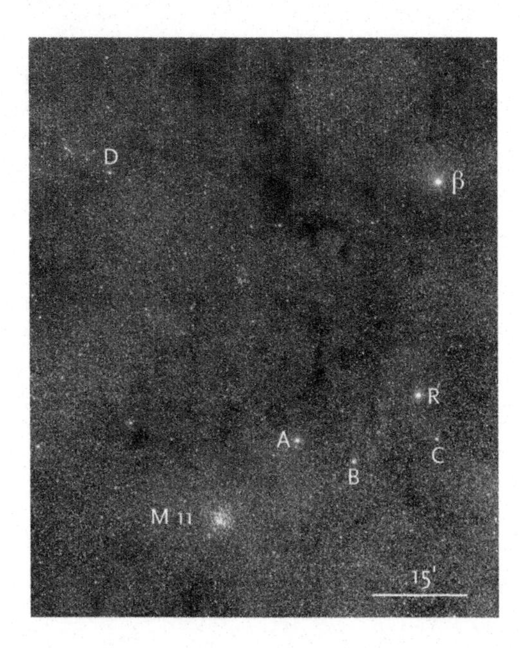

Among the stars within the cluster, only one is magnitude 9; all the others are fainter and when viewed through binoculars melt into the haze. Some of them show up when there are excellent observing conditions with the aid of averted vision. There are approximately 500 stars up to magnitude 13, and 50 of these are brighter than magnitude 11. The total number of the stars in the area of the cluster exceeds 2,000! M 11 is in the same field of view as Beta, so it is not hard to find.

The cluster is approximately 6,000 light years from us, so its actual diameter is 45 light years. It lies in the denser part of the summer Milky Way, which is extremely rich in stars; therefore, the sight of it and its surroundings will charm even the most demanding observer.

By astronomical standards M 11 is a young cluster. It is estimated to be only 250 million years old. It is one of the densest open clusters known to us. In its central region, which measures 15 light years in diameter, the density of stars reaches as much as 8 stars per cubic light year, which means that the average distance between two neighboring stars is less than 1 light year. The density of stars in this open cluster is not much lower than in an average globular cluster. Try to envision the following. Standing on a hypothetical planet in the central part of the cluster one could gaze in the sky and see a few hundred magnitude 1 stars, and at least 40 would be between 3 to 50 times brighter than our Sirius! What a sight that would be!

Open cluster M 26 (8m0/14′) is smaller and much fainter than its northern neighbor. Through binoculars it can barely be seen as a few arc minutes large spot of light. In order to observe it you have to choose a really clear and dark night; otherwise, you will not be able to see it at all. Even if you perform your observations with a larger telescope, you could see only approximately 20 stars in the area of the cluster, none of them brighter than magnitude 9. The cluster is easy to find, since it is in the same field of view with Alpha and less than a degree east-southeast of Delta (see detailed chart on previous page). It is estimated to be 5,000 light years from us, which means that its real diameter is only 12 light years.

SERPENS (The Serpent)

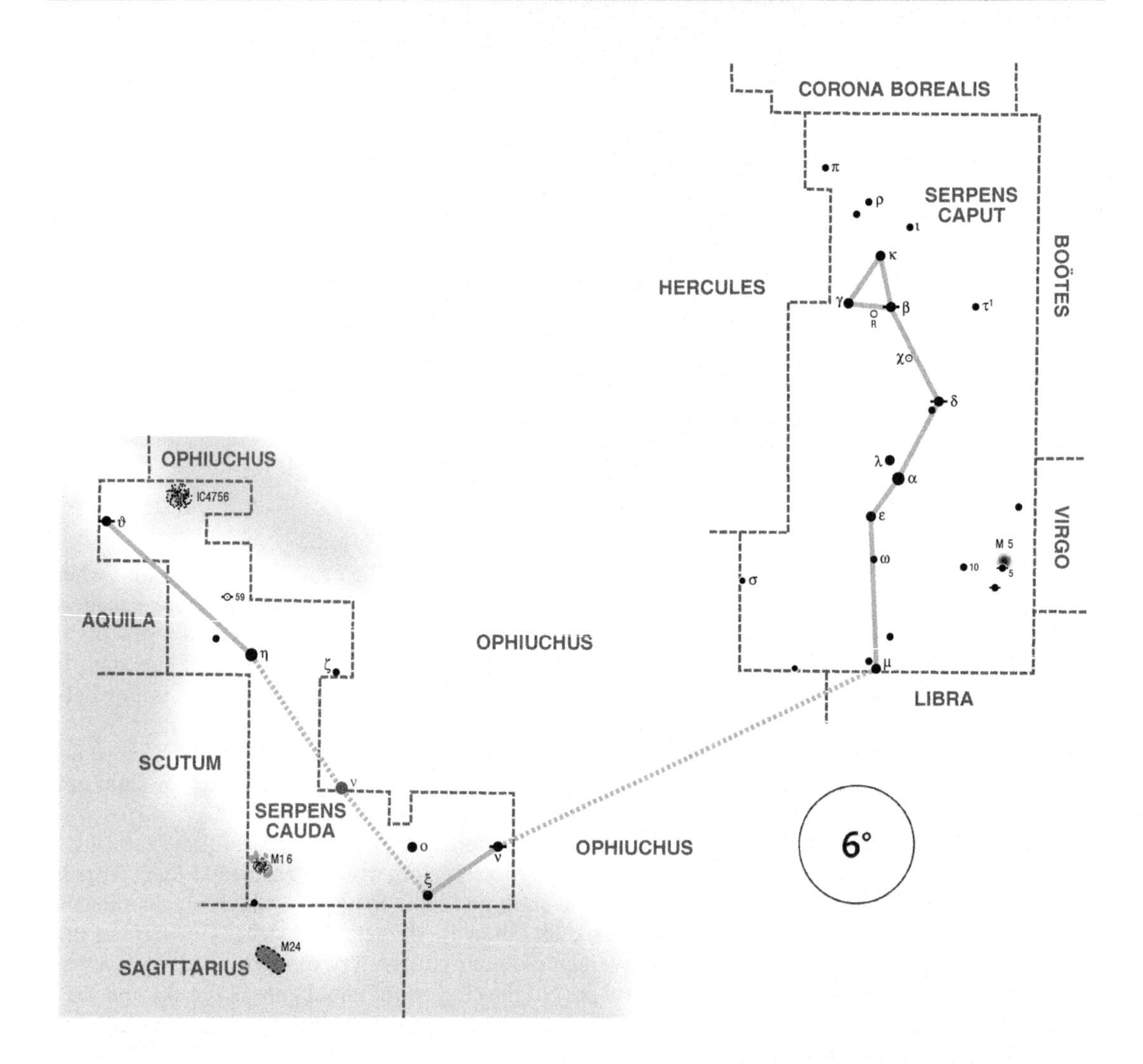

The very large (extended) Serpens has its beginning under the spring constellation Corona Borealis, east of Bootes. From there it crosses Ophiuchus and ends in the middle of the summer Milky Way at Aquila and Scutum. Serpens is the only constellation that is divided in two (Ophiuchus – The Serpent Holder – cuts it in half). The part known as Serpens Caput (The Serpent's Head) lies west of Ophiuchus, while the second part, known as Serpens Cauda (The Serpent's Tail), is found to the east of it.

SERPENS CAPUT CULMINATES		
end April at 2 a.m.	end May at midnight	end June at 10 p.m.

SERPENS CAUDA CULMINATES		
end May at 2 a.m.	end June at midnight	end July at 10 p.m.

The brightest stars within Serpens Caput are Alpha (2m6), Mu (3m5), Beta (3m6), and Epsilon (3m7).

The yellow-orange Alpha is the 103rd brightest star in the sky and is 73 light years from us. Its luminosity is 34 times that of the Sun.

The surroundings of Beta are especially rich in stars and are attractive for observing with wide-field binoculars. Besides the bright Gamma (3m8) and Kappa (4m1) there are at least a dozen magnitude 6 stars. Beta itself is a double, but for observers with binoculars it is very hard to separate the two stars, since the companion is of magnitude 10.0. The stars are 31 arcsec (p. a. 264°, 1999) apart, which means that it is possible to split them through binoculars. The fainter star is yellowish. It is a dwarf with luminosity of just one-sixth that of the Sun. Anyone who tries to spot the fainter star through binoculars should attempt this only when there are excellent observing conditions and when the constellation is high in the sky.

Variable star R and Beta appear together in the same field of view of binoculars (Figures 11.12 and 11.13). R is a red giant, a typical long-period variable of the Mira type. It changes its brightness from magnitude 5.2 to 14.4 in a period of 356.41 days. The highest and the lowest brightness are just extreme values that have been reached by the star only a few times during the last 180 years in which it has been regularly observed. During most of its maximums its brightness has reached between magnitude

Figure 11.12 Long-period variable star R at maximum (*bottom*) and minimum brightness. The *bright star* on the *right* is Beta. The apparent separation between them is approximately 1 degree

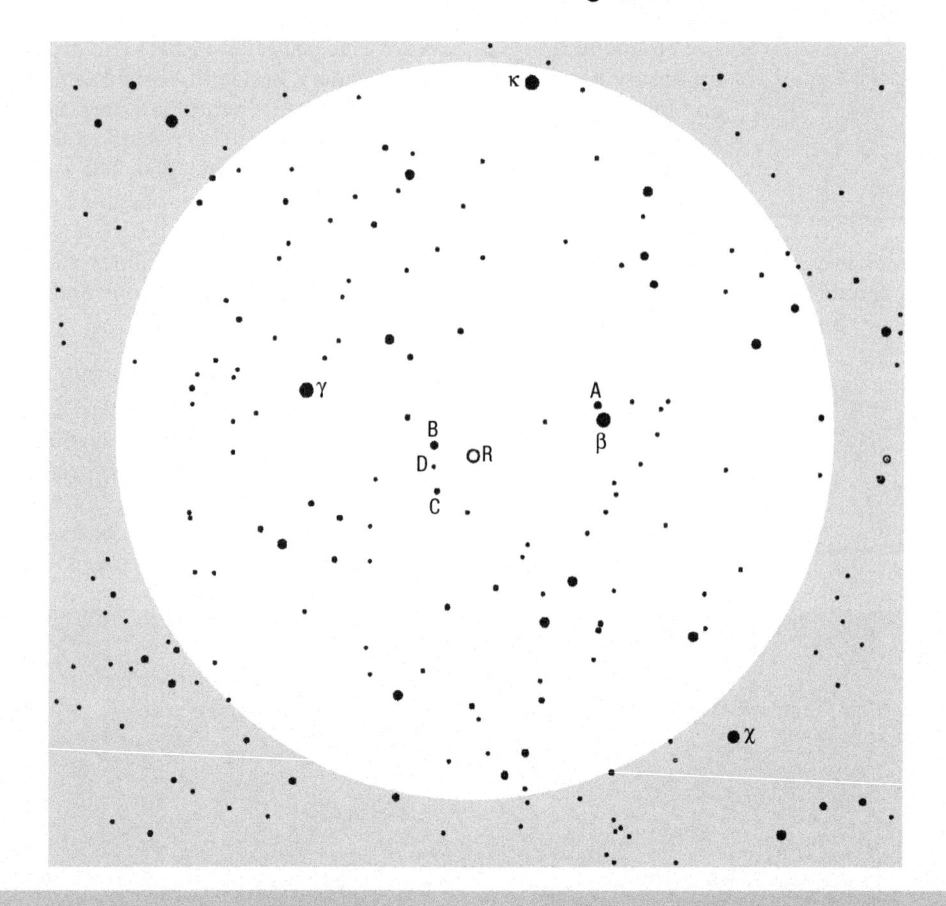

Figure 11.13 Variable R is easy to find, since it is together in the same field of view of binoculars with Beta. In its vicinity we can find some suitable comparison stars: A (6m8), B (7m4), C (8m4), and D (9m2)

6 and 7, and 13 at minimum. At its brightest it can occasionally be seen with the naked eye, but it is always clearly visible through binoculars. When its brightness falls toward the minimum, it is out of reach of binoculars. The star reached one of its maximums on January 18, 2008. From this date and the known period, one can easily calculate the dates of the next maxima. As all red giants R is also distinguished for its orange-reddish color that becomes even more vivid when the star is nearing its minimum, but unfortunately this is the period that it is not seen through binoculars.

R Serpentis is a typical red giant with a diameter 100 times bigger than that of the Sun. Its luminosity surpasses our star by about 250 times. The star is approximately 600 light years from us.

Globular cluster M 5 (5m6/23′) is easy to see, but finding it can be a much harder task, since it lies in a part of the sky in which there are not a lot of bright stars (Figure 11.14). Through binoculars it is seen as a bright spot of light with a distinct brighter central region. Its size depends on observing conditions. With excellent conditions, the spot of light extends to at least 10 arcmin. The star leading to the cluster is Alpha. If the star is moved to the extreme northeastern edge of the field of view, the magnitude 5 star labeled as 10 Serpentis appears on the extreme southwestern edge. Move the binoculars a mere 2 degrees west and the cluster appears in the field. M 5 is one of the brightest and most beautiful globular clusters in the sky and truly appears splendid when viewed through larger amateur telescopes. Only Omega Centauri, NGC 104 in Tucana, M 22 in Sagittarius, and M 13 in Hercules are brighter than M 5.

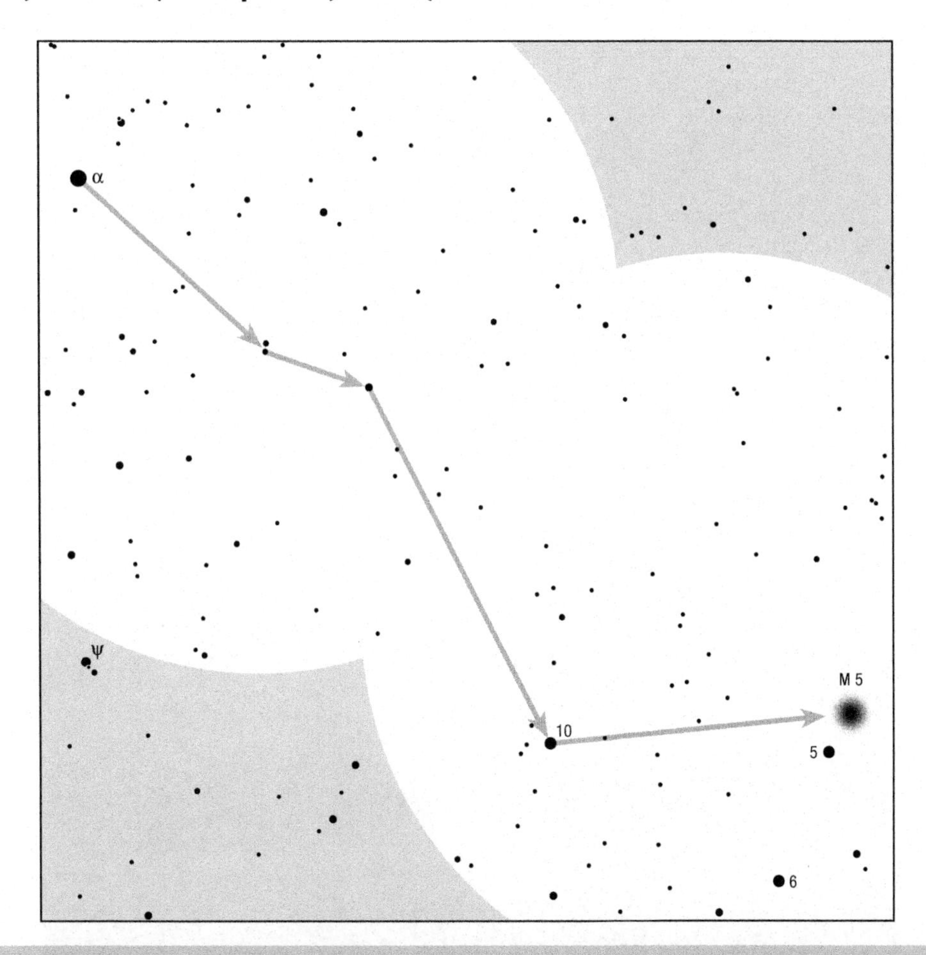

Figure 11.14 The way to the globular cluster M 5 leads us from the bright Alpha over a rather empty area of the sky to the star with designation 10

Like almost all globular clusters M 5 is very old. It is estimated be over 10 billion years of age. Its origins date back to the same time as our galaxy formed.

M 5 is approximately 24,500 light years from us (this is slightly less than the distance to the globular cluster M 13 in Hercules). It contains at least half a million stars. The largest and brightest of these are real giants. Their luminosity is over 2,000 times that of our Sun. From this distance our star would hardly even be seen even through the largest telescopes; it would appear as a faint star of magnitude 19!

West of Aquila and east of Ophiuchus, in the area of the Great Rift (the dark zone of opaque clouds of gases and dust that divides the Milky Way into two), lies the second part of the constellation Serpens, known as the Serpens Cauda. The brightest stars here are the yellow-orange Eta (3m2), Xi (3m5), and Theta (4m0).

Theta lies on the extreme northeastern edge of the constellation and represents the end of the serpent's tail. It is a binary star, seen through binoculars as a close pair. The two stars with magnitudes 4.6 and 4.9 are 23 arcsec apart (p. a. 104°, 2006). Unfortunately, both are white. Theta is approximately 130 light years from us. The real separation between the stars is 900 AU – enough to fit in 15 Solar Systems!

Figure 11.15 Globular cluster M 5 and the star 5 Serpentis

The splendid open cluster IC 4756 (4m6/52′) lies in the extreme northern part of the constellation. The star leading to the cluster is Theta. If the star is moved to the extreme eastern edge of the field of view, the cluster appears on the western edge. On a clear, Moonless night approximately 80 stars are scattered within a field of slightly less than 1 degree in diameter.

The brightest ones are of magnitude 7. The background of the cluster is also very rich; plenty of Milky Way stars can be found there. With perfect observing conditions the number of stars will increase to 130! Viewing IC 4756 with binoculars, with their large field of view and low magnification, offers greater success than looking through larger telescopes, where the impression of the cluster is completely lost. Next to it one can see its smaller neighbor, the open cluster NGC 6633, which already lies in Ophiuchus (see page 360).

Close to the triple border between Serpens Cauda, Scutum, and Ophiuchus lies one of the most wonderful celestial objects – M 16, or the Eagle Nebula (6m0/15′) (Figure 11.16). This is an open cluster, the stars of which are scattered among a gigantic cloud of interstellar gases and dust, a wonderful interlacing of stars and emission nebula covered by dark, opaque areas of dust. The cluster is composed of over a thousand stars. But – and this is the moment of truth – all we can see through binoculars is a modest open cluster with approximately 40 stars, the brightest of which are a mere magnitude 8. The clouds of interstellar matter that is spread around the stars can only be glimpsed through binoculars when there are excellent observing conditions. On a perfect night, the densest and brightest part of the nebula might appear in the area southeast of the cluster. The true beauty of the intricate structure of interstellar matter is unfortunately seen only on long-exposure images, amateur as well as professional. The nebula measures as much as 30 × 35 arcmin, which is slightly more than the full Moon.

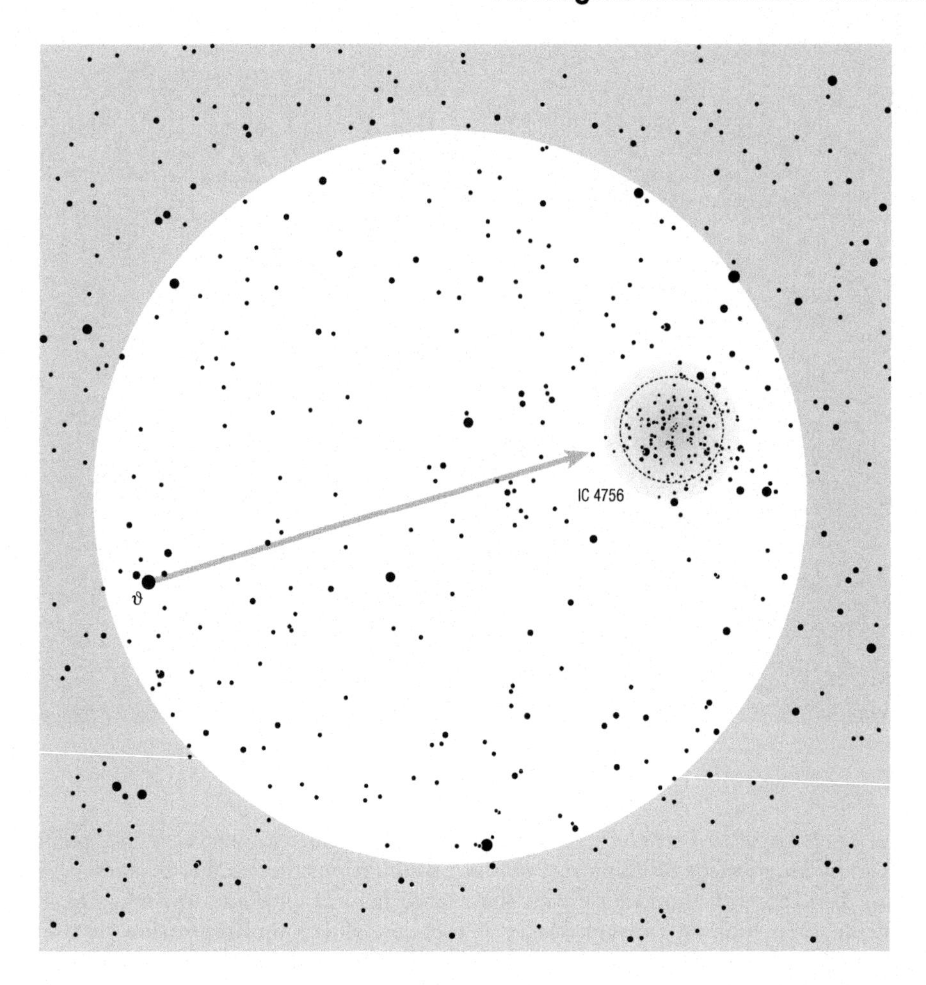

The cluster is hard to find, since there are no bright stars in its vicinity. The best starting point is Nu Ophiuchi (Figure 11.17). If the star is moved to the extreme northwestern edge of the field of view of binoculars, a magnitude 6 star appears on the extreme southeastern edge. This star lies only 2 degrees west of the cluster. We can also find it by starting at the Sagittarius Star Cloud (M 24 – page 420), which is clearly seen with the naked eye. Just 2 degrees above this lies M 18, 1 degree north lies M 17, and some 2 degrees further on north lies M 16.

In the nebula M 16 are numerous new stars that were born from interstellar matter. In the open cluster, which lies in the northwestern part of the nebula, there are young giants with high surface temperatures and gigantic luminosity. Measurements show that M 16 is approximately 7,000 light years from us and that its diameter is as much as 70 light years. The brightest part of the nebula measures 25 light years.

Figure 11.16 The Eagle Nebula, or M 16

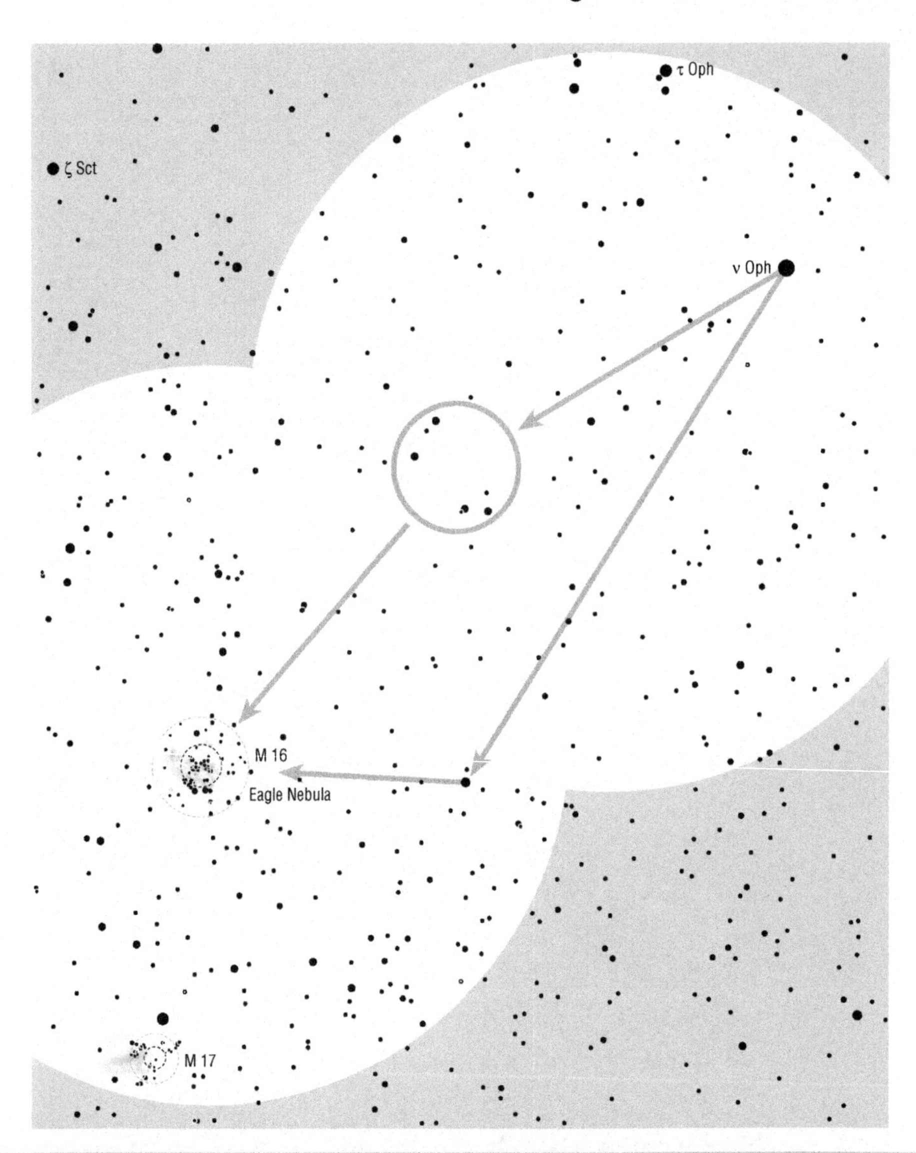

Figure 11.17 The star leading to the open cluster M 16 in Serpens Cauda is Nu Ophiuchi, which lies on the border of the two constellations

SEXTANS (The Sextant)

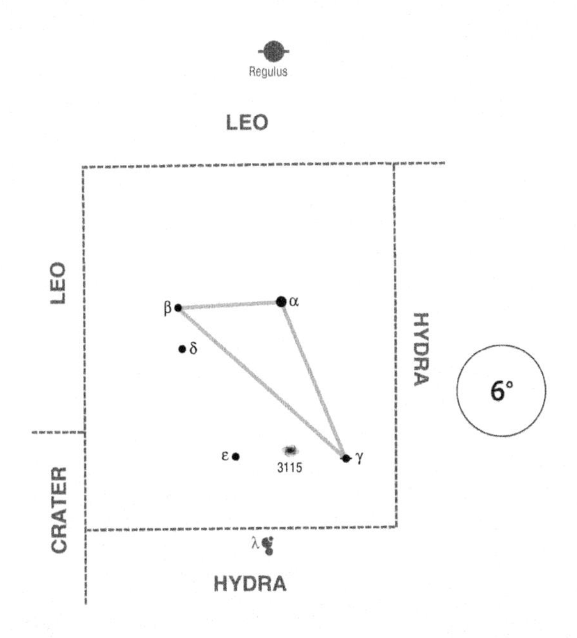

CONSTELLATIONS CULMINATE		
end January at 2 a.m.	end February at midnight	end March at 10 p.m.

Sextans is a faint spring constellation that lies below Leo. The only star that is clearly visible with the naked eye is Alpha (4m5). All others are, when there are average observing conditions, on the border of visibility and in light-polluted areas are completely lost.

With excellent observing conditions, the lenticular galaxy NGC 3115 (9m2/ 8′3 × 3′2) can be glimpsed through binoculars (Figure 11.18). It is seen edge on, and because of its appearance it has been given the nickname "The Spindle Galaxy." The leading stars are Gamma Sextantis, which is a rather faint magnitude 5 star, or the brighter one Lambda Hydrae of magnitude 4, which is easy to recognize, since it has two stars of magnitude 5 and 6 right near it. With the aid of Gamma or Lambda we move to the pair of magnitude 6 stars labeled 17 and 18 Sextantis. The galaxy lies slightly more than 1 degree to the northwest. Between them we should see a close pair of magnitude 7 stars. In the field of view of binoculars the galaxy is seen as a several arcmin long line of faint light.

As mentioned previously, you should attempt searching for such a dim object only when there are excellent observing conditions and when the constellation is at culmination (at its highest point in the sky). You should also make sure that you have plenty of time for your observations. Once you have recognized the key stars in the vicinity of the galaxy and you believe the galaxy is in the center of the field of view, you should wait for your eyes to adjust to night vision and then concentrate on watching the spot where the galaxy should be. If you still cannot see it, try with averted vision or try gently shaking the binoculars.

It is interesting to note that even though NGC 3115 is somewhat brighter and smaller than spiral galaxy M 65 in Leo, it is harder to see, since it is some 20 degrees lower and we have to always observe it though a thicker layer of the atmosphere. Probably, it would not be seen through binoculars at all if it did not have a bright and condensed nucleus.

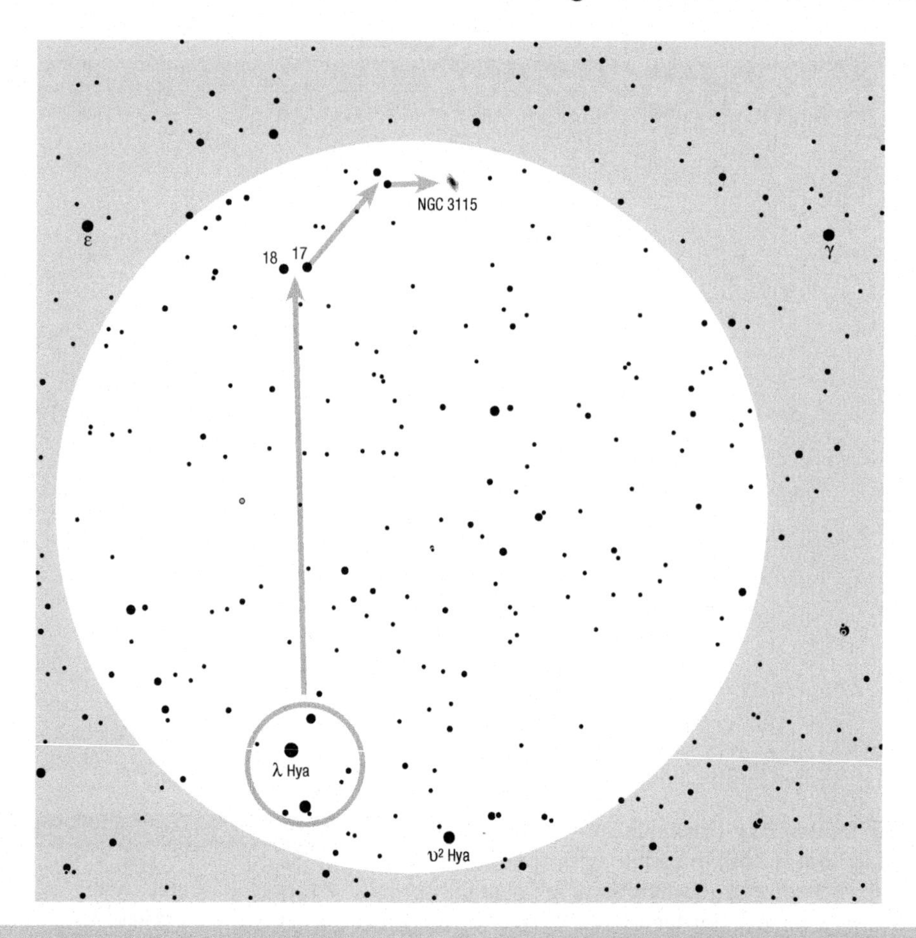

Figure 11.18 In the field of view of binoculars the key stars that will lead you to the faint galaxy NGC 3115 are always clearly visible: the leading star Lambda Hydrae with its nearby stars of magnitudes 5 and 6, the pair of magnitude 6 stars labeled as 17 and 18 Sextantis, and the close pair of magnitude 7 stars east of the galaxy. However, the faint line of the galaxy's light will only be seen when observing conditions are excellent

Observing such faint objects as the two galaxies just mentioned may not be extremely enjoyable in itself. However, you should be pleased that you can see them at all! In this case you will have pushed your binoculars to their limits.

The distance to NGC 3115 has not been determined precisely. The best estimates place it at approximately 30 million light years from us. If this is true, its real diameter is 35,000 light years, so it is a mid-sized galaxy.

TAURUS (The Bull)

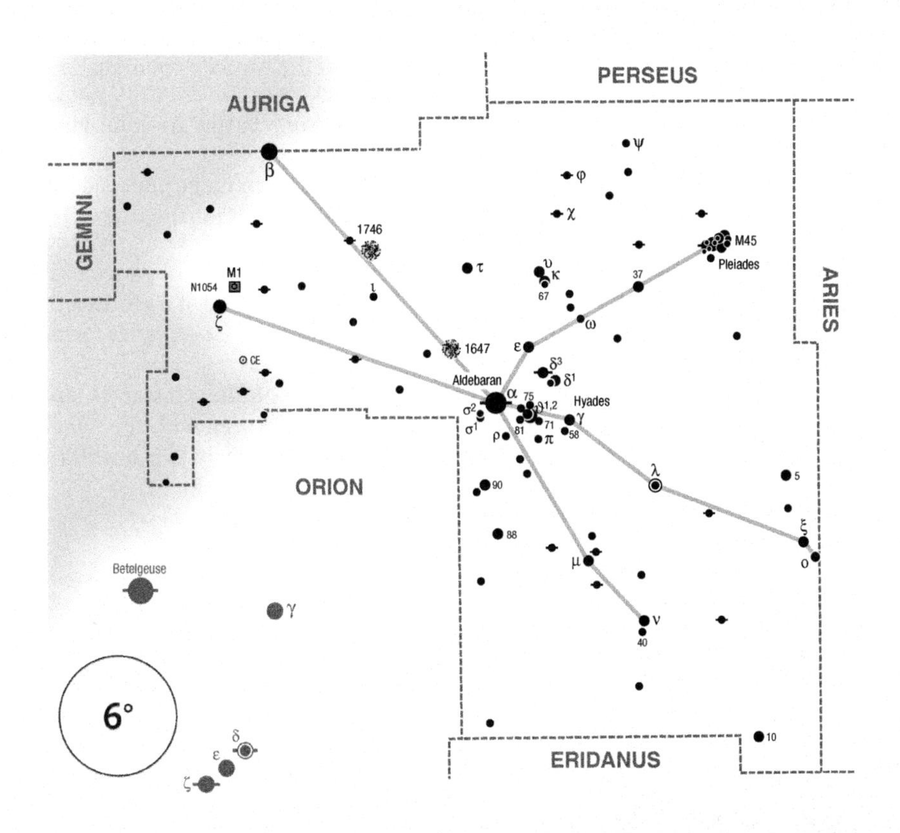

CONSTELLATIONS CULMINATE		
end October at 2 a.m.	end November at midnight	end December at 10 p.m.

Taurus is a large and remarkable winter constellation of the zodiac. It contains at least three truly splendid celestial treasures: two wonderful open clusters (Hyades and Pleiades) and the Crab Nebula, a remnant of an exploded star. Besides, the brightest star in the constellation – Aldebaran (0m9) – is a dazzling deep orange color. The other brighter stars are Beta (once known as Gamma Aurigae; 1m6), Eta (2m8), and Zeta (3m0).

The bright Aldebaran, which is one of six stars that forms the asterism called the Winter Hexagon, cannot be missed. It lies in the same line with Orion's Belt. Aldebaran is the 14th brightest star in the sky; in fact it is a red giant, 65 light years from us. Its luminosity is about 140 times that of the Sun. The color of the star glitters when viewed through binoculars. Its diameter is 38 times that of the Sun, but the surface temperature reaches a mere 3,400 K. Like all red giants Aldebaran is also slightly variable, but its brightness varies by a mere 0.2 magnitude. The star has four faint apparent companions. The brighter is of magnitude 11.3 and is 133 arcsec (p. a. 31°, 1997) away; however, it is not visible through binoculars.

Aldebaran lies in the vicinity of the ecliptic and is one of the few first magnitude stars that is often in conjunction with the Moon. (A conjunction is when two or more celestial objects appear near one another in the sky. If the Moon or a planet covers another celestial body, such as a planet or a star, the event is called an occultation.) There is a very interesting and revealing story related to the occultation of the Moon and Aldebaran in year 509. According to ancient records the event was visible from Athens. More than 1,000 years later this report was studied by Edmond Halley. He discovered that the occultation could not have happened, unless Aldebaran changed its position in the sky. This discrepancy led Halley to compare the positions of some of the brighter stars with the positions noted in ancient reports and books. He discovered that Aldebaran, Sirius, and Arcturus noticeably changed their positions. This is how Halley discovered the proper motion of the stars (their true motion through space). Modern measurements show that Aldebaran changed its position by as much as 7 arcmin in the last 2,000 years, which is almost a quarter of the Moon's apparent diameter.

Beta is the 27th brightest star in the sky, 130 light years from Earth. Its luminosity is that of approximately 270 Suns.

Lambda is an eclipsing binary. The star changes its brightness from magnitude 3.4 to 3.9 within a period of 3.9529478 days. A suitable comparison star is the rather distant northeastern Gamma (3m7).

Open cluster Hyades (0m5/5°5) extends west of Aldebaran to Gamma and back to Epsilon (Figure 11.19). Its brightest stars form a clearly visible V-shape. On older drawings of the constellation, Hyades forms the outline of a bull's head, with Aldebaran as its eye.

The diameter of the densest part of the cluster is approximately 6 degrees and under a dark sky is clearly seen with the naked eye. But it appears at its most beautiful when viewed through binoculars with a wide

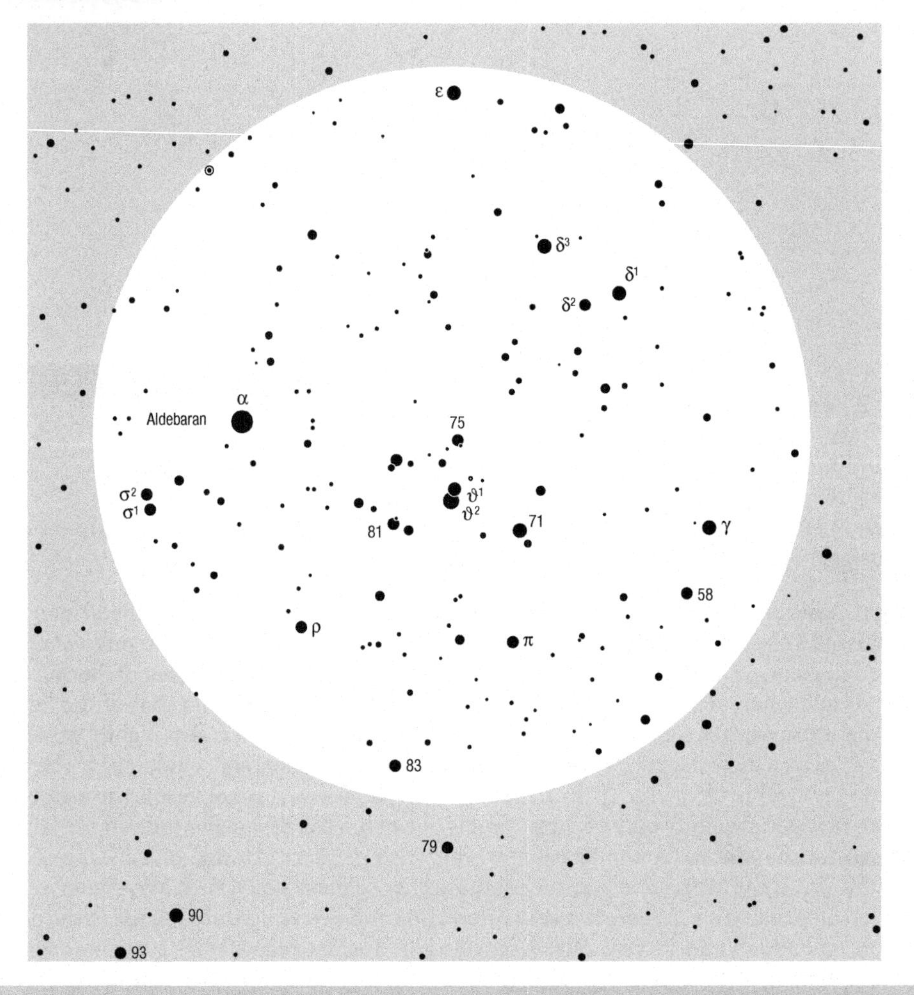

Figure 11.19 Hyades

field of view and low magnification. The cluster is one of the closest to us, since it is only 150 light years away, which leads us to conclude that Aldebaran is not a member. In the area of Hyades, there are approximately 28 stars that can be seen with the naked eye, over 160 are up to magnitude 9 and can be seen through binoculars in good observing conditions, and on a perfect winter night, when the limiting magnitude of the binoculars reaches magnitude 11, we can see more than 400 stars! Unfortunately, their number rises with fainter magnitudes. In total, there are over a thousand stars, but they are increasingly faint. Less than half of them are members of the cluster.

In such a crowd there are always numerous widely separated double stars suitable for viewing through binoculars, such as Theta (the brightest star in the cluster) or Sigma, which is even more interesting, for it is an apparent double. Stars with magnitude 4.7 and 5.1 are 437.5 arcsec apart (p. a. 195°, 2002). They are not gravitationally bound, since they lie too far from each other; it just appears so when we look at them from Earth because both lie in the same direction. Sigma-2, which is the brighter star of the pair, is a member of Hyades.

Theta is a double that consists of magnitude 3.4 and 3.9 stars, which are 337 arcsec apart (p. a. 348°, 1998). The brighter star is white, while the fainter is yellow-orange in color.

Among the brighter stars we should not neglect to mention Epsilon (3m5) and Delta-1 (3m8), both of which are also yellow-orange in color, and Gamma, which has a yellowish tone.

The real diameter of the prominent central group is approximately 10 light years, while the entire cluster extends across 75 light years in diameter!

In Hyades, there are no clouds of gases and dust between the stars. This means that the cluster is rather mature – the best estimations make it 800 million years old. The stars in Hyades move with a velocity of approximately 40 km/s in the direction of Betelguese in Orion. They were closest to us approximately 800,000 years ago. In about 50 million years, they will be so far away that they will be seen only as a faint group of stars about 20 arcmin in diameter through binoculars.

Although they are prominent, the Hyades are pale in comparison with Pleiades (1m2/2°), another open cluster in the constellation Taurus. In the night sky these stars are so clearly seen that they are known even to those who do not dabble in astronomy. At least 6 or 7 bright stars, gathered in a region only a bit larger than 1 degree, are clearly seen with the naked eye. This is why the cluster was already well known in antiquity and why the beauty of the Pleiades is celebrated by poets from everywhere and from all periods. In the Messier catalog, the cluster is noted under the number 45.

If the Pleiades seem charming with the naked eye, they are truly breathtaking when viewed through binoculars! The nine brightest stars are joined by numerous fainter ones, making the field of the cluster literally overflowing with stars. If the observer with binoculars feels otherwise slightly deprived because of the low magnification, when observing the Pleiades the binoculars' low magnification becomes an advantage. The large field of view offers a panoramic view of this splendid group (Figure 11.20)!

With larger amateur telescopes and really excellent observing conditions we lose the magnificent feel of this cluster but notice that the stars are sunk in a faint nebula. Interestingly, this haze is not the remnant of matter from which the stars in the cluster were born but is a cloud of gases and dust through which the stars are traveling merely by coincidence during their journey through interstellar space. The nebula, which is illuminated by the stars, is clearly visible on long-exposure images, even ones taken with amateur equipment.

Investigations of the cluster have shown that it contains approximately 500 stars. The brightest are all hot and white: Eta or Alcyone (2m8), Atlas (3m6), Electra (3m7), Maia (3m9), Merope (4m2), Taygeta (4m3), Pleione (5m1), Celaeno (5m4), and Asterope (5m6). Alcyone is the 145th brightest star in the sky and has a luminosity equaling 700 times that of our Sun.

Figure 11.20 The soft nebula in Pleiades

The Pleiades are 370 light years from us. This makes them one of the closest open clusters. The cluster's real diameter measures 20 light years, but the nine brightest stars are gathered within an area that measures only 7 light years in diameter. By cosmic criteria this group of stars is very young, which is why there are no red giants found in it. The cluster is estimated to be only approximately 20 million years old.

The Pleiades have been used for testing eyesight from antiquity onward. In good observing conditions, an observer with average sight can see 6 or 7 stars. What about people with excellent eyesight? Maestlin, Kepler's teacher, saw 14 stars; he had drawn 11 stars on charts of the Pleiades before the telescope was invented! The well-known English observer Denning also saw 14 stars in the cluster. Madame Airy (the wife of a well-known nineteenth-century English Astronomer Royal) could see 12, while the English amateur astronomer William Dawes, who was in his time known to have extraordinary eyesight, saw 13 stars in the cluster. The Austrian astronomer Carl von Littrow maintained that he could see as many as 16! In 2001, the Slovene meteor observers (MBK Team) traveled to Arizona in the United States to observe the annual meteor shower known as the Leonids. When they returned they reported on the exceptionally dark sky in which they could see 23 members of the Pleiades! How is this possible? As we mentioned in the first section of this book, when we described the limiting magnitude, the ultimate limit of the eyesight is magnitude 8. This means that we should (theoretically) be able to see as many as 36 stars in the Pleiades!

We can use the Pleiades for another very interesting and useful exercise – to find out the limiting magnitude of our binoculars or telescope. On Figure 11.21 the eastern part of the Pleiades around Atlas, Pleione, and Alcyone are shown with stars up to magnitude 13, so this view is fit for testing even moderate-size telescopes. But when you test the optics, be sure that the observing conditions are perfect or at least excellent. Otherwise you will be testing observing conditions!

Figure 11.21 Detailed chart of the eastern part of the Pleiades with magnitudes of stars up to 13, written without decimal point

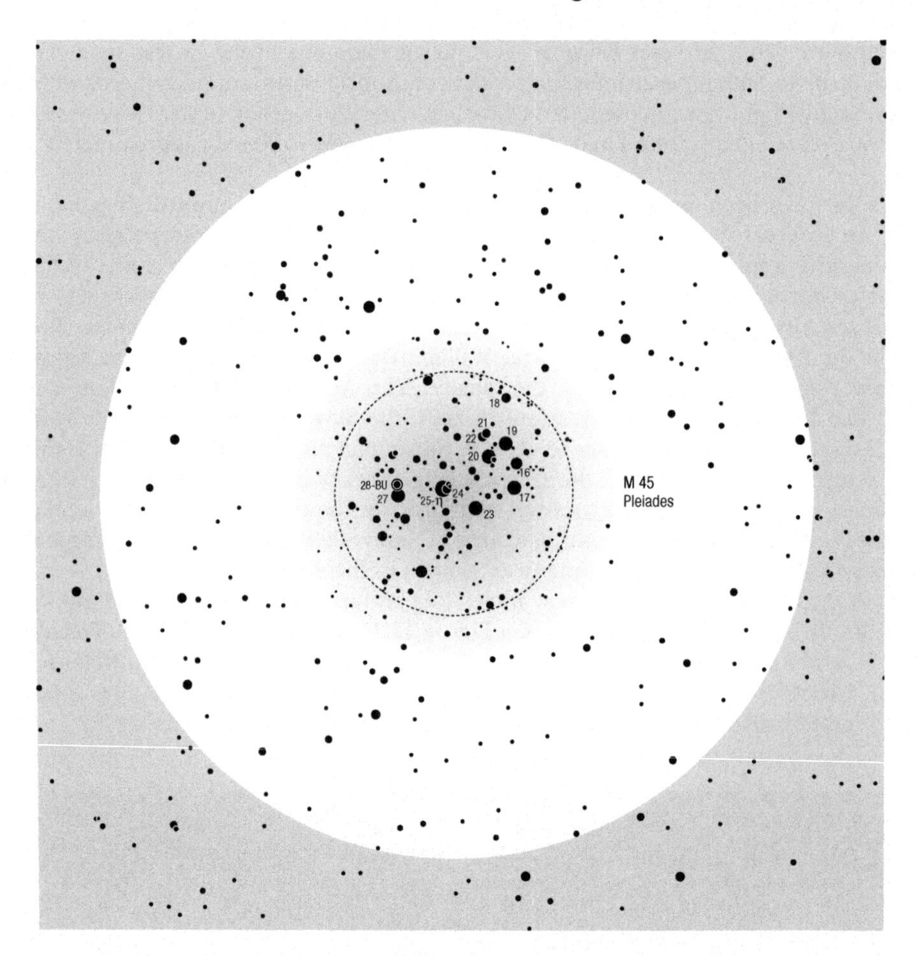

Open cluster NGC 1647 (6m4/45′) is easy to find as it lies in the same field of view with the bright Aldebaran, approximately 3 degrees northeast. If we move Aldebaran to the southwestern edge of the field of view, the cluster should be at its center. It contains about 160 stars; however, all of them are faint. The brightest in the region of the cluster is magnitude 6.6, but it is not a true member. On good nights we could see approximately 15 stars, on excellent approximately 25, but when there are perfect observing conditions their number increases up to 40. The cluster is completely resolvable when viewed through binoculars. It is some 1,800 light years from us, so its real diameter is 21 light years.

Open cluster NGC 1746 (6m/42′) is a large but loose association of faint stars. Through binoculars we see approximately 20 of them, on a prefect night maybe more. The brightest stars are only magnitude 8. The cluster is fully resolvable when viewed through binoculars.

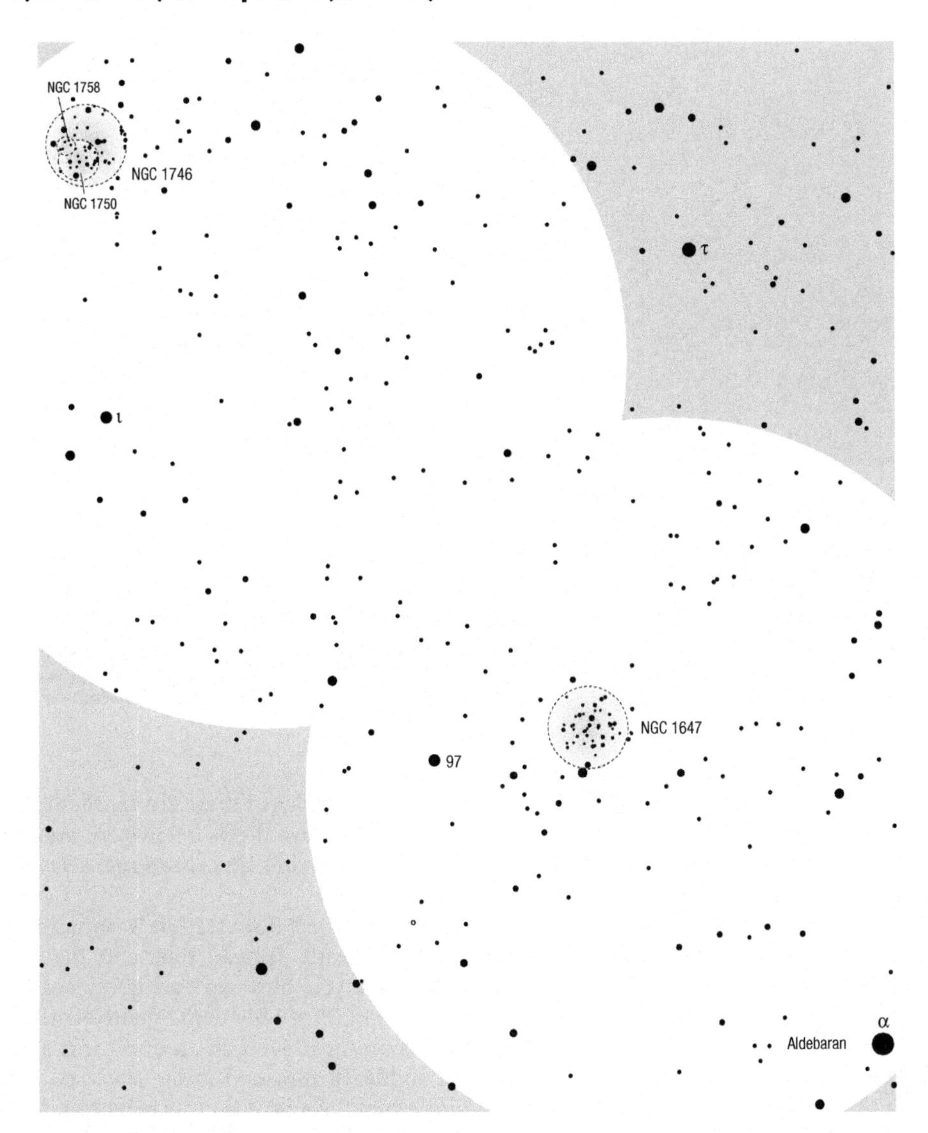

NGC 1746 is a rather close neighbor of the previously described NGC 1647. If you move from NGC 1647 one diameter of the field of view to the northeast, you will catch a glimpse of the first NGC 1746 stars.

The cluster NGC 1746 is not a jewel in the sky, but it has its hidden treasures that are unfortunately not seen through binoculars. In the same area there are actually three overlapping clusters: NGC 1746, NGC 1750, and NGC 1758. The brightest stars that we can see through binoculars come from all three clusters. NGC 1746 and NGC 1758 are rather loose groups of stars, while NGC 1750 is a slightly more compact and rich association of stars, but these are all faint. This interesting group is more suitable for observation through larger telescopes.

The famous Crab Nebula or M 1 (8m4/6′ × 4′) is a remnant of a supernova explosion that was observed by Chinese astronomers in 1054. It is not hard to find, since it lies slightly over 1 degree to the northwest of the bright star Zeta. They both appear together in the same field of view of binoculars. It is much harder to notice the faint nebula. The tenuous spot

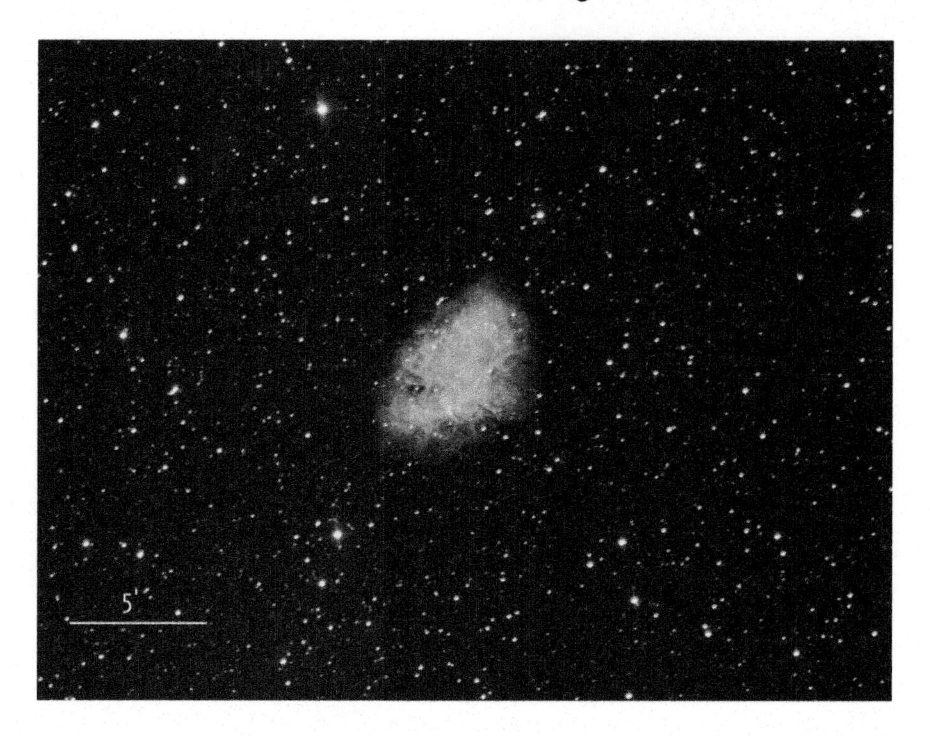

of light, an arc minute or perhaps two in diameter, shows up only when there are excellent observing conditions, on a clear and dark winter night, when the temperature drops below zero and the air is almost completely dry. Although the nebula is hardly visible, it is such an exceptional celestial object that we will devote a bit more attention to it here.

In one of the previous chapters, in which the life of the stars was discussed, it was mentioned that massive stars are not satisfied with a peaceful and slow death. Instead they end their life in a magnificent explosion, during which the outer parts of the star blow up and its pieces are flung into interstellar space, while the core shrinks into a small and incredibly dense neutron star or black hole. This rather rare event is called a supernova explosion. In 1054, such an event was seen in the constellation Taurus. The incredibly bright star that suddenly appeared in the sky was named the "Guest Star" by Chinese chroniclers. At the beginning, when the star was at its brightest, it could be seen even in daylight, but after three weeks its brightness slowly decreased, and after a year or so it fell below the visibility of the naked eye.

The nebula close to Zeta Tauri was first seen by the English physicist and amateur astronomer John Bevis, who noticed it in 1731. It was independently discovered by the famous French comet hunter Charles Messier in 1758, while he was trying to find Halley's Comet, which was due to return to perihelion that year. The nebula that misled him in his comet search became the first entry in his famous list of nebulous objects that confuse comet hunters.

Since it was discovered, the Crab Nebula has been regularly observed by astronomers. In 1844, they noticed its filamentary crablike shape, which gave the nebula its name – the Crab Nebula (Figure 11.22). Approximately 80 years later (1921), astronomers discovered that the gases in the nebula are expanding at high velocity in all directions. This led to the idea that the Crab Nebula might be a result of a tremendous explosion. In 1942, Walter Baade estimated that the explosion could have occurred 760 years ago. In the same year, the astronomer J. Oort and a professor of Asian languages, J. Duyvendak, linked the Crab Nebula to the Chinese records of the "Guest Star."

Accurate measurements have shown that different parts of the nebula move with different velocities. The rough average distance given for gases moving in the outer parts is 0.2 arcsec per year. If we

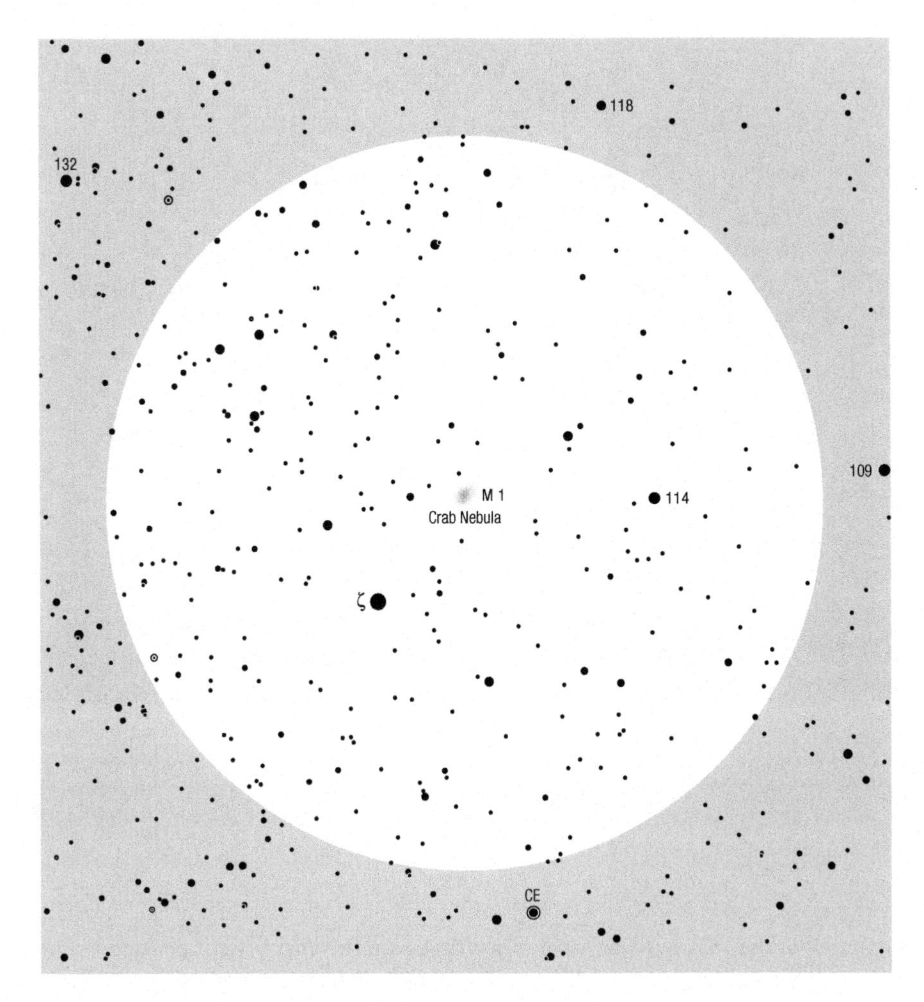

assume that the Crab Nebula is 6,300 light years from us, then the true velocity of matter is approximately 960 km/s, or 83 millions km/day. This means that in the next four months the gases in the nebula will have expanded by one diameter of our Solar System!

As early as 1948, astronomers discovered that the Crab Nebula is also a strong source of radio waves. In fact, it is the fourth strongest in the sky. Radio waves that come from the nebula are polarized.

Astronomers were looking for the star that illuminated the nebula for a long time, with no luck. At the same time as the radio nature of the nebula was discovered, Walter Baade studied two stars in the central region of the nebula. Both could have been candidates for the central star. Spectroscopic investigations have since shown that the southwestern star is most probably the actual remnant of a nova. This is a hot, blue dwarf magnitude 16 star. Astronomers wondered if it was possible that this faint star had illuminated the nebula, which was as much as 1,000 times brighter than the star. Calculations have shown that the surface temperature of the star would have to be at least 500,000 K and its luminosity approximately 100 times that of the Sun for this little star to illuminate the entire nebula. At such high temperatures, the star should emit most of its radiation in the ultraviolet part of the spectrum. If all this is true, it would explain the dimness of the star in visible light, as well as its strong influence on the surrounding gases. However, new, even more surprising facts came to light that led astronomers to think this hypothesis was faulty.

Figure 11.22 This image of the Crab Nebula depicts all three main components of the supernova remnant: the central pulsar (*arrow*), the almost uniform synchrotron light emitted by the charged particles trapped in the strong magnetic field (synchrotron nebula), and the network made of brighter filaments that glow in the light of the hot and cold hydrogen and oxygen atoms

In the beginning of the 1950s, it was ascertained that even the visible light emitted by the nebula is strongly polarized. What does this mean? If we view the nebula through a polarization filter, which lets light through on only a single plane, the appearance of the nebula will change greatly as we turn the filter around. For a normal nebula that emits nonpolarized light, the nebula will remain the same regardless of how the polarization filter is turned. The polarized light emitted by the Crab Nebula is proof that it is also home to an extremely strong magnetic field.

The data gathered from this observation led to the theory proposed in 1953 by Russian astronomer I. S. Schklovsky and simultaneously (but independently from him) by Dutch astronomers J. Oort and T. Walraven. The radiation of the nebula that became known as the synchrotron nebula (this is the central, brightest part of the Crab Nebula that shines in an almost uniform blue light and that can be seen through binoculars and amateur telescopes) originates from the acceleration and deceleration of the fast electrons within a strong magnetic field. On Earth, we can see synchrotron radiation of electrons in a strong magnetic field in cyclotrons (accelerators of charged particles). Of course, astronomers immediately started asking themselves from where such a strong magnetic field came in the nebula. But even before they tried to answer this question new facts emerged. In 1968, they discovered the true nature of the small star that Walter Baade considered to be a nova remnant. At that time radio astronomers discovered that this weak and small star emits radio pulses 30 times per second, regular as clockwork.

A year prior to this discovery radio astronomers had noticed regularly repeating signals from approximately 30 pulsating stars, but this one in the Crab Nebula had by far the shortest period. Today, it is clear to us that these objects are neutron stars – collapsed star cores that have approximately the mass of the Sun and a typical diameter as small as 20 km. Back in 1934, Walter Baade and

Fritz Zwicky had announced the possible existence of such exotic objects that supposedly emerge as a result of massive stars exploding. In these situations the core of the star supposedly collapses under its own weight and becomes so dense that the electrons and protons merge into neutrons. These (according to the principles of quantum mechanics) represent the so-called degenerated fermion fluid, and only its pressure can stop the further collapse of the core (Figure 11.23).

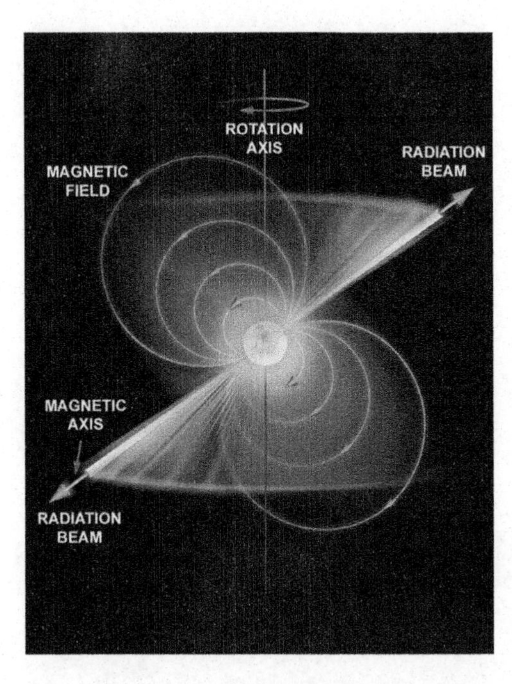

Figure 11.23 As all other stars the neutron star also rotates around its axis; however, in this case, the magnetic axis is tilted against the rotational one. Electrons are accelerated along the lines of the magnetic field, which is the strongest and densest near the magnetic poles. The electrons emit electromagnetic radiation in narrow beams around the magnetic axis. We see the pulsar every time the beam points toward Earth

In 1967, Thomas Gold suggested that the repeating radio signals from these stars were not a consequence of radial pulsating (expansion and shrinking), but that they emerge due to the extremely quick spinning of the star around its axis. Soon afterward it was clear that anything else but a neutron star would immediately disperse in all directions if it spun around its axis as fast as 30 times a second.

The pulsar is a truly interesting object. Its radiation is a consequence of the reciprocal influence of its strong magnetic field (with the density of 10^9 Tesla, which is 10^{12} times denser than Earth's magnetic field) and the charged particles surrounding it. The magnetic field accelerates the electrons up to almost the speed of light. As a result they emit electromagnetic radiation in a narrow beam along the direction of movement. The radiation leaves the pulsar in two narrow beams of light in line with the magnetic lines of force. Because the pulsar revolves the electromagnetic beams also spin with it. In this way it is similar to a lighthouse on the coast. We can see it when the narrow beam of electromagnetic radiation shines upon us. The beam emitted by the pulsar in the Crab Nebula hits Earth in its every turn. If it didn't do this, the true nature of the object in the heart of the Crab Nebula could have remained a secret for a very long time.

The typical wavelengths of synchrotron radiation can be found within radio wavelengths, which is why almost all pulsars were discovered by radio astronomers. The pulsar in the Crab Nebula is interesting for astronomers also because it is one of the two that emit electromagnetic radiation across the entire spectrum and can thus be seen through optical telescopes as well as radio telescopes.

For years astronomers have been trying to figure out the structure and composition of the nebula. Detailed images produced by the Hubble Space Telescope reveal its gentle, filamentary structure. The bright filaments of matter, heated to a high temperature by the strong ultraviolet radiation of the

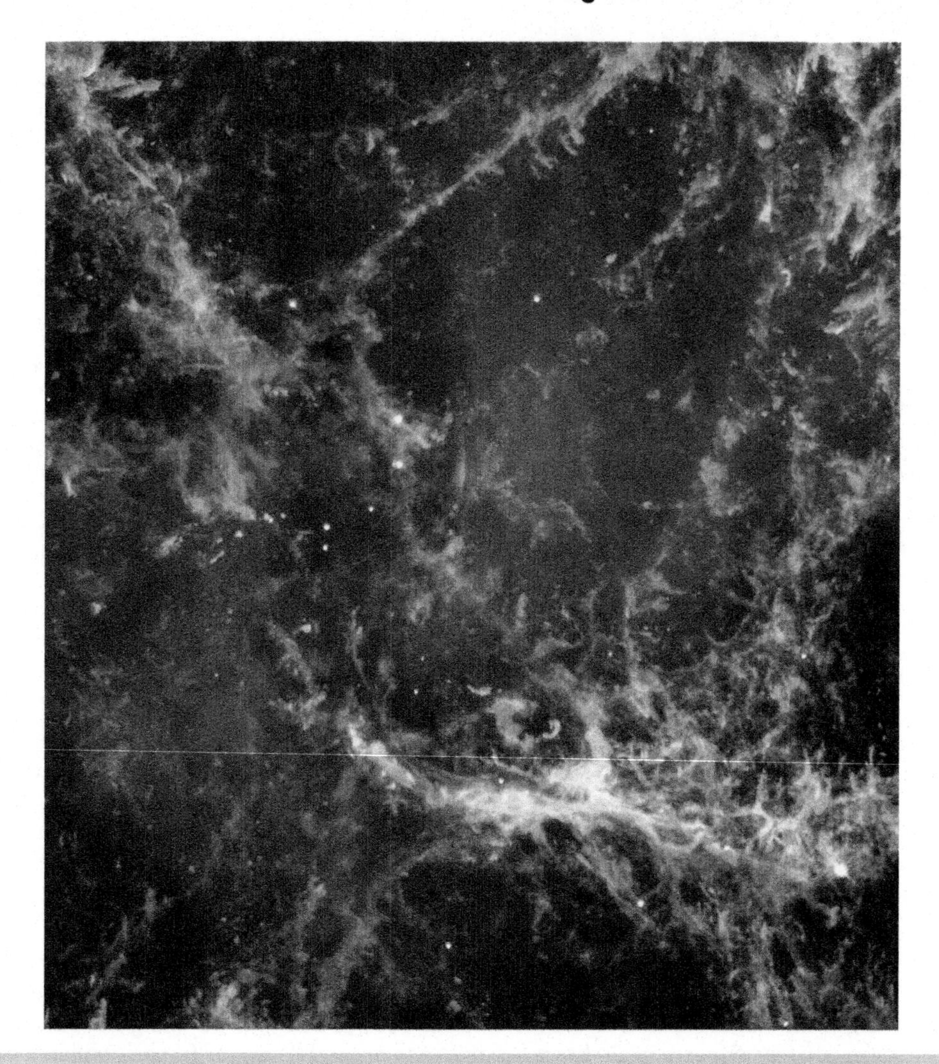

Figure 11.24 These stunning details in the filamentary structure of the Crab Nebula were taken with the Hubble Space Telescope. Gases in these filaments emit light due to the ionization and recombination at discrete wavelengths; this occurs because of the quantum energy levels in atoms. The strongest visible light comes from atoms of hydrogen, nitrogen, sulfur, and oxygen

synchrotron nebula, emit light. On images obtained from Hubble it could be seen that the web structure is much more complex than we previously thought. The individual filaments of gases differ from each other not merely in their form but also in temperature. This unique spider web of gases also includes much more dust than we anticipated possible. Until recently, astronomers were of the opinion that the surroundings in the nebula are too hostile for any complex molecules to form. This is why they were surprised when the infrared spectrum and optical observations from Earth showed that the dust particles in the Crab Nebula somehow survived all disasters. The dust is present throughout the entire nebula and it is especially dense in its coldest parts.

The chemical composition of the filaments of the Crab Nebula represented a mystery to astronomers for a long time. Because the star from which the nebula originated had to have at least eleven times the mass of our Sun, the fusion of hydrogen nuclei into helium nuclei had to take place in its core in so-called CNO cycles. When such a massive star ends its life cycle its remains should be extremely rich in nitrogen. However, there is not as much of this element in the Crab Nebula as one would expect. As an addition to the unusual chemical composition of the nebula we should also mention the recent discovery of a group of denser clusters of gases that are almost in line with the poles of the pulsar. In these, astronomers noticed a strong radiation of ionized argon atoms. The radiation is much stronger than at other known supernova remnants.

The density of the gases in the synchrotron nebula is only one-trillionth of the density of air, and this is considered to be almost a perfect vacuum by Earth standards. The mass of the visible part of the Crab Nebula was estimated to equal three Sun's masses. The remaining mass is hidden in the extremely fragile but large halo that surrounds the fibrous structure of the nebula and emits hydrogen H-alpha light.

Astronomers are studying the nebula with great interest, since its great luminosity and its rather recent origins make it the best example they have of a supernova remnant, even though it is 6,300 light years away. Studying the events in the Crab Nebula has contributed greatly to our understanding of the final phases in the lives of stars. Our knowledge about it is increasing each year, but still we only have a rough idea of what is happening inside it. And with every new observation, new questions are raised that cannot be answered with today's observational techniques.

Neutron Stars

If we wished to turn our Sun into a typical neutron star, we would have to compress it into a ball with a diameter of approximately 30 km across. Earth's corresponding diameter would be a mere 300 m!

From these data we can conclude that the average density of such a star is unimaginably high and reaches as much as 10^{18} kg/m^3 or 10^{15} kg/dm^3. This number tells us that a sugar cube made of neutron star matter would weight as much as 1 billion tons! If we put this sugar cube on one side of the scales we would need the entire human race to stand on the other side just to balance it!

The fastest known pulsar is PSR J1748-2446ad. Its rotation period is 1.4 ms, which means that it spins around its axis 716 times per second!!!

TRIANGULUM (The Triangle)

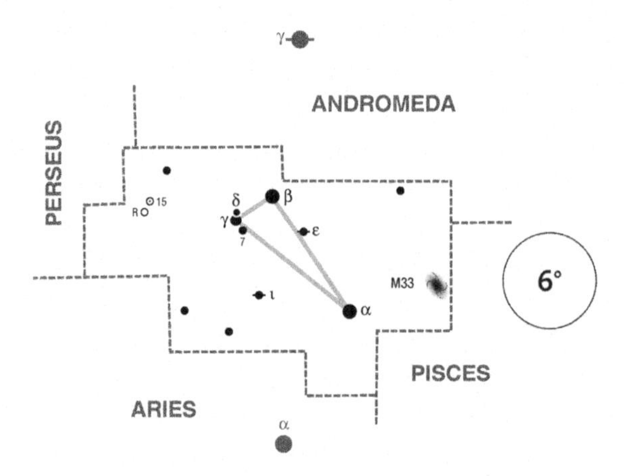

The small constellation Triangulum lies southeast of Perseus, below Andromeda. The three brightest stars, Beta (3m0), Alpha (3m4), and Gamma (4m0) represent the corners of an extended triangle that lies midway between Alpha Arietis and Gamma Andromedae.

CONSTELLATIONS CULMINATE		
mid September	mid October	mid November
at 2 a.m.	at midnight	at 10 p.m.

Alpha is 65 light years from us and is slightly larger and brighter than our Sun. Its luminosity is that of 13 Suns.

Iota is a splendid double in which the magnitude 5.3 and 6.7 stars are 4.0 arcsec apart (p. a. 69°, 2006). There is a distinct color contrast between them – the brighter one is strong yellow and the fainter is blue. Both stars are also spectroscopic binaries. This system is 200 light years from us. Unfortunately, the stars cannot be separated through binoculars.

The constellation is adorned by the spiral galaxy M 33 (5m7/ 68′7 × 41′6), which was first noticed by Charles Messier in 1764 (Figure 11.25). In his observation diary, Messier described it as a pale, almost uniform spot of light that was slightly brighter in the center and contained no stars. This is how it is seen through modern binoculars. Of course, its appearance depends greatly on observing conditions. With average observing conditions the galaxy is not seen at all, or only its faint nucleus can be noticed. But on a really dark night, far away from light-polluted areas, we will see a large, distinctive, yet faint oval patch of light, brighter in the center, and in our mind we might be even able to glimpse its spiral structure. This is definitely a sight worth seeing!

The galaxy is easy to find with binoculars, since it is together in the same field of view with Alpha. The spot of light is 4 degrees west and slightly to the north. On excellent viewing nights it can be glimpsed with the naked eye, so this is an object we can use to see the deepest into space (and into the past!) without an optical aid.

Astronomers were not able to resolve individual stars within the galaxy until the beginning of the twentieth century. Then they realized that this is not a nebula in our galaxy, but a distant spiral galaxy of the type Sc, with a small central nucleus and very loose arms, seen face on. Today we know that M 33 is a member of our Local Group of galaxies and that following Andromeda's Galaxy and our own it is the third largest. Its mass is only one-sixth of that of our galaxy. It is 3 million light years from us. Its apparent diameter is slightly over half a degree. Even though its integral brightness is high, its surface brightness is rather low. On long-exposure images M 33 extends for over 1 degree, so it is a frequent target of amateur astrophotographers. Its actual diameter is 50,000 light years, and its mass is estimated to be that of between 10 and 40 billion Suns.

R Trianguli is a yellow-orange pulsating variable star of the Mira type, which changes its brightness from 5m4 to 12m6 within a period of 267 days. The star reached one of its maximums on April 2, 2008. From this date and the known period, one can easily calculate the dates of the next maxima. When the star is at its brightest, it is seen with the naked eye. When it is at its faintest, it is far beyond the reach of binoculars. When its brightness is on the decline, its color becomes increasingly intense, but unfortunately the star is too faint to notice this when viewed through binoculars. You should be careful not to mistake this variable for another orange and slightly variable star labeled 15 Trianguli, which lies less than 1 degree to the northwest (see Figure 11.27 for details).

Figure 11.25 Spiral galaxy M 33

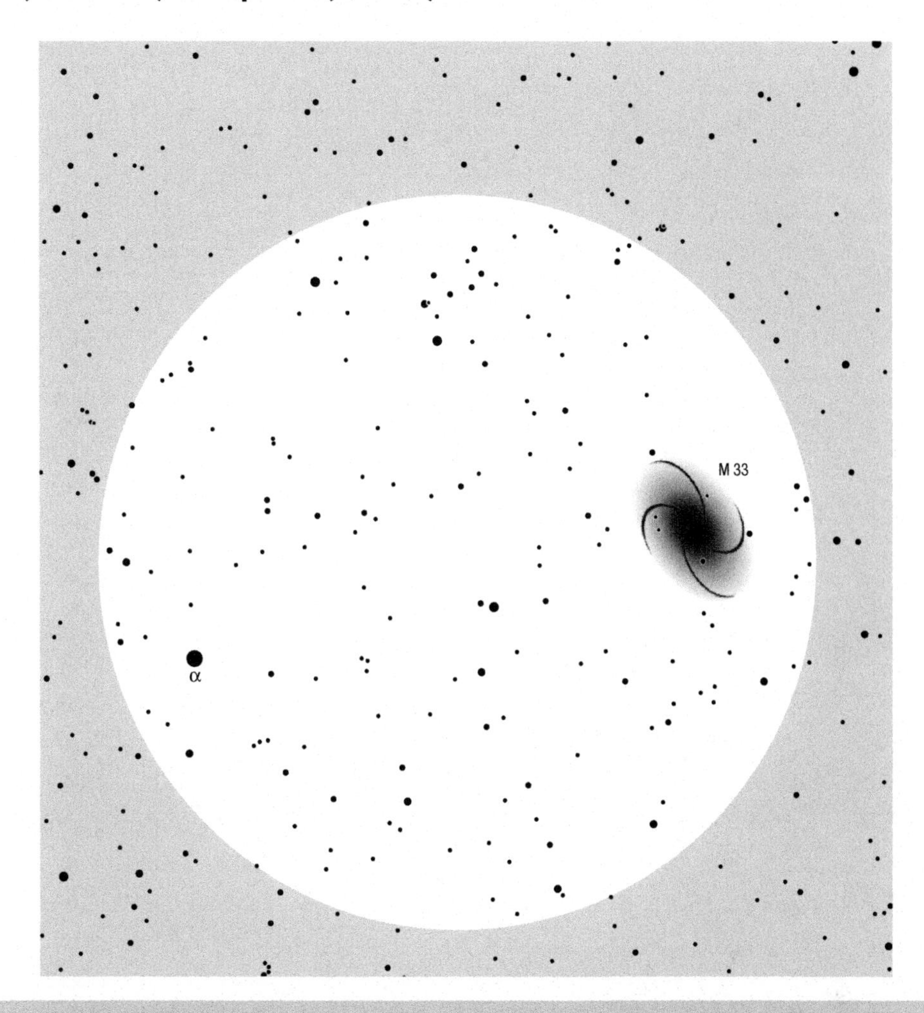

Figure 11.26 The spiral galaxy M 33 is easy to find, as it is together in the same field of view with Alpha

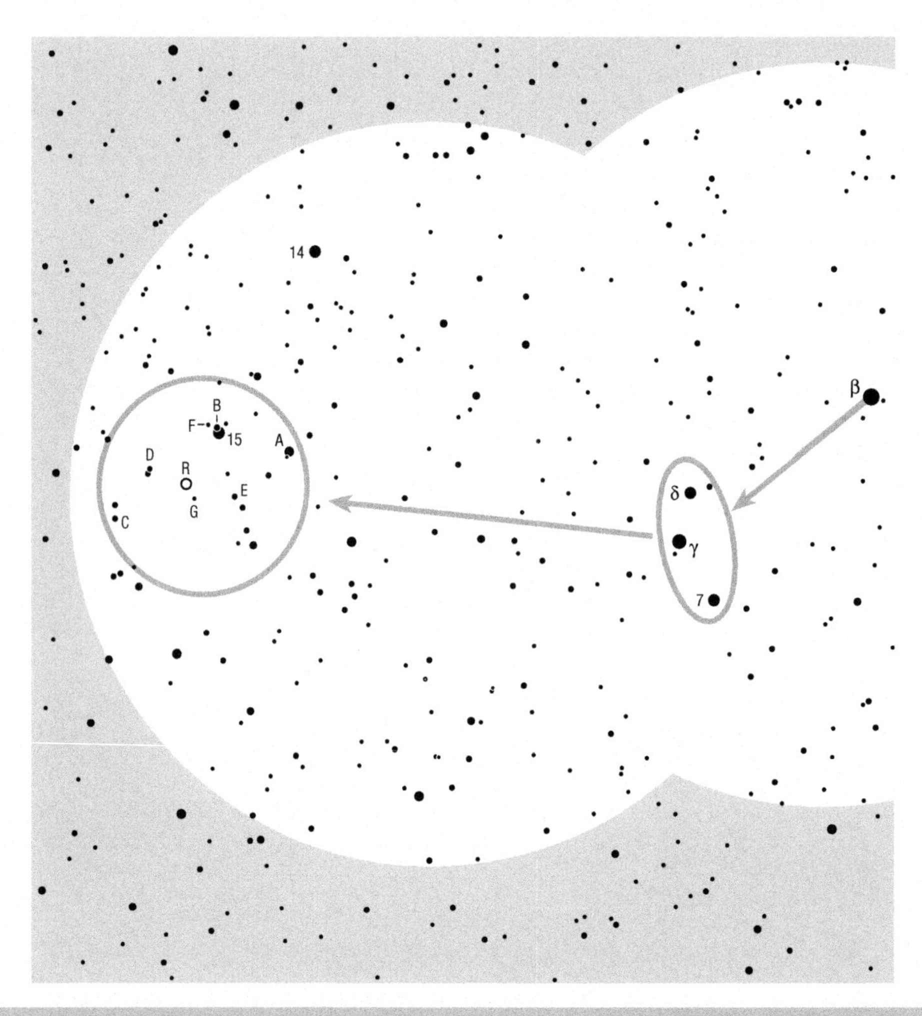

Figure 11.27 Surroundings of the variable star R with some comparison stars: A (5m8), B (6m7), C (7m4), D (8m0), E (8m4), F (9m2), and G (9m5)

URSA MAJOR (The Great Bear)

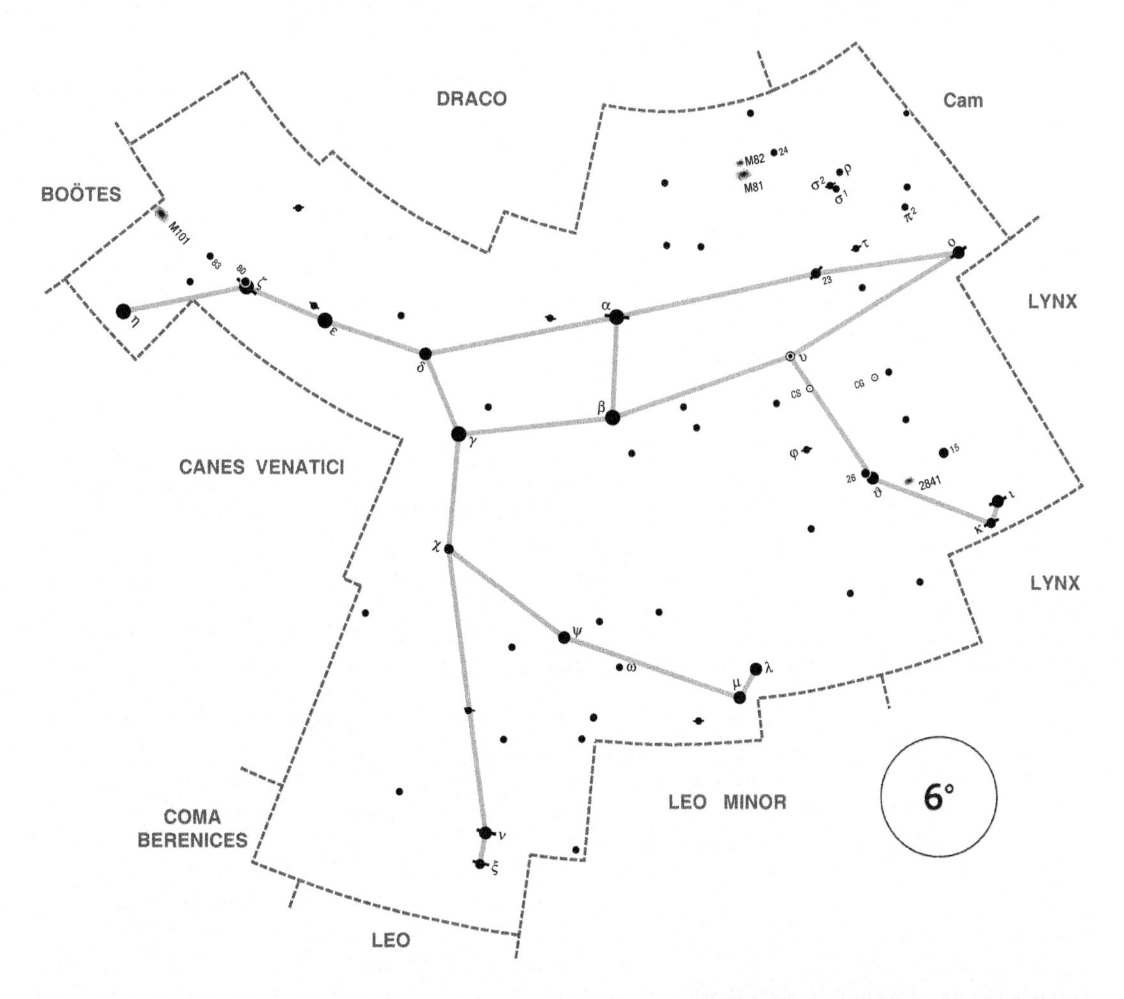

Ursa Major is certainly the most famous constellation in the northern skies, even though it does not contain any first magnitude stars in it. The main asterism known as the Big Dipper, which consists of seven stars, is of such a distinct shape that it cannot be missed. People who are not astronomers often think it is an independent constellation.

CONSTELLATION CULMINATES		
end February at 2 a.m.	end March at midnight	end April at 10 p.m.

On April evenings, the stars of Ursa Major can be seen from the majority of the populated world. The entire constellation can be admired from the North Pole all the way down to the Tropic of Capricorn at 23.5 degrees southern latitude. Only from locations below 40 degrees south – the southern part of South Africa, the southern island of New Zealand, and Tasmania – can none of its seven brightest stars be seen.

The brightest stars in the constellation are Epsilon, or Alioth; Alpha, or Dubhe; Eta, or Alkaid (all 1m8); Zeta, or Mizar (2m2); Beta, or Merak (2m3); Gamma, or Phecda (2m4); Psi; Mu (both 3m0); Iota (3m1); Theta (3m2); Delta, or Megrez; Omicron (both 3m3); Lambda (3m4); Nu (3m5); Kappa (3m6); and Xi (3m7).

For mid-northern observers, the constellation is circumpolar and extremely suitable for the orienting oneself in the sky as well as a starting point for recognizing other constellations. Besides Alpha and Beta are well known pointers to the north celestial pole. If we extend the line between Beta to Alpha by five times, we arrive at Polaris.

Epsilon is the 33rd brightest star in the sky. It is 81 light years from us. Its luminosity is 93 times that of the Sun.

Eta is the 38th brightest star in the sky, 100 light years from Earth. Its luminosity surpasses our Sun by 130 times.

Beta is the 80th brightest star in the sky. Its light travels 80 years in order to reach us. Its luminosity is that of 52 Suns.

Gamma is the 85th brightest star in the sky. It is 84 light years from us, and its luminosity is 55 times that of the Sun.

Mu (3m1) and Lambda represent a nice color contrast. Lambda is white, while Mu is orange. The distance between them is 1.5 degrees, so they appear together in the same field of view of binoculars.

Xi is a close double star with components of magnitude 4.3 and 4.8, only 1.7 arcsec apart (p. a. 238°, 2006). Both are yellow and represent an interesting couple in mid-sized telescopes. When viewed through binoculars they are seen as one.

Alpha is the 36th brightest star in the sky. It is an interesting triple star. One companion is of magnitude 7.0 and is as far as 6.3 arcmin away from Alpha (p. a. 204°, 1991). The stars are easily resolved when viewed through binoculars. Alpha has a distinct golden yellow color, and its fainter companion is bluish. However, Alpha itself is a close binary, with an orbital period of approximately 45 years. Its close companion is of magnitude 4.9 and is a mere 0.4 arcsec away (p. a. 89°, 2004), so they are a difficult pair to separate even in large amateur telescopes. This triple system is 124 light years from us.

Mizar, or Zeta, is the 70th brightest star in the sky, 78 light years away. This is probably the most famous double star. The companion known as Alcor, or 80 Ursae Majoris, is a magnitude 4.0 star. The apparent distance between them is 11.8 arcmin (p. a. 71°, 1991), so we can see them as two stars even with the naked eye. When we look at Mizar through a small telescope, we notice that it is also a double, with stars of magnitude 2.2 and 3.9 separated by 14.3 arcsec (p. a. 153°, 2005). They cannot be split when viewed through binoculars. The true distance between the stars is so great that it takes a few millennia to complete one orbit around their common center of gravity. Mizar was the first double star discovered with a telescope (Riccioli, 1650). But there is even more! Both stars are spectroscopic binaries, which is something astronomers discovered from the shifts of absorption lines in their spectra.

In Ursa Major, there are two interesting galaxies to be found: M 81 and M 82. These are the brightest members of a smaller group approximately 12 million light years from us. This group is most probably the closest to our Local Group (Figure 11.28).

M 81 (6m9/21′ × 10′) is so bright that is seen through binoculars even with only fair observing conditions. In the field of view we see an approximately 10 arcmin large spot of light, noticeably brighter in its center. The size and details depend greatly on observing conditions. The better the observing conditions and the higher the galaxy is in the sky, the more we are able to see. On long-exposure images, taken with large professional telescopes, M 81 is one of the most magnificent spiral galaxies, entirely symmetrical and extraordinarily rich (Figure 11.30). The clearly seen spiral arms, composed of millions of faint stars and countless gas and dust clouds, are wound tightly around its bright and condensed core. The real diameter of M 81 is some 75,000 light years, so it is just half as large as our galaxy.

Figure 11.28 Galaxies M 81, M 82, and NGC 3077 (the latter is not visible through binoculars)

With excellent observing conditions we should see a fainter, few arc minutes big, and elongated spot of light approximately 38 arcmin north of M 81. This is M 82 (8m4/9′ × 4′), one of the most unusual galaxies seen with amateur instruments. On long-exposure images it is shown as a spindle of light, covered in darker stripes that give the feeling of extraordinary motion. Even the largest telescopes cannot reveal a spiral structure or individual stars within the galaxy; thus, it was classified as an irregular galaxy. However, the newest research has shown that M 82 is wrapped in gases and dust that were pulled out of it by its larger and more massive neighbor M 81, only 150,000 light years away. Behind the dense clouds of gases and dust, a relatively normal spiral galaxy is hidden. M 82 is thus among the smallest known spirals, with a diameter of merely 28,000 light years (Figure 11.31).

The fantastic spiral galaxy M 101 (7m9/22′) is found 5.5 degrees east of Mizar (Zeta Ursae Majoris), and the two therefore appear together in the same field of view of binoculars (Figure 11.32). Its integral brightness is as much as magnitude 7.9. But we should also take a look at the data related to its apparent diameter, which is almost as big as that of the full Moon. If the first number gets us excited, the second quickly sobers a skilled observer. Such a large celestial object has extremely low surface brightness, so it is much harder to see than a small magnitude 9 galaxy with a bright nucleus. In order to observe M 101 we have to choose a dark, Moonless night, and the galaxy should be close to its highest point in the sky. In the field of view of binoculars we can see a large but faint spot of light with a brighter central part. The size of the spot of light and how many details can be seen depend mainly on observing conditions (and the quality of the binoculars, of course).

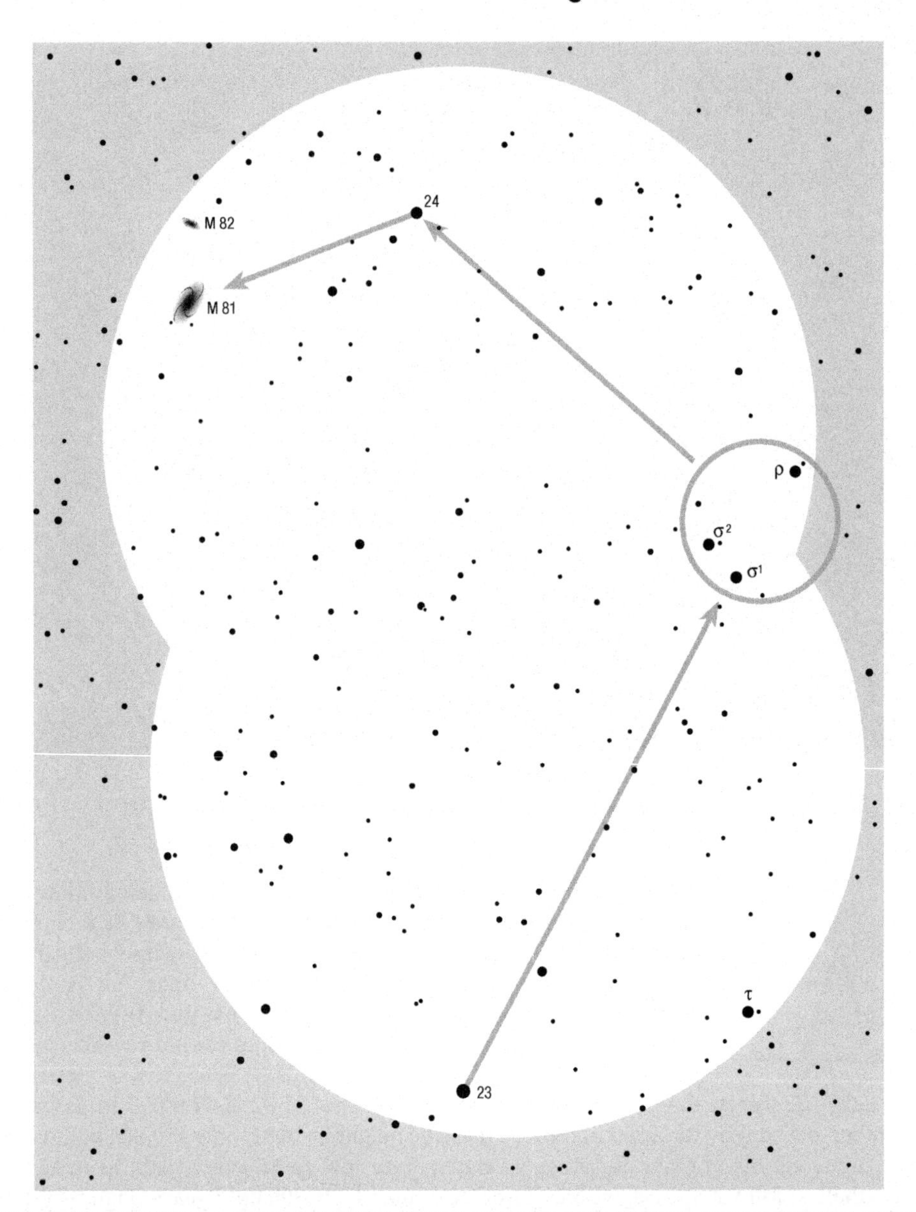

Figure 11.29 Galaxies M 81 and M 82 lie approximately 10 degrees northwest of Alpha. First you should find star 23 (3m7), which is clearly seen with the naked eye. Star 23 lies in the same field of view of binoculars as the trio Sigma–1, Sigma–2, and Rho (all magnitude 5 stars). In the same field of view, you can also find the star labeled 24 (4m6); both galaxies are only 2 degrees east and slightly south of it

The galaxy is approximately 27 million light years from us. Its actual diameter is almost 170,000 light years, which places it among the largest known spiral galaxies.

M 101 is seen face on and therefore presents a perfect object for all amateur astrophotographers. On long-exposure images, taken even with a 200-mm objective lens, its spiral structure is already clearly shown.

Figure 11.30 The wonderful spiral galaxy M 81

For skilled observers here is another faint galaxy NGC 2841 (9m3/8′1 × 3′8), located less than 2 degrees southwest of the bright Theta (Figure 11.33). The galaxy should be observed only when there are excellent observing conditions and when it is close to culmination. In the field of view close to a magnitude 8 star, we can see a faint, elongated spot of light, at least a few arc minutes in length. Your eyes should be completely

Figure 11.31 The odd spiral galaxy M 82

Figure 11.32 Spiral galaxy M 101

adapted to night vision, and if you know the exact position of the galaxy, you can use averted vision. Through binoculars you will only see the central part of the galaxy with its bright core. The wonderful spiral arms are visible only on long-exposure images.

The galaxy is approximately 31 million light years from us. Its real diameter measures 130,000 light years.

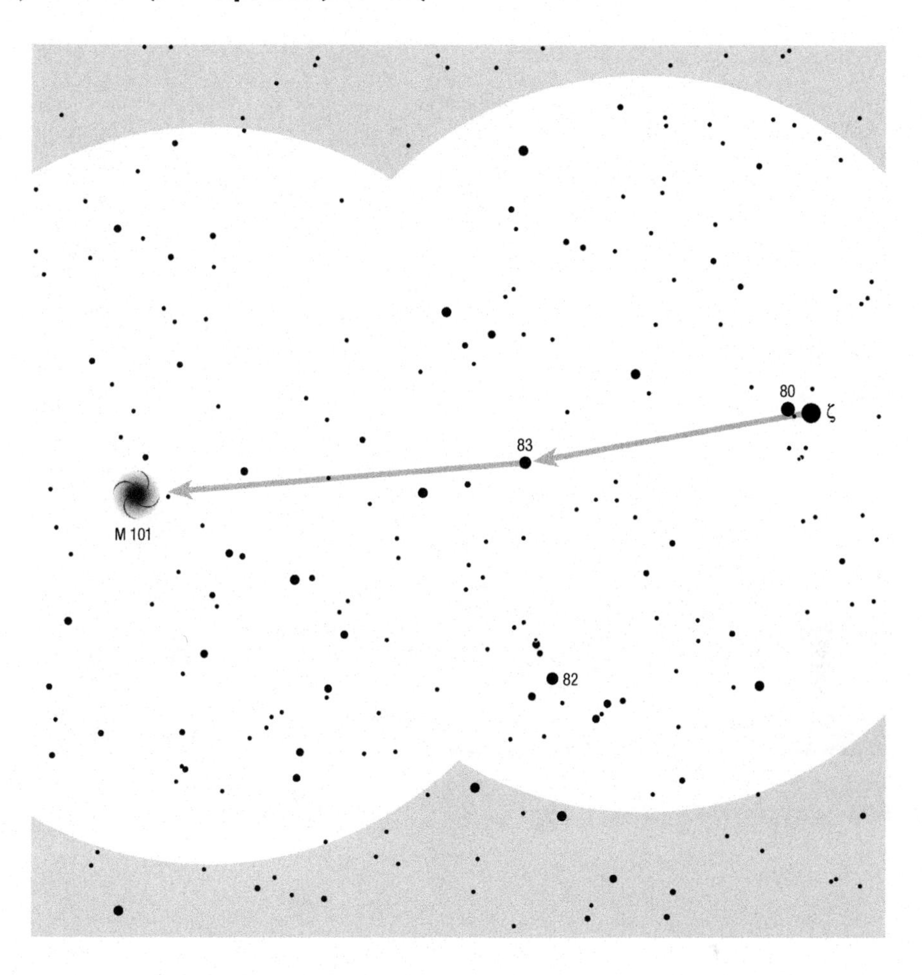
80
ζ
83
M 101
82

Figure 11.33 Spiral galaxy NGC 2841

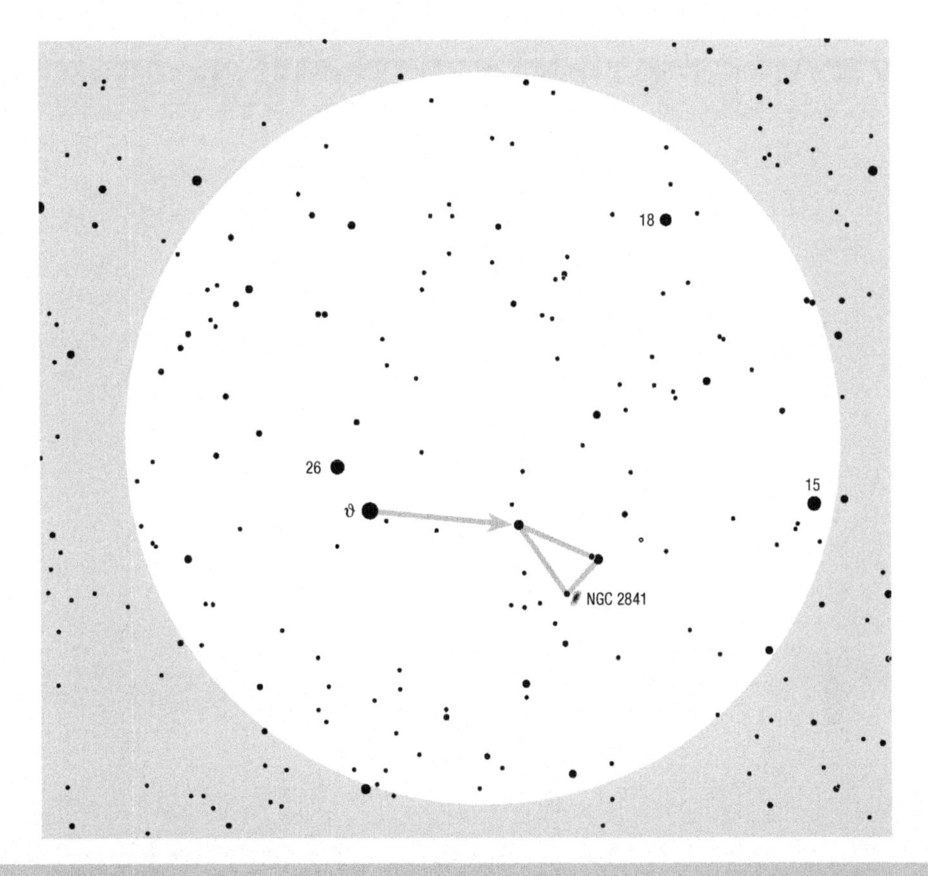

Figure 11.34 The position of the faint spiral galaxy NGC 2841. The star leading to it is Theta. In the field of view we can clearly see both magnitude 6 stars above the galaxy as well as the magnitude 8 star in the vicinity. The galaxy itself shows as a faint stripe of light only when there are excellent observing conditions

URSA MINOR (The Little Bear)

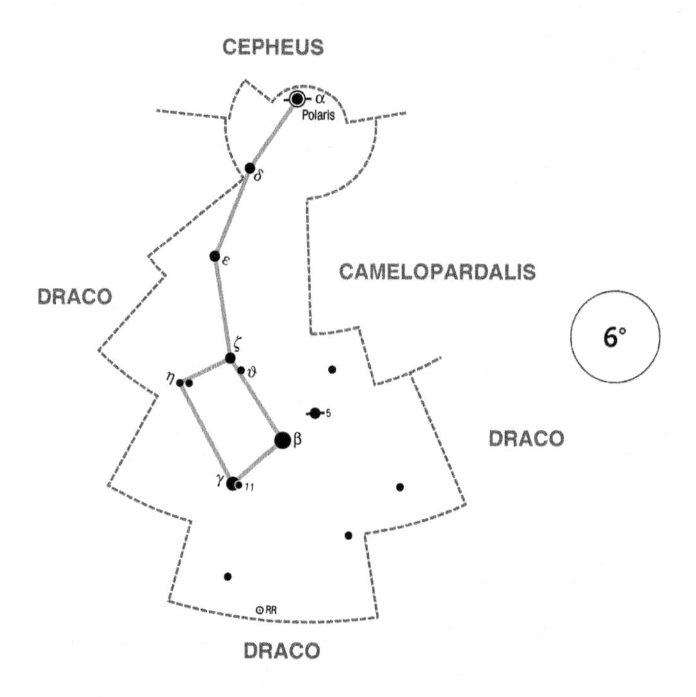

CONSTELLATION CULMINATES		
mid April at 2 a.m.	mid May at midnight	mid June at 10 p.m.

Ursa Minor is one of the rare constellations known to almost everybody. The brightest star in it is Polaris (2m0), an important orientation point and the famous guiding star for travelers and sailors.

The star pattern extending from Polaris in the direction of Mizar in Ursa Major cannot be mistaken; it looks like a smaller and slightly deformed Big Dipper (the best known asterism in Ursa Major). The other brighter stars are Beta, or Kochab (2m1), and Gamma (3m0), which are known by their common nickname the "Guardians of the Pole." Other stars in the constellation are of magnitude 4 or fainter.

The yellow-orange Beta is the 58th brightest star in the sky. It is 127 light years from us. Its luminosity is 170 times that of the Sun.

When viewed from mid-northern latitudes the Northern Star, or Polaris, is approximately 45 degrees above the northern point of the horizon. We can also find it by using the last side of the Big Dipper, and extending the connecting line between Alpha and Beta Ursae Majoris by approximately five times. We cannot miss it, for it is the only bright star in that part of the sky. It is an important star due to the fact that it is less than a degree away from the celestial North Pole, a fixed spot around which it appears that the celestial sphere and the stars revolve. And while all stars move in smaller or larger circles around the celestial North Pole day and night, throughout the entire year Polaris appears to be constantly almost in the same spot when viewed with the naked eye (Figure 11.35).

Astronomers, who like to be very precise, soon established that the celestial poles are not entirely motionless but change their position in the heavens due to the precession of Earth's axis. Precession

Figure 11.35 The apparent motion of the stars around the north celestial pole

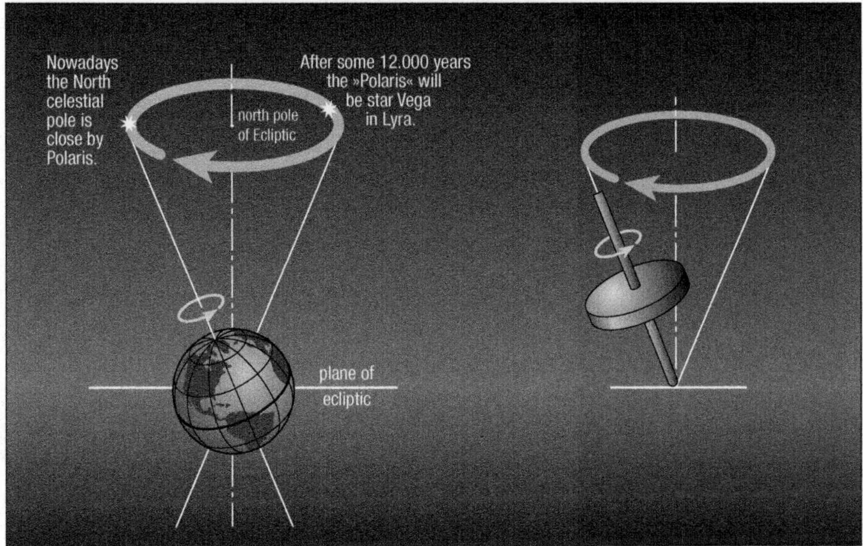

Figure 11.36 The precession of the Earth's axis and the precession of the spinning top

occurs because axis is tilted by approximately 23.4 degrees against the vertical plane perpendicular to the orbital plane around the Sun. The reason is the gravitational influence of the Sun and Moon. A similar occurrence can be seen at a spinning top that spins around a tilted axis (Figure 11.36).

The celestial poles, therefore, move among the stars and make an entire circle (with the diameter of 47 degrees) in approximately 25,800 years. Currently, the northern celestial pole is in the vicinity of the Polaris. Approximately 5,000 years ago (when the great Egyptian pyramids were built), the role of the "North Star" was taken by Thuban (Alpha Draconis), since it was only 2 degrees away from the northern celestial pole. In the time of the ancient Greeks, the North Star was represented by the Beta Ursae Minoris. Since it was only 7 degrees away from the pole it was the closest to it. Our Polaris assumed its role as the North Star in the fifteenth century, when it was approximately 4 degrees from the pole.

Nowadays it is less than 1 degree from the pole, and the distance is still decreasing. In 2102, it will be a mere 28 arcmin away. After this date the distance will slowly increase, and around year 3000 it will have to cede its prominent position to another star – Delta Cephei.

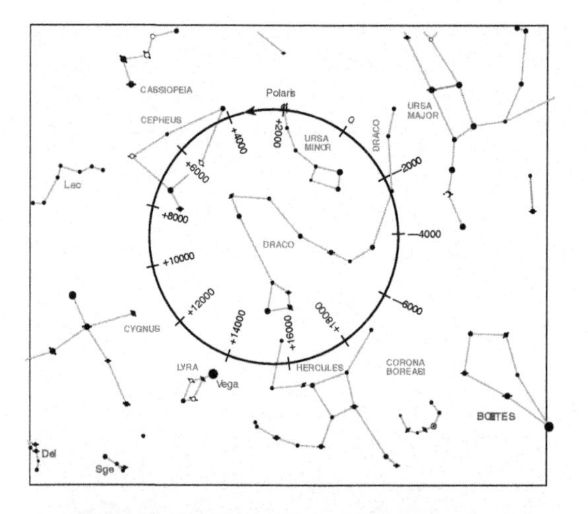

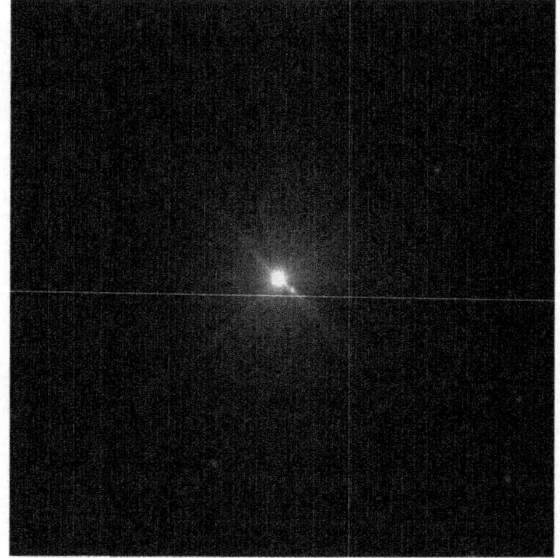

Figure 11.36A Polaris and its 9.1 magnitude companion

Polaris is the 46th brightest star in the sky. It is a variable, a Cepheid, but the changes in brightness are just one-tenth of a magnitude, which is not noticeable with average amateur equipment. The star is 430 light years from Earth. Its luminosity surpasses our Sun by 2,200 times. It is actually a triple star, but it cannot be separated through binoculars. However, even through small amateur telescopes the magnitude 9.1 companion can be seen. The two stars are 18.6 arcsec apart (p. a. 233°, 2005). The orbiting period is in excess of 72,000 years, and the actual separation between them measures at 2,000 AU (we could place 33 Solar Systems between the two stars). Polaris itself is a spectroscopic binary. In this close pair the stars are only 5 AU apart (the same as the distance between the Sun and Jupiter), and it takes 30.5 years for the pair to complete one orbit around each other.

Gamma and the star labeled 11 Ursae Minoris (5 m) are a wide pair that can be seen with the naked eye. The stars are not gravitationally bound, since they seem close together in the sky merely by coincidence when viewed from Earth. They are 11 arcmin apart (p. a. 270°). When viewed through binoculars we can notice a distinct color contrast. Gamma is white, and the star 11 is yellow-orange.

VELA (The Sails)

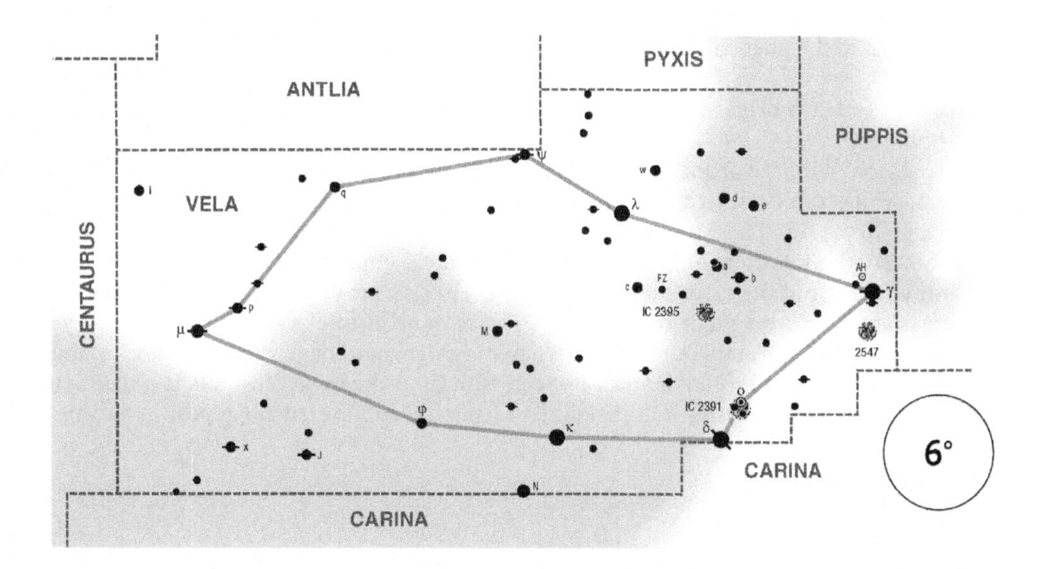

CONSTELLATION CULMINATES		
in begin. January at 2 a.m.	in begin. February at midnight	in begin. March at 10 p.m.

Vela lies under Antlia, even further south, so from the mid-northern latitudes only its extreme northern part can be seen. But if our observing site is 32 degrees north or further south, we could see the entire constellation, which is quite prominent. Vela is a part of the formerly largest constellation Argo Navis. It used to be exceptionally large, so modern astronomers divided it into Carina, Puppis, and Vela. Puppis is the furthest north of them and is clearly seen from mid-northern latitudes. Carina is the most southern and cannot be seen from the mid-northern latitudes. Even observers from southern Europe and the southern United States can see only its most northern parts with the bright Canopus, and even for them the star rises a mere 7 degrees above the horizon at culmination. Therefore, Carina is not described in this book.

As mentioned earlier, the astronomers split Argo Navis; however, they left the old names and designations of the stars. This means that there is only one Alpha, one Beta, one Gamma, and so on, in the three new constellations, which, in turn, means that the brightest star in Vela is Gamma (1m7). In brightness, Gamma is followed by Delta (1m9), Lambda (2m2), Kappa (2m5) and Mu (2m7). All of them except Lambda are always below the horizon when viewed from mid-northern latitudes.

Gamma is the 32nd brightest star in the sky and is a very interesting multiple star. The brightest components are of magnitudes 1.8 and 4.1, separated by 41 arcsec (p. a. 219°, 2002) and can be split in binoculars. The brighter star of this pair is famous for being the brightest Wolf–Rayet star in the sky. The other two fainter components can also be resolved in binoculars, but you will have great difficulty in seeing the faint stars of magnitudes 7.3 and 9.4 in the vicinity of the bright primary one. The stars are 840 light years away from us. Their common luminosity surpasses our Sun by 10,000 times, but the main part of this is contributed by the Wolf–Rayet star.

Delta is the 43rd brightest star in the sky. It is 80 light years away from Earth. Its luminosity is 77 times that of the Sun.

Lambda is the 68th brightest star in the sky. It is 570 light years from us. The star is a yellow-orange giant, the luminosity of which is over 3,000 times that of the Sun. In southern localities, where Lambda is high in the sky at culmination, it is as bright as Mizar in Ursa Major.

Kappa is the 90th brightest star in the sky. It is 540 light years from Earth. The star is a blue-white subgiant with luminosity that surpasses our Sun by 2,100 times. There is an interesting fact related to this star: it is only a few degrees from the south celestial pole of the planet Mars. Thus, future human expeditions on the Red Planet will have a good orientation point if their electronic equipment breaks down.

Yellow Mu is the 111th brightest star in the sky. It is 116 light years from us. Its luminosity is 80 times that of the Sun.

The entire region of the Velorum Milky Way is extremely rich in star fields. There are three bright open clusters, which can be seen through binoculars from the southern latitudes, where the constellation is high in the sky. These are NGC 2547 (4m7/74′), IC 2391 (2m5/50′), and IC 2395 (4m6/20′). Observers from southern Europe or the southern United States could try to find any of them on a clear and dark winter night, but the easiest target is IC 2391. This is a group of rather bright stars gathered around the variable Omicron Velorum (3m5–3m7). It is quite visible even with the naked eye from southern sites. Of about 300 stars in that area, 60 are brighter than magnitude 11 and 25 of these are brighter than magnitude 9. Seven stars can be seen with the naked eye. Yes, this is a true southern jewel! The star leading to the cluster is the bright Delta; the star and the cluster appear in the same field of view.

VIRGO (The Virgin)

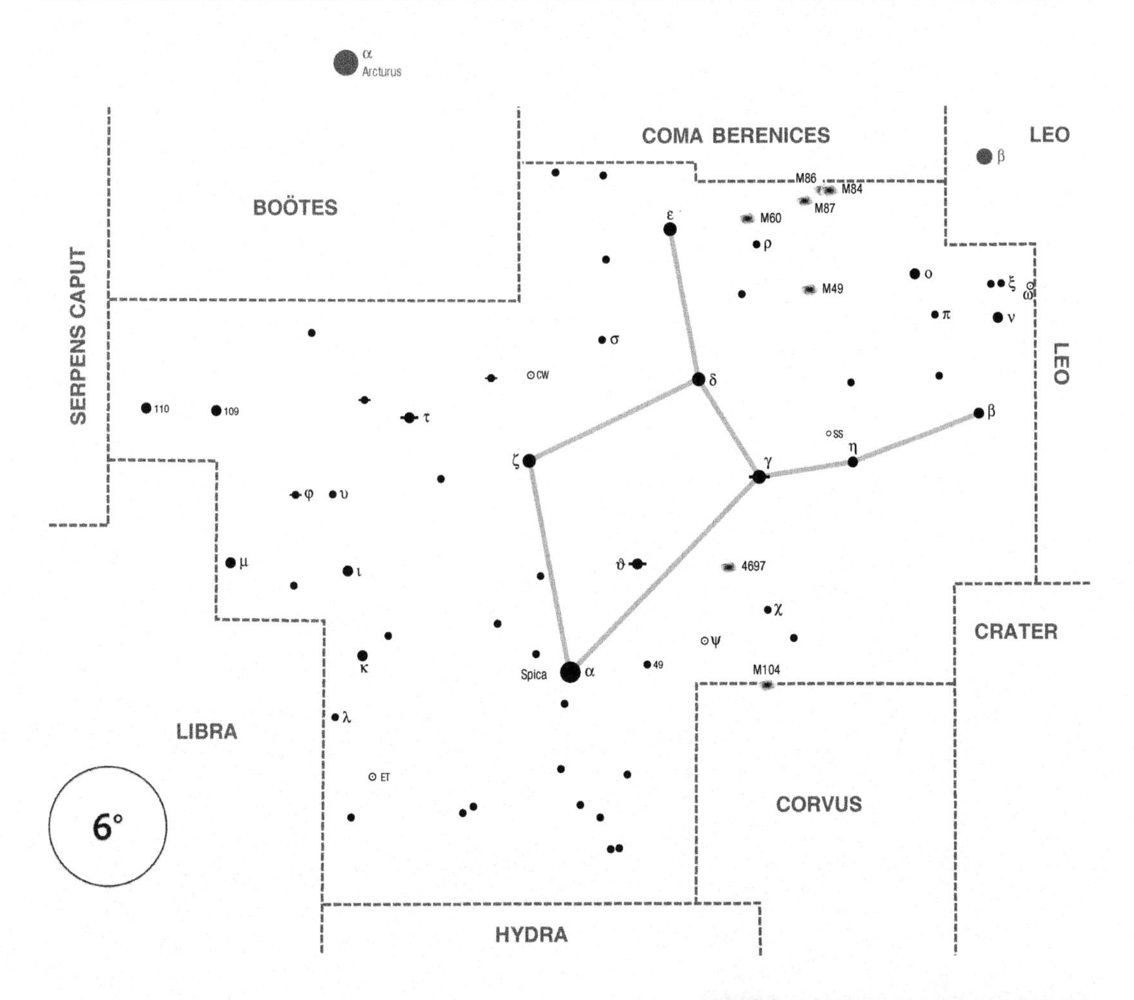

CONSTELLATION CULMINATES		
mid March at 2 a.m.	mid April at midnight	mid May at 10 p.m.

Virgo is a spring constellation of the zodiac and is one of the largest in the sky. The brightest star is Spica (1m0). Its name means "corn ears" in Latin.

In brightness Spica is followed by Gamma (2m7), Epsilon (2m8), Zeta, Delta (both 3m4), and Beta (3m6).

Spica is the 15th brightest star in the sky. Its color is blue white, and it has a high surface temperature – approximately 20,000 K. The star is 263 light years away and has the luminosity of approximately 2,000 Suns.

Gamma, or Porrima, is the 120th brightest star in the sky. It is a splendid binary (the stars are gravitationally bound) with an orbital period of 169 years. In 2005, when they were at their closest, the two stars with equal brightness (3.5 magnitude) were only 0.3 arcsec apart. During the next 85 years, the apparent distance will increase; however, the Porrima will return within the range of amateur telescopes no earlier than in 2020. The last data on their distance is from 2006 (0.5 arcsec, p. a. 82°). The stars are only 38.7 light years from us; their common luminosity is 8 times that of the Sun. Porrima cannot be separated through binoculars.

Epsilon is the 144th brightest star in the sky and 102 light years from us. Its luminosity is that of approximately 54 Suns.

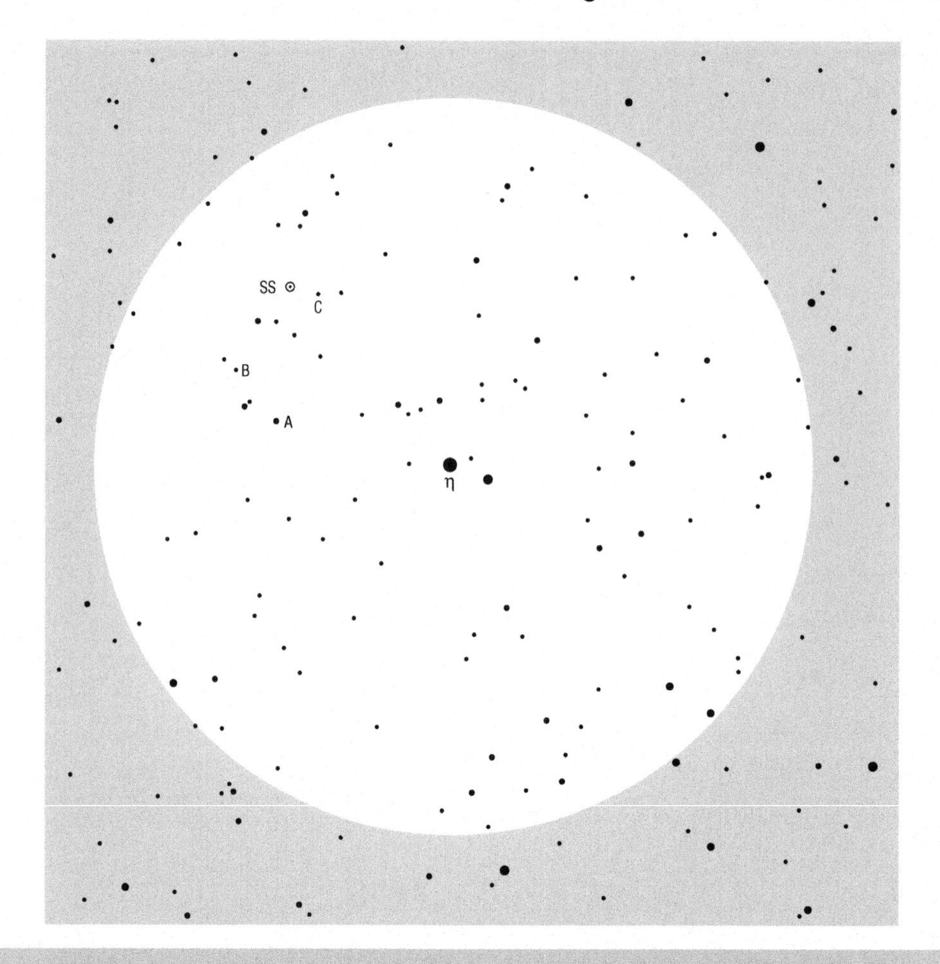

Figure 11.37 The surroundings of the variable SS, with comparison stars: A (7m7), B (8m9), and C (9m5)

The pulsating variable star SS Virginis is one of numerous variables found in this constellation (Figure 11.37). It is mentioned here because of its distinctive orange-reddish color. The star changes its brightness between magnitudes 6 and 9.6 and is visible throughout the entire period (364 days) through binoculars. SS Virginis is found less than 2 degrees northeast of the bright Eta. The star reached one of its maximums around January 1, 2008. In winter, the constellation is visible after midnight and culminates at around 6 o'clock in the morning. The pulsation period is just a day shorter than one year, which means that the star will be visible in the morning sky for the next couple of decades.

On Virgo's shared border with the constellation Coma Berenices, we can find the largest galaxy cluster visible through amateur telescopes. This prominent association is approximately 60 million light years away and consists of some 2,000 members. Some of the galaxies in this area are also visible through binoculars.

We start our search for the galaxies with the bright Epsilon. If this star is moved to the eastern edge of the field of view, you will catch the sight of a V-shape pattern created by five rather bright stars. The brightest of these is labeled 33 Virginis. Northwest of this group is the clearly visible pair of Rho and 27 Virginis. 34 Virginis can be seen in the same field of view with this pair. The elliptical galaxy M 60 (8m8/7′2 × 6′2) is located between them.

Through binoculars it is seen as a tenuous spot of light a few arc minutes across. The galaxy can only be seen when there are excellent observing conditions and when it is at its highest point in the sky. When you are convinced that the part of the sky in which the galaxy should be is in the middle of the field of view, wait for as long as it takes for your eyes to fully adapt to night vision. If you cannot see the galaxy directly, try with averted vision. The appearance and size of the galaxy depends greatly on observing conditions. If they are poor, try at a different time!

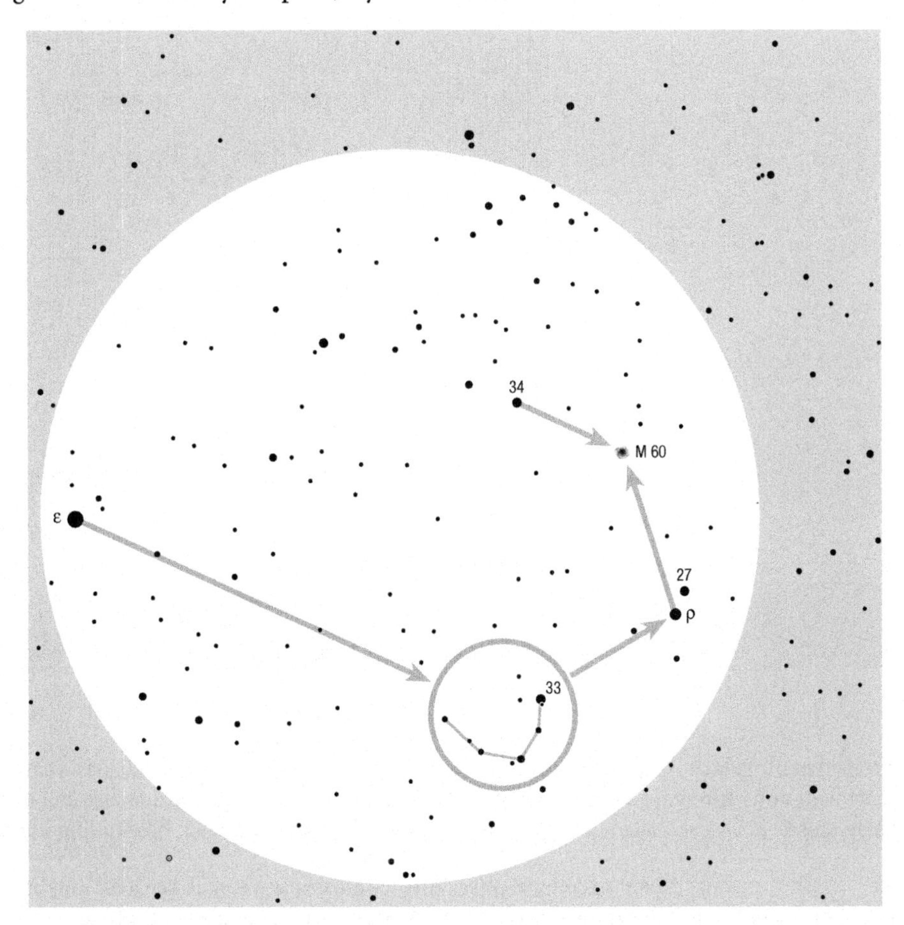

M 60 is one of the largest known elliptical galaxies. In mass, it is similar to M 49 but slightly smaller. The diameter of M 60 is estimated to measure at 120,000 light years across, and it has a mass of a thousand billions suns!

If we choose the familiar stars 34, 27, and Rho as guides, we will find a group of magnitude 8 and 9 stars that form a recognizable pattern approximately 4 degrees to the west. Just below the magnitude 8 star (see chart) we will notice a larger, more or less uniform spot of light. This is the elliptical galaxy M 87 (8m6/7′2 × 6′8), one of the biggest and brightest members of the cluster (with a diameter exceeding 120,000 light years). It might prove a slight disappointment when viewed through binoculars, but you should be aware that the light falling onto your eyes at this very moment left the galaxy 60 million years ago, at the time dinosaurs became extinct.

This galaxy is known for two unusual characteristics, which are visible only on long-exposure images. M 87 is surrounded by thousands of globular clusters, more than 10,000, according to some modern data. Just for comparison: in our galaxy a total of 150 can be found. The second characteristic is the presence of a narrow jet

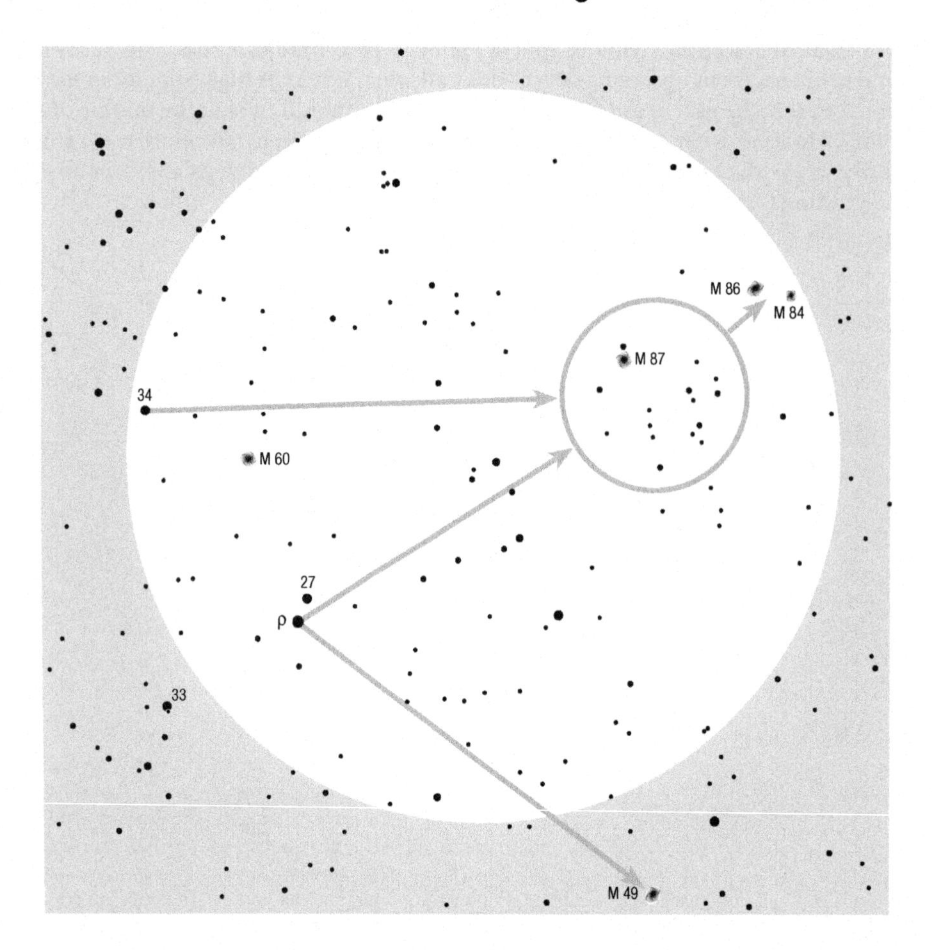

of fast electrons that the galaxy is emitting from its center. The jet is approximately 4,000 light years long and 400 light years wide and represents a strong source of X-rays. Its true nature and origin are still not known. It is very likely that M 87 is a so-called active galaxy, with a massive galactic black hole hiding in its center.

With excellent observing conditions we will see (or maybe it might be better to say glimpse) two more faint ovals – elliptical galaxies M 84 (9m1/5′) and M 86 (8m9/7′5 × 5′5) about a degree and a half west of M 87. Here you will have to use averted vision. Look at the chart with elliptical galaxy M 87 for more details (Figure 11.38).

The diameter of M 84 is estimated at 90,000 light years. Astronomers are not certain yet whether M 86 is a true member of the Virgo cluster or whether it merely lies in the same direction when viewed from Earth and is in fact closer to us. Some measurements suggest that the galaxy is only 20 million light years away (the cluster is – as mentioned before – 60 million light years away).

M 49 (8m4/8′9 × 7′4) is totally different from the previously mentioned galaxies. This is one of the largest elliptical galaxies known to us. With excellent observing conditions it is clearly visible through binoculars and is evidently of an oval shape. We start our search at Rho (see the chart on page 489) or at the rather bright star Omicron (magnitude 4; see chart on page 492). If this star is moved to the extreme western edge of the field of

Figure 11.38 The central region of the Virgo cluster with the bright elliptical galaxy M 87, which is clearly seen through binoculars, and another two elliptical members M 84 and M 86, which are on the border of visibility. All of the other galaxies in the image are not seen through binoculars but can be glimpsed through amateur telescopes

view, M 49 will appear on its eastern edge. As the faint objects are not visible on the edge of the field of view, shift your binoculars and center them on the two magnitude 6 stars that are always clearly visible. The galaxy lies between them, slightly closer to the southern star.

M 49 is 5 degrees from the center of the Virgo cluster, to which it belongs. Its mass is estimated to be that of a thousand billion Suns, and it has a diameter of approximately 160,000 light years. It is 60 million light years from us.

On the border of the constellation Corvus lies the spiral galaxy M 104 (8m3/ 8′9 × 4′1), popularly known as the Sombrero. This galaxy is seen edge on and is divided into two by a dark stripe consisting of clouds of obscuring dust and gases. The Sombrero, that appears very attractive on images, is visible through binoculars, but (unfortunately) only as a few arc minutes long stripe of light. Despite its promising brightness (8m3), the galaxy lies rather low for mid-northern observers, so we need excellent observing conditions in order to see it.

The Sombrero lies in a rather empty part of the sky. Therefore, the best leading star is Spica (see the chart on page 493). If we move it to the extreme eastern edge of the field of view, we will see the characteristic pattern comprised of four brighter stars on its western part. The upper three are magnitude 6, while the lower, with designation 49 Virginis, is as bright as magnitude 5. Now move this star to the extreme eastern edge of the field of view, and on the west you will see another magnitude 5 star – Psi. One additional shift to the west, and the Sombrero will appear in the field of view. Its stellar

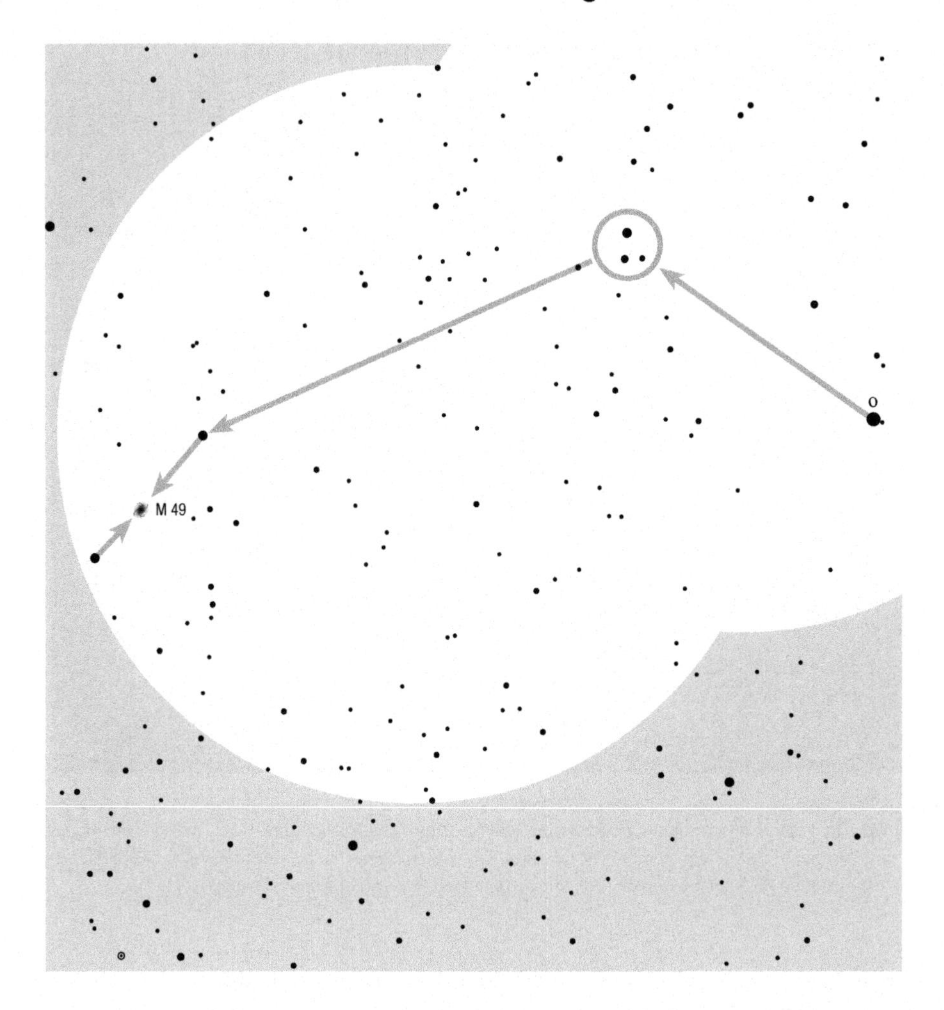

neighborhood is very picturesque and easy to recognize. The most noticeable is a string of three closer stars, separated by a mere 3.5 arcmin. The brightest is the westernmost one that shines with a magnitude of 8.

In fact, the Sombrero is a gigantic galaxy with a diameter of 130,000 light years. Its light travels 50 million years through space before it reaches us. The mass of the galaxy is estimated to equal 1,300 billion Suns. Astounding to even contemplate, isn't it?

At the very end of this constellation and for all those who like a challenge there is another elliptical galaxy – NGC 4697 (9m3/6′0 × 3′8), which is on the border of visibility through binoculars. It is situated approximately 5 degrees west of the rather bright star Theta. Another way to find the galaxy is to start at Psi and move across the magnitude 7 and 6 stars to where the galaxy should be located. Do not expect too much in the field of view. The galaxy is seen merely as an arc minute long, dim line of light. In order to find this faint object you have to choose a night with excellent observation conditions, and the galaxy should be close to its highest point in the sky. Your eyes have to be completely adapted to night vision. In your search you should use averted vision. It always helps if you gently shake the binoculars, for your eyes are better at noticing moving objects. This truly works! If you do not see the galaxy, try at a different time. Maybe your eyes are tired from staring at the chart or through binoculars, or maybe the conditions are not as perfect as you think.

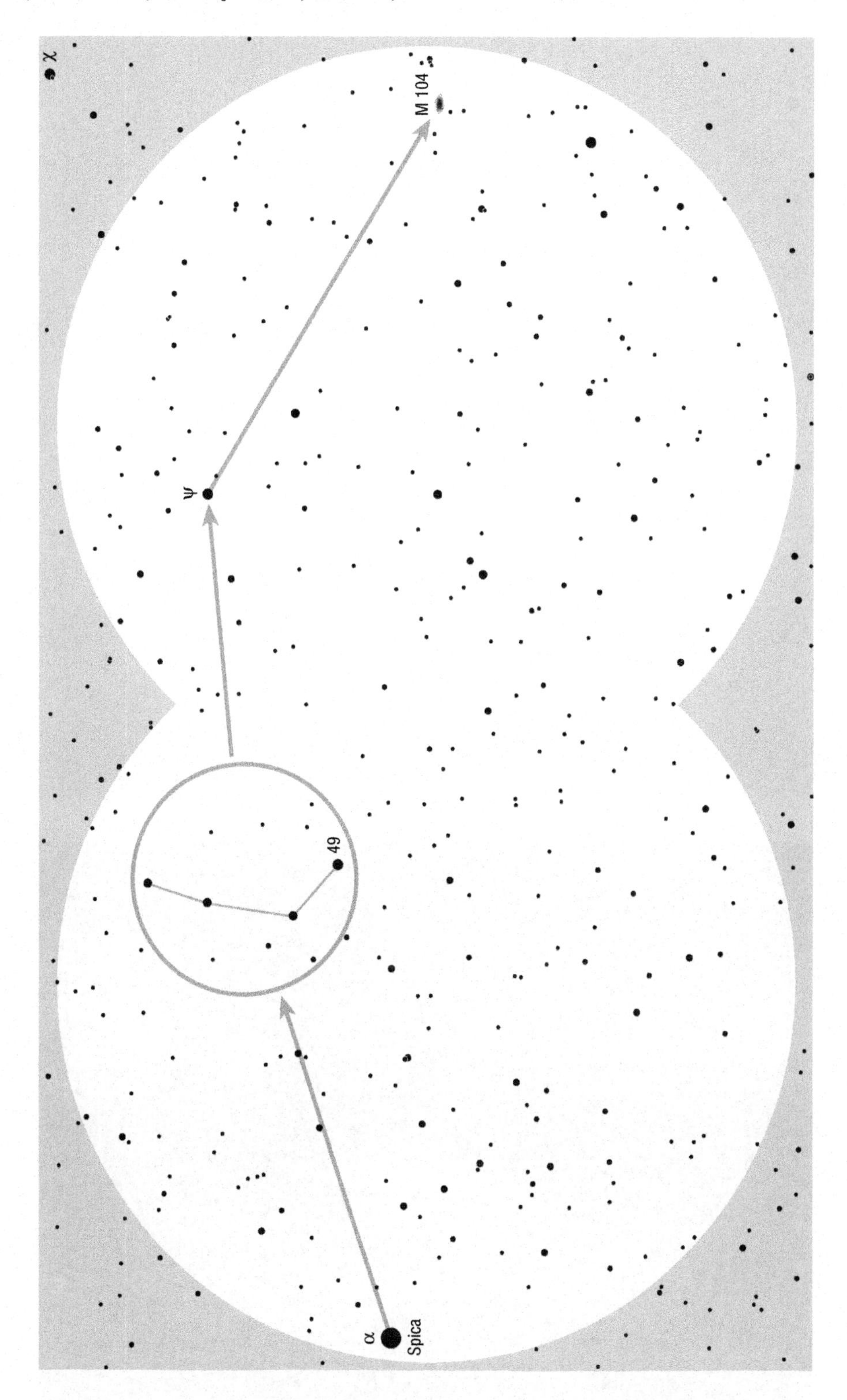
χ
M 104
ψ
49
α
Spica

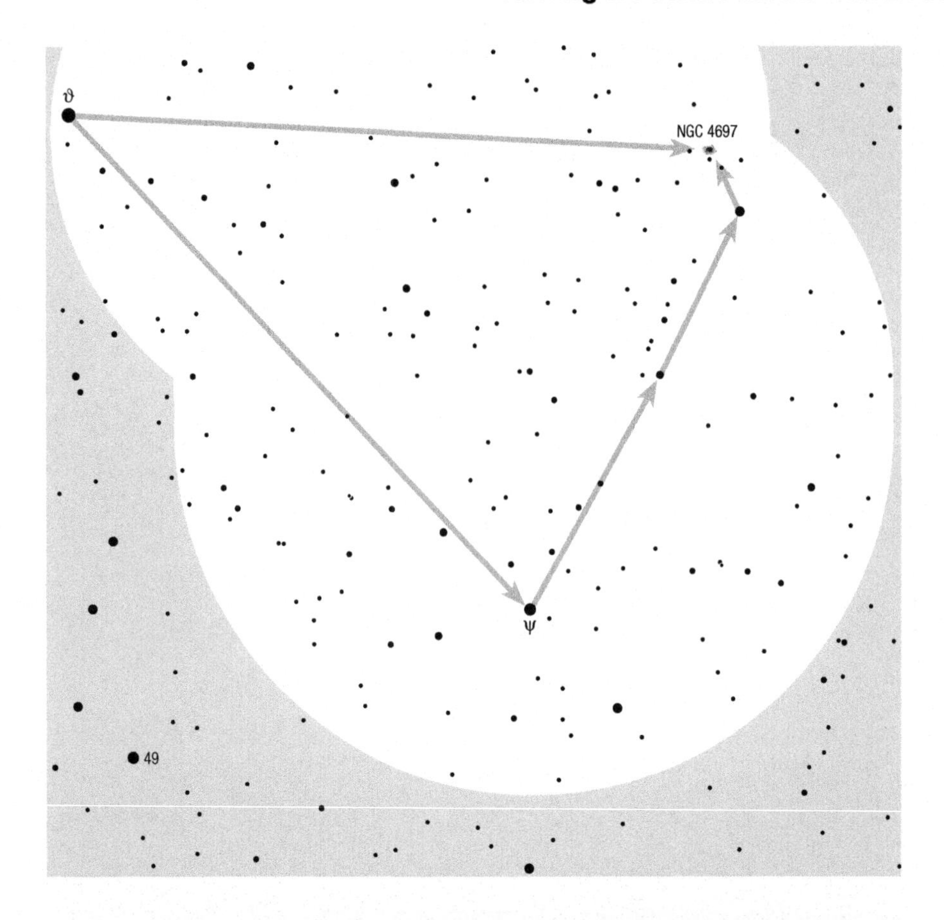

Figure 11.39 The Sombrero Galaxy

VULPECULA (The Fox)

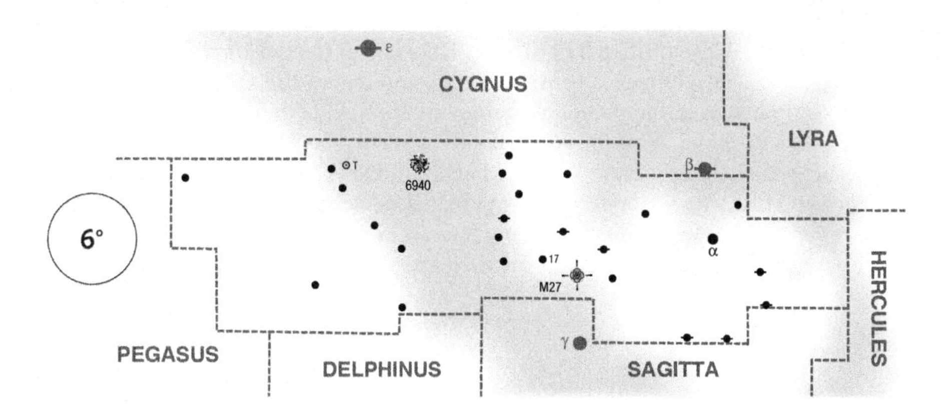

Vulpecula is a small and unremarkable summer constellation, lying between Cygnus, Lyra, and Sagitta. The easiest way to find it is with the aid of Beta Cygni and Gamma Sagittae. Vulpecula lies between the two. The brightest star is Alpha, but even this is only magnitude 4.4.

CONSTELLATION CULMINATES		
mid June at 2 a.m.	mid July at midnight	mid August at 10 p.m.

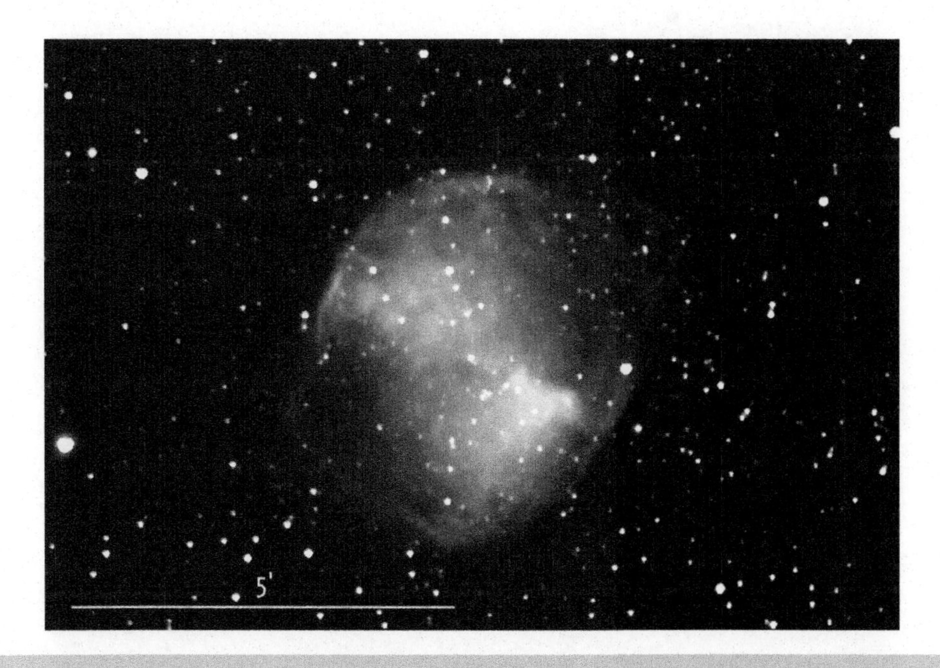

Figure 11.40 Planetary nebula M 27, photographed with amateur equipment

In this constellation we find the M 27 (7m4/ 8′0 × 5′7), popularly known as the Dumbbell Nebula, a planetary nebula that is one of the largest and closest of its kind (Figures 11.40 and 11.41). It is clearly visible through binoculars, but merely as a spot of faint light approximately 6 arcmin across. Its size can vary greatly, depending on observing conditions. With excellent conditions we might even glimpse the irregularities in the otherwise unified bright spot. Here is a perfect example of how the adjustment of your eyes to night vision influences the size and visibility of celestial objects. You should try catching the planetary nebula in the field of view of binoculars, then turn the lights on or walk into a lit room. Wait for your eyes to adjust to the brightness. Then walk back to the binoculars and watch the field of view at least for half an hour. What you are going to experience will certainly convince you that we are not exaggerating in this book when we say that your eyes need to be adjusted to night vision. You will notice that only half an hour separates you from seeing absolutely nothing to seeing many details. It is half an hour well spent!

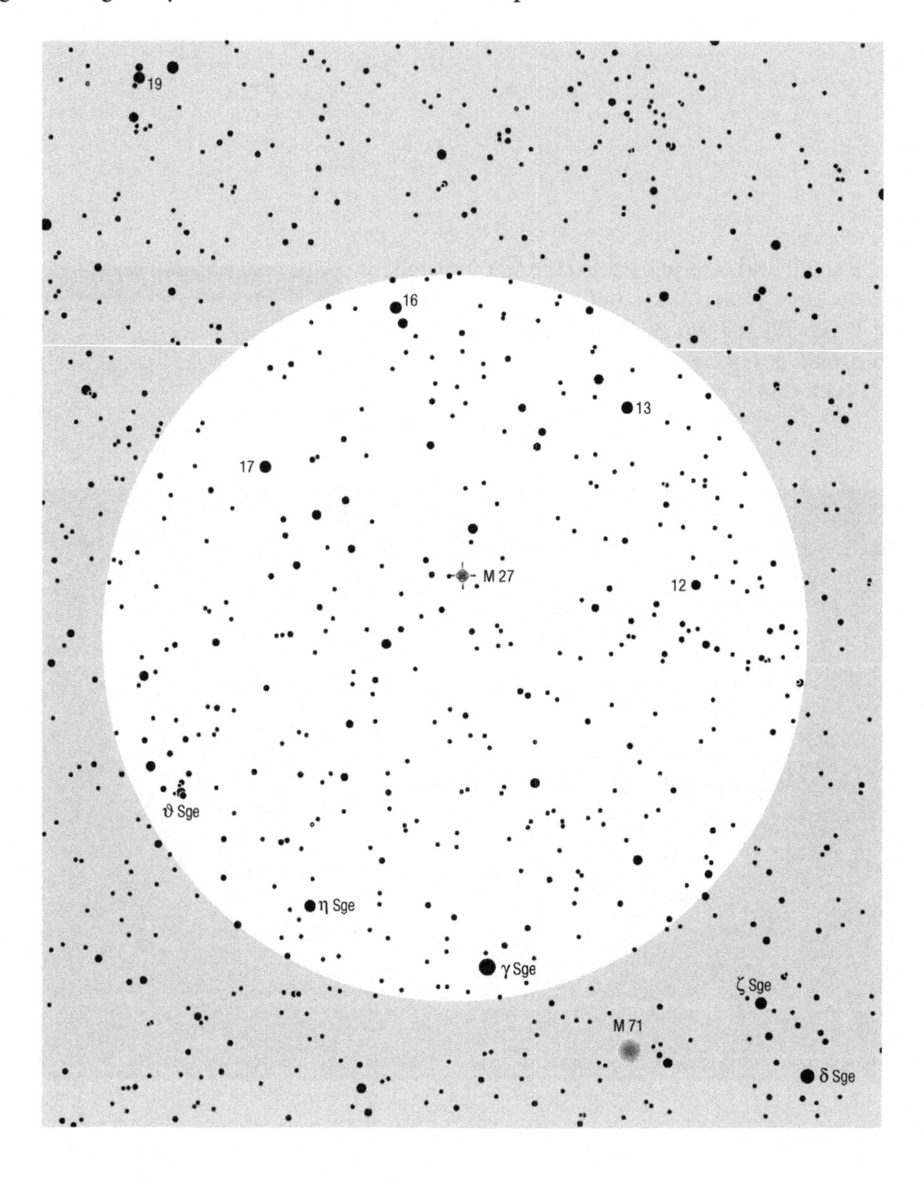

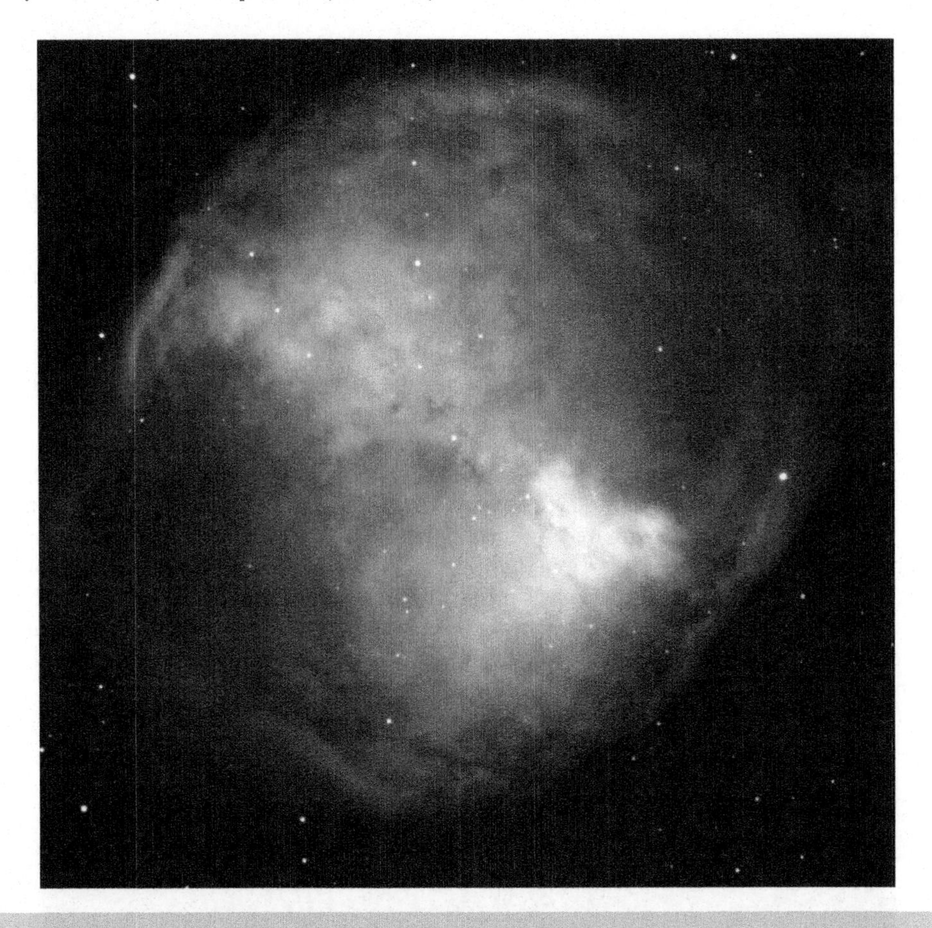

Figure 11.41 Professional portrait of M 27. The image is 5 arcmin wide.

The star leading to the nebula is the Gamma Sagittae. This star is located approximately 3 degrees to the south; thus, it appears in the same field of view of binoculars. The Dumbbell obtained its name from its shape, which unfortunately is not seen through binoculars. However, it can be seen through mid-sized amateur telescopes and of course on photographs, even amateur ones.

The planetary nebula is approximately 1,250 light years from us, which means that its real diameter is 2.5 light years. The gases in the nebula are expanding in all directions with the speed of approximately 30 km/s. If we assume that the speed of expansion has stayed consistent over time, we can calculate that the nebula is approximately 48,000 years old. The central star from which the nebula emerged is an extremely hot blue subgiant with a surface temperature of approximately 85,000 K and is one of the hottest stars known to us. However, since it is extremely small, its luminosity is only half that of our Sun. In our sky, the star shines with magnitude 13.5 and can thus not be seen through binoculars. With its strong ultraviolet radiation, it excites the gases in an otherwise cold surrounding nebula. These shine in glorious colors and reveal the nebula's composition. Green and red prevail. The red light is emitted mainly by ionized hydrogen atoms, while the green is emitted by twice-ionized oxygen atoms.

Open cluster Collinder 399 (3m6/60′), sometimes named Brocchi's Cluster, is so bright and large that it was known even before the invention of the telescope. The Arabic astronomer Al Sufi described it in his book as early as A.D. 964. The cluster contains approximately 40 bright stars, which are scattered across a large area of the sky. This is why the cluster was never ascribed a Messier or an NGC number. However, this is also the reason why it is best seen through binoculars. The pattern of the brightest stars is reminiscent of a reversed coat hanger, and this gave the cluster its other popular nickname – the "Coathanger cluster."

It is still uncertain if the stars of Collinder 399 are close together only when seen from Earth, or whether they form a true cluster. Data acquired by ESA's Hipparcos satellite shows that there is no evidence that the stars are connected, and therefore they would not have a common origin. The best way to find it through binoculars is to follow the instructions on how to find the variable star U Sagittae, described on page 408.

Vulpecula lies in the summer Milky Way, so the entire area is full of stars, groups, and open clusters. When there are excellent observing conditions we can find the open cluster NGC 6940 (6m3/25′) through binoculars. It lies in the northeastern part of the constellation and is in the field of view seen as a 20 arcmin large spot of light from which a few of the brightest magnitude 9 members shine. In total, the cluster has approximately 140 members, out of which about 50 are up to magnitude 11 and can be seen through binoculars when there are perfect observing conditions. A suitable leading star is the bright Epsilon Cygni. If this star is moved to the extreme northeastern edge of the field of view, the cluster will appear on the southwestern edge. Between them a rather good checking point is represented by a close pair of clearly visible magnitude 6 stars. NGC 6940 is approximately 2,500 light years from us, so its actual diameter is 18 light years.

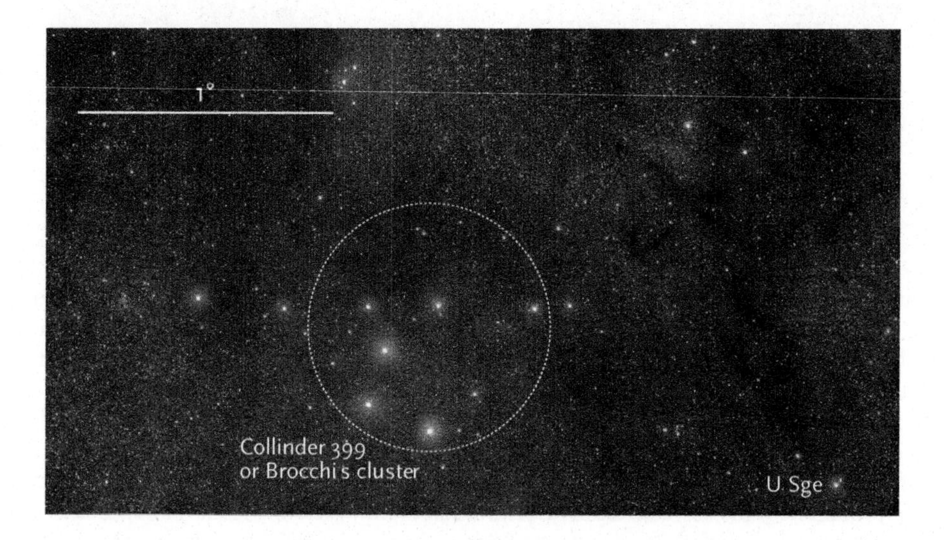

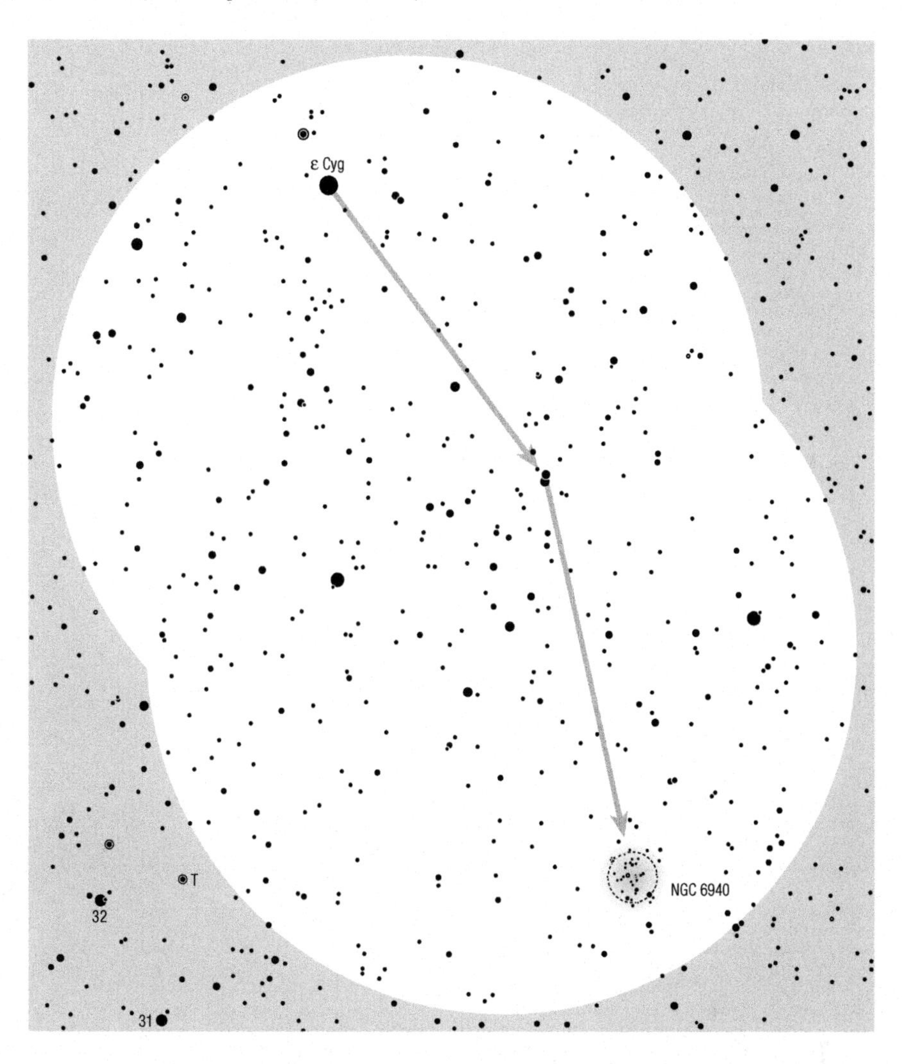
ε Cyg
NGC 6940
T
32
31

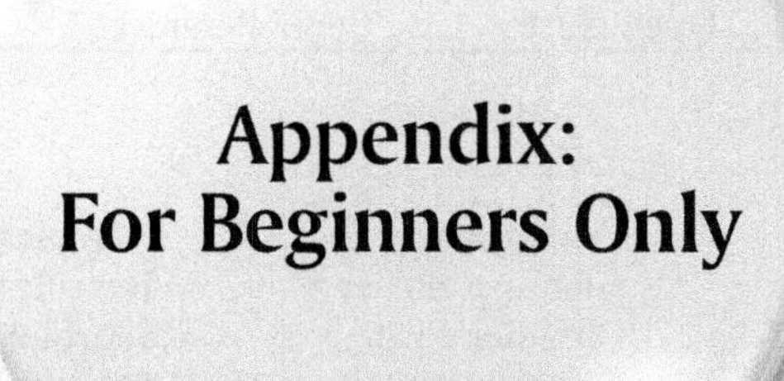

Appendix: For Beginners Only

All celestial objects described in the book are, in perfect observing conditions, seen in 10×50 binoculars and therefore seen in any small amateur telescope. The author has seen them all through such binoculars. The observational data were gathered for several years on Bloška planota (Slovenia), which at that time belonged to Class 2 of the Bortle scale. But in spite of the very dark and calm sky you have to be an experienced observer in order to see many of them. What does it means to be an experienced observer, after all?

An experienced observer has seen many celestial objects through different types and sizes of the telescopes. That is why he or she is well aware that binoculars have to be mounted on a stable tripod in order to make the best use of their potentials. An experienced observer always waits for his or her eyes to become adapted for night vision. An experienced observer always takes enough time to observe the chosen object. The key word here is *enough*. For some easy and bright double stars it is enough to observe them for five minutes, but to see the faintest celestial objects on the border of visibility of the binoculars or as many details as possible in some open cluster or elusive disperse nebula, you need – after your eyes fully adapt to night vision – half or maybe a whole hour.

An experienced observer is patient like a cat and does not run from object to object. An experienced observer makes essential observations only when the chosen object is near its culmination. And last but not least, an experienced observer uses all observing techniques.

For all of you, who are only beginning your observational career and do not see yet how the sky lies, here is a list of 45 of the most easily seen and attractive celestial objects, which are also easy to find. If you read carefully the instructions beside the descriptions you will start your hobby with little hassle and gather some observing experiences easily and with great enjoyment. And then you can go on! Objects are listed seasonally, as they appear high in the evening sky.

Spring Objects

Name	Type	Constellation	Page Reference
M 81/82	Spiral galaxies	Ursa Major	xxx
Mizar	Double star	Ursa Major	xxx
M 101	Spiral galaxy	Ursa Major	xxx
M65/66	Spiral galaxies	Leo	xxx
Mellote 111	Open cluster	Coma Berenices	xxx

Summer Objects

Kappa Her	Double star	Hercules	3xxx
M 13	Globular cluster	Hercules	xxx
M 4	Globular cluster	Scorpius	xxx
M 6	Open cluster	Scorpius	xxx
M 7	Open cluster	Scorpius	xxx
Nu Dra	Double star	Draco	xx
IC 4665	Open cluster	Ophiuchus	xxx
M 8	Bright nebula	Sagittarius	xxx
M 20	Bright nebula	Sagittarius	xx
M 22	Globular cluster	Sagittarius	xxx
M 24	Star cloud	Sagittarius	xxx
Delta Lyr	Double star	Lyra	xxx
Theta Ser	Double star	Serpens	xxx
M 27	Planetary nebula	Vulpecula	xxx
Cr 399	Open cluster	Vulpecula	xxx
Albireo	Double star	Cygnus	xxx
M 39	Open cluster	Cygnus	xxx

Autumnal Objects

Mu Cep	Variable star	Cepheus	xxx
M 15	Globular cluster	Pegasus	xxx
NGC 253	Spiral galaxy	Sculptor	xxx
M 31	Spiral galaxy	Andromeda	xxx
NGC 752	Open cluster	Andromeda	xxx
M 33	Spiral galaxy	Triangulum	xxx
Mira	Variable star	Cetus	xxx
Alpha Per	Open cluster	Perseus	xxx
Algol	Variable star	Perseus	xxx
Double Cluster	Open clusters	Perseus	xxx

Winter Objects

Pleiades	Open cluster	Taurus	xxx
Hyades	Open cluster	Taurus	xxx
M 42	Bright nebula	Orion	xxx
M 36/37/38	Open cluster	Auriga	xxx
M 35	Open cluster	Gemini	xxx
M 41	Open cluster	Canis Major	xxx
M 44	Open cluster	Cancer	xxx
M 67	Open cluster	Cancer	xxx

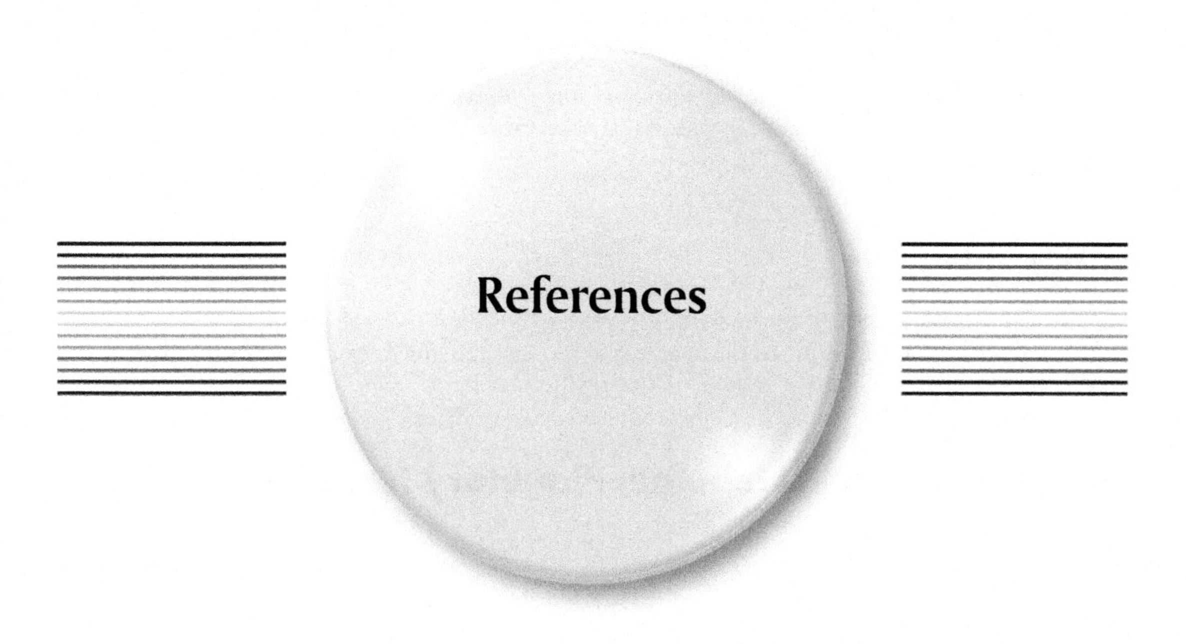

Detailed Star Charts in the Book

All detailed star charts are drawn by author. References for basic data (on the charts as well as in the text) are from:

Single Stars

The Deep Sky Browser J2000.0. On-line: http://messier45.com/cgi-bin/dsdb/dsb.pl

The Hipparcos Space Astronometry Mission, ESA. On-line: www.rssd.esa.int/index.php?project=HIPPARCOS

Ridpath, I. 1998. *Norton's Star Atlas and Reference Handbook (Epoch 2000.0).* Edinburgh: Addison Wesley Longman Limited, 1998

SIMBAD Astronomical Database – SIMBAD project is operated at CDS, Strasbourg, France

Tirion, W. 1989. *SkyAtlas 2000.0.* Cambridge, MA: Sky Publishing Corporation

Tirion, W., B. Rappaport, and G. Lovi. 1989. *Uranometria 2000.0, Volume I and II.* Richmond: Willmann–Bell, Inc.

Wikipedia. On-line: http://en.wikipedia.org/wiki

Double and Multiple Stars

Mason, D. B., G. L. Wycoff, and W. I. Hartkopf. *The Washington Double Star Catalog.* On-line: http://ad.usno.navy.mil/wds

Variable Stars

AAVSO (The American Association of Variable Star Observers). On-line: www.aavso.org/ Light curves are provided by the AAVSO, with visual observations taken from the AAVSO International Database.

Open Clusters

The Deep Sky Browser J2000.0. On-line: http://messier45.com/cgi-bin/dsdb/dsb.pl

WEBDA database – WEBDA project is operated at the Institute for Astronomy of the University of Vienna. On-line: www.univie.ac.at/webda/navigation.html

Globular Clusters, Nebulae, Planetary Nebulae and Galaxies

The Deep Sky Browser J2000.0. On-line: http://messier45.com/cgi-bin/dsdb/dsb.pl

Ridpath, I. 1998. *Norton's Star Atlas and Reference Handbook (Epoch 2000.0).* Edinburgh: Addison Wesley Longman Limited, 1998

SEDS (Students for the Exploration and Development of Space), on-line: www.seds.org/MESSIER

SEDS. On-line: http://seds.org/~spider/ngc/ngc.html

Tirion, W. 1989. *SkyAtlas 2000.0.* Cambridge, MA: Sky Publishing Corporation

Tirion, W., B. Rappaport, and G. Lovi. 1989. *Uranometria 2000.0, Volume I and II.* Richmond: Willmann-Bell, Inc.

Wikipedia. On-line: http://en.wikipedia.org/wiki

Books

1. Avsec, F., and M. Prosén, *Astronomija za 4. razred gimnazije (Astronomy for Grammar Schools).* Ljubljana: DZS, 1971 (in Slovene)
2. Burnham, Jr., R., *Burnham's Celestial Handbook Vol. 1–3.* New York: Dover Publications, Inc., 1978
3. Consolmagno, G., and D. M. Davis, *Turn Left at Orion: A Hundred Night Sky Objects to See in a Small Telescope – and How to Find Them.* Cambridge: Cambridge University Press, 2000
4. Hack, M., and C. Lamberti (eds.), *Corso di Astronomia.* Milano: Fabri Editori, 1985
5. Learner, R., *Astronomy Through the Telescope.* New York: Van Nostrand Reinhold Company, 1981
6. Moore, P., *Exploring the Night Sky with Binoculars.* Cambridge: Cambridge University Press, 1990
7. Seronik, G., *Binocular Highlights.* Cambridge, MA: Sky Publishing, New Track media LLC, 2006
8. Shklovskii, S., *Stars, Their Birth, Life, and Death.* San Francisco, CA: W. H. Freeman and Company, 1978
9. Tonkin, S., *Binocular Astronomy.* London: Springer-Verlag London Limited, 2007

Journal Articles

1. Arsov, Z., "O spektroskopiji (About Spectroscopy)", *Spika,* June, 1998 (in Slovene)
2. Downs, H., "Modeling the Universe in Your Mind", *Sky & Telescope,* October, 1993

3. Fonović, M., "R Severne krone (R Coronae Borealis)", *Spika,* May, 1997 (in Slovene)
4. Fonović, M., "Planetarne meglice (Planetary Nebulae)", *Spika,* July/August, 1998 (in Slovene)
5. Fonović, M., "Vizualno opazovanje spremenljivih zvezd (Visual Observing of Variable Stars)", *Spika,* September, 1998 (in Slovene)
6. Fonović, M., "Nove (Novae)", *Spika,* July/August, 1999 (in Slovene)
7. Fonović, M., "Naša Galaksija (Our Galaxy)", *Spika,* November, 2004 (in Slovene)
8. Fonović, M., "Bele pritlikavke (White Dwarfs)", Spika, January, 2005 (in Slovene)
9. Fonović, M., "Barve in spektri zvezd (Colors and spectra of stars)", Spika, July/August, 2005 (in Slovene)
10. Galičič, M., "Pulzarji (The Pulsars)", *Spika,* February, 1993 (in Slovene)
11. Gomboc, A., "Temna snov (The Dark Matter)", *Spika,* September, 1995 (in Slovene)
12. Gomboc, A., "O črnih luknjah in zvezdah, ki jih srečajo (About Black Holes and Stars meet them)", *Spika,* April, 2004 (in Slovene)
13. Gomboc, A., "Subrahmanyan Chandrasekhar (1910–1995)", *Spika,* December, 1995 (in Slovene)
14. Guštin, A., "Binokularji (Binoculars)", *Spika,* March, 1993 (in Slovene)
15. Guštin, A., "Kako daleč so zvezde? (How Far are the Stars?)", *Spika,* February, 1993 (in Slovene)
16. Guštin, A., "Mejniki sodobne astronomije (Boundary Stones of Modern Astronomy)", *Gea (Special edition),* December, 2006 (in Slovene)
17. Habing, H., and P. Murdin, "Being Around at the Death", *Nature.* August 19, 1993
18. Hearnshaw, J. B., "Origins of the Stellar Magnitude Scale", *Sky & Telescope,* November, 1992
19. Kajdič, P., "Življenje zvezd (The Life of Stars)", *Spika,* January, 2000 (in Slovene)
20. Kilar, B., "Prehod nebesnih teles čez meridian opazovališča (The Transit of Celestial Bodies over the Meridian of Observing Site)", *Spika,* May, 1998 (in Slovene)
21. Lake, G., "Cosmology of the Local Group", *Sky & Telescope,* December, 1992
22. MacRobert, A., "The Outer Limits: How Far Can You See?", *Sky & Telescope,* April, 1983
23. MacRobert, A., "Dealing with Dew", *Sky & Telescope,* June, 1995
24. MacRobert, A., "Caring for Optics", *Sky & Telescope,* July, 1997
25. Maddox, J., "The Future History of the Solar System", *Nature,* December 15, 1994
26. Marschall, L. A., S. J. Ratcliff, and T. J. Balonek, "Parallax You can See", *Sky & Telescope,* December, 1992
27. Mezgec, I., "Barve neba in Sonca (Colours of the Sky and Sun)", *Spika,* September, 1994 (in Slovene)
28. Mezgec, I., "Optični pojavi v atmosferi (Optical Events in the Atmosphere)", *Spika,* December, 1994 (in Slovene)
29. Nemanič, V., "Človeško oko (The Human Eye)", *Spika,* January, 1993 (in Slovene)
30. Parker, W. J., "Anatomy of a Crab", *Sky & Telescope,* January, 1995
31. Roth, J. and R. W. Sinnott, "Our Nearest Celestial Neighbors", *Sky & Telescope,* October, 1996
32. Schilling, G., "Jan Oort Remembered", *Sky & Telescope,* April, 1993
33. Steffey, P. C., "The Truth About Star Colors", *Sky & Telescope,* September, 1992
34. Strnad, J., "Hertzsprung-Russellov diagram (Hertzsprung-Russell Diagram)", *Spika,* March, 1995 (in Slovene)
35. Strnad, J., "Velike oddaljenosti v vesolju (Great distances in Universe)", *Spika,* Februar, 1997 (in Slovene)
36. Širca, S., "Nevtrini s Sonca I in II (Neutrinos from the Sun)", *Spika,* December, 1993 and January, 1994 (in Slovene)
37. Širca, S., "Nevtrini s Sonca: nova generacija detektorjev (Neutrinos from the Sun: New Generation of Detectors)", *Spika,* November, 1997 (in Slovene)
38 Špenko, T., "Velemesta na zahodu (The Metropolises on the West)", *Spika,* June, 2000 (in Slovene)
39. Špenko, T., "Bogastvo Strelca (The Treasures of Sagittarius)", *Spika,* July/August, 2000 (in Slovene)
40. Trimble, V. and S. Parker, "Meet the Milky Way", *Sky & Telescope,* January, 1995
41. Zwitter, T., "Kefeide (The Cepheids)", *Spika,* January, 1997 (in Slovene)

Picture Credits

All pictures are reproduced by permission of their copyright owners.
Page number A: above B: below C: centre L: left R: right

5 A - Tomaž Perme; **5** B, **7-11** - Primož Kalan; **12** – Igor Žiberna; **19** - Primož Kalan; **22** A – Jurij Stare; **27** - Brane Vasiljevič; **32-33** - Aleš Arnšek (pictures in the background of illustrations); **48** – Jurij Stare and Srečko Lavbič; **49, 61** L - Jurij Stare; **61** R - Alain Klotz; **70** - AAVSO (The American Association of Variable Star Observers); **72** - Aleš Arnšek (pictures in the background of illustration); **77, 80** - Image of the Sun in the background of illustration: SOHO/EIT consortium (SOHO is a project of international cooperation between ESA and NASA); **82** - Jurij Stare; **83** A – Tone Špenko (pictures in the background of illustrations); **83** B - Mark McCaughrean (Max-Planck-lnstitute for Astronomy), C. Robert O'Dell (Rice University), and NASA; **84** - NASA/ESA/STScI/ J. Hester and P. Scowen (Arizona State University); **85** - ToneŠpenko (image of the Sun in the background of illustration); **87-90** – Image of the Sun in the background of illustrations: SOHO/EIT consortium (SOHO is a project of international cooperation between ESA and NASA); **93** - NASA and The Hubble Heritage Team (AURA/STScI); **93** - NASA/CXC/M. Weiss; **94** - J, Hester/Arizona State University/NASA; **96** -NASA/ESA/Felix Mirabel; **97** - David A. Hardy (www.astroart.org/STFC); **98** A - Jurij Stare; **98** B - Ernst Paunzen/lnstitute for Astronomy of the University of Wienna; **99** - Image of the Sun in the background of illustration: SOHO/EIT consortium (SOHO is a project of international cooperation between ESA and NASA); **103** - illustration based on Richard Powell's An Atlas of The Universe (www.atlasoftheuniverse.com/); **104** - ESA/AOES Medialab; **110** A - Jurij Stare; **110** B - Herman Mikuž/Črni Vrh Observatory/www.observatorij.org; **110** B (inset), **111, 112** - Srečko Lavbič; **113** - NASA/Andrew Fruchter and the ERO Team (Sylvia Baggett (STScI), Richard Hook (ST-ECF), Zoltan Levay (STScI)); **114** A – Tone Špenko; **114** B, **115** A - Jurij Stare; **115** B - Herman Mikuž/Črni Vrh Observatory; **116, 117** A, **117** B - Jurij Stare; **117** BBL - SDSS (Sloan Digital Sky Survey); **117** BBR - Robert Gendler; **118** A - NASA, ESA, and The Hubble Heritage Team (STScl/AURA); **118** B - NASA, ESA, and The Hubble Heritage Team (STScl/AURA)-ESA/Hubble Collaboration; Acknowledgment: B. Whitmore (STScI); **119** - Simon Krulec; **123** - Eddie Trimarchi (Tin Shed Observatory);

128 - NASA/JPL-Caltech; **129** - illustration based on Richard Powell's An Atlas of The Universe (www.atlasoftheuniverse.com/); **130** - NASA/JPL-Caltech; **131 A** - Naval Research Laboratory/Image processing by N. E. Kassim, D. S. Briggs, T. J. W. Lazio, T. N. LaRosa, C. A. Gross and J. Imamura/ original data from the NRAO VLA courtesy of A. Pedlar, K, Anantharamaiah, M. Goss and R. Ekers; **131 B** - NASA/UMass/D. Wang et al.; **133 A** - NASA/JPL/lnfrared Astrophysics Team; **133 B** - ESO (European Southern Observatory); **134 A** - Prof. Andrea Ghez (UCLA); **137** - Herman Mikuž/Črni Vrh Observatory; **138** - Josch Hambsch and Rober Gendler; **139** - NASA, ESA, the Hubble Heritage (STScl/ AURA)-ESA/Hubble Collaboration, A. Evans (University of Virginia, Charlottesville/NRAO/Stony Brook University); **142 A** - Primož Kalan; image of Jupiter: NASA/JPL/University of Arizona; **142 B** - ToneŠpenko; **144** - Primož Kalan; image of our galaxy: NASA/JPL-Caltech; **145** - Jurij Stare; **146** - Volker Springel; **147** -NASA/WMAP Science Team; **149** – Tone Špenko; **150** - ESO (European Southern Observatory); **151** - NASA, M. Clampin (STScI), H. Ford (JHU), G. illingworth (UCO/Lick Observatory), J. Krist (STScI), D. Ardila (JHU), D. Golimowski (JHU), the ACS Science Team, J. Bahcall (IAS) and ESA; **152** - Inset image of quasar's surroundings: Tone Špenko; **160 A** - Matt BenDaniel; **160 B, 161** - Javor Kac; **163** - H. Fukushima, D. Kinoshita and J. Watanabe (National Astronomical Observatory of Japan); **173** - Tone Špenko; **177,179** - Jurij Stare; **180** - Tone Špenko; **184** - Jurij Stare; **187** - Boštjan Guštin; **188**-Jurij Stare; **190** - Tone Špenko; **194** - Jurij Stare; **198** - Boštjan Guštin; **200, 210, 214, 216** -Jurij Stare; **219** - Herman Mikuž/Črni Vrh Observatory; **220** - Jurij Stare; **222** - Srečko Lavbič; **224, 227** - Jurij Stare; **229 A** - NASA, H. E. Bond and E. Nelan (STScI, Baltimore, Md.); M. Barstow and M. Burleigh (University of Leicester, U.K.); and J. B. Holberg (University of Arizona); **229 B** - Earth by Apollo 17: GSFC/NASA/Apollo 17; image of Jupiter: NASA/JPL/University of Arizona; image of Uranus: NASA/E. Karkoschka, University of Arizona; Image in the background: Jurij Stare; **231, 241, 242** - Jurij Stare; **244** - Alain Klotz; **245** - Robert Gendler and Stephane Guisard; **253** - ToneŠpenko; **254** - AAVSO (The American Association of Variable Star Observers); **256** - Images of the Sun on the illustration: SOHO/EIT consortium (SOHO is a project of international cooperation between ESA and NASA). Image of quiet Sun: Tone Špenko; **257** - Tone Špenko; **259** - ESO (European Southern Observatory); **263** - Srečko Lavbič; **265, 267, 276, 280, 281** - Jurij Stare; **283-286 B** - Tone Špenko; **286 A** - ESA/Hubble European Space Agency Information Centre (M. Kornmesser, L. L. Christensen); **291** - NASA, ESA, HEIC, and The Hubble Heritage Team (STScl/ AURA); **298, 303, 309 B** - Jurij Stare; **309 A, 317** - AAVSO (The American Association of Variable Star Observers); **320, 323** - Jurij Stare; **327 A** - AAVSO (The American Association of Variable Star Observers); **327 B, 329** - Tone Špenko; **336** - Jurij Stare; **337** - The Hubble Heritage Team (AURA/ STScl/NASA); **338** - Srečko Lavbič; **345 A** - Jurij Stare; **345 B** -Tone Špenko; **347, 349** - Jurij Stare; **350, 356 A** - ToneŠpenko; **356 B** - Jurij Stare; **360** - ToneŠpenko; **362** -Herman Mikuž; **368** - Jurij Stare; **369 A** - NASA, C. R. O'Dell and S. K. Wong (Rice University); **369 B** - C. R. O'Dell/Rice University/NASA; **371** - Jurij Stare; **372** - Herman Mikuž/Črni Vrh Observatory; **373** - Robert Gendler; **379** - Primož Kalan; **380, 383, 385** - Jurij Stare; **387** - Dave Erickson/www.hbastro.com; **392, 398** - Tone Špenko; **400** - Jurij Stare; **402** - Tone Špenko; **404** - David A. Kodama; **410** - Tone Špenko; **412-413** - Boštjan Guštin; **414, 416** - Tone Špenko; **417** - Jurij Stare; **418** - Boštjan Guštin; **420, 421 A** -Tone Špenko; **421 B** - Jurij Stare; **427** - Tone Špenko; **432** - S. Binnewies, R. Sparenberg and V. Robering/Capella Observatory; **434** - Jurij Stare; **435** - Steve Crouch; **436, 440** - Jurij Stare; **442** - AAVSO (The American Association of Variable Star Observers); **443, 445** - Tone Špenko; **448** - Jurij Stare; **449** - Tone Špenko; **451, 458 A** - Jurij Stare; **458 B** - ToneŠpenko; **459** – DSS (Digitized Sky Survey); **462** - Jurij Stare; **464** - Herman Mikuž/Črni Vrh Observatory; **465** - Bill Saxton, NRAO/AUl/NSF; **466** - NASA and The Hubble Heritage Team (STScI/AURA); **468** - ToneŠpenko; **470** - Jurij Stare; **475** - Srečko Lavbic; **477** - Jurij Stare; **478** - Srečko Lavbič; **480** - ToneŠpenko; **483** - Brane Vasiljevič; **484** - Tone Špenko; **491** - Jurij Stare; **494, 495** - Srečko Lavbič; **497** - ESO/VLT; **498** - Tone Špenko.

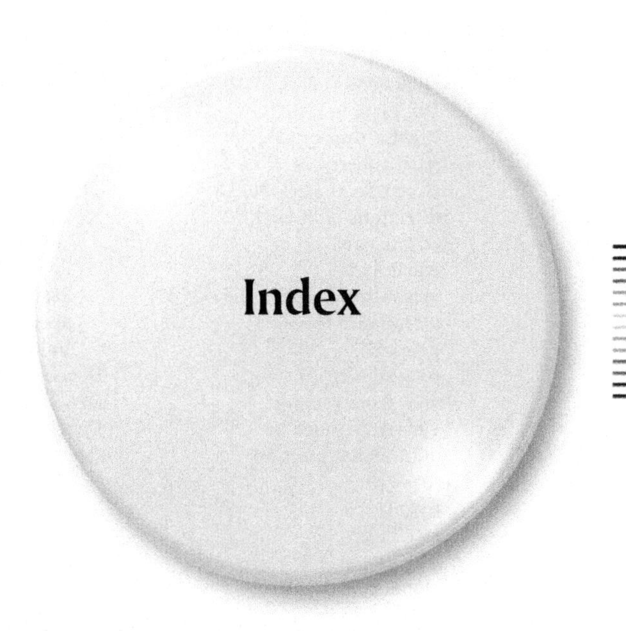

Index

Index of Persons

Index of described binocular objects

Abrreviation: n+oc - nebula + open cluster

Stars

Double or Multiple Stars

Variable Stars

X-Ray Source

Open Clusters

Globular Clusters

Bright Nebulae

Dark Nebulae

Planetary Nebulae

Milky Way Galaxy

Galaxies